22nd International Conference on Computational Statistics (COMPSTAT 2016)

Oviedo, Spain
23 – 26 August 2016

Editors:

Ana Colubi
Angela Blanco
Cristian Gatu
Gil Gonzalez-Rodriguez

ISBN: 978-1-5108-3759-1

Printed from e-media with permission by:

Curran Associates, Inc.
57 Morehouse Lane
Red Hook, NY 12571

Some format issues inherent in the e-media version may also appear in this print version.

Printed by Curran Associates, Inc. (2017)

For permission requests, please contact International Statistical Institute (ISI) at the address below.

International Statistical Institute (ISI)
P.O. Box 24070
2490 AB The Hague
The Netherlands

Phone: 31-70-3375737
Fax: 31-70-3860025

www.isi-web.org

Additional copies of this publication are available from:

Curran Associates, Inc.
57 Morehouse Lane
Red Hook, NY 12571 USA
Phone: 845-758-0400
Fax: 845-758-2633
Email: curran@proceedings.com
Web: www.proceedings.com

Proceedings of

COMPSTAT 2016

22nd International Conference on Computational Statistics

August 23-26, 2016

Auditorio Príncipe Felipe, Oviedo, Spain

22nd International Conference on Computational Statistics

COMPSTAT 2016

23-26 August 2016
Oviedo, Spain

Editors:
Ana Colubi, Angela Blanco and Cristian Gatu.

Technical Editors:
Gil Gonzalez-Rodriguez and Angela Blanco.

Preface

The 22nd International Conference on Computational Statistics, COMPSTAT 2016, is held in Oviedo, Spain, from August 23rd to August 26th 2016. It is locally organized by members of the University of Oviedo assisted by active Spanish researchers. The COMPSTAT is an initiative of the European Regional Section of the International Association for Statistical Computing (IASC-ERS), a society of the International Statistical Institute (ISI). COMPSTAT is one of the best-known world conferences in Computational Statistics, regularly attracting hundreds of researchers and practitioners.

The first COMPSTAT conference took place in Vienna in 1974, and the last two editions took place in Limassol in 2012 and Geneve in 2014. It has gained a reputation as an ideal forum for presenting top quality theoretical and applied work, promoting interdisciplinary research and establishing contacts amongst researchers with common interests.

Keynote lectures are addressed by Prof. Gerard Biau, Universit Pierre et Marie Curie, Paris, France, Prof. Alastair Young, Imperial College, London, UK and Prof. Hans-Georg Mueller, University of California Davis, United States.

From more than 450 submissions received for COMPSTAT, 360 have been retained for presentation in the conference. The conference programme has 41 contributed sessions, 8 invited sessions, 3 keynote talks, 30 organized sessions and 3 tutorials. There are approximately 430 participants.

The Proceedings are published in an electronic book comprising 34 papers. The participants can find an electronic copy in a USB stick placed in their conference bags or download it at the conference web page. All the papers submitted have been evaluated through a rigorous peer review process. Those papers that have been accepted for publication in the Proceedings have been evaluated thoroughly by at least 2 referees. This ensures a high quality proceedings volume in the main areas of computational statistics.

The organization would like to thank the editors, authors, referees and all participants of COMPSTAT 2016 who contributed to the success of the conference. Our gratitude to sponsors, scientific programme committee, session organizers, local universities, the city of Oviedo, and many volunteers who have contributed substantially to the conference. We acknowledge their work and support.

The COMPSTAT 2016 organizers invite you to the next edition of the COMPSTAT, which will take place in Iasi, Romania in 2018. We wish the best success to Cristian Gatu the Chairman of the 23rd edition of COMPSTAT.

Ana Colubi
Organiser and Chairperson of the SPC.

Scientific Program Committee

Ex-officio:

COMPSTAT 2016 organiser and Chairperson of the SPC: Ana Colubi.

Past COMPSTAT organiser: Manfred Gilli.

Next COMPSTAT organiser: Cristian Gatu.

IASC-ERS Chairman: Alessandra Amendola.

Members:

Marc Genton, Salvatore Ingrassia, Jean-Michel Poggi, Igor Pruenster, Juan Romo, Tamas Rudas and Stefan Van Aelst.

Consultative Members:

Representative of the IFCS: Christian Hennig.

Representative of the ARS of IASC: Chun-houh Chen.

Representative of ERCIM WG CMS: Erricos Kontoghiorghes.

COMPSTAT 2016 Proceedings Management Committee:

Ana Colubi, Angela Blanco and Cristian Gatu.

Local Organizing Committee:

Ana M. Aguilera, Gil Gonzalez-Rodriguez, M. Dolores Jimenez-Gamero, Agustin Mayo, Domingo Morales and M. Carmen Pardo.

Additional Referees:

Andreas Alfons, Laura Anderlucci, Andriy Andreev, Jan Beirlant, Rosa Maria Benito, Christophe Biernacki, Danilo Bolano, Papa Ousmane Cisse, Francisco de Carvalho, Christel Faes, Christian Francq, Pierre Geurts, Jose M. Gozalvez-Zafrilla, Francesca Greselin, Arthur Gretton, Markus Haas, Marie Huskova, Tadashi Imaizumi, Antonio Irpino, O. Alaoui Ismaili, Atsushi Iwai, James Keraita, Takafumi Kubota, Antonio Lijoi, Jesus Lopez-Fidalgo, Elizabeth Ann Maharaj, Alina Matei, Yuichi Mori, Alam Moudud, Makiko Oda, Akinori Okada, Shunji Ouchi, Vygantas Paulauskas, Gonzalo Perera, Stephen Pollock, Luc Pronzato, Fernando Quintana, Ana B. Ramos-Guajardo, Marco Riani, Javier Roca, Peter Rousseeuw, Juan Eloy Ruiz-Castro, Sergio Scippacercola, Alwin Stegeman, Dilek Teker, Makoto Tomita, Henghsiu Tsai, Chen-Hsiang Yang, Mariangela Zenga.

Sponsors

http://www.cronosaction.com

http://www.iasc-isi.org

http://www.cost.eu

http://www.oviedo.es

http://www.elsevier.com

http://www.uniovi.es

http://www.alquisa.es

http://CMStatistics.org

Contents

Davide Martinetti and Ghislain Geniaux
ProbitSpatial R package: fast and accurate spatial probit estimations 1

Aldo Corbellini, Luigi Grossi and Fabrizio Laurini
Robustness for multilevel models with the forward search 13

Kohei Adachi and Nickolay T. Trendafilov
Some mathematical notes on comprehensive factor analysis 25

Claude Manté and Saikou Oumar Kidé
Approximating the Rao's distance between negative binomial distributions. Application to counts of marine organisms. 37

Anne De Moliner, Camelia Goga and Hervé Cardot
Estimation of total electricity consumption curves of small areas by sampling in a finite population 49

Jan Ámos Víšek
Coping with level and different type of contamination by SW-estimator 59

Lida Mavrogonatou and Vladislav Vyshemirsky
Sequential importance sampling for online Bayesian changepoint detection 73

Fumio Ishioka and Koji Kurihara
Detection of space–time clusters for radiation data using spatial interpolation and scan statistics 85

Massimiliano Giorgio, Maurizio Guida, Fabio Postiglione and Gianpaolo Pulcini
A Bayesian approach for the transformed gamma degradation process 99

Ricardo A. Collado and Germán G. Creamer
Time series forecasting with a learning algorithm: an approximate dynamic programming approach 111

Masatoshi Nakamura, Yoshimichi Ochi and Masashi Goto
Trees Garrote for Regression Analysis 123

S.X. Lee and G.J. McLachlan
On mixture modelling with multivariate skew distributions 137

Alexander Pulido-Rojano and J.Carlos García–Díaz
A modified control chart for monitoring the multihead weighing process 149

A. Hitaj, F. Hubalek, L. Mercuri and E. Rroji
On multivariate extensions of the Mixed Tempered Stable distribution 159

Jocelyn Chauvet, Catherine Trottier, Xavier Bry and Frédéric Mortier
Extension to mixed models of the Supervised Component-based Generalised Linear Regression 169

Brodinova *et al.*
Evaluation of robust PCA for supervised audio outlier detection 183

Xavier Bry, Théo Simac and Thomas Verron
Supervised-Component based Cox Regression 195

Hirohito Sakurai and Masaaki Taguri
Test of mean difference in longitudinal data based on block resampling approaches 205

W. James Murdoch and Mu Zhu
Expanded alternating optimization for matrix factorization and penalized regression 217

Angela Alibrandi *et al.*
NPC to assess effects of maternal iodine nutrition and thyroid status on children cognitive development 231

Baldvin Einarsson *et al.*
Using intraclass correlation coefficients to quantify spatial variability of catastrophe model errors **243**

Ana Tavares *et al.*
Detection of exceptional genomic words: a comparison between species **255**

Wataru Sakamoto
Cluster detection of disease mapping data based on latent Gaussian Markov random field models **267**

M. Ivette Gomes, Helena Penalva, Frederico Caeiro and Manuela Neves
Non-reduced versus reduced-bias estimators of the extreme value index — efficiency and robustness **279**

Yoshitake Takebayashi, Takafumi Kubota and Hiroe Tsubaki
Risk profiles for severe mental health difficulty: classification and regression tree analysis **291**

Carlos G. Maté and Javier Redondo
Forecasting financial time big data using interval time series **303**

Michel Philipp, Achim Zeileis and Carolin Strobl
A toolkit for stability assessment of tree-based learners **315**

Pierre Michel and Badih Ghattas
Variable Importance in Clustering Using Binary Decision Trees **327**

Takafumi Kubota, Hitoshi Fujimiya and Hiroyuki A. Torii
Time series changes of the categorical data using the text data regarding radiation **339**

Thomas Michael Bartlett, Levy Boccato
A Multimomental ARMA model: initial formulation and a case study **349**

Tommi Salminen, Lasse Koskinen and Arto Luoma
Joint modeling of inflation and real interest rate dynamics with application to equity-linked investment **361**

Yoshiro Yamamoto and Sanetoshi Yamada

Visualization of cross tabulation by the Association rules by using the Correspondence analysis **373**

Giurghita and Husmeier
Inference in nonlinear systems with unscented Kalman filters **383**

Guillaume Sagnol, Hans-Christian Hege and Martin Weiser
Using sparse kernels to design computer experiments with tunable precision **397**

ProbitSpatial R package: fast and accurate spatial probit estimations

Davide Martinetti, *INRA Avignon, France*, davide.martinetti@paca.inra.fr
Ghislain Geniaux, *INRA Avignon, France*, ghislain.geniaux@avignon.inra.fr

Abstract. This package meets the emerging needs of powerful and reliable models for the analysis of spatial discrete choice data. Since the explosion of available and voluminous geospatial and location data, older estimation techniques cannot withstand the course of dimensionality and are restricted to samples with less than a few thousand observations. The functions contained in ProbitSpatial allow fast and accurate estimations of Spatial Autoregressive and Spatial Error Models under Probit specification. They are based on the full maximization of likelihood of an approximate multivariate normal distribution function, a task that was considered as prodigious just few years ago. Extensive simulation and empirical studies proved that these functions can readily handle sample sizes with as many as several millions of observations, provided the spatial weight matrix is in convenient sparse form, as is typically the case of large data sets, where each observation neighbours only a few other observations. SpatialProbit relies amongst others on Rcpp, RcppEigen and Matrix packages to produce fast computations for large sparse matrixes. Possible applications of spatial binary choice models include spread of diseases and pathogens, plants distribution, technology and innovation adoption, deforestation, land use change, amongst many others.

Keywords. Spatial Statistics, Probit, Discrete Choice model, R package

1 Introduction

`ProbitSpatial` library [24] contains a set of tools for estimating and testing different types of spatial probit models. Those models belong to the growing family of econometrics methods that deals with observations showing some kind of spatial or network dependence. In particular, these choice models are focused on data that have a binary dependent variable, such as the choice of adopting a new farming technology [9], increasing tax rates in a district [5], reopening of a damaged infrastructure [23], location of suppliers plants [20] or of R&D labs [3], tree harvesting choice [15], defoliation [17], deforestation [6] or plants distribution [11] and land use changes [8, 28, 30, 39, 42].

In spatial models, the interdependence between observations not only is viewed as a violation of the independence hypothesis, but also as an actual feature of the problem, and its presence and intensity is of major interest, especially in spatial econometric and social network analysis. Despite there exists an extensive literature on continuous-dependent-variable models with spatial dependence, began three decades ago with the seminal paper of Anselin [2], the case of spatial regression models with binary or multinomial dependent variables has received limited attention [34], mainly due to its complexity. A recent and comprehensive review by Calabrese et al. [7] collects and compares through a Monte-Carlo experiment all significant contributions in this field: the expectation-maximization (EM) algorithm by McMillen [26], the Bayesian Gibbs sampler by LeSage [21], the recursive importance sampling (RIS) algorithm by Beron and Vijverberg [5], the generalized method of moments (GMM) algorithm by Pinkse and Slade [33] and its linearised (linearised GMM) version by Klier and McMillen [20]. Also, on the same subject, the unpublished paper of Pace and LeSage [32] that uses Geweke-Hajivassiliou-Keene (GHK) approximations of a multivariate normal probability with sparse variance-coavariance matrix (GHK and RIS are basically the same) and the contribution by Diallo and Geniaux [13] on linearised GMM. From this review, we observed that only RIS and Gibbs sampler estimators perform reasonably well in terms of accuracy, but they are unfeasible for large samples ($n \gg 1000$). On the other hand, GMM-type estimators scale well w.r.t. sample size, but their accuracy is really poor, especially when the value of the spatial dependence parameter is high.

For spatial binary-dependent variables, the most used techniques are spatial logit [13, 20] and spatial probit [5, 21, 26, 32, 33, 40] models. `ProbitSpatial` package contains functions for estimating probit models, in which the disturbances of the regression model are supposed to follow a multivariate normal (MVN) distribution and are based on the maximisation of the corresponding likelihood, a *prodigious task* according to Wang et al. [38]. In contrast with the logit specification, that assumes multivariate logistic distribution of the error terms, the probit model is particularly attractive for the flexibility of the MVN covariance structure. Nevertheless, this flexibility comes at a cost: there exists no closed form expression for computing MVN probabilities, that are eventually expressed as a multiple integral. For dealing with this problem, there exists several proposals than we can group into three main families: numerical integration methods, simulation methods and analytical approximation methods. We propose to use a MVN computation method that belongs to the last family and that improve the univariate conditioning approximation method proposed by Mendell and Elston in [29], described by Kakamura in [18] and recently reviewed by Connors et al. in [12]. We will show that this method allows to compute the log-likelihood of a spatial probit model accurately and rapidly.

In Section 2 we will present two methods for spatial probit parameter estimations, with special emphasis on the algorithmic implementation and the corresponding use in the `ProbitSpatial` `R` library. In Section 3 we will compare the two techniques on a set of simulated Monte Carlo experiments against existing functions in `R`. We will focus then on the accuracy and estimation time of the different methods.

2 ProbitSpatial library

Spatial binary-choice regression models are used to analyse sample data that are associated with specific locations in space and that represent binary outcomes (usually actions or choices). We deal with spatial regression models of the following form (the notations follow LeSage and Pace [22]):

$$\begin{aligned} \boldsymbol{y} &= \rho W\boldsymbol{y} + \boldsymbol{X}\boldsymbol{\beta} + \boldsymbol{u}\,, \\ \boldsymbol{u} &= \lambda M\boldsymbol{u} + \boldsymbol{\epsilon}\,, \\ \boldsymbol{\epsilon} &\sim \mathcal{N}(\boldsymbol{0}, \sigma^2 I_n)\,, \end{aligned} \tag{1}$$

where $\boldsymbol{y}$ represents an $n \times 1$ vector of binary dependent variables, $\boldsymbol{X}$ an $n \times k$ matrix of independent variables, I_n is the identity matrix of size n and $\boldsymbol{\beta}$ ($k \times 1$ vector), ρ and λ (both scalars in $[-1, 1]$) are parameters to be estimated. The two $n \times n$ matrices W and M are known as spatial weight matrices and contain the information on the spatial relationship between observations. Spatial weight matrices are usually constructed as a function of the distance between observations or other contiguity measures (shared borders, shared department, etc.). Typically $w_{ij} = 1$ (or $m_{ij} = 1$), if observations i and j are contiguous, while $w_{ij} = 0$ (or $m_{ij} = 0$) otherwise (it follows immediately that both W and M are symmetric and by convention, the diagonal is set to be zero). It is common practice to row-standardise the spatial weight matrices, i.e. $w_{ij}/(\sum_j w_{ij})$.

The model in Eq. (1) is often referred as spatial autoregressive model with spatial autoregressive disturbances (or SARAR$(1, 1)$), when both the parameters ρ, a.k.a. spatial lag parameter, and λ, a.k.a.

spatial error parameter, are different from zero. If $\lambda = 0$, then we will speak of a spatial autoregressive model (SAR, or SARAR(1, 0)) model, while if $\rho = 0$, we will speak of a spatial error model (SEM, or SARAR(0, 1)). The SAR and SEM models are used to explain spatial dependences of different nature[1]. In this work we will focus in particular on spatial autoregressive and spatial error models. In particular, since the disturbances $\boldsymbol{\epsilon}$ in Eq. (1) follow a multivariate normal distribution, then we are in the case of a SAR probit or SEM probit specification.

The SAR probit model can be reformulated from Eq. (1) as

$$\boldsymbol{y} = (I_n - \rho W)^{-1}(\boldsymbol{X\beta} + \boldsymbol{\epsilon}), \tag{2}$$

and hence the variance of the error term $\boldsymbol{v} = (I_n - \rho W)^{-1}\boldsymbol{\epsilon}$ can be written as

$$\Sigma = E(\boldsymbol{v}\boldsymbol{v}') = \sigma^2((I_n - \rho W)^{-1}((I_n - \rho W)^{-1})^t). \tag{3}$$

Consistent and efficient estimates of the $\boldsymbol{\beta}$ and ρ parameters are obtained by maximising the corresponding likelihood function, that takes the form of a multivariate normal (MVN) probability, namely an n-dimensional integral, as follow:

$$\begin{aligned} L(\beta, \rho) &= \Phi_n(\boldsymbol{x} \in \boldsymbol{A} \mid \Sigma) \qquad (4)\\ &= \frac{1}{(2\pi)^{n/2}\,|\Sigma|^{\frac{1}{2}}} \int_{A_1}\int_{A_2}\cdots\int_{A_n} e^{-\left(\frac{1}{2}\boldsymbol{x}^t\Sigma^{-1}\boldsymbol{x}\right)}, \end{aligned}$$

where $\boldsymbol{A} = \{A_i\}_{i\in\{1,\ldots,n\}}$ and

$$A_i = \begin{cases} [((I_n - \rho W)^{-1}\boldsymbol{X\beta})_i, +\infty) & \text{, if } y_i = 0,\\ (-\infty, ((I_n - \rho W)^{-1}\boldsymbol{X\beta})_i] & \text{, if } y_i = 1. \end{cases}$$

The SEM probit model can be reformulated from Eq. (1) as

$$\boldsymbol{y} = \boldsymbol{X\beta} + (I_n - \lambda M)^{-1}\boldsymbol{\epsilon}, \tag{5}$$

and hence the variance of the error term $\boldsymbol{v} = (I_n - \lambda M)^{-1}\boldsymbol{\epsilon}$ can be written as

$$\Sigma = E(\boldsymbol{v}\boldsymbol{v}') = \sigma^2((I_n - \lambda M)^{-1}((I_n - \lambda M)^{-1})^t). \tag{6}$$

Consistent and efficient estimates of the $\boldsymbol{\beta}$ and λ parameters are obtained by maximising the following likelihood function, that correspond again to the n-dimensional integral

$$\begin{aligned} L(\beta, \rho) &= \Phi_n(\boldsymbol{x} \in \boldsymbol{A} \mid \Sigma) \qquad (7)\\ &= \frac{1}{(2\pi)^{n/2}\,|\Sigma|^{\frac{1}{2}}} \int_{A_1}\int_{A_2}\cdots\int_{A_n} e^{-\left(\frac{1}{2}\boldsymbol{x}^t\Sigma^{-1}\boldsymbol{x}\right)}, \end{aligned}$$

where $\boldsymbol{A} = \{A_i\}_{i\in\{1,\ldots,n\}}$ and

$$A_i = \begin{cases} [(\boldsymbol{X\beta})_i, +\infty) & \text{, if } y_i = 0,\\ (-\infty, (\boldsymbol{X\beta})_i] & \text{, if } y_i = 1. \end{cases}$$

As previously mentioned, MVN probabilities cannot be computed exactly, since there exists no closed formula for solving the integrals in Eqs. (4) and (7) as long as Σ is different from the identity matrix (i.e. the case of independent observations). We hence resort to use a modified version of the analytic

[1]See Chapter 1 of the book of LeSage and Pace [22] for more details on spatial autoregressive and spatial error models.

approximation proposed by Mendell and Elston [29]. The idea is to rewrite the MVN probabilities as the product of univariate conditional probabilities (we adopt the notation of Trinh and Genz [37]):

$$\begin{aligned}
\Phi_n(\boldsymbol{x} \in \{A_1, \dots, A_n\} \mid \Sigma) &= P(x_1 \in A_1, \dots, x_n \in A_n) \\
&= P(x_1 \in A_1) \cdot P(x_2 \in A_2 \mid x_1 \in A_1) \\
&\quad \cdot P(x_3 \in A_3 \mid \{x_1 \in A_1, x_2 \in A_2\}) \dots \\
&\quad \cdot P(x_n \in A_n \mid \{x_1 \in A_1, x_2 \in A_2, \dots, x_{n-1} \in A_{n-1}\}) \\
&= P(x_1 \in A_1) \cdot \prod_{i=2}^{n} P(x_i \in A_i \mid \{x_1 \in A_1, \dots, x_{i-1} \in A_{i-1}\}) .
\end{aligned}$$

The algorithm is based on the Cholesky decomposition of the variance-covariance matrix $\Sigma = CC^t$, where C is a lower triangular matrix (the decomposition always exists, since Σ is symmetric and positive semi-definite by definition). The exponent of the integrand of Eq. (4) takes then the form

$$\boldsymbol{x}^t \Sigma^{-1} \boldsymbol{x} = \boldsymbol{x}^t (C^{-1})^t C^{-1} \boldsymbol{x}$$

and we use the transformation $\boldsymbol{x} = C\boldsymbol{z}$ (or, equivalently, $\boldsymbol{z} = C^{-1}\boldsymbol{x}$), where $\boldsymbol{x}^t\Sigma^{-1}\boldsymbol{x} = \boldsymbol{z}^t\boldsymbol{z}$, with $d\boldsymbol{x} = |C|\, d\boldsymbol{z} = \sqrt{|\Sigma|} d\boldsymbol{z}$. Taking advantage of the lower triangular structure of the Cholesky decomposition of Σ, the integral intervals $A_i = (a_i, b_i)$ are transformed according to $\boldsymbol{a} \leq C\boldsymbol{z} \leq \boldsymbol{b}$. The new integral limits $(\boldsymbol{a'}, \boldsymbol{b'})$ can be computed iteratively, while the values of z_i, that cannot be computed directly, are approximated using their truncated expected values:

$$\tilde{z}_i = \frac{\phi(a'_i) - \phi(b'_i)}{\Phi(b'_i) - \Phi(a'_i)} , \tag{8}$$

where ϕ and Φ represent the standard normal univariate density and cumulative distribution functions, respectively. This is the key intuition beyond the Mendell-Elston approximation method, i.e. to replace the $\tilde{z}_i$ values by truncated expected values ($\tilde{z}_i$ is then the average value of a random variable z_i that follows the truncated univariate distribution). The algorithm iteratively substitutes the univariate conditional probabilities with the value of Eq. (8), starting from the first random variable and uses the obtained approximation for computing the next step. The algorithm ends when the probability of the last random variable is computed and the final result is given by the following approximation:

$$\Phi_n(\boldsymbol{x} \in \{A_1, \dots, A_n\} \mid \Sigma) = P(x_1 \in A_1, \dots, x_n \in A_n) \sim \prod_{i=1}^{n} (\Phi(b'_i) - \Phi(a'_i)) .$$

The iterations of the Mendell and Elston algorithm follow the order of the variables as given by the problem, but there exists evidence that different re-orderings can lead to better approximations [12, 16, 36]. If there were an exact solution to the integral, these orderings would not change the value of the probability, as long as integration limits and rows and columns of Σ are rearranged accordingly, but since we are using an approximation that runs iteratively over the n observations, it is advisable to control the propagation of the approximation error, especially during the first steps of the iteration. For that, Gibson et al. [16] propose to sort the observations in such a way that the outermost integrals correspond to those observations that have the smallest expected values. A recent study by Connors et al. [12] corroborates this hypothesis.

This numerical approximation can be directly used to compute the likelihood associated to SAR and SEM probit models through Eqs. (4) and (7). Nevertheless, there are a few further precautions that we adopted in order to reduce the computation time of the algorithm and improve the quality of the approximation:

- massive use of sparse matrices and sparse linear algebra, especially for the computations relative to the spatial weight matrixes W and M, that have dimension $n \times n$, but usually high sparsity (lot of zero entries);

- the inverse matrixes $(I_n - \rho W)^{-1}$ and $(I_n - \lambda M)^{-1}$ are approximated (when $n > 1000$) with their Taylor expansion $(I_n - \rho W)^{-1} \sim I_n + \rho W + \rho^2 W^2 + \rho^3 W^3 + \ldots$ and similarly for $(I_n - \lambda M)^{-1}$. This is possible since W (resp. M) is row-standardised and the spatial lag parameter ρ (resp. spatial error parameter λ) is contained in the $[-1, 1]$ interval, hence all of the eigenvalues of ρW (resp. λM) have absolute value smaller than 1.

- the most time-consuming part of the entire algorithm is the Cholesky decomposition of the sparse $n \times n$ matrix Σ. The length of this part depends strongly on the number of non-zero elements of Σ, that in turns depends on the sparsity of W and on the order of the Taylor approximation of $(I_n - \rho W)^{-1}$ or $(I_n - \lambda M)^{-1}$ detailed in the previous point. There exists several algorithms that allow to reduce the number of non-zero elements in the Cholesky factor, based on a clever reordering of the rows and columns of Σ. We chosed to perform the Approximate Minimum Degree algorithm (a.k.a. AMD, see [1]) before the Cholesky decomposition, since it has been proved to be really fast and efficient: it slightly worsens the accuracy of the MVN probability, but busts the computation time by several order of magnitude.

- the reordering of the observations proposed by Gibson et al. [16] should in theory be applied at each iteration of the Mendell-Elston algorithm. This is obviously unfeasible for large samples. We prefer instead to reorder only once, before the Cholesky decomposition.

- one may consider the use of the precision matrix $P = \Sigma^{-1}$ instead of the variance-covariance matrix Σ, as suggested by Pace and Barry [31] and LeSage and Pace [22]. This proposal makes perfectly sense, since P is usually sparser than Σ, and allows a faster Cholesky decomposition. Obviously, the accuracy of the approximation is reduced. Further details on the implementation of the precision-matrix version can be found in [32].

For the estimation of the parameters $(\hat{\boldsymbol{\beta}}, \hat{\rho})$ we propose two optimisation procedures:

FL Maximisation of the approximated full log-likelihood by means of a multi-dimensional optimisation algorithm and the corresponding gradient functions (not reported here, for details see [25]);

CL Maximisation of the log-likelihood conditional on ρ: at each step i of the optimization, for a given value of ρ_i, a set of $\hat{\boldsymbol{\beta}}_i$ parameters is estimated by means of a Standard Probit estimation[2]. The corresponding log-likelihood is computed and the optimizer will try to minimize the log-likelihood by searching the optimal $\hat{\rho} \in [-1, 1]$. In this way, we can use one-dimensional optimization, that is way faster than a multi-dimensional one[3].

To summarise, our proposal consists of four estimators, listed her with the corresponding R commands:

FLUC Full log-likelihood with univariate conditional approximation of MVN probabilities and variance-covariance matrix;
`SpatialProbitFit(f,d,W,DGP,method="full-lik",varcov="varcov")`

CLUC Conditional log-likelihood with the univariate conditional approximation of MVN probabilities and variance-covariance matrix;
`SpatialProbitFit(f,d,W,DGP,method="conditional",varcov="varcov")`

FLUP Full log-likelihood with univariate conditional approximation of MVN probabilities and precision matrix;
`SpatialProbitFit(f,d,W,DGP,method="full-lik",varcov="precision")`

CLUP Conditional log-likelihood with univariate conditional approximation of MVN probabilities and precision matrix;
`SpatialProbitFit(f,d,W,DGP,method="conditional",varcov="precision")`

[2]We perform standard probit estimations with the `speedglm.wfit` function in `R` from package `speedglm` [14].
[3]We use the `optimize` function from the `stats` library in `R`, that performs golden-section search [35].

where `f` is the regression formula to be estimated , `d` is the database containing the variables, `W` is a spatial weight matrix of class `dgCMatrix` (see [4]) and `DGP` indicates if the data generating process is either `SAR` or `SEM`.

3 Comparison with existing R libraries and other methods found in the literature

In this section we compare the performances of existing algorithms for the estimation of spatial probit models in R. We will consider the following implementations:

M1: `ProbitSpatial`, our package;

M2: `spatialprobit` R package by Wilhelm and Godinho de Matos [41];

M3: `McSpatial` R package by McMillen [27];

M4: EM estimator by McMillen [26], as coded in [7];

M5: Gibbs estimator by LeSage [21], as coded in [7];

M6: RIS estimator by Beron and Vijverberg [5], as coded in [7];

M7: GMM estimator by Pinkse and Slade [33], as coded in [7];

M8: LGMM estimator by Klier and McMillen [20], as coded in [7].

An extensive comparison with multiple sets of parameters is beyond the scope of this presentation. We will consider a simple SAR DGP as in Eq. (2), with $\boldsymbol{\beta} = (4, -2, 1)$, $\rho = 0.5$, W the spatial weight matrix with the 3 first nearest neighbours of n observations randomly distributed in the unit square, X_1 the intercept term, $X_2 \sim \mathcal{N}(2, 2)$ and $X_3 \sim \mathcal{N}(0, 1)$. We will look at the performances of the different estimators in terms of both accuracy of the estimated parameters and in terms of estimation time. The tests are performed over samples of increasing size ($n \in \{100, 500, 1000, 5000, 10000, 50000\}$). Whenever an estimator takes more than 5 minutes in average, we drop it from the comparison. In the Table 1 we report the mean bias of the estimated $\beta_1, \beta_2, \beta_3$ and ρ parameters, as well as the mean estimation time over 100 repetitions with randomly generated samples. All simulations are performed using `R` on a MAC OS X 10.6.8, with a dual-core 2.66 Ghz processor and 8 GB of RAM.

The results contained in Table 1 show that the `ProbitSpatial` package is definitively outperforming existing libraries in `R`. By looking at the first two columns we can see that `ProbitSpatial` does clearly better than the functions contained in `spatialprobit` [41] (they use Bayesian estimation), both in terms of accuracy than in terms of time. In particular, it is worth noting that the `spatialprobit` package tends to underestimate the spatial dependence parameter. By comparing column one and three, we can observe that the performance in terms of accuracy of `ProbitSpatial` and `McSpatial` are similar. This is due to the fact that both packages are based on likelihood maximisation. Nevertheless, `ProbitSpatial` is more suitable for large samples.

4 Other features of the library

The function `SpatialProbitFit` used for the estimation of SAR probit models can also be used to fit spatial error models (SEM). Furthermore, it allows to set the following parameters:

DGP either `SAR` or `SEM`;

method either `conditional` or `full-lik`. The first one implies a conditional estimation, the second one corresponds to full-likelihood estimation, with gradient functions. Further details can be found in [25].

Table 1. Mean bias and estimation time (in seconds) of the eight estimators for increasing sample sizes.

n		M1	M2	M3	M4	M5	M6	M7	M8
100	β_1	1.40	-0.38	1.43	-1.15	1.44	0.76	780.87	1.73
	β_2	-0.63	0.28	-0.65	0.64	-0.64	-0.35	-317.78	-0.71
	β_3	0.41	-0.08	0.42	-0.28	0.43	0.14	226.63	0.54
	ρ	-0.02	-0.26	-0.01	-0.16	-0.10	0.00	-0.05	0.02
	t	**0.10**	**4.79**	**0.13**	**11.17**	**13.46**	**29.94**	**2.32**	**0.01**
500	β_1	0.59	-1.04	0.59	-1.56	0.18	-0.18	56.36	0.67
	β_2	-0.27	0.54	-0.27	0.79	-0.07	0.12	-27.51	-0.32
	β_3	0.16	-0.26	0.16	-0.37	0.06	-0.13	14.87	0.14
	ρ	0.00	-0.22	-0.00	-0.14	-0.08	0.01	0.00	0.13
	t	**0.24**	**8.15**	**1.45**	**762.34**	**1278.48**	**1273.64**	**276.93**	**0.32**
1000	β_1	0.12	-1.18	0.12					0.18
	β_2	-0.04	0.61	-0.04					-0.06
	β_3	0.01	-0.31	0.01					0.02
	ρ	-0.01	-0.22	-0.01					0.12
	t	**0.41**	**11.70**	**10.26**					**3.13**
5000	β_1	0.07	-1.20						
	β_2	-0.03	0.61						
	β_3	0.01	-0.31						
	ρ	-0.01	-0.22						
	t	**2.68**	**44.52**						
10000	β_1	0.00	-1.29						
	β_2	0.00	0.65						
	β_3	0.01	-0.33						
	ρ	0.00	-0.22						
	t	**6.22**	**86.41**						
50000	β_1	0.01							
	β_2	0.00							
	β_3	0.00							
	ρ	0.00							
	t	**58.40**							

varcov either `varcov` or `precision`. Should the likelihood function be computed using the variance-covariance matrix (`varcov`) or the precision matrix (`precision`)? Default is `varcov`.

control a list of control parameters on the order of approximation of the Taylor expansion $(I_n - \rho W)^{-1} \sim I_n + \rho W + \rho^2 W^2 + \rho^3 W^3 + \ldots$, the tolerance of the optimizer and the pruning tolerance for certain matrices. See documentation of the function in the package.

It is worth noting that the output of the `SpatialProbitFit` is a model of class `SpatialProbit` containing information on the estimated parameters, the value of the log-likelihood and other model characteristics.

For the `SpatialProbit` class, the following `methods` are available

effects returns the marginal effects of the model (average direct, average indirect and average total effects as in LeSage and Pace [22]).

fitted returns the fitted values of the model in one of the following forms:

- `link`: the value of the latent variable;
- `response` the value of the probability;
- `binary` 0/1 vector of probability being greater than `cut` (default for `cut` is 0.5).

predict returns in-sample predict values of a new matrix of covariates. Same available forms as `fitted`;

residuals returns the generalised residuals of the model;

summary returns descriptive statistics of the data and model, the estimation time, the standard errors of the estimated parameters (computed with likelihood-ratio tests or with the variance covariance matrix if the option `covar = TRUE`) and the confusion matrix with the accuracy of the estimated model.

Finally, we also included two functions for simulation purposes: `generate_W` for generating a spatial weight matrix from a set of coordinates and `sim_binomial_probit` for generating the data.

5 Conclusions and forthcoming versions

Comprehensive tests presented in [25] and the ones presented here confirm that the `ProbitSpatial` package is, at the moment, the best option in R for fitting spatial binary choice models, under both SAR and SEM specifications. The improvement with respect to existing libraries is both in terms of accuracy than in terms of estimation time. It is also the only feasible option for large samples with more then a few thousand observations.

For the forthcoming version of the package we plan to implement:

- an estimator for the SARAR$(1, 1)$ model;
- an estimator for the spatial Durbin model, where spatial autoregression affects the covariates;
- a prediction function for out-of-sample observations (see [19] for details on best linear unbiased prediction in spatial framework);
- fix bugs of the current version.

Acknowledgement

Davide Martinetti has received the support of the EU in the framework of the Marie-Curie FP7 COFUND People Programme, through the award of an AgreenSkills fellowship under grant agreement 267196. The research reported in this work has been partially supported by projects URBANSIMUL and EPIDEC.

Bibliography

[1] P.R. Amestoy, T.A. Davis and I.S. Duff. (1996) *An approximate minimum degree ordering algorithm.* SIAM Journal on Matrix Analysis and Applications, **17**, 886–905.

[2] L. Anselin. (1982) *A note on small sample properties of estimators in a first-order spatial autoregressive model*, Environment and Planning, **14**, 1023–1030.

[3] C. Autant-Bernard. (2006) *Where do firms choose to locate their R&D? A spatial conditional logit analysis on French data*, European Planning Studies, **14**, 1187–1208.

[4] D. Bates and M. Maechler. (2016) *Matrix: Sparse and Dense Matrix Classes and Methods.* R package version 1.2–4. `http://CRAN.R-project.org/package=Matrix`.

[5] K.J. Beron and W.P.M. Vijverberg (2004). *Probit in a spatial context: a Monte Carlo analysis*, in Advances in Spatial Econometrics. Methodology, Tools and Applications, Luc Anselin, Raymond J.G.M. Florax, and Sergio J. Rey (eds.), Berlin: Springer, 169–195.

[6] C. Brun, A.R. Cook, J.S. Huay Lee, S.A. Wich, L. Pin Koh and L.R. Carrasco. (2015) *Analysis of deforestation and protected area effectiveness in Indonesia: a comparison of Bayesian spatial models*, Global Environmental Change, **31**, 285–295.

[7] R. Calabrese and J.A. Elkink. (2014) *Estimators of binary spatial autoregressive models: a Monte Carlo study.* Journal of regional science, **54**, 664–687.

[8] C. Carrión–Flores and E.G. Irwin. (2004) *Determinants of residential land-use conversion and sprawl at the rural-urban fringe.* American Journal of Agricultural Economics, **86**, 889–904.

[9] A. Case. (1992) *Neighborhood influence and technological change.* Regional Science and Urban Economics, **22**, 491–508.

[10] C.E. Clark. (1961) *The greatest of a finite set of random variables*, Operations Research, **9**, 145–162.

[11] Y.C. Collingham, R.A. Wadsworth, B. Huntley and P.E. Hulme. (2000) *Predicting the spatial distribution of non-indigenous riparian weeds: issues of spatial scale and extent.* Journal of Applied Ecology, **37**, 13–27.

[12] R.D. Connors, S. Hess and A. Daly. (2014) *Analytic approximations for computing probit choice probabilities.* Transportmetrica A: Transport Science, **10**, 119–139.

[13] A. Diallo and G. Geniaux. (2015) *Spatial logit for large samples with local spatial lag and regional spatial random effects using linearized GMM: an application to land use models.* Journal of Regional Science, submitted.

[14] M. Enea. (2014) *speedglm: Fitting Linear and Generalized Linear Models to large data sets.* R package version 0.2–1.0. `http://CRAN.R-project.org/package=speedglm`.

[15] M. Fortin, S. Delisle-Boulianne and D. Pothier. (2013) *Considering spatial correlations between binary response variables in forestry: an example applied to tree harvest modeling.* Forest Science, **59**, 253–260.

[16] G.J. Gibson, C.A. Glasbey and D.A. Elston. (1994) *Monte Carlo evaluation of multivariate normal integrals and sensitivity to variate ordering.* In: I.T. Dimov, B. Sendov and P.S. Vassilevski (eds.), Advances in Numerical Methods and Applications, 120–126.

[17] P.J. Heagerty and S.R. Lele. (1998) *A composite likelihood approach to binary spatial data.* Journal of the American Statistical Association, **93**, 1099–1111.

[18] W.A. Kamakura. (1989) *The estimation of multinomial probit models: a new calibration algorithm.* Transportation Science, **23**, 253–265.

[19] H.H. Kelejian and I.R. Prucha. (2007) *The relative efficiencies of various predictors in spatial econometric models containing spatial lags.* Regional Science and Urban Economics, **37**, 363–374.

[20] T. Klier and D.P. McMillen. (2008) *Clustering of auto supplier plants in the United States: Generalized method of moments spatial logit for large samples.* Journal of Business & Economic Statistics, **26**, 460–471.

[21] J.P. LeSage. (2000) *Bayesian estimation of limited dependent variable spatial autoregressive models.* Geographical Analysis, **32**, 19–35.

[22] J. LeSage and R.K. Pace. (2008) *An Introduction to Spatial Econometrics*, Chapman and Hall/CRC Eds..

[23] J.P. LeSage, R.K. Pace, N. Lam, R. Campanella and X. Liu. (2011) *New Orleans business recovery in the aftermath of hurrican Katrina.* Journal of the Royal Statistical Society A, **174**, 1007–1027.

[24] D. Martinetti and G. Geniaux. (2016) *ProbitSpatial: Probit with Spatial Dependence, SAR and SEM Models. R package version 1.0.* `http://CRAN.R-project.org/package=ProbitSpatial`.

[25] D. Martinetti and G. Geniaux. (2016) *Approximate likelihood estimation of spatial probit models.* Regional Science and Urban Economics, submitted. https://urbansimul.paca.inra.fr/urbansimul/pdf/recherche/

[26] D.P. McMillen. (1992) *Probit with spatial autocorrelation.* Journal of Regional Science, **32**, 335–348.

[27] D. McMillen. (2013) *McSpatial: Nonparametric spatial data analysis.* R package version 2.0, `http://CRAN.R-project.org/package=McSpatial`.

[28] D. McMillen and M.E. Soppelsa. (2015) *A conditionally parametric probit model of microdata land use in Chicago.* Journal of Regional Science, **55**, 391–415.

[29] N. Mendell and R. Elston. (1974) *Multifactorial qualitative traits: genetic analysis and prediction of recurrence risks.* Biometrics, **30**, 41–57.

[30] D.K. Munroe, J. Southworth and C.M. Tucker. (2004) *Modeling spatially and temporally complex land-cover change: the case of Western Honduras.* The Professional Geographer, **56**, 544–559.

[31] R.K. Pace and R. Barry. (1997) *Quick computation of regressions witha spatially autoregressive dependent variable.* Geographical Analysis, **29**, 232–237.

[32] R.K. Pace and J.P. LeSage. (2011) *Fast maximum likelihood estimation of the spatial probit model capable of handling large samples.* Available at SSRN: `http://ssrn.com/abstract=1966039` or `http://dx.doi.org/10.2139/ssrn.1966039`.

[33] J. Pinkse and M.E. Slade. (1998) *Contracting in space: an application of spatial statistics to discrete choice models.* Journal of Econometrics, **85**, 125–154.

[34] J. Pinkse and M. Slade. (2010) *The future of spatial econometrics.* Journal of Regional Science, **50**, 103–117.

[35] R Core Team. (2013) *R: A language and environment for statistical computing.* R Foundation for Statistical Computing, Vienna, Austria, `http://www.R-project.org/`.

[36] M.J. Schervish. (1985) *Algorithm AS 195: multivariate normal probabilities with error bound.* Journal of the Royal Statistical Society Series C, **33**, 81–94, correction, **34**, 103–104.

[37] G. Trinh and A. Genz. (2015) *Bivariate conditioning approximations for multivariate normal probabilities.* Statistics and Computing, **25**, 989–996.

[38] X. Wang, K.M. Kockelman. (2009) *Maximum simulated likelihood estimation with spatially correlated observations: a comparison of simulation techniques.* Transportation Statistics, Brian Sloboda (Ed.) JD Ross Publications, 173–189.

[39] Y. Wang, K.M. Kockelman and X. Wang. (2011) *Anticipation of land use change through use of geographically weighted regression models for discrete response.* Transportation Research Record, **2245**, 111–123.

[40] H. Wang, E.M. Iglesias and J.M. Wooldridge. (2013) *Partial Maximum likelihood estimation of spatial probit models.* Journal of Econometrics, **172**, 77–89.

[41] S. Wilhelm and M. de Matos. (2013) *Estimating spatial probit models in R.* R Journal, **5**, 130–43.

[42] T.-Y. Zhong, X.-J. Huanga, X.-Y Zhang, K. Wang. (2011) Temporal and spatial variability of agricultural land loss in relation to policy and accessibility in a low hilly region of Southeast China. Land Use Policy, **287**, 762–769.

Robustness for multilevel models with the forward search

Aldo Corbellini, *University of Parma, Italy*, aldo.corbellini@unipr.it
Luigi Grossi, *University of of Verona, Italy*, luigi.grossi@univr.it
Fabrizio Laurini, *University of Parma, Italy*, fabrizio.laurini@unipr.it

Abstract. Robustness of standard regression models have been studied quite extensively. When repeated measures are available, the methodological framework is generalized to multilevel models, for which little is known in term of robustness, even in the simplest case of ANOVA. We present a sequential forward search algorithm for multilevel models that allows robust and efficient parameters estimation in presence of outliers, and it avoids masking and swamping. The influence of outliers will be monitored at each step of the sequential procedure, which is the key element of the forward search. There are peculiar features when the forward search is applied to multilevel models. Such features pose new computational challenges, as some restrictions, that make the sub-models identifiable at every step, are required. The method is illustrated by an application to real data where exports of coffee to European countries are modeled and analyzed to identify outliers that might be linked to potential frauds. Preliminary results on simulated data have highlighted the benefit of adopting the forward search algorithm, which can reveal masked outliers, influential observations and show hidden structures.

Keywords. Multilevel Analysis, Outliers, Robust Methods

1 Introduction and motivation

In this paper a robust approach to the study of multilevel models is suggested. Multilevel models are a particular case of Linear Mixed Models (LMM) which are particularly attractive when repeated measures of variables are collected on a sample of individuals [7]. Parameters of these models are obtained through the GLS estimator which can be strongly affected by the presence of outliers and influential observations. It is then crucial to monitor the effect that highly influential observations could have on the final estimates and apply robust estimators. Many influence diagnostics have been proposed to detect outliers in longitudinal data analysis (see [7], ch. 6, for a review), but they are all based on the leave-k-out approach and can be strongly affected by the so called "masking" effect when the real number of outliers is greater than k. A monitoring method, which has revealed to be particularly effective to cope with the masking effect, is the forward search which has been originally proposed for linear regression models (see [1]), but it has been extended to many other fields thanks to its great flexibility [3].

The out-performance of the forward search and the statistical properties under the regression model are illustrated in the recent papers [2] and [8]. The key of the success of the forward search is that it offers a sequential algorithm such that, at each step, an efficient maximum likelihood estimate is carried out. In general, after fitting a parametric model with a maximum likelihood estimator, a set of residuals is computed. Being a sequential algorithm, it suffers from heavy load of computational time and efficient routines must be adopted to get results in short time.

In this paper we extend the forward search approach to multilevel models to monitor the influence of outliers on estimated coefficients and to suggest a way to obtain a robust estimator. The correlation structure of longitudinal data is more complex than that of regression models for cross-sectional data and this opens new challenging issues we try to address in this paper. In particular, we suggest an original method to increase the size of a basic subset which is tailored on LMM features and enables us to take into account the identifiability problems which could arise from the application of the forward search to longitudinal data.

The paper is structured as follows. In the next section the data used to apply the robust forward search approach are presented. The description of the data comes before the introduction of the model, which is specified in sections 3 and 4 together with the coefficients estimator, because the construction of the model is conditional to the trend and seasonal pattern of the data. The extension of the forward search to LMM and the main related issues are discussed in Section 5. The application of the forward search to simulated data and to real trade data is illustrated in section 6. Section 7 is devoted to the final discussion of results and possible further extensions.

2 Data, their features and economic characteristics

Motivation of the case study

Regression models have widely been used to understand, model and predict many trade data concerning commodities. Among others, coffee represents one of the most actively traded commodities on earth. Existing modern robust approaches using liner models are presented in [5].

The commercial policy of the European Union (EU) is generally based on uniform principles and common agreements with third countries. The infringement of such principles by traders operating in the EU market can have big negative impact on the EU economy and the EU budget.

One of the most common consequences is the considerable and often systematic under/over declaration of the invoice value which produces groups of outliers and clusters with linear structures, which are clearly visible in scatter plots of the traded value and volume. Nevertheless, such patterns can also emerge from perfectly regular trade, for example in presence of different product quality levels, peculiar market conditions, or episodic meteorological events.

Additionally, since the foundation of the Economic and Monetary Union (EMU) in Europe, it was expected to foster price transparency and convergence of goods and services among Member States. From a theoretical point of view, this means reducing deviations from the law of one price which states that market arbitrage should enforce broad parity in prices across individual goods. Once prices are converted to a common currency, the same good should be sold for the same price in different countries.

For all these reasons, detecting outliers for trade data is particularly important. That might highlight frauds or market inefficiencies. On the other hand, outlier detection has to be very accurate and it has to minimize the number of false detection. As robust estimators are usually less efficient than non robust estimators, tests for outlier detection based on very robust estimators usually detect a high number of false outliers [6]. Therefore, a robust and efficient algorithm, like the forward search, must be used to correctly identify true outliers.

Description of the data-set

Coffee production takes place mainly in developing countries. One of the biggest producers is Brazil. Coffee's consumption, instead, is well spread all over European countries, none of which produces it. The coffee price is characterized by relatively low price elasticity of supply because new plants require more than two years to become productive. Likewise demand is characterized by low price elasticity, because it

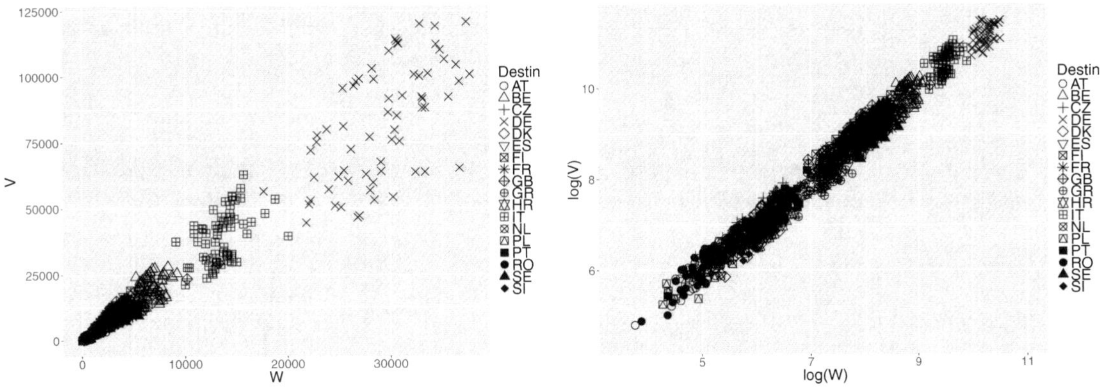

Figure 1. Left panel: Scatterplot of the data with symbol for each destination. Largest importers are Germany and Italy. Right panel: scatterplot of the log-log data

is quite stable and changes only if big movements in price arise independently from any income increase. As a consequence, coffee price time series show a very high variability and outliers are very likely to appear.

In this work we consider the coffee traded from the origin Brazil to destination of 18 countries in Europe. For each destination country the monthly value (in Euro) and the monthly quantity (in tons) of coffee sold from Brazil, to the specific country, are recorded. Monthly figures of sales are recorded for approximately 5 years. We stick on the 18 countries that traded with Brazil every month in these 5 years span. Therefore, our dataset represents a case of a balanced design in the framework of longitudinal data analysis. However, our results and methods are not sensitive to deviations from that balanced structure, although minor adjustments might be required to handle some missing data.

To make notation specific, we have that for each destination country (group) it is available the value of coffee imported from Brazil, denoted as $V_{i,t}$ with $i = 1, \ldots, g$ and $t = 1, \ldots, ng$. Additionally, the quantity imported by each country is denoted by $W_{i,t}$. From the above notation, it is already considered the option to have a different number of replicates for each country as the number of replicates ng might be different for each country.

A snapshot of the whole set of data is given in Figure 1 where each destination country is plotted with a different symbol. The left panel of Figure 1 is the prototype of scatterplot used by [5] for their analysis, where destinations were ignored.

To highlight the monthly time evolution of the trades, we also show, in Figure 2, the logged time series of V and W for Brazil versus Italy. A clear relationship is visible between the two series, but the presence of a long-term trend is questionable. The seasonality is quite clear in June and December, when the minimum and maximum yearly values happen, while it does not look particularly strong in other months.

3 Linear Mixed Models (Multilevel) for repeated measures

We consider the unified theory of linear mixed model as multilevel models, using the notation of [9], Chapter 4. The LMM has fixed component $X\beta$ and the random effect u written in the form

$$y = X\beta + Zu + \epsilon, \tag{1}$$

where Z is a kind of indicator matrix; an example of the structure of Z is postponed to Section 4. The model is well specified under a number of subsequent assumptions. Denoting with $\mathbf{I}$ the identity matrix

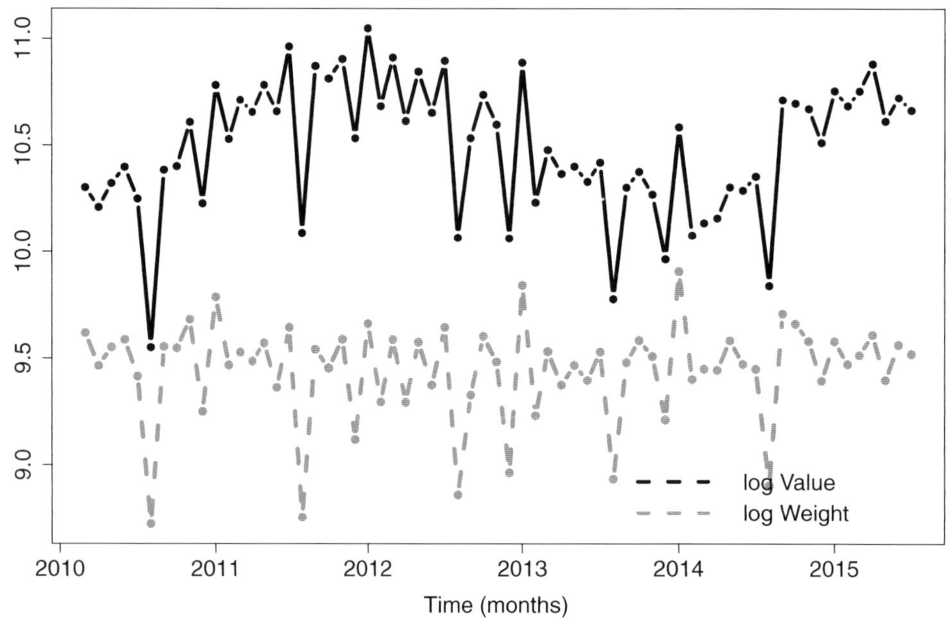

Figure 2. Time series of relevant quantities, in log scale, for destination Italy

and with **0** either a vector or a matrix with zeros (clear from the context), the random component u and the error term ϵ have the following features:

$$E\begin{bmatrix} u \\ \epsilon \end{bmatrix} = \begin{bmatrix} \mathbf{0} \\ \mathbf{0} \end{bmatrix}, \quad \text{Cov}\begin{bmatrix} u \\ \epsilon \end{bmatrix} = \begin{bmatrix} G & \mathbf{0} \\ \mathbf{0} & R \end{bmatrix} \quad \text{with } G = \sigma_u^2\mathbf{I} \text{ and } R = \sigma_\epsilon^2\mathbf{I}.$$

Model (1) is very flexible, despite its simplicity. For instance it offers a very convenient framework when repeated measures, which are generally dependent, have to be analyzed. For repeated measures it can be used as a "simple" random intercept model (also called a constant correlation model) where all components of β, but the intercept, are common among groups. Representation (1) is appropriate even if regression coefficients are assumed to vary among groups. Model (1) can also be used when groups are correlated to each others, by considering a richer structure for either R or G, or both. In the sequel we review some existing results of LMM and illustrate our robust procedure under the simplest specification of LMM, i.e. the uniform correlation model with only a random intercept.

One way to derive an estimate of β is to rewrite the LMM as a linear model with correlated errors having form

$$y = X\beta + \epsilon^\star \quad \text{where} \quad \epsilon^\star = Zu + \epsilon.$$

Under this representation, denoting with T the transposition of a matrix, clearly we have

$$\text{Cov}(\epsilon^\star) = V = ZGZ^T + R.$$

For the linear model with correlated errors, regression parameters can be estimated via the Generalized Least Squares (GLS) which can be written as

$$\tilde{\beta} = (X^TV^{-1}X)^{-1}X^TV^{-1}y. \tag{2}$$

For y having a general distribution, (2) can be shown to be the best linear unbiased estimator for β. Alternatively, if y is multivariate normal then the right-hand side of (2) is both the maximum likelihood (ML) estimator and the uniformly minimum variance unbiased estimator.

When random effects are included, then a unified way to see the overall estimation problem is to rewrite it through the notion of best linear unbiased prediction (BLUP). It turns out that (2) is also the BLUP of β and the BLUP of u is given by

$$\tilde{u} = GZ^T V^{-1}(y - X\tilde{\beta}).$$

Both GLS and BLUP estimators require elements of covariance matrices. There is a large and varied literature on estimation of covariance matrices in mixed models. In recent years, with the advent of better computing algorithms, ML or restricted ML (REML) have become the most common strategies for estimating the parameters in covariance matrices. An advantage in using REML compared to ML is a better accuracy in small samples.

Whatever the method used, once parameters of V and G are estimated, they are plugged into the BLUP yielding to estimated BLUP (EBLUP). Formally, EBLUP are given by introducing $\hat{\sigma}_u$ and $\hat{\sigma}_\epsilon$. Subsequently, estimated covariance matrices $\hat{V}$ and $\hat{G}$ lead to estimated BLUP given by

$$\hat{\beta} = (X^T\hat{V}^{-1}X)^{-1}X^T\hat{V}^{-1}y \text{ and } \hat{u} = \hat{G}Z^T\hat{V}^{-1}(y - X\hat{\beta}).$$

EBLUPs are important to obtain the fitted values of the model given by

$$\hat{y} = X\hat{\beta} + Z\hat{u}. \tag{3}$$

Estimated BLUPs have two sources of variability, namely estimation of the fixed and random effects pair β and u and estimation of covariance matrices G and V. This is somehow critical as both should be taken into account when making inference.

All statistical results discussed so far are key elements to derive the residuals from model (1), which are computed by taking the observed y and the fitted values from (3). Residuals are the building bricks to set up the forward search, whose details will be discussed next.

4 Model specification for trade coffee data

Given the data and our time-varying variables, discussed in Section 2, we will be using the following notation. It is often customary to take a log-log relationship in economics, due to its connection with elasticity. The resulting plot of such a joint transform is sketched in the right panel of Figure 1. From that plot emerges that there is a sort of "common slope" driving the relationship between the log variables, justifying the choice of the random intercept model.

Matching the general notation of LMM discussed in Section 3, let $y = \log V$ and the X matrix built using $\log W$, a linear trend and a set of seasonal dummies, assuming that the seasons will begin in January, and drop in December. So, for g Countries, if we have ng observations for the first Country, we could sketch the X matrix as

$$X = \begin{bmatrix} 1 & \log W_{\text{Country 1,1}} & 1 & 1 & 0 & \cdots & 0 \\ 1 & \log W_{\text{Country 1,2}} & 2 & 0 & 1 & \cdots & 0 \\ \vdots & \vdots & \vdots & \vdots & \vdots & \vdots & \\ 1 & \log W_{\text{Country 1,ng}} & ng & 0(?) & 0(?) & \cdots & 1(?) \\ \vdots & \vdots & \vdots & \vdots & \vdots & \vdots & \\ 1 & \log W_{\text{Country g,1}} & 1 & 1 & 0 & \cdots & 0 \\ \vdots & \vdots & \vdots & \vdots & \vdots & \vdots & \end{bmatrix}$$

The question mark associated to the last row of the first country is related to the last recording of value and quantity, for which the occurring month is "arbitrary".

As stated in Section 3, the matrix Z is, in general, a kind of indicator matrix. Under the special case of random intercept model, then Z has a number of columns equal to the number of groups. Specifically, for the random intercept model we have

$$Z = \begin{bmatrix} \mathbf{1}_{ng} & \mathbf{0}_{ng} & \cdots & \mathbf{0}_{ng} \\ \mathbf{0}_{ng} & \mathbf{1}_{ng} & \cdots & \mathbf{0}_{ng} \\ \vdots & \vdots & \vdots & \vdots \\ \mathbf{0}_{ng} & \mathbf{0}_{ng} & \cdots & \mathbf{1}_{ng} \end{bmatrix}$$

5 The forward search for LMM

The forward search is a sequential procedure which is based on a set of algorithms such that we start from an outlier-free subset of size $m^\star < n$ (the Basic Subset, BSB) and, at every step, the BSB is increased by including units closer to the selected model. In general, inclusion is such that at every step we move from a subset of size m to a subset of size $m+1$, with iterations until all n data are included. When outliers or other influential observations enter the subset, through the monitoring of relevant statistics, sharp movements are recorded. For regression models, this procedure is illustrated in [1] and made precise in [8].

There are several differences from regression to LMM. We start by highlighting that it is common to have time series even for simple random effects models, thus the selection of units belonging to $m^\star$ is made accordingly. To be specific, a sensible approach is to build the initial subset by taking contiguous observations in each group (here represented by all destination countries). Using proper notation, tailored to the set of available regressors (one explanatory variable, one time index for the trend and eleven dummies for the monthly seasonality), we have that, for each destination country i, with $i = 1, \ldots, g$, a coherent set of observations is given by fixing a time index $t^{(i)}$, and then selecting contiguous observations, the number of which is driven by k, the number of columns of the X matrix. Specifically, we select an arbitrary set of observations for each group $m_{t^{(i)},t^{(i)}+1,\ldots,t^{(i)}+k+1}$ which ensures that the model is identifiable. A similar choice is made for all groups but, in general, $t^{(i)} \neq t^{(j)}$, with $j = 1, \ldots, g$ and $i \neq j$. An initial subset M1 given by

$$\text{M1} = \bigcup_{i=1}^{g} \{m_{t^{(i)},t^{(i)}+1,\ldots,t^{(i)}+k+1}\},$$

is then used to fit the random effect model. After fitting the model the following algorithm is performed

1. Squared residuals $r^2_{i,t} = (y_{i,t} - \hat{y}_{i,t})^2$, for all units are computed (even for those that did not contribute to the fitting). With this notation $y_{i,t}$ and $\hat{y}_{i,t}$ denote a single element from the vector expressed in equations (1) and (3) respectively.
2. Squared residuals are sorted yielding to $r^2_{i,t}[\star_i]$, with the argument $[\star_i]$ denoting the that the sorting is kept separated for each group.
3. The median of $r^2_{i,t}[\star_i]$ is computed for each i and then stored.

Steps 1 to 3 above are then repeated by changing, for each group, the time index $t^{(i)}$ and leading to sets M2, M3, ..., all having the same size. This procedure is repeated 10000 times, as suggested by [3], since choosing among all possible subsets is unfeasible for almost all practical applications.

For each group the observations that lead to the smallest median of sorted residuals are those contributing to the BSB $m^\star$. Hence, not necessarily the time index to build $m^\star$ is identical for all groups, but inside each group, the contiguity of observation is at this stage guaranteed. This last restriction, however, might be relaxed when having an X matrix with a time index as explanatory variable, since the sorting of the data is irrelevant in the fitting when the time dependence is explicitly modelled with a time trend variable.

As stated above, once fitted the model, computing residuals during the forward search is quite similar to standard regression. The fitted values (3) are obtained after getting $\hat{\beta}$, $\hat{\sigma}_u$ and $\hat{\sigma}_\epsilon$ from the function `lmer` of library lme4 ([4]). Plugging such estimates into BLUP give estimates of $\hat{u}$ and $\hat{y}$.

Some discussion is needed when fitted values (3) and residuals are computed for all units, i.e. also for units that did not contribute to the estimation step. Fitted values and residuals are obtained by considering all entries of X (similarly to the regression model) but using only units that contributed to the fit to extract entries of Z (unlike the regression model). This last restriction is needed to ensure the conformability of product of matrices.

The key difference with regression model, however, comes when moving from m to $m+1$ because of the presence of groups in the data (here represented by each destination). The move from m to $m+1$ requires, once again, the sorting of squared residuals.

Our forward search algorithm is quite flexible and relatively general, as it does not require any group membership balancing during the procedure. In other words, at every step of the forward search, all groups must have only the minimum number of observations, say $m_{t^{(i)},t^{(i)}+1,\ldots,t^{(i)}+k+1}$, such that the full model is identifiable.

As stated above, the inclusion of new units, i.e. the move from size m to $m+1$, is based on the squared residuals computed for all data. As for the choice of the BSB, once residuals have been computed, the sorting is made separately for every group. Therefore, the difference compared to the forward search in regression, is that the sorting is not made on the whole dataset, but split by group. In practice, there are squared residuals sorted for the first group, then squared residuals sorted for the second group, and so forth.

The $(m+1)$-th observation joining the dataset is the one for which the squared sorted residuals not belonging to the m-th step is smallest. As a consequence, if observations belonging to the same group are well described by the model, and consequently have all small residuals, then the units forming the subset at steps $m+1$, $m+2$, ..., will belong to the same group. Therefore, at each step of the forward search, the size of each group belonging to the subset of size m can be, potentially, very different. As stated above, when discussing the choice of the BSB, the time index in the inclusion is not considered, as there is an explanatory variable taking the time trend into account, so that the fit does not require, necessarily, observations to be contiguous.

The peculiar feature of the forward search is that the inclusion of outliers or the inclusion of influential observations is highlighted by sharp peaks moving from m to $m+1$. This is also true for LMM. Influential observations, hidden structures, or outliers display similar pattern in many diagnostics summaries, like plot of residuals and standardized estimates of random effects.

6 Forward search for simulated and trade coffee data

An example with simulated data, where two outliers were added, is reported in Figure 3. The two outliers are the two central lowest extreme observations in the log-log transform (see the right panel of Figure 3) and, for the sake of the illustration, they have been assigned to Germany (DE). The time series of the simulated data, contaminated with two outliers is illustrated in Figure 4. Their effect is visible in the $\log V$ variable, but less visible, in this scale, for the $\log W$.

The structure is quite different from our example coffee data, but a similar pattern with some heteroschedasticity can be seen.

After running the forward search we monitor several graphical diagnostics. One of the most important plot in this framework is the monitor of estimates of standardized random effects, which is sketched in Figure 5. There is evidence, from findings in Figure 5, that toward the last steps of the forward search, influential observations were included into the procedure, showing sharp peaks. Not all groups, in the simulated data, were affected in the same way by the inclusion of outliers. In general the random components is absorbing influential observations, so outliers are likely to appear in the estimated random

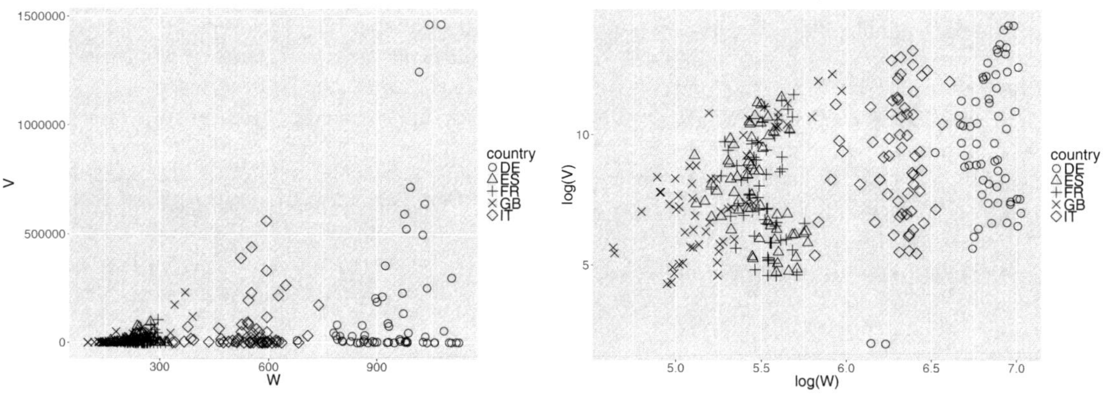

Figure 3. Scatterplot of the simulated data with five groups (left panel). Simulated data have also two outliers. The right panel is the associated log-log transform, with the two outliers located in the lowest part of the plot

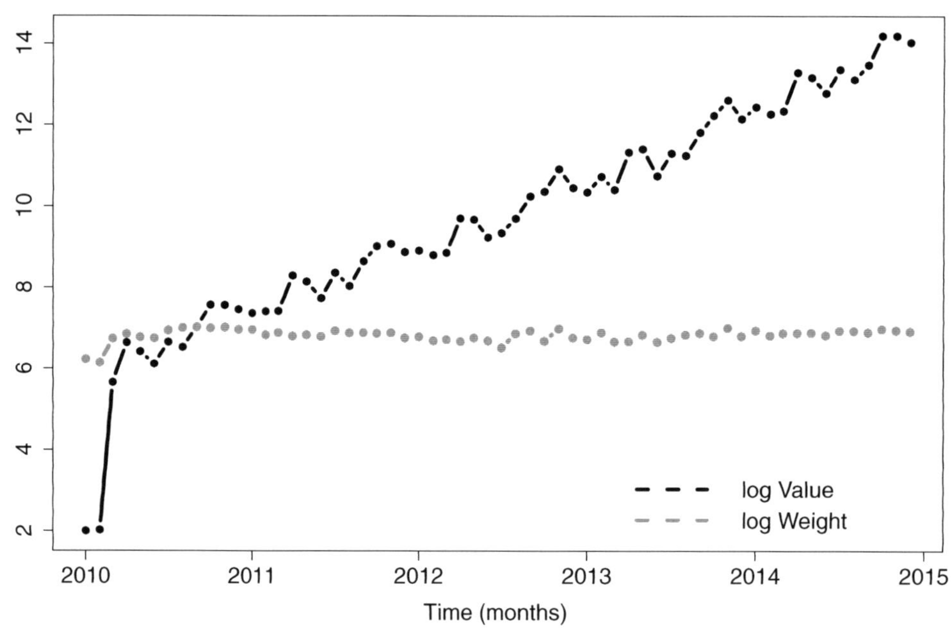

Figure 4. Time series of the simulated data for the country with two outliers. The outliers were added in both variables

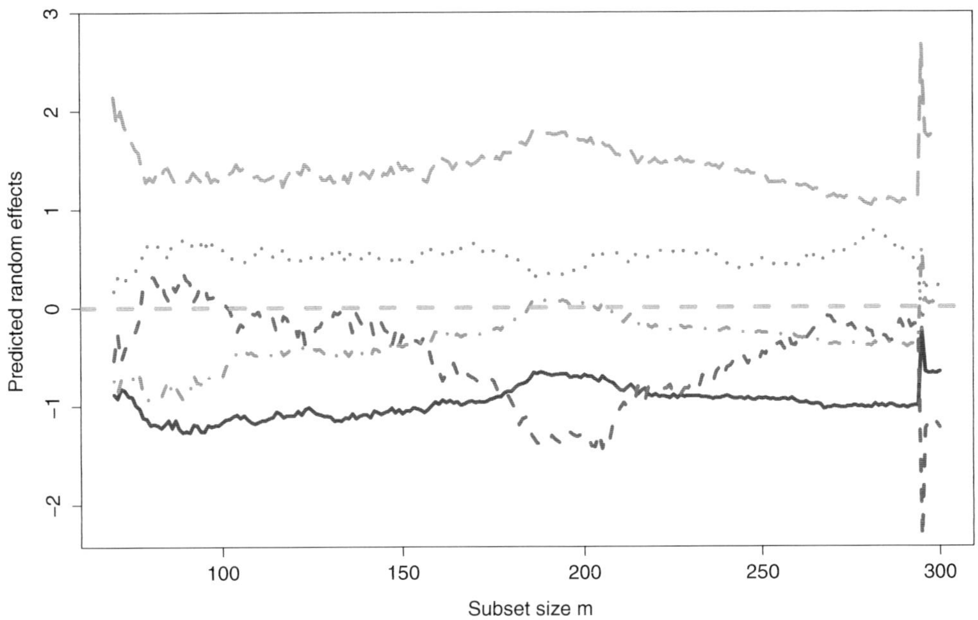

Figure 5. Estimates of random effects on the simulated and contaminated data during the forward search. Peaks toward the end of procedure shows the steps when outliers are included

components. There are peaks also in the residuals, but since the random effects are fewer, the diagnostic plot is simpler to interpret.

For the coffee data, introduced and discussed in Section 2, running the forward search lead to results which do not show any influential observations in the data. A snapshot of such a summary is visible by exploring the estimated standardized random effects during the forward search. Since there are 18 destination countries, it is hard to understand whether or not outliers, or hidden structures, are into the dataset.

Robust procedures, often, tend to spot outliers even when they are not present (false signal). The forward search does not suffer from that unappealing feature. However, in the future, we will perform more simulations, with cleaned and contaminated data, for better understanding the false discovery rate of the forward search, and compare it with other robust techniques.

7 Conclusions and discussion

We have introduced the forward search for LMM where correlation is induced by repeated measures. Previous strategies were restricted mostly to uncorrelated data and regression. The benefits of the introduced robust and efficient technique are, essentially, two-folds: when outliers are present, they are properly highlighted and estimates are unaffected by their presence. When data are outlier-free or when no hidden structure is inside the data, the forward search does not flag observations as outliers (false signals). An interesting avenue for future research is to add some outliers in the real data for understanding the effect of influential observations on the coffee data. A preliminary study on contaminated coffee data, not reported here for conciseness, suggested to introduce a small fraction of outliers in order to have a meaningful set of diagnostics. This is a subject that will be studied carefully in the nearest future.

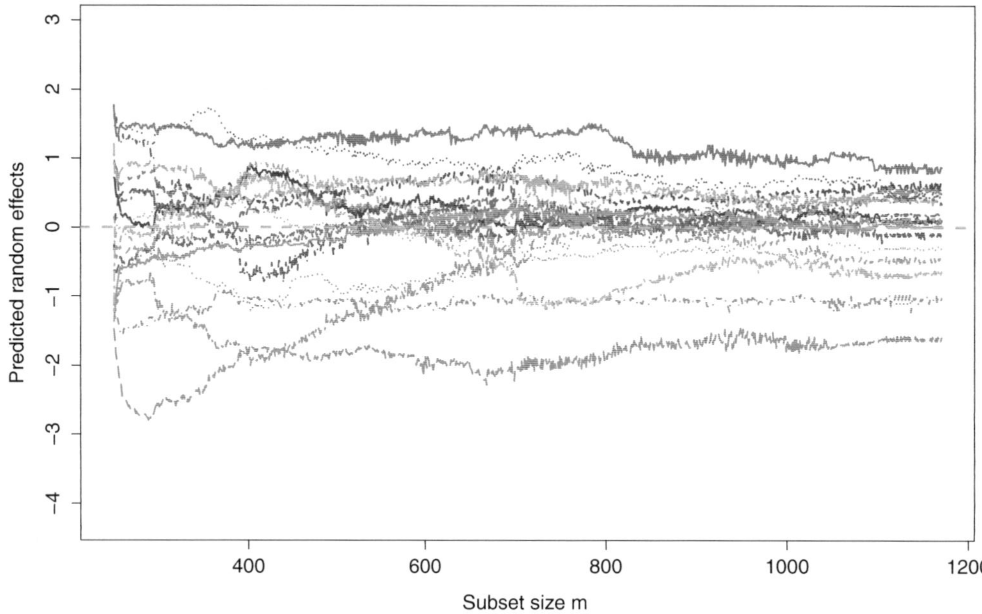

Figure 6. Estimates of random effects on the coffee trade data during the forward search. A quite smooth behaviour is visible, denoting lack of evidence of suspicious trades

Future research is toward several directions. The first avenue is to consider more general models than the random intercepts. In this paper, we have taken "for granted" that the explanatory variables had always to be included, but this might be questionable, and even dependent on the step of the forward search.

Simulations with more structured outliers must be carried out so that there might be more confidence and support for interpretations. A key feature is to build proper "test-based" inference to create a sequential procedure for testing for outliers, as explained by [8]. This last feature is, probably, the most challenging from the computational point of view, as it requires many forward search run in parallel.

Acknowledgements

We thank two anonymous referees and Marco Riani for helpful comments on the preliminary version of the manuscript. This research was partly supported by the project MIUR PRIN "MISURA"– Multivariate models for risk assessment.

Bibliography

[1] Atkinson, A.C. and Riani, M. (2000), *Robust Diagnostic Regression Analysis*. Springer–Verlag, New York.

[2] Atkinson, A.C., Riani, M. and Cerioli, A. (2010), The forward search: Theory and data analysis. *Journal of the Korean Statistical Society*, **39**, 117–134.

[3] Atkinson, A.C., Riani, M. and Cerioli, A. (2004), *Exploring Multivariate Data with the Forward Search*. Springer–Verlag, New York.

[4] Bates, D., Mächler, M., Bolker, B. and Walker, S. (2015), Fitting Linear Mixed-Effects Models Using lme4, *Journal of Statistical Software*, **67**, 1–48.

[5] Cerioli, A. and Perrotta, D. (2013) Robust clustering around regression lines with high density regions. *Advances in Data Analysis and Classification*, **8**, 5–26.

[6] Grossi, L. and Laurini, F. (2009) A robust forward weighted Lagrange multiplier test for conditional heteroscedasticity, *Computational Statistics & Data Analysis*, **53**, 6, 2251–2263.

[7] Liu X. (2016), *Methods and Applications of Longitudinal Data Analysis*, Academic Press, London.

[8] Riani, M., Atkinson, A.C. and Cerioli, A. (2009), Finding an unknown number of multivariate outliers. *Journal of the Royal Statistical Society, Series B*, **71**, 447–466.

[9] Ruppert, D., Wand, M.P. and Carroll, R.J. (2003) *Semiparametric regression*. Cambridge University Press, New York

Some mathematical notes on comprehensive factor analysis

Kohei Adachi, *Osaka University*, adachi@hus.osaka-u.ac.jp
Nickolay Trendafilov, *Open University*, nickolay.trendafilov@open.ac.uk

Abstract. In the currently prevalent model of factor analysis (FA), specific factors and errors are not dissociated, though they had been separated in the original conception of FA. Thus, an FA model with specific factors dissociated from errors is considered here, whose least squares procedure is referred to as comprehensive FA. The main goal includes showing that the model part of comprehensive FA and residuals are identifiable, common and specific factors are undetermined but have some constancy, and comprehensive FA has clear relationships with principal component analysis.

Keywords. Specific factors, Errors, Original factor analysis, Principal component analysis.

1 Introduction

The currently prevalent model of factor analysis (FA) can be expressed as

$$x = \Lambda f + \Psi u \tag{1}$$

for $p \times 1$ observed variable vector x whose expectation $E[x]$ equals the $p \times 1$ zero vector 0_p. Here, f and u are $m \times 1$ and $p \times 1$ latent variable vectors, respectively, with the elements of f called common factors and those of u called unique factors, and $m < p < n$. On the other hand, $\Lambda(p \times m)$ and $\Psi(p \times p)$ are fixed parameter matrices, where Λ contains factor loadings, and Ψ is the diagonal matrix, the squares of whose diagonal elements are called uniqueness [5].

According to [8] original conception of FA, Ψu is decomposed into an unsystematic error and the factor specific to the corresponding variable: $\Psi u = \Delta s + e$, with $e(p \times 1)$ an error vector, $s(p \times 1)$ the specific factor vector, and Δ a $p \times p$ diagonal matrix [4, 6]. The decomposition allows us to rewrite (1) as

$$x = \Lambda f + \Delta s + e. \tag{2}$$

Each diagonal element of Δ^2 is called specificity in that it stands for to what degree an observed variable is explained by the corresponding specific factor in s. The expectations for f, s, and e are assumed as $E[f] = 0_m, E[s] = E[e] = 0_p$, and

$$E[ff'] = I_m, E[ss'] = I_p, E[fs'] = O_{m\times p}, \tag{3}$$

$$E[fe'] = O_{m\times p}, E[se'] = O_{p\times p}, E[ee'] = \Omega^2. \tag{4}$$

with Ω a $p \times p$ diagonal matrix, $O_{m\times p}$ the $m \times p$ matrix of zeros, and I_m the $m \times m$ identity matrix.

It is desirable that the specificity Δ^2 in (2) is estimated rather than the error-perturbed uniqueness Ψ^2 in (1). However, the model (2) has only been mentioned so far without a parameter estimation procedure presented. In order to allow the estimation, we consider modifying (2) in this paper.

The modifications are as follows:
[1] Common and specific factors are treated as fixed parameters rather than variables;
[2] No distributional assumptions are made for errors: (4) is not assumed.
By **[1]**, our version of (2) can be written in the matrix form

$$X = F\Lambda' + S\Delta + E = ZB' + E \ . \tag{5}$$

Here, X is the n-observations $\times$ p-variables column-centered data matrix, F is the $n \times m$ matrix of common factor scores, S is the $n \times p$ one of specific factor scores, and E is the $n \times p$ error matrix, with $Z = [F, S](n \times (m+p))$ and $B = [\Lambda, \Delta](p \times (m+p))$ the block matrices of parameters to be estimated. The matrix versions of (3) can be expressed as

$$\frac{1}{n}F'F = I_m, \frac{1}{n}S'S = I_p, \frac{1}{n}F'S = 0_{m\times p} \text{ summarized as } \frac{1}{n}Z'Z = I_{p+m}. \tag{6}$$

Owing to **[2]**, i.e., no constraint for the errors in E, we may simply consider minimizing the squared sum of errors

$$f(Z, B) = ||E||^2 = ||X - (F\Lambda' + S\Delta)||^2 = ||X - ZB'||^2 \tag{7}$$

subject to (6). Here, Z and E being column-centered (i.e., the matrix version of $E[f] = 0_m$ and $E[s] = E[e] = 0_p$) is not be considered, since it is satisfied by the solution of the minimization problem as described later.

Indeed, the algorithms for the minimization have already been presented by [1, 3, 7, 9, 11]. However, it has not been recognized that the minimization is underlain by the model (5): in the above literature, the underlying model has not been considered and (7) is rather described as $f(F, \Lambda, U, \Psi) = ||X - (F\Lambda' + U\Psi)||^2$ with U an $n \times p$ matrix of unique factor scores: the algorithms have been presented rather for estimating the unique factors/uniqueness.

We refer to minimizing (7) subject to (6) as comprehensive FA (Comp-FA) in this paper. Its purpose is to present new results on the Comp-FA solution as theorems with some lemmas. They includes [A] the identifiability of Λ, Δ, ZB', and E, [B] the identifiability of $S_{XF}, S_{XS}, S_{FE}, S_{SE}$, and S_{EE} with S_{XF} denoting the covariance matrix between the columns of X and those of the optimal F, [C] the indeterminacy of F and S, and [D] the relationships to principal component analysis (PCA). Before presenting main results, we describe the preliminary ones in the next section. Throughout the paper, we suppose that the rank of XB equals the number of variables:

$$\text{rank}(XB) = p, \tag{8}$$

which implies $\text{rank}(X) = \text{rank}(B) = p$, the existence of $(X'X)^{-1}$, and $BB^+ = I_p$ with B^+ the Moore-Penrose inverse of B.

2 Preliminary Results

The Comp-FA solution is given by the matrices Z and B that minimizes (7) subject to (6). Both matrices cannot be jointly obtained in an explicit form, but if one of them is given, the solution of the other matrix can be obtained explicitly. In this section, we first present a useful decomposition of the loss function (7) as a theorem, which is followed by describing the properties of the solution already known.

Theorem 2.1. *Under the constraint (6), the loss function (7) can be decomposed as*

$$f(Z, B) = ||X - ZS'_{XZ}||^2 + n||S_{XZ} - B||^2. \tag{9}$$

Here, $S_{XZ} = n^{-1}X'Z = n^{-1}X'[F, S] = [S_{XF}, S_{XS}]$ with its blocks $S_{XF} = n^{-1}X'F$ and $S_{XS} = n^{-1}X'U$ the covariance matrices of observed variables to common factors and unique ones, respectively.

Proof. (7) can be rewritten as $||X - ZS'_{XZ} + ZS'_{XZ} - ZB'||^2 = ||X - ZS'_{XZ}|| + h - 2trC$ with $h = ||ZS'_{XZ} - ZB'||^2$ and $C = (X - ZS'_{XZ})'(ZS'_{XZ} - ZB')$. Here, we can expand C and use (6) to have $C = X'ZS'_{XZ} - X'ZB' - S_{XZ}Z'ZS'_{XZ} + S_{XZ}Z'ZB' = nS_{XZ}S'_{XZ} - nS_{XZ}B' - nS_{XZ}S'_{XZ} + nS_{XZ}B' = O_{p\times p}$. Further, (6) leads to $h = n||S_{XZ} - B||^2$. □

From (9), we can find that, for given Z, the optimal $B = [\Lambda, \Delta]$ is given by

$$\hat{B} = [S_{XF}, \text{diag}(S_{XS})], \text{ i.e., } \hat{\Lambda} = S_{XF} \text{ and } \hat{\Delta} = \text{diag}(S_{XS}). \tag{10}$$

The optimal Z for given $\hat{B}$ is given by

$$\hat{Z} = Z_1 + Z_2 = n^{1/2}K_1L_1' + n^{1/2}K_2L_2' = X\hat{B}L_1\Theta^{-1}L_1' + n^{1/2}K_2L_2' \tag{11}$$

with $Z_1 = n^{1/2}K_1L_1' = X\hat{B}L_1\Delta^{-1}L_1'$ and $Z_2 = n^{1/2}K_2L_2'$. Here, $K_1(n \times p), L_1((p+m) \times p)$, and Θ (a $p \times p$ diagonal matrix) is given through the singular value decomposition (SVD) of $n^{-1/2}X\hat{B}$:

$$n^{-1/2}X\hat{B} = K_1\Theta L_1'. \tag{12}$$

The remaining matrices $K_2(n \times m)$ and $L_2((p+m) \times m)$ form $K = [K_1, K_2]$ and $L = [L_1, L_2]$ satisfy $K'K = L'L = LL' = I_{p+m}$. In (11), we can find that Z_1 is uniquely determined, but Z_2 and $\hat{Z}$ are not uniquely determined.

Using the fact that (12) leads to the eigenvalue decomposition (EVD) of $\hat{B}'S_{XX}\hat{B}$ defined as

$$\hat{B}'S_{XX}\hat{B} = L_1\Theta L_1' \tag{13}$$

and the transpose of (12) post-multiplied by (11) provides

$$S_{XZ} = \hat{B}'^{+}L_1\Theta L_1', \tag{14}$$

the following algorithm can provide the solution of $B = [\Lambda, \Delta]$ [1]:
Step 1. Initialize $\hat{B}$ so as to satisfy (8)
Step 2. Perform EVD (13) to obtain S_{XZ} with (14)
Step 3. Update $\hat{B}$ with (10)
Step 4. Finish if convergence is reached; otherwise, back to Step 2.

In [1], it has been proved that (11) is column-centered if X is so.

3 Identifiability of Comp-FA Model

We suppose that $\hat{B}$ resulting in (10) also satisfies (8) with $\text{rank}(X\hat{B}) = p$. This section is started with the theorem which shows the identifiability of the loadings in Λ and the specificities in Δ^2 obtained through the four steps in the last section.

Theorem 3.1. *If the diagonal elements of Θ in EVD (13) have distinct values, then S_{XZ} in (14) is uniquely determined for given S_{XX} and $\hat{B}$. Further, S_{XZ} gives the unique $\hat{B}$ with (10).*

Proof. If Θ has distinct diagonal elements, then L_1 and Θ are uniquely determined for given S_{XX} as in (13). Further, $\hat{B}$ satisfying (8) provides the unique $\hat{B}^+$. Those L_1, Θ, and $\hat{B}^+$ uniquely determine S_{XZ} as in (14). Further, it gives the unique $\hat{B} = [S_{XF}, \text{diag}(S_{XS})]$ with (10), since of $S_{XZ} = [S_{XF}, S_{XS}]$. □

The following lemma and theorem show that the model part $F\Lambda' + S\Delta = ZB'$ in (5) can be identified:

Lemma 3.2. *The covariance matrix S_{XZ} in (14) is also expressed as*

$$S_{XZ} = S_{XX}\hat{B}L_1\Theta^{-1}L_1'. \tag{15}$$

Proof. Pre-multiplying the left and the right hand sides of (11) by $n^{-1}X'$ leads to $S_{XZ} = S_{XX}\hat{B}L_1\Theta^{-1}L_1' + n^{-1/2}X'K_2L_2'$. Here, $X'K_2L_2'$ is found to disappear as $X'K_2L_2' = \hat{B}'^{+}\hat{B}'X'K_2L_2' = \hat{B}'^{+}L_1'\Delta K_1'K_2L_2' = O_{p\times(m+p)}$ using (12) and $K_1'K_2 = O_{p\times m}$. □

Theorem 3.3. *The solution for* $ZB' = F\Lambda' + S\Delta$ *is expressed as*

$$\hat{Z}\hat{B}' = \sqrt{n}K_1L_1'\hat{B}' = X\hat{B}L_1\Theta^{-1}L_1'\hat{B}' = XS_{XX}^{-1}S_{XZ}'\hat{B}' = X\hat{B}S_{XZ}'S_{XX}^{-1}. \quad (16)$$

Proof. Pre-multiplying both sides of (12) by $(X'X)^{-1}X'$ leads to $n^{-1/2}\hat{B} = (X'X)^{-1}X'K_1\Delta L_1'$, which implies $\hat{B}L_2 = O_{p\times m}$ since of $L_1'L_2 = O_{p\times m}$. This fact and (11) lead to $\hat{Z}\hat{B}' = (n^{1/2}K_1L_1' + n^{1/2}K_2L_2')\hat{B}' = n^{1/2}K_1L_1'\hat{B}' = X\hat{B}L_1\Theta^{-1}L_1'\hat{B}'$. Further, (15) leads to $S_{XX}^{-1}S_{XZ}\hat{B}' = \hat{B}L_1\Theta^{-1}L_1'\hat{B}'$ and this being symmetric implies $S_{XX}^{-1}S_{XZ}\hat{B}' = \hat{B}S_{XZ}'S_{XX}^{-1}$, which complete (16). □

As found in (16), $F\Lambda' + S\Delta = ZB'$ is identifiable and its solution is a linear combination of the columns in X.

4 Properties of Residuals

The properties of the resulting residuals in $\hat{E} = X - \hat{Z}\hat{B}'$ are considered in this section. Theorem 3.3 shows that $\hat{E}$ is also identifiable and a linear combination of the columns of X. It implies that $\hat{E}$ is column-centered and its inter-variable covariance matrix is expressed as $S_{EE} = n^{-1}\hat{E}'\hat{E}$, which can be rewritten as follows:

Theorem 4.1. *The* $p\times p$ *covariance matrix* $S_{EE} = n^{-1}\hat{E}'\hat{E}$ *for residuals is expressed using* $\mathit{offd}(S_{XS}) = S_{XS} - \mathit{diag}(S_{XS})$ *as*

$$S_{EE} = S_{XX} - \hat{\Lambda}\hat{\Lambda}' - \mathit{offd}(S_{XS})\hat{\Delta} - \hat{\Delta}\mathit{offd}(S_{XS}) - \hat{\Delta}^2. \quad (17)$$

Proof. Using $\hat{E} = X - \hat{Z}\hat{B}'$ and (6), we have $S_{EE} = S_{XX} - S_{XZ}\hat{B}' - \hat{B}S_{XZ}' + \hat{B}\hat{B}'$. Here, $\hat{B}$ is substituted by (10) to provide $S_{XZ}\hat{B}' = [S_{XF}, S_{XS}][\hat{\Lambda}, \hat{\Delta}]' = \hat{\Lambda}\hat{\Lambda}' + S_{XS}\hat{\Delta} = \hat{\Lambda}\hat{\Lambda}' + \{\hat{\Delta} + \text{offd}(S_{XS})\}\hat{\Delta} = \hat{\Lambda}\hat{\Lambda}' + \hat{\Delta}^2 + \text{offd}(S_{XS})\hat{\Delta}$. Using this and $\hat{B}\hat{B}' = \hat{\Lambda}\hat{\Lambda}' + \hat{\Delta}^2$ in $S_{EE} = S_{XX} - S_{XZ}\hat{B}' - \hat{B}S_{XZ}' + \hat{B}\hat{B}'$ leads to (17). □

In general, (17) is not a diagonal matrix, which implies that the residuals are correlated between variables.

Theorem 4.2. *The p-residuals* $\times(m+p)$*-factors covariance matrix,* $S_{EZ} = [S_{EF}, S_{ES}] = n^{-1}\hat{E}'\hat{Z} = [n^{-1}\hat{E}'\hat{F}, n^{-1}\hat{E}'\hat{S}]$*, is expressed as* $S_{EZ} = [O_{p\times m}, \mathit{offd}(S_{XS})]$*, i.e.,*

$$S_{EF} = O_{p\times m} \text{ and } S_{ES} = \mathit{offd}(S_{XS}) \quad (18)$$

Proof. Using (6) and (10), we have $S_{EZ} = n^{-1}(X - \hat{Z}\hat{B}')'\hat{Z} = S_{XZ} - \hat{B} = [S_{XF}, S_{XS}] - [\hat{\Lambda}, \hat{\Delta}] = [O_{p\times m}, \text{offd}(S_{XS})]$. □

The equation $S_{EF} = O_{p\times m}$ in the theorem shows that the residuals are uncorrelated with common factors, while $S_{ES} = \text{offd}(S_{XS})$ implies that the residual for each variable is uncorrelated with the specific factor for that variable, but correlated with the specific factors for the other variables.

The following lemma and theorem concern the amount of residuals:

Lemma 4.3. *The post-multiplication of the p-variables* $\times$ $(m+p)$*-factors covariance matrix by its transpose equals the sample covariance matrix:*

$$S_{XZ}S_{XZ}' = S_{XX} \quad (19)$$

Proof. Using (14), we have $S_{XZ}S'_{XZ} = \hat{B}'^{+}L_1\Theta^2L'_1\hat{B}^{+}$, which is rewritten as $\hat{B}^{+\prime}\hat{B}'S_{XX}\hat{B}\hat{B}^{+} = S_{XX}$ using (13) and $\hat{B}\hat{B}^{+} = I_p$. □

Theorem 4.4. *The amount of residuals, i.e., the resulting value of loss function (7), is expressed as*

$$||\hat{E}||^2 = n\ tr(S_{XX} - \hat{\Lambda}\hat{\Lambda}' - \hat{\Delta}^2) = n||offd(S_{XS})||^2 \tag{20}$$

Proof. The first identity in (20) is given by using $troffd(S_{XS})\hat{\Delta} = 0$ in $f(\hat{Z}, \hat{B}) = ntrS_{EE}$ with (17). The second identity follows from (9): the term $||X - ZS'_{XS}||^2$, in which $\hat{Z}$ is substituted, disappears as

$$||X - S'_{XZ}||^2 = ntrS_{XX} - 2ntrS_{XZ}S'_{XZ} + ||\hat{Z}S'_{XZ}||^2 = 0 \tag{21}$$

using (6) and (19). That is, the resulting value of (7) or (9) amounts to $n||S_{XZ} - \hat{B}||^2$ which is found to equal $n||\text{offd}(S_{XS})||^2$ using (10). □

The theorem shows that the amount of residuals is proportional to the sum of the squared covariances of each variable to the specific factors for the other variables. It stands for the deviation of a solution from the FA assumption that each of specific factors is specific to the corresponding variable.

5 Factor Scores as Higher Rank Approximation

The $n \times p$ matrix $ZB' = F\Lambda' + S\Delta$ was shown to be identified in Section 3, while $Z = [F, S](n \times (m+p))$ cannot be identified as found in (11). Their clear difference is in the numbers of columns and their ranks with $rank(Z) = p + m > rank(ZB') = p = rank(X)$: the rank of the factor score matrix Z is higher than that of the data matrix X.

Differently from X and ZB', the two matrices XB and Z_1 in (11) are $n \times (m+p)$ as is Z, and have the following relationships to Z. First, XB is the target matrix approximated by Z, which is shown by that Comp-FA loss function (7) can be rewritten as

$$f(Z, B) = ||Z - XB||^2 + ||X||^2 + n||B||^2 - ||XB||^2 - n(p+m). \tag{22}$$

In its right-side hand, only $||Z - XB||^2$ is relevant to Z [2]. It shows that the resulting $\hat{Z}$ is a higher rank approximation of XB with $rank(Z) = p + m > rank(XB) = p$. Second, $\hat{Z}$ is equidistant from $Z_1 = X\hat{B}L_1\Theta^{-1}L'_1$ with

$$||\hat{Z} - Z_1||^2 = ||Z_2||^2 = ||n^{1/2}K_2L'_2||^2 = nm \tag{23}$$

[1]. Further, Z_1 can be expressed as in the next theorem:

Theorem 5.1.

$$Z_1 = X(X'X)^{-1}X'\hat{Z}. \tag{24}$$

Proof. We can rewrite (15) into $S_{XX}^{-1}S_{XZ} = \hat{B}L_1\Theta^{-1}L'_1$. Its use in (11), i.e., $Z_1 = X\hat{B}L_1\Theta^{-1}L'_1$ implies (24). □

This theorem shows that the columns of Z_1 are the projection of those of $\hat{Z}$ onto the column space of X. Equations (23) and (24) allows us to draw the cone in Figure 1, where matrices are depicted as arrows. This figure shows that the matrix $\hat{Z} = [\hat{F}, \hat{S}]$ containing common and specific factor scores forms the cone whose central axis is Z_1 with the distance of $\hat{Z}$ to Z_1 being $(nm)^{1/2}$. In other words, $\hat{Z}$ exists somewhere on the circumference of the circle whose center is Z_1 and radius is $(nm)^{1/2}$.

Theorem 5.2. *It holds*

$$X = \hat{Z}S'_{XS} = \hat{Z}(\hat{Z}'\hat{Z})^{-1}\hat{Z}'X \tag{25}$$

Proof. Equation (21) implies $X = \hat{Z}S'_{XS} = \hat{Z}(n^{-1})\hat{Z}'X$. Here, we can use (6) to provide (25). □

This theorem shows that the column space of X is included in that of $\hat{Z}$.

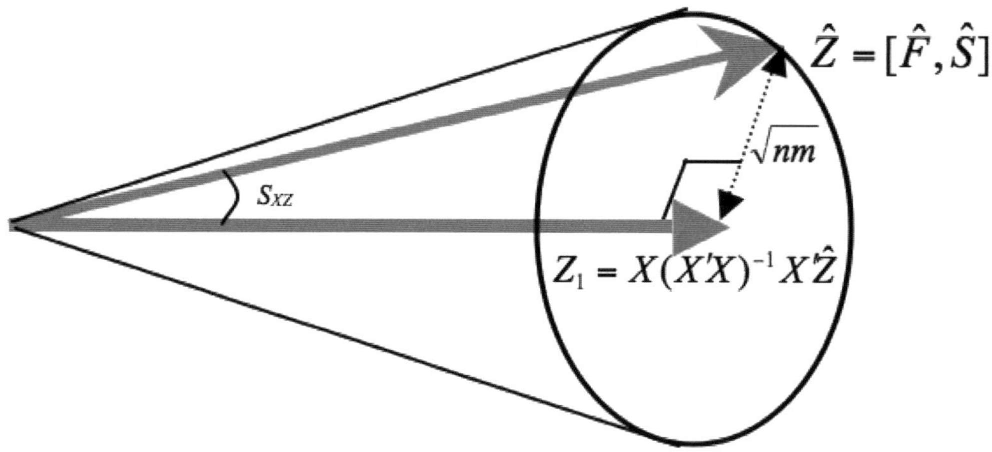

Figure 1. Cone formed by common and specific factor scores

6 Comparisons to PCA

It can be considered that an established procedure should have clear relationships to a related one, which may be principal component analysis (PCA) for FA. How PCA is similar to/different from FA is inconclusive in the prevalent framework of FA with (1), but is clarified by Comp-FA. The difference is simply whether specific factors are considered: the loss function of PCA can be defined as

$$f_{PC}(F, \Lambda) = ||X - F\Lambda'||^2, \tag{26}$$

by excluding the specific factor part $S\Delta$ from the Comp-FA loss function (7) [10].

We can further show clear inequalities holding between PCA and FA solutions, as shown below. There, F_{FA} and A_{FA} are used for the FA solutions of F and A, while F_{PC} and A_{PC} denote the PCA ones with

$$\frac{1}{n}F'_{PC}F_{PC} = I_m \tag{27}$$

which can be supposed without loss of generality.

In the next theorem, the fact is used that the resulting value of the PCA loss function (26) is expressed as

$$||X - F_{PC}\Lambda'_{PC}||^2 = ntr(S_{XX} - tr\Lambda_{PC}\Lambda'_{PC}). \tag{28}$$

and [10] FA-like PCA is considered in which

$$g(S^*, \Delta^*) = ||X - (F_{PC}\Lambda'_{PC} + S^*\Delta^*)||^2 \tag{29}$$

is minimized over S^* and Δ^* subject to

$$F'_{PC}S^* = O_{p\times m},\ n^{-1}S^{*'}S^* = I_p,\ \Delta^* \text{ being diagonal.} \tag{30}$$

Theorem 6.1. *The amount of residuals (28) for PCA cannot be less than that for FA (20):*

$$tr(S_{XX} - tr\Lambda_{PC}\Lambda'_{PC}) \geq tr(S_{XX} - \Lambda_{FA}\Lambda'_{FA} - \hat{\Delta}^2) \tag{31}$$

implying $tr\Lambda_{PC}\Lambda'_{PC} \leq tr\Lambda_{FA}\Lambda'_{FA} + tr\hat{\Delta}^2$.

Proof. As compared with PCA, EFA-like PCA has additional parameters S^* and Δ^*, which implies the function (29) value cannot exceed (28):

$$ntr(S_{XX} - tr\Lambda_{PC}\Lambda'_{PC})||X - (F_{PC}\Lambda'_{PC} + S^*\Delta^*)||^2. \tag{32}$$

Further, S^*, Δ^*, and F_{PC} with (27) and (30) can be substituted into S, Δ, and F in (6), respectively: S^*, Δ^*, and F_{PC} meet the Comp-FA model (5) and (6), but they are not optimal and thus the (29) value cannot be less than the Comp-FA loss function value (20):

$$||X - (F_{PC}\Lambda'_{PC} + S^*\Delta^*)||^2 \geq ntr(S_{XX} - \Lambda_{FA}\Lambda'_{FA} - \hat{\Delta}^2). \tag{33}$$

Inequalities (32) and (33) imply (31). □

The relationships of loadings between PCA and FA are shown next:

Theorem 6.2.

$$tr\Lambda_{FA}\Lambda'_{FA} \leq tr\Lambda_{PC}\Lambda'_{PC} \tag{34}$$

Proof. Since the PCA solution minimizing (26) is known as the best lower rank approximation, the resulting value of (26), i.e., (28) cannot exceed $f_{PC}(F_{FA}, \Lambda_{FA}) = ||X - F_{FA}\Lambda'_{FA}||^2$, which is rewritten using (10) as $n(trS_{XX} - 2trS_{XF}\Lambda'_{FA} - tr\Lambda_{FA}\Lambda'_{FA}) = n(trS_{XX} - tr\Lambda_{FA}\Lambda'_{FA})$. It is not less than (28), which implies (34). □

This theorem shows that the absolute values of PCA loadings tend to be larger than those of FA ones for a data set.

7 Illustration

In order to illustrate Comp-FA and its properties discussed so far, we carried out Comp-FA for the column-standardized version of Harman's [4] 12-observations × 5-variables socio-economic data matrix. The resulting loadings in Λ and specificities in Δ^2 are shown in Table 1(A) together with D_{EE} indicating the diagonal elements of (17), i.e., the variances of the residuals. On the other hand, Table 1(B) and (C) present the maximum likelihood solution for the prevalent FA model (1) and the PCA solution, respectively. Every loading matrix Λ has been rotated by the varimax method. The loadings are mutually similar among the three solutions and allow the equivalent interpretation of factors/components. Let us compare Table 1(A) against (B) for the parameters other than Λ. For example, in (B), the uniqueness for the variable employment is found to be 0.040. In this value, the specificity and the amount of residuals are compounded, while they are dissociated in (A) with their values 0.026 and 0.003, respectively. Such a specificity for each variable is not estimated in PCA without considering specific factors: only amounts of the residuals remaining unexplained by components are given by D_{EE} in (C). Comparing (A) and (C), we can confirm that the amounts of residuals and the absolute values of loadings for Comp-FA tend to be smaller than those for PCA, which are suggested in Theorems 6.1 and 6.2, although exceptional values are also found in Table 1.

Table 1. Solutions of Comp-FA, Prevalent FA, and PCA for the socio-economic data.

Variable	(A) Comp-FA				(B) Prevalent FA			(C) PCA		
	Λ		Δ^2	D_{EE}	Λ		Ψ^2	Λ		D_{EE}
Population	0.995	0.018	0.007	0.002	0.999	0.015	0.000	0.994	0.011	0.000
School	0.016	0.878	0.229	0.000	-0.003	0.900	0.190	-0.004	0.941	0.013
Employment	0.977	0.128	0.026	0.003	0.970	0.132	0.040	0.981	0.132	0.000
Services	0.425	0.787	0.200	0.000	0.427	0.796	0.184	0.451	0.823	0.014
House	0.002	0.983	0.033	0.001	0.008	0.960	0.078	-0.001	0.968	0.004

As shown in Theorem 3.3, the model-part $\hat{Z}\hat{B}'$ is identifiable, which implies that the residual matrix $\hat{E} = X - \hat{Z}\hat{B}'$ is also so. Table 2 presents the matrices $\hat{Z}\hat{B}'$ and $\hat{E}$ resulting in Comp-FA. The former elements have a far larger variance than the latter ones, with $\|\hat{Z}\hat{B}'\|^2 = 59.93$ and $\|\hat{E}\|^2 = 0.07$ whose sum equals $\|X\|^2 = 60$. It shows that the Comp-FA model is fitted very well to the data set.

Table 3 shows the matrix S_{XS} containing the covariances between observed variables and specific factors. The squares of the diagonal elements in Table 3 equal to the specificities Δ^2 in Table 1(A). As explained in Section 4, the largeness of the off-diagonal elements in S_{XS} stands for the deviation of the solution from the FA assumption that a specific factor is specific to the corresponding variable: the factor is uncorrelated to the other variables. Since both variables and factors are standardized, the covariances in Table 3 are also correlation coefficients. It allows us to easily find that no off-diagonal element shows a substantially large correlation. We can thus ascertain that all observed variables meet the FA assumption fairly well.

Table 2. Observation × variables matrices of the model-part and residuals.

Obs.	(A) Model-part					(B) Residuals				
	Po	Sc	En	Se	Ho	Po	Sc	En	Se	Ho
1	-0.205	0.776	0.184	1.351	1.327	0.040	0.018	-0.044	0.005	-0.015
2	-1.525	-0.318	-1.532	-0.996	-1.147	-0.067	0.001	0.073	-0.011	-0.001
3	-0.896	-1.532	-1.086	-1.013	-1.323	0.033	-0.012	-0.036	0.005	0.010
4	-0.731	1.251	-0.544	0.176	1.322	-0.010	0.011	0.011	-0.002	-0.010
5	-0.714	0.789	-0.581	0.169	1.317	0.033	0.005	-0.036	0.006	-0.005
6	0.491	-1.833	0.338	-0.570	-0.822	0.104	-0.003	-0.113	0.017	0.002
7	-1.514	-0.013	-1.645	-1.006	-0.173	-0.017	-0.011	0.018	-0.002	0.009
8	0.914	0.034	0.764	-0.546	-0.491	-0.046	0.000	0.049	-0.006	-0.001
9	1.159	0.649	0.845	0.544	0.138	-0.048	-0.031	0.052	-0.006	0.027
10	1.032	1.346	1.052	2.449	1.289	-0.012	-0.026	0.014	-0.002	0.023
11	0.986	-1.076	0.85	-0.377	-0.82	0.033	0.000	-0.036	0.006	0.000
12	1.003	-0.071	1.355	-0.181	-0.616	-0.044	0.047	0.048	-0.009	-0.040

Table 3. Covariances between observed variables and specific factors (in bold).

Variable	Specific Factor				
	Po	Sc	En	Se	Ho
Population1	**0.081**	-0.045	0.000	0.002	0.015
School	-0.008	**0.478**	0.017	-0.007	0.000
Employment	0.000	0.049	**0.162**	-0.002	-0.016
Services	0.000	-0.007	-0.001	**0.447**	0.002
House	-0.007	0.000	-0.014	0.006	**0.181**

Acknowledgements

The first author is supported by Grant (C)-26330039 from the Japan Society for the Promotion of Science. The second author is supported by a grant RPG-2013-211 from The Leverhulme Trust, UK.

8 Conclusions

In this paper, we presented some mathematical results on the least squares solution for the comprehensive factor analysis (Comp-FA) model $X = F\Lambda' + S\Delta + E$, in which the specific factors in S are separated from the unsystematic errors in E with the columns of $[F, S]$ standardized and mutually orthogonal.

Main results for the solutions of F, Λ, S, and Δ are as follows: the loadings in Λ, the specificities in Δ^2, and the model part $F\Lambda' + S\Delta$ are identifiable and it is a linear combination of the columns in data matrix X (Theorem 3.1 and Theorem 3.3), although the matrix $[F, S]$ containing common and specific factors is not unique: its rank is higher than the data matrix X and the column space of $[F, S]$ includes that of X (Theorem 5.2).

The identifiability of the model part implies that of the residuals in $\hat{E} = X - \hat{F}\hat{\Lambda}' - \hat{S}\hat{\Delta}$. The results for $\hat{E}$ are summarized as follows: residuals are correlated among variables (Theorem 4.1). The residual for each variable is uncorrelated with common factors and the specific factor for that variable, but correlated to the specific factors for the other variables (Theorem 4.2). The amount of residuals is proportional to the sum of the squares of the covariances between each variable and the specific factors for the other variables: the amount stands for the deviation from the FA assumption that a specific factor is specific to each variable (Theorem 4.4).

The Comp-FA model clarifies the relationships of FA to principal component analysis (PCA): the incorporation of the specific factor part $S\Delta$ into PCA leads to FA. Owing to it, FA shows the better fit to a data set than PCA, while PCA tend to provide the loadings of larger absolute values than FA, as shown in Theorem 6.1 and Theorem 6.2.

Bibliography

[1] Adachi, K. (2012). Some contributions to data-fitting factor analysis with empirical comparisons to covariance-fitting factor analysis. *Journal of the Japanese Society of Computational Statistics*, 25, 25-38.

[2] Adachi, K. (2014). A matrix-intensive approach to factor analysis. *Japanese Journal of Statistics*, 44, 363-382 (in Japanese).

[3] de Leeuw, J. (2004). Least squares optimal scaling of partially observed linear systems. In *Recent developments of structural equation models: Theory and applications*, Eds. by K. van Montfort, J. Oud and A. Satorra, 121-134. Kluwer Academic Publishers.

[4] Harman, H. H. (1976). *Modern factor analysis* (3rd Edition Revised). The University of Chicago Press.

[5] Mulaik, S. A. (2010). *Foundations of factor analysis*, 2nd Edition. Boca Raton: CRC Press.

[6] Reyment R. and Jöreskog, K. G. (1996). *Applied factor analysis in the natural sciences.* Cambridge University Press.

[7] Sočan, G. (2003). *The incremental value of minimum rank factor analysis.* PhD Thesis, University of Groningen.

[8] Spearman, C. (1904). General Intelligence, objectively determined and measured. *American Journal of Psychology*, 15, 201-293.

[9] Stegeman, A. (2016). A new method for simultaneous estimation of the factor model parameters, factor scores, and unique parts. *Computational Statistics and Data Analysis*, 99, 189-203.

[10] Trendafilov, N. T., Unkel, S., and Krzanowski, W. (2013). Exploratory factor and principal component analyses: Some new aspects. *Statistics and Computing*, 23, 209-220.

[11] Unkel, S. and Trendafilov, N. T. (2010). Simultaneous parameter estimation in exploratory factor analysis: An expository review. *International Statistical Review*, 78, 363-382.

Approximating the Rao's distance between negative binomial distributions. Application to counts of marine organisms.

Claude Manté, *Aix-Marseille Université, Université du Sud Toulon-Var, CNRS/INSU, IRD, MIO, UM 110, F13288 Marseille Cedex 09, France*, claude.mante@mio.osupytheas.fr
Saikou Oumar Kidé, *Institut Mauritanien de Recherches Océanographiques et des Pêches, Laboratoire de Biologie et Ecologie des Organismes Aquatiques- BP 22 Nouadhibou Mauritania*, saikoukide@gmail.com

Abstract. While the negative binomial distribution is widely used to model catches of animals, it is noteworthy that the parametric approach is ill-suited from an exploratory point of view. Indeed, the "visual" distance between parameters of several distributions is misleading, since on the one hand it depends on the chosen parametrization and on the other hand these parameters are not commensurable (*i. e.* they measure quite different characteristics). Consequently, we settle the topic of comparing abundance distributions in a well-suited framework: the Riemannian manifold $NB(D_{\mathcal{R}})$ of negative binomial distributions, equipped with the Fisher-Rao metrics. It is then possible to compute an intrinsic distance between species. We focus on computational issues encountered in computing this distance between marine species.

Keywords. Information geometry, abundance distributions, geodesic, cut point

1 Introduction

The statistical analysis of counts of living organisms brings information about the collective behavior of species (schooling, habitat preference, *etc*), possibly associated with their biological characteristics (growth rate, reproductive power, survival rate, *etc*). This task can be implemented in an exploratory setting (see for instance [8, 7] and the references therein), but parametric distributions are also widely used for modeling populations abundance. Thus, the negative binomial (NB) distribution is commonly used to model catches of animals [2, 10, 12, 9].

This distribution is especially relevant for this purpose, because [9]:

1. it arises as a Gamma-Poisson mixture, whose parameters depend on the more or less aggregative behavior of the species, and on the efficiency of the trawl for catching it

2. it arises as the limit distribution of the Kendall's [6] birth-and-death model; in this setting, the parameters depend on the demography of the species (reproductive power, mortality, immigration rate)
3. in addition, it is a natural model for collections (of animals, for instance).

But it is noteworthy that the parametric approach is ill-suited from an exploratory point of view: the "visual" distance between parameters of several NB distributions is misleading, because on the one hand it depends on the chosen parametrization and, on the other hand, these parameters are not commensurable in general (they are associated with completely different characteristics of the species, in the setting of different statistical models). Considering the Riemannian manifold $NB(D_{\mathcal{R}})$ of negative binomial distributions (NB) equipped with the Fisher-Rao metrics, we can compute intrinsic distances between species, on the basis of their counts. Then, the "visual" distance between species approximated through Multidimensional Scaling of the table of Rao's distances (for instance) is a sound dissimilarity measure between species.

2 Notations

Consider a Riemannian manifold $\mathfrak{M}$, and a parametric curve $\alpha : [a, b] \to \mathfrak{M}$; its first derivative with respect to "time" will be denoted $\dot{\alpha}$. A geodesic curve γ connecting two points p and q of $\mathfrak{M}$ will be alternatively denoted $p \curvearrowright q$, and $p \curvearrowright q \oplus q \curvearrowright r$ will denote the broken geodesic [1] connecting p to r with a "stopover" at q. A probability distribution $\mathfrak{L}^i$ will be identified with its coordinates with respect to some chosen parametrization; for instance, we will write $\mathfrak{L}^i \equiv (\phi^i, \mu^i)$.

We also consider for any $x \in \mathfrak{M}$ the local norm $\|V\|_g(x)$ associated with the metrics g on the tangent space $T_x\mathfrak{M}$:

$$\forall V \in T_x\mathfrak{M}, \ \|V\|_g(x) := \sqrt{V'.g(x).V}. \tag{1}$$

Finally, the length of a curve α traced on $\mathfrak{M}$ will be denoted $\Lambda(\alpha)$.

3 The Rao's distance

In a seminal paper, Rao [11] noticed that, equipped with the Fisher information metrics denoted $\mathfrak{g}(\bullet)$, a family of probabilities depending on p parameters can be considered as a p-dimensional Riemannian manifold. The associated Riemannian (Rao's) distance between the distributions with parameters $\theta^{(1)}$ and $\theta^{(2)}$ is given by:

$$D_{\mathcal{R}}\left(\theta^{(1)}, \theta^{(2)}\right) := \int_0^1 \sqrt{\dot{\gamma}'(t).\mathfrak{g}(\gamma(t)).\dot{\gamma}(t)}dt \tag{2}$$

where γ is a **segment** (minimal length curve) connecting $\theta^{(1)} = \gamma(0)$ to $\theta^{(2)} = \gamma(1)$. As any Riemannian distance, $D_{\mathcal{R}}$ is **intrinsic** (*i.e.* it is coordinates-free).

Riemannian geometry in a nutshell

Definition 1.
[1] Consider the differentiable manifold $\mathfrak{M}$, and the set $\mathcal{X}(\mathfrak{M})$ of vector fields on $\mathfrak{M}$. A linear connection (or covariant derivative) $\boldsymbol{D}$ on $\mathfrak{M}$ is a bilinear map

$$\begin{cases} \boldsymbol{D} : \mathcal{X}(\mathfrak{M}) \times \mathcal{X}(\mathfrak{M}) \to \mathcal{X}(\mathfrak{M}) \\ (X, Y) \mapsto \boldsymbol{D}_X Y \end{cases}$$

which is linear in X and a derivation on Y.

According to the fundamental theorem of Riemannian geometry [1], there is a unique symmetric connection ∇ compatible with a fixed metrics $\mathfrak{g}$ (the so-called Levi-Civita or Riemann connection), giving in our case the Rao's distance.

Definition 2.
[1, 5] Let $\gamma : [0,1] \to \mathfrak{M}$ be a curve traced on $\mathfrak{M}$, and $\boldsymbol{D}$ be a connection on $\mathfrak{M}$. γ is a geodesic with respect to $\boldsymbol{D}$ if its acceleration $\boldsymbol{D}_{\dot{\gamma}(t)}\dot{\gamma}(t)$ is null $\forall t \in]0,1[$. In other words, ***a geodesic has constant speed*** *in the local norm (1):*

$$\|\dot{\gamma}\|_{\mathfrak{g}} := \|\dot{\gamma}(\bullet)\|_{\mathfrak{g}}(\gamma(\bullet)) = \sqrt{\dot{\gamma}'(\bullet) . \mathfrak{g}(\gamma(\bullet)) . \dot{\gamma}(\bullet)}.$$

Corollary 1.
Let $\gamma : [0,1] \to \mathfrak{M}$ be a geodesic, and $[a,b] \subseteq [0,1]$. Then

$$\int_a^b \sqrt{\dot{\gamma}'(t) . \mathfrak{g}(\gamma(t)) . \dot{\gamma}(t)} dt = (b-a) \|\dot{\gamma}\|_{\mathfrak{g}}.$$

Geodesics on a p-dimensional Riemannian manifold with respect to ∇ are solutions of the Euler-Lagrange equation [5, 1, 3]:

$$\forall 1 \le k \le p,\ \ddot{\gamma}_k(t) + \sum_{i,j=1}^{p} \Gamma_{i,j}^k \dot{\gamma}_i(t) \dot{\gamma}_j(t) = 0 \tag{3}$$

where each coefficient of ∇ (some "Christoffel symbol" $\Gamma_{i,j}^k$) only depends on $\mathfrak{g}$, and is defined in coordinates by:

$$\Gamma_{i,j}^k := \sum_{m=1}^{p} \frac{\mathfrak{g}^{\mathrm{im}}}{2} \left(\frac{\partial \mathfrak{g}_{mj}}{\partial \theta_k} + \frac{\partial \mathfrak{g}_{mk}}{\partial \theta_j} - \frac{\partial \mathfrak{g}_{jk}}{\partial \theta_m} \right) \tag{4}$$

where $\mathfrak{g}^{\mathrm{im}}$ (resp. $\mathfrak{g}_{mk}$) is some entry of $\mathfrak{g}^{-1}$ (resp. $\mathfrak{g}$).

To determine the shortest curve between two points of $\mathfrak{M}$, one applies the following result.

Lemma 1.
[5, 1] Let $\mathfrak{M}$ be an abstract surface, and $p, q \in \mathfrak{M}$. Suppose that $\alpha : [a,b] \to \mathfrak{M}$ is a curve of minimal length connecting p to q. Then, α is a geodesic.

Nevertheless, building the segment connecting p to q is not straightforward, since the lemma above only shows that a segment is a geodesic. But a geodesic is not necessarily a segment...

Theorem 1.
[1] Let $p = \alpha(0)$ be the initial point of a geodesic. Then there is some $0 < t_0 \le +\infty$ such that α is a segment from p to $\alpha(t)$ for every $t \le t_0$ and for $t > t_0$ thereafter never again a segment from p to any $\alpha(t)$ for $t > t_0$. This number t_0 is called the cut value of α and $\alpha(t_0)$ is called the cut point of α. There are only two possible reasons (which can occur simultaneously) for $\alpha(t_0)$ to be to be the cut point of α:

- there is a segment from p to $\alpha(t_0)$ different from α

- $\alpha(t_0)$ is the first conjugate point on α to p (i.e. $t_0 \dot{\alpha}(0)$ is a critical point of the exponential map, defined hereunder).

Remark 1.
No matter the cause of the phenomenon, the main point for us is that if t_0 is a cut value of α, $\forall t \le t_0$, $D_{\mathcal{R}}(p, \alpha(t)) = t$ while $\forall t > t_0$, $D_{\mathcal{R}}(\alpha(t_0), \alpha(t)) < t - t_0$.

Definition 3.
[1] Let $\mathfrak{M}$ be a Riemann manifold and $x \in \mathfrak{M}$. The exponential map of $\mathfrak{M}$ at x is $\exp_x : W_x \to \mathfrak{M}$, defined on some neighborhood W_x of 0 in the tangent space $T_x\mathfrak{M}$ by:

$$\exp_x(V) := \alpha_{\mathcal{B}(V)}(\|V\|)$$

where $\mathcal{B}(V)$ is the projection of V onto the unit ball and $\alpha_{\mathcal{B}(V)}$ is the unique geodesic in $\mathfrak{M}$ such that $\alpha_{\mathcal{B}(V)}(0) = x$ and $\dot{\alpha}_{\mathcal{B}(V)}(0) = \mathcal{B}(V)$.

Remark 2.
If $\alpha := p \frown q$ is a segment and $V_0 := \dot{\alpha}(0)$, because of uniqueness of geodesics, $\exp_p(V_0) := \alpha_{\mathcal{B}(V_0)}(1) = q$; reciprocally, if $V_1 := \dot{\alpha}(1)$, $\exp_q(V_1) := \alpha_{\mathcal{B}(V_1)}(1) = p$ (compare Figures 1 & 2).

4 The geometry of $NB(D_{\mathcal{R}})$

The most classical parametrization of the NB distribution is given by

$$P(X = j; (\phi, p)) = \begin{pmatrix} \phi + j - 1 \\ j \end{pmatrix} p^j (1-p)^\phi \ j \geq 0 \tag{5}$$

with $(\phi, p) \in \mathbb{R}^+ \times]0, 1[$; ϕ is the index parameter (denoted k by [2] and many other authors). Nevertheless, because of its orthogonality, we chose instead the parametrization used by Chua and Ong [4]:

$$P(X = j; (\phi, \mu)) = \begin{pmatrix} \phi + j - 1 \\ j \end{pmatrix} \left(\frac{\mu}{\mu + \phi}\right)^j \left(1 - \frac{\mu}{\mu + \phi}\right)^\phi, \ j \geq 0 \tag{6}$$

$(\phi, \mu) \in \mathbb{R}^+ \times \mathbb{R}^+$; here, μ is the mean of the distribution. In these coordinates, the information matrix is:

$$\mathfrak{g}(\phi, \mu) = \begin{pmatrix} G_{\phi\phi} & 0 \\ 0 & G_{\mu\mu} \end{pmatrix}$$

with $G_{\mu\mu} = \frac{\phi}{\mu(\mu+\phi)}$, while the expression of $G_{\phi\phi}$ is more complicated:

$$G_{\phi\phi} = -\frac{\mu + \phi(\mu + \phi)\left(\left(\phi/\mu+\phi\right)^\phi - 1\right)\psi^1(\phi)}{\phi(\mu + \phi)} \tag{7}$$

where ψ^1 is the Trigamma function (first derivative of the logarithmic derivative of $\Gamma(\bullet)$). One will find in [3] the closed-form expression of the Rao's distance for a number of probability families. These authors reported that when the index parameter of two NB distributions **is the same** the Rao's distance is given, in the parametrization (5), by:

$$D_{NB(p)}\left((\phi, p^1), (\phi, p^2)\right) := 2\sqrt{\phi}\ \arccos\left(\frac{1 - \sqrt{p^1 p^2}}{\sqrt{(1-p^1)(1-p^2)}}\right). \tag{8}$$

Of course, if μ^1 (resp. μ^2) is the mean of $\mathfrak{L}^1 = NB(\phi, p^1)$ (resp. $\mathfrak{L}^2 = NB(\phi, p^2)$), we have necessarily:

$$D_{\mathcal{R}}(\mathfrak{L}^1, \mathfrak{L}^2) \leq D_{NB(p)}(\mathfrak{L}^1, \mathfrak{L}^2). \tag{9}$$

Due to the complexity of (7), $D_{\mathcal{R}}(\mathfrak{L}^1, \mathfrak{L}^2)$ cannot be obtained in a closed-form. It must be computed by finding the numerical solution of the Euler-Lagrange equation (3), completed in the parametrization (6) by the boundary conditions

$$\left\{\gamma(0) = (\phi^1, \mu^1), \gamma(1) = (\phi^2, \mu^2)\right\}. \tag{10}$$

Geodesics can be as well be computed by solving (3) under the alternative constraints

$$\left\{\gamma(0) = (\phi^1, \mu^1), \dot{\gamma}(0) = V \in \mathbb{R}^2\right\}. \tag{11}$$

This solution is associated with the exponential map at (ϕ^1, μ^1).

5 Approximating $D_{\mathcal{R}}\left(\mathfrak{L}^1, \mathfrak{L}^2\right)$

In this section, $\mathfrak{L}^i \equiv \left(\phi^i, \mu^i\right)$ denotes a NB distribution parametrized in the (6) system, but our purpose could be extended to any parametric family.

Firstly, all the Christoffel symbols (4) were calculated from the expression (7) of $G_{\phi\phi}$, with the help of *Mathematica*. Then, the differential equation (3) was numerically solved under the the boundary conditions (10), for the estimated parameters of a number of marine organisms. In most case a solution could be found in an acceptable time (four CPU minutes, at most), with a good numerical precision (15 digits), but was each one of the geodesics found a segment? And what about failures in computation? We indeed had to face two different problems: a theoretical one and a computational one.

Theoretical issue

Suppose a solution $\gamma = \mathfrak{L}^1 \curvearrowright \mathfrak{L}^2$ of (3) under the boundary condition (10) has been found; according to Corollary 1, a straightforward approximation of $D_{\mathcal{R}}\left(\mathfrak{L}^1, \mathfrak{L}^2\right)$ should be $\|\dot{\gamma}\|_{\mathfrak{g}}$. But notice that $\|\dot{\gamma}\|_{\mathfrak{g}}$ is only an **upper bound**, which is attained only when there is **no cut point** in $\gamma([0,1])$ (cf. Theorem 1). That is why we need some test to detect a possible cut point on some geodesic curve (see Section 5). Suppose now a cut point $\left(\phi^{c(1,2)}, \mu^{c(1,2)}\right)$ has been detected on γ. Then, it is natural [1] to supersede γ by the broken geodesic

$$\left(\phi^1, \mu^1\right) \curvearrowright \left(\phi^{c(1,2)}, \mu^{c(1,2)}\right) \oplus \left(\phi^{c(1,2)}, \mu^{c(1,2)}\right) \curvearrowright \left(\phi^2, \mu^2\right)$$

whose length is shorter than $\Lambda(\gamma)$, provided $\left(\phi^{c(1,2)}, \mu^{c(1,2)}\right) \curvearrowright \left(\phi^2, \mu^2\right)$ is also a segment.

Computational issues

Wet met various numerical problems in computing $\mathfrak{L}^1 \curvearrowright \mathfrak{L}^2$:

(P1) no solution was found (due to time limit, singularities, *etc*)

(P2) an unsuitable solution was found: for some $t \in [0,1]$, $(\phi(t), \mu(t)) \notin \mathbb{R}^+ \times \mathbb{R}^+$

(P3) the boundary condition (10) was not fulfilled with a satisfactory precision.

Simple configurations

When none of these issues is met, we first check that there is no cut point on γ. Then, the canonical solution is acceptable, and we can write:

$$D_{\mathcal{R}}\left(\mathfrak{L}^1, \mathfrak{L}^2\right) \approx \Lambda(\gamma) = \|\dot{\gamma}\|_{\mathfrak{g}}. \tag{12}$$

If a cut point $\left(\phi^{c(1,2)}, \mu^{c(1,2)}\right)$ is detected on γ, and if $\left(\phi^{c(1,2)}, \mu^{c(1,2)}\right) \curvearrowright \left(\phi^2, \mu^2\right)$ is free of cut point, we adopt as an upper bound for $D_{\mathcal{R}}\left(\mathfrak{L}^1, \mathfrak{L}^2\right)$:

$$\Lambda\left(\left(\phi^1, \mu^1\right) \curvearrowright \left(\phi^{c(1,2)}, \mu^{c(1,2)}\right)\right) + \Lambda\left(\left(\phi^{c(1,2)}, \mu^{c(1,2)}\right) \curvearrowright \left(\phi^2, \mu^2\right)\right).$$

Intricate configurations

When (P1) or (P2) is met, we consider that the best achievable solution would consist in breaking $\gamma = \mathfrak{L}^1 \curvearrowright \mathfrak{L}^2$ by inserting a well-placed "stopover". But since γ is undetermined, how should $\left(\phi^{S(1,2)}, \mu^{S(1,2)}\right)$ be chosen? We propose two heuristics for approaching γ:

1. compute a "rough solution" $\widetilde{\gamma_R}$ to the original problem, contenting ourselves with low-precision (here: 5 digits), and substitute $\widetilde{\gamma_R}$ for γ to search for $\left(\phi^{S(1,2)}, \mu^{S(1,2)}\right)$
2. when $\widetilde{\gamma_R}$ cannot be obtained, merely use instead $\widetilde{\gamma_L}(t) := t\left(\phi^1, \mu^1\right) + (1-t)\left(\phi^2, \mu^2\right)$.

In the second case, after fixing a convenient sampling rate $\frac{1}{N}$, the stopover naturally corresponds to the shortest broken geodesic:

$$\begin{cases} \left(\phi^{S(1,2)}, \mu^{S(1,2)}\right) = \widetilde{\gamma_L}\left(\frac{k_L}{N}\right) \\ k_L := \arg \min_{1 \le k \le N-1} \left(\Lambda\left(\left(\phi^1, \mu^1\right) \curvearrowright \widetilde{\gamma_L}\left(\frac{k}{N}\right)\right) + \Lambda\left(\widetilde{\gamma_L}\left(\frac{k}{N}\right) \curvearrowright \left(\phi^2, \mu^2\right)\right)\right). \end{cases} \tag{13}$$

In the first case two eventualities must be considered:

1. a cut point $\left(\phi^{c(1,2)}, \mu^{c(1,2)}\right)$ is detected on $\widetilde{\gamma_R}([0,1])$; then $\left(\phi^{S(1,2)}, \mu^{S(1,2)}\right) = \left(\phi^{c(1,2)}, \mu^{c(1,2)}\right)$
2. if no cut point is detected, proceed like in (13):

$$\begin{cases} \left(\phi^{S(1,2)}, \mu^{S(1,2)}\right) = \widetilde{\gamma_R}\left(\frac{k_R}{N}\right) \\ k_R := \arg \min_{1 \le k \le N-1} \left(\Lambda\left(\left(\phi^1, \mu^1\right) \curvearrowright \widetilde{\gamma_R}\left(\frac{k}{N}\right)\right) + \Lambda\left(\widetilde{\gamma_R}\left(\frac{k}{N}\right) \curvearrowright \left(\phi^2, \mu^2\right)\right)\right). \end{cases} \tag{14}$$

Boundary problems

(P3) is easy to solve, since it merely corresponds to $\gamma(0) \neq \mathfrak{L}^1$ or $\gamma(1) \neq \mathfrak{L}^2$. We just have to add to formulas (12), (13) or (14) the corrective boundary error term

$$BE(\gamma) := \left\|\gamma(0) - \mathfrak{L}^1\right\|_{\mathfrak{g}}\left(\mathfrak{L}^1\right) + \left\|\gamma(1) - \mathfrak{L}^2\right\|_{\mathfrak{g}}\left(\mathfrak{L}^2\right) \tag{15}$$

given by formula (1). Finally, we can write:

$$D_{\mathcal{R}}\left(\mathfrak{L}^1, \mathfrak{L}^2\right) \le \Lambda(\gamma) + BE(\gamma) \tag{16}$$

whatever the selected geodesic (broken, or not) may be.

Locating a (N, ϵ)- cut point on some geodesic γ

For that purpose, the unit interval is first divided into N intervals: $[0,1] = \bigcup_{i=1}^N \delta_i$, with $\delta_i := [\frac{i-1}{N}, \frac{i}{N}[$. Suppose there exists a cut point $\gamma(t_c)$ on γ, such that $t_c \in \delta_{i_c}$. Consider the set

$$\mathfrak{C}_N(\gamma) := \left\{\gamma_1 := \gamma\left(\frac{1}{N}\right), \cdots, \gamma_k := \gamma\left(\frac{k}{N}\right), \cdots, \gamma_{N-1} := \gamma\left(\frac{N-1}{N}\right)\right\} \subset \mathfrak{M}$$

and, for each $1 \le i \le N$ the geodesic $\alpha_i := \gamma_{i-1} \curvearrowright \gamma_i$ obtained by solving (3) **under the constraints**

$$\{\alpha_i(0) = \gamma_{i-1},\ \alpha_i(1) = \gamma_i\}.$$

Because of the uniqueness of segments, Corollary 1 and Remark 1, $\forall i < i_c$, $\frac{\|\dot\gamma\|_{\mathfrak{g}}}{N} = \Lambda(\alpha_i) = \|\dot\alpha_i\|_{\mathfrak{g}}$. On the contrary, when $i \ge i_c$, the distance between γ_{i-1} and γ_i **along** γ is $\frac{\|\dot\gamma\|_{\mathfrak{g}}}{N}$ yet, while $\|\dot\alpha_i\|_{\mathfrak{g}}$ should be smaller. More precisely, if the resolution $\frac{1}{N}$ is small enough (for instance, smaller than the injectivity radius [1] of $\mathfrak{M}$), $\gamma_{i-1} \curvearrowright \gamma_i$ is a segment and we may write:

$$\begin{cases} \forall i < i_c,\ \frac{\|\dot\gamma\|_{\mathfrak{g}}}{N} - \|\dot\alpha_i\|_{\mathfrak{g}} = 0 \\ \forall i \ge i_c,\ \frac{\|\dot\gamma\|_{\mathfrak{g}}}{N} - \|\dot\alpha_i\|_{\mathfrak{g}} > 0. \end{cases}$$

Thus, after fixing ϵ (small), we can locate possible cut points, with a precision depending on (N, ϵ).

Definition 4.

We will say that $\gamma_{i_c} \in \mathfrak{C}_N(\gamma)$ is a (N, ϵ)- cut point on γ if

$$i_c = \arg \min_{1 \le i \le N-1} \left(\left|\frac{\|\dot\gamma\|_{\mathfrak{g}}}{N} - \|\dot\alpha_i\|_{\mathfrak{g}}\right| > \epsilon\right).$$

6 The MEEZ data

The Mauritanian coast, situated on the Atlantic side of the northwestern African continent, embeds a wide long continental shelf of about 750 km and 36000 km^2 with an Exclusive Economic Zone (MEEZ) of 230000 km^2. This study focuses on the analysis of abundance of fish and invertebrates data collected during annual scientific trawl surveys performed by oceanographic vessels on the continental shelf ($<$ 200 m depth), from 1987 to 2010. All the species (fish and invertebrate) captured in a given station were identified, counted and then recorded on the database. In addition, each station has been characterized by supplementary environmental variables: bathymetry, sedimentary type of the substrate, latitude and longitude. The counts of species collected were then fitted by NB distributions. For that purpose, it was necessary to determine homogeneous regions (habitats) in the MEEZ; it was found that the optimal number of habitats is four. Then the counts of each species were separately fitted in each one of these regions, and it was observed that in each one of the habitats, only a reduced number of species could be satisfactorily fitted by some NB distribution; other species were discarded. For further details on the data or estimation methods, see [9, 7].

7 Results

Geodesics: a bestiary

We illustrate hereunder the diversity of cases encountered in computing $D_{\mathcal{R}}(A,B)$. From now, the approximation parameter are fixed to $(N,\epsilon) = (10, 0.01)$. All the figures displayed will be composed of three panels. On the left one, we superimposed the final solution to the rough geodesic (when it could be computed). On both the other panels, we investigated the structure of broken geodesics in the neighborhood of a stopover S, with the help of the exponential map. We determined first $\gamma_1 = A \curvearrowright S$ (resp. $\gamma_2 = B \curvearrowright S$) by solving equation (3) under the constraints (10). We afterward considered $\{V_i(\theta_k) := \rho(\theta_k) \cdot \mathcal{B}(\dot{\gamma_i}(0)) : i = 1, 2\}$, where the angle of the rotation ρ is (in degrees) $\theta_k \in \{0, \pm 0.1, \pm 0.2, \pm 0.3\}$. Equation (3) was then solved under the constraints (11) with $V = V_i(\theta_k)$, giving rise to two bundles of seven geodesics; remember that for $\theta = 0$, $\exp_A(V_1(0)) = S = \exp_B(V_2(0))$ (see Remark 2). In all these plots, the red point will be "A" and the black one will be "B", while the stopover is represented by the big gray point.

On Figure 1, we represented the geodesic $\gamma_1 := A \curvearrowright B$, with $A \equiv (0.7767, 11.2078)$ and $B \equiv (0.7767, 87.268)$ in the system (6). It corresponds to a simple configuration: no (N,ϵ)- cut point was found, and we can see on the left panel that there is practically no difference between the segment and the sampled rough geodesic. We stress that the stopover S is in this case quite unnecessary; it was introduced only for illustration. On the central panel, the segment $A \curvearrowright S$ has been extrapolated with the exponential map, as well as the other geodesics of the bundle. On the right panel the segment $B \curvearrowright S$ and the corresponding bundle of geodesics have been extrapolated in the same way. We can see that there is practically no difference between extrapolations of $A \curvearrowright S$ and $B \curvearrowright S$, the segment γ_1 and the rough geodesic $\widetilde{\gamma_{1,R}}$. Notice finally that in this (artificial) case, the distance $D_{NB(p)}$ (8) given by [3] can be computed. In the (5) system, $A \equiv (0.7767, 0.935191)$, $B \equiv (0.7767, 0.991178)$ and we have, in compliance with (9):

$$1.7 \approx D_{\mathcal{R}}(A,B) < D_{NB(p)}(A,B) = 1.783.$$

On Figure 2, we plotted geodesics connecting two species: *Rhizoprionodon acutus*, coded RIAC70, and *Anguilla sp*, coded ANSP50. Even if $RIAC70 \curvearrowright ANSP50$ could not be determined, the rough geodesic $\widetilde{\gamma_R}$ could be computed. A (N,ϵ)- cut point was detected in the second position of the sampled curve, and used as a stopover S to compute the final broken segment. But we can see of the central and the right panels that neither $A \curvearrowright S$ nor $B \curvearrowright S$ could be extrapolated to obtain a geodesic connecting A to B.

Figure 1. A geodesic without any cut point $\left(\widetilde{D_{\mathcal{R}}} \approx 1.70 \approx D_{\mathcal{R}}\right)$. Left panel: the rough geodesic (cyan suits) is superimposed to the segment; A is the red point, B the black one and the stopover is represented by the gray point. Right panels: plot of the two bundles of geodesics issued from A or B. Red curve: $\theta = 0$; dashed curves: $\theta \neq 0$. The header corresponds to the parameters of the distributions.

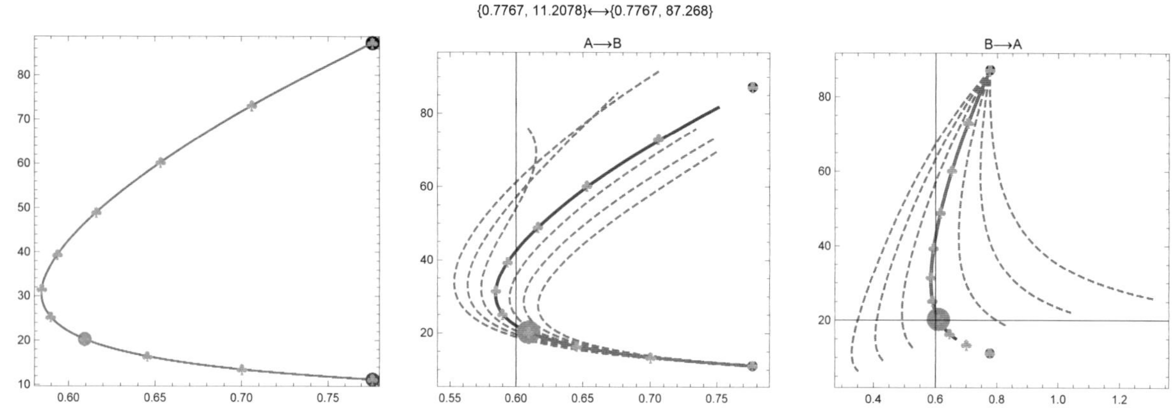

Figure 2. A geodesic with a cut point; $\widetilde{D_{\mathcal{R}}}\,(RIAC70, ANSP50) \approx 3.98$ and $D_{\mathcal{R}}\,(ANSP50, RIAC70) \approx 3.90$. Same graphical conventions as in Figure 1.

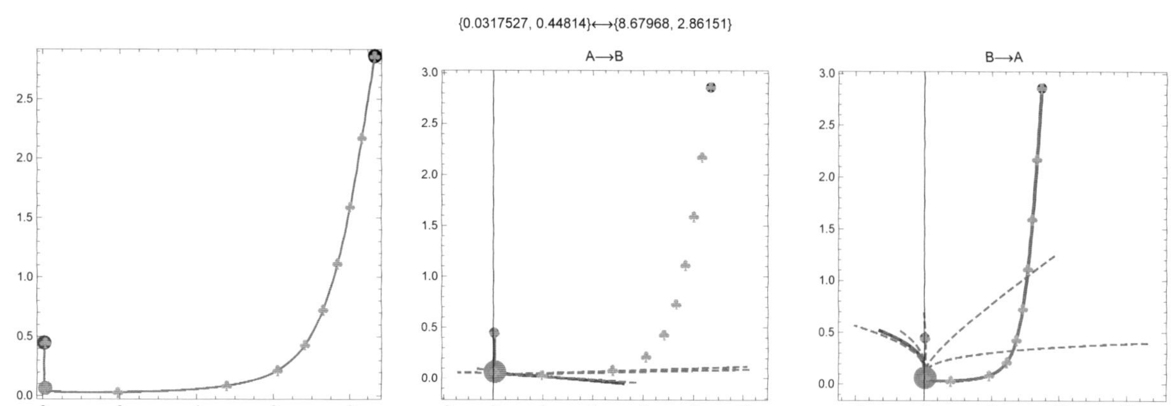

Figure 3. A broken geodesic between two species; $D_{\mathcal{R}}(HISP00, SCAN40) \approx 29.62$. Same graphical conventions as in Figure 1.

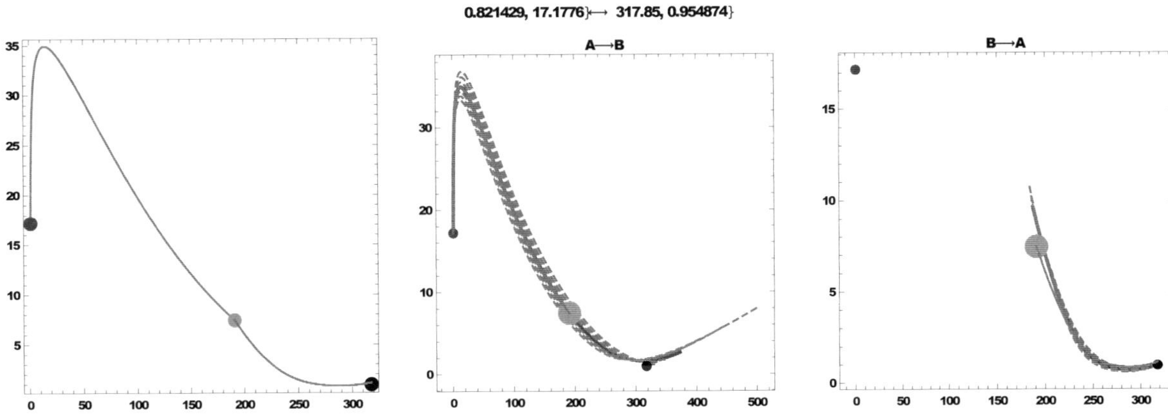

Figure 4. A broken geodesic between two species; $D_{\mathcal{R}}(HISP00, TRTR20) \approx 30.60$. Same graphical conventions as in Figure 1.

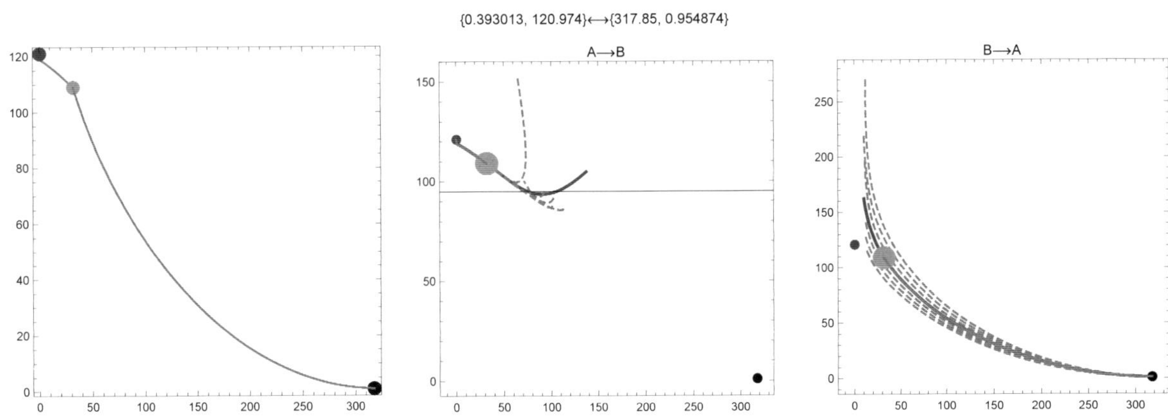

Now, what about the worst cases, when linear interpolation was unavoidable? There were 4 such pairs of species in the habitat $C4$ (see Table 1). Notice first that all these pairs were associated with a particular species, of parameters $(317.85, 0.954874)$: this is *Hippocampus sp*, coded HISP00, the less aggregative species in this habitat.

Let us start with $\{HISP00, SCAN40\}$, whose processing is represented on Figure 3 (SCAN40 is the code of *Scorpaena angolensis*). In this case, neither of the geodesics could be computed, and we used in last resort linear interpolation in the space of parameters. The obtained curve is rather smooth, and one could probably find a genuine segment close to this broken geodesic, with enough computation time.

Another example: $D_{\mathcal{R}}(HISP00, TRTR20)$, where TRTR20 is the code of *Trachurus trecae*. This case, displayed on Figure 4, is quite different: the structure of the geodesics near the stopover S looks like the structure of geodesics in the neighborhood of a cut point (see Figure 2). But notice S was found by traveling across $\widetilde{\gamma_L}(\bullet)$, and one cannot claim it is a realistic first guess for $HISP00 \curvearrowright TRTR20$.

Return to the exploratory setting

Remember that the MEEZ could be split into four homogeneous regions (see Section 6), named $\{C_1, \cdots, C_4\}$. From the estimates of the parameters of the N_h species kept in C_h, it is possible to tabulate the Rao's distance between species and process the resulting table with methods designed for non-Euclidean distances (Multidimensional Scaling, Isomap, *etc*), as proposed by Rao [11] himself. Because of the computational cost of Rao's distances, we were forced to select, for each habitat, a sub-sample of species representing as well as possible the whole (landmark species, say). Thus, in $C4$ (like in other habitats), species were first split into two categories: very aggregative and moderately aggregative. We focused on the second category, keeping for computation the 30-species set (amongst the 121 species correctly fitted, while 301 species were observed) obtained by gathering isolated species and species constituting the vertices of the convex envelope of non-isolated species (see Figure 6 of [9]).

Global statistics

It is interesting to tally the various configurations encountered in different habitats: simple or intricate, and the presence of possible (N, ϵ)-cut points on the obtained geodesics. In the intricate case, it is also interesting to tally the cases where linear interpolation was unavoidable. The obtained results are gathered in Table 1. More than 70% of the configurations (88% in $C4$) were simple (*i.e.* the canonical solution was accepted), and (N, ϵ)- cut points were quite rare. In the intricate cases, the rough solution was generally accepted (more than 90% of the cases). We can thus claim that the obtained upper bounds given by Formula(16) were mostly tight approximations of true Rao's distance.

Table 1. Global results obtained in the four habitats of the MEEZ

Habitat	Number of species (well-fitted)	Simple configurations	Intricate (Rough, Linear)	Cut points
$C1$	30	356	(75,4)	1
$C2$	19	124	(46,1)	2
$C3$	26	227	(88,10)	1
$C4$	26	288	(33,4)	1

Acknowledgments

We thank the Mauritanian Institute of Oceanographic Research and Fisheries (IMROP) and the Department of Cooperation and Cultural Action of the Embassy of France in Mauritania for their support for this study. We also thank all scientists who contributed to field surveys and data collection.

Bibliography

[1] Berger, M. (2003) *A Panoramic View of Riemannian Geometry.* Springer Verlag, Berlin.

[2] Bliss, C. I. and Fisher, R. A. (1953) *Fitting the Negative Binomial distribution to biological data.* Biometrics, **9**, 176-200.

[3] Burbea, J. and Rao, C. R. (1986) *Infomative geometry of probality spaces.* Expo. Math., **4**, 347-378.

[4] Chua, K. C. and Ong, S. H. (2013). *Test of mispecification with application to Negative Binomial distribution.* Computational Statistics, **28**, 993-1009.

[5] Gray, A. (1999) *Modern differential geometry of curves and surfaces with Mathematica* (2nd ed.). CRC Press, London.

[6] Kendall, D.G. (1948). *On some modes of population growth leading to R. A. Fisher's logarithmic series distribution.* Biometrika, **35**, 6-15.

[7] Kidé, S. O., Manté, C., Dubroca, L., Demarcq, H., Mérigot, B. (2015) *Spatio-Temporal Dynamics of Exploited Groundfish Species Assemblages Faced to Environmental and Fishing Forcings: Insights from the Mauritanian Exclusive Economic Zone.* PLoSONE, **10, 10**, e0141566. doi:10.1371/journal.pone.0141566

[8] Manté, C., Durbec, J.P. and Dauvin, J. C. (2005) *A functional data-analytic approach to the classification of species according to their spatial dispersion. Application to a marine macrobenthic community from the Bay of Morlaix (western english channel).* J. Appl. Statist., **32, 8**, 831-840.

[9] Manté, C., Kidé, O. S., Yao, A. F. and Mérigot, B. (2016) *Fitting the truncated negative binomial distribution to count data. A comparison of estimators, with an application to groundfishes from the Mauritanian Exclusive Economic Zone.* Environmental and Ecological Statistics, DOI 10.1007/s10651-016-0343-1.

[10] O'Neil, M. F. and Faddy, M. J. (2003) *Use of binary and truncated negative binomial modelling in the analysis of recreational catch data.* Fisheries Research, **60**, 471-477.

[11] Rao, C. R. (1992) *Information and accuracy attainable in the estimation of statistical parameters.* In: *Breakthroughs in Statistics: Foundations and Basic Theory.* Springer, New York, 235-247.

[12] Vaudor, L., Lamouroux, N. and Olivier, J. M. (2011) *Comparing distribution models for small samples of overdispersed counts of freshwater fish.* Acta Oecologica, **37, 3**, 170-178.

Estimation of total electricity consumption curves of small areas by sampling in a finite population

Anne De Moliner, *Institut de Mathématiques de Bourgogne, Université de Bourgogne-Franche-Comté, EDF Lab Paris-Saclay*, anne.de-moliner@edf.fr
Camelia Goga, *Institut de Mathématiques de Bourgogne, Université de Bourgogne-Franche-Comté*, camelia.goga@u-bourgogne.fr
Hervé Cardot, *Institut de Mathématiques de Bourgogne, Université de Bourgogne-Franche-Comté*, herve.cardot@u-bourgogne.fr

Abstract. Many studies carried out in the French electricity company EDF are based on the analysis of the total electricity consumption curves of groups of customers. These aggregated electricity consumption curves are estimated by using samples of thousands of curves measured at a small time step and collected according to a sampling design. Small area estimation is very usual in survey sampling. It is often addressed by using implicit or explicit domain models between the interest variable and the auxiliary variables. The goal here is to estimate totals of electricity consumption curves over domains or areas. Three approaches are compared: the first one consists in modeling the functional principal scores with linear mixed models. The second method consists in using functional linear regression models and the third method which is mon-parametric is based on regression trees for functional data. These methods are evaluated on a dataset of French consumption curves.

Keywords. big data, energy, functional data, functional principal component analysis, mixed models, regression trees.

1 Introduction and context

Many studies carried out by the EDF company are based on the analysis of the total or the mean electricity consumption of groups of clients sharing some common characteristics. In what follows, these groups will be called domains. In particular, there is a growing need to estimate the mean electricity consumption curves for small geographic areas such as regions, cities or districts, ... in order to create new services for local authorities.

The individual electricity consumption is recorded by means of smart meters at a very fine time scale (often half-hourly). In view of this new setting, the relevant variables, such as the consumption curve, may be considered as realizations of functional variables depending on a continuous time index t that is in the $[0, T]$ rather than as multivariate vectors. Totals or means of the consumption curves are estimated by using a sample of thousands of individuals selected from the whole population of EDF customers. The total or mean curve estimation under various sampling designs and estimators as well as the construction of confidence bands have been addressed in the recent works of [5], [6] and [7].

We are concerned here with the small area estimation in a functional context. Small area estimation is very common in surveys and many authors have addressed this issue in a non-functional setting. The very recent book of [9] gives a thorough review of the existing methods. When the domain of interest is small, the direct estimators which are built only on the individuals belonging to the domain are not very efficient. To improve the results, auxiliary information is used and estimators are built using an implicit or explicit modeling of the link between quantities of interest and the auxiliary information.

Many auxiliary variables are available at the EDF company and some of them, such as the billing information, may be known both at the unit and the domain levels. This information will be used to build models in order to enhance the estimations and also to provide reasonable estimations for non-sampled domains (e.g domains with no units in the sample). Note that we consider here only multivariate auxiliary information which do not depend on time.

In this paper, we compare three methods to estimate mean curves over domains: linear mixed models combined with principal components analysis, functional linear regression implemented using the calibration algorithm according to the idea given by [1] and a non-parametric approach based on regression trees for functional data.

The paper is organized as follows: in Section 2, we introduce some notations and hypotheses about the survey sampling framework, definition of the domains and auxiliary information. We present, in Section 3, three estimation methods and we compare in Section 4 the proposed estimators on a French consumption curve dataset. Section 5 sums up our conclusions.

2 Notations and framework

In this section we introduce some notations about survey sampling framework, domains definition, auxiliary information and functional data.

Let U be a population of interest of known size N. To each unit i of the population a (load) curve defined over a time interval $[0, T]$ is associated: for each unit i we have a function of time $y_i(t), t \in [0, T]$, where the continuous index t represents time. In practice, the curves are not observed continuously for $t \in [0, T]$ but only for a set V of measurement instants $0 = t_1 < t_2 < ... < t_v = T$ which are supposed to be the same for all units and equispaced. We also assume that there are no missing values.

The population U is decomposed into D disjointed domains U_d of known sizes $N_d, d = 1, ..., D$. Our goal is to estimate the mean curve μ_d over each domain, i.e.

$$\mu_d(t) = \frac{1}{N_d} \sum_{i \in U_d} y_i(t), \quad t \in [0, T]. \tag{1}$$

Let $1_{d,i} = 1_{U_d(i)}$ denote the indicator equal to one if the unit i belongs to domain d and zero otherwise. Let $d(i)$ be the domain to which unit i belongs.

Let X_i denote an auxiliary information vector known for each unit $i \in U$ and let $Z_{d(i)}$ denote an additional auxiliary information vector, known only at the domain level. In order to apply linear methods easily, this information is grouped into a single vector $X_i^* = (X_i, Z_{d(i)})$. Let $\bar{X}_d^*$ be the mean of variables X_i^* over the domain d. For sake of simplicity, only multivariate variables which do not vary over time are considered.

Let us consider the following functional superpopulation model between the interest variable y and the auxiliary variable X_i^* :

$$y_i(t) = f_{d(i)}(X_i^*, t) + \epsilon_i(t), \quad i \in U_d, \quad t \in [0, T] \tag{2}$$

where $f_{d(i)}$ is an unknown regression function to be estimated, which may vary from one domain to another, and $\epsilon_i(t)$ is a zero mean noise process.

From the population U, a random sample s of size n is drawn using a probability sampling design assumed to be non-informative, meaning that the selection of the individuals is not depending on the values of y. Let s_d be the intersection of the domain U_d and the sample s and n_d be the size of s_d. The domain size n_d can be equal to zero for one or more domains. In practice, data collection must often respect strong technical constraints and sometimes its main purpose is non statistical (technical tests or power grid management), so it is often hard to consider that the sample has been drawn according to a proper sampling design. For this reason, we use model-based inference rather than design-based inference.

In order to assess the performances of our methods, we use as a benchmark the model

$$y_i(t) = \mu_{d(i)}(t) + \epsilon_i(t), \quad \forall i \in U_d, \tag{3}$$

with ϵ_i a noise process with mean zero. The mean load curve estimator under this model is the mean of the curves belonging to s_d, i.e. $\hat{\mu}_d^0(t) = \frac{\sum_{i \in s_d} y_i(t)}{n_d}$. Obviously, this estimator can not be calculated for non-sampled domains (i.e. $n_d = 0$) and moreover it is extremely unstable for small domain sample sizes.

3 Estimation methods

In this section, we present three approaches for estimating the domain mean curve: linear mixed model on principal component scores, functional linear regression models and regression trees adapted to curves.

Linear mixed model on principal component scores

In this section, we adapt the unit level linear mixed models frequently used in small area estimation. Our solution consists in using a functional Principal Component Analysis (PCA) in order to transform our curve estimation problem into a multivariate one. More precisely, we perform a functional principal components analysis (see [10]) and we decompose the curve y_i into the space spanned by the first K principal components following the Karhunen-Loeve expansion:

$$y_i(t) = \mu(t) + \sum_{k=1}^{K} f_{k,i}\zeta_k(t) + \nu_i(t), \quad i \in U, \tag{4}$$

where $\zeta_k(t)$ denotes the functional principal component k, $k = 1, \ldots, K$, $\nu_i(t)$ the reminder term and $f_{k,i}$ the score of unit i on component k.[4] Using (4), the domain mean μ_d can then be approximated as follows

$$\mu_d(t) \simeq \mu(t) + \sum_{k=1}^{K} \left(\frac{1}{N_d} \sum_{i \in U_d} f_{k,i} \right) \zeta_k(t).$$

The functional principal components $\zeta_k(t)$ are unknown and they can be estimated by $\hat{\zeta}_k(t)$ as suggested in [4]. Thus, to estimate μ_d, we need to estimate μ and the mean of principal scores over the domain d, i.e. $\overline{f}_{k,d} = \frac{1}{N_d} \sum_{i \in U_d} f_{k,i}$. Let consider for that the following unit level linear mixed model on $f_{k,i}$, $k = 1, \ldots, K$ as in [9]:

$$f_{k,i} = \beta_k' X_i^* + u_{k,d(i)} + \epsilon_{k,i}, \quad \forall i \in U_d,$$

where $\beta_k' X_i^*$ is the (functional) fixed effect of the auxiliary information, $u_{k,d(i)}$ the (functional) random effect of the domain $d(i)$ distributed normally with mean 0 and variance $\sigma_{d,k}^2$ and $\epsilon_{k,i}$ the residual

[4] Here, the PCA is not used as a dimension reduction method but only in order to transform our problem into uncorrelated mean of real variables estimation problems so we keep a number K of principal components as large as possible.

distributed normally with mean 0 and variance $\sigma^2_{\epsilon,k}$. This model is a parametric case of the general model considered in (2). Using the same lines as in [9], Chapter 7, we estimate β_k by EBLUP (Empirical Best Linear Unbiased Prediction) and deduce then the estimator $\widehat{\overline{f}}_{k,d}$ of $\overline{f}_{k,d}$. Finally, the mean μ_d is estimated by

$$\hat{\mu}_d(t) = \hat{\mu}(t) + \sum_{k=1}^{K} \widehat{\overline{f}}_{k,d} \hat{\zeta}_k(t), \tag{5}$$

with $\hat{\mu}(t)$ the sample mean and $\hat{\zeta}_k(t)$ the estimated principal component, $k = 1, \dots, K$.

Functional linear regression estimator

In this section, we assume that the following superpopulation model holds for the entire population:

$$y_i(t) = \mu(t) + \beta(t)X_i^* + \epsilon_i(t), \quad \forall i \in U, \tag{6}$$

with ϵ_i a noise process with mean zero.

It is the same model as before but without random effects: it is therefore assumed that, conditionally to the auxiliary information, the distribution of curves $y(t)$ is the same over all the domains (with no area-specific effects to be taken into account). We are in the usual context of functional regression (regression of a functional variable on real explanatory variables). This problem can easily be addressed by projecting the curves on an adapted basis (for example the principal components basis as before or B-splines). By discretizing the curves over the V measurement instants and then fitting an Ordinary Least Square estimator for each instant, we build the following estimator

$$\hat{\mu}_d(t) = \hat{\mu}(t) + \hat{\beta}(t)\bar{X}_d^*, \tag{7}$$

with $\hat{\mu}(t)$ and $\hat{\beta}(t)$ the Ordinary Least Square parameters estimated over the whole sample.

The estimation of this functional model can be computationally heavy for large datasets (many domains or many measures for each curve) so we follow the approach proposed by [1] to compute our domain mean curve estimators: under the model-assisted framework, the author proposes to estimate quickly many domain totals using a Generalized Regression Estimator (see [11]) by remarking that, as proved in [8], this estimator is equal to the calibration estimator so we can estimate multiple totals by only calculating an unique weight vector for each domain. In addition to that, our regression estimator (for one principal component or one basis vector) is equal to the Generalized Regression Estimator for the Simple Random Sampling (with modified weights). Following this idea, we can implement our estimators by only calculating an unique weight vector for each domain and using a projection of our curves on the basis of our choice.

Regression tree for functional data (Courbotree)

In this section, we propose to predict the curve of each non-sampled unit and then, to derive the estimator of the domain mean curve as follows (see [13]):

$$\hat{\mu}_d(t) = \frac{1}{N_d} \left(\sum_{i \in s_d} y_i(t) + \sum_{i \in U_d - s_d} \hat{y}_i(t) \right). \tag{8}$$

In order to compute the prediction $\hat{y}_i(t)$ for each unit $i \in U_d - s_d$, we use a regression tree approach. This approach, as suggested by [3], is a nonparametric method consisting in splitting iteratively the dataset into two parts until obtaining classes such as the dependent variable y_i is as homogeneous as possible within each class according to some criterion. In order to do that, we use the auxiliary information X_i supposed to be known for each unit i from the population U and the homogeneity criterion is based on the Euclidean norm (we use the "courboTree" approach as in [12]).

Finally, the prediction $\hat{y}_i(t)$ for each unit $i \in U_d - s_d$ is given by the mean of the consumption curves for the individuals that belong to the same class as i.

In practice, as the levels of the consumption curves are extremely various, the regression tree performs badly on raw curves so we must pre-process the data: we divide each curve by a level proxy (for example the consumption over the previous year) and in a post-processing step we multiply the estimated curve by this proxy level. Moreover, we recommend to smooth the curve in the preprocessing step (using a mobile average of order 5).

4 An illustration on a real dataset

We have applied the methods presented previously on a real load curves dataset in order to assess their performances and to compare the ability of these methods to give good estimations.

Description of the data set

The test population consists in 1904 consumption curves of French households recorded at a daily time step from October 2011 to March 2012 (177 points) with no missing values. This population is formed by eight domains, corresponding to geographic areas.

For each unit of the test population, we have the following auxiliary information: contract power, tariff option and consumption over the previous year. For each domain, we know the electric heating rate and the mean surface of the housing.

Test protocol

In order to assess the quality of our estimators, we draw a large number $B = 2000$ of random samples of consumption curves and, on each sample, we estimate the domain mean curve by using the suggested estimators. Then we build quality indicators to compare the estimated mean curves to the real ones.

In our simulations, the 8th domain is always non-sampled (in order to assess the quality of the methods for non-sampled domains). For each simulation, we use a simple random sampling to draw 200 units from the 7 sampled domains.

Quality indicators

We build separate quality indicators on sampled and non-sampled domains. Let $Y_d(t)$ be the mean curve of domain d at time t and $\hat{Y}_d(t)$ be its estimate for a given method. We denote by $E_{MC}[\hat{Y}_d(t)] = \frac{1}{B}\sum_{b=1}^{B} \hat{Y_d}^b(t)$ the Monte Carlo expectation of estimate $\hat{Y}_d(t)$ with $\hat{Y_d}^b(t)$ the mean load curve obtained on the sample selected in the run b, with $b = 1, ..., B$.

For given domain d and time t we define the following Relative Bias indicator as

$$RB(\hat{Y}_d(t)) = 100\frac{|E_{MC}[\hat{Y}_d(t)] - Y_d(t)|}{Y_d(t)}. \tag{9}$$

Then, for a given domain d and time t, we define the following Mean Square Error indicator as

$$MSE_{MC}(\hat{Y}_d(t)) = \frac{1}{B}\sum_{b=1}^{B}(\hat{Y_d}^b(t) - Y_d(t))^2. \tag{10}$$

This indicator is the sum of the square bias and the variance. Estimators showing the smallest MSE are preferred. In order to facilitate the comparison between estimators, we used a derived measure, easier to read, named Relative Efficiency, obtained by dividing the mean of the MSE of an estimator for a given domain by the mean of the MSE for the benchmark estimator (the sample mean for the same domain, see Eq. (3)): [5]

$$RE(\hat{Y}_d(t)) = 100\frac{\overline{MSE}_{MC}(t)(\hat{Y}_d(t))}{\overline{MSE}_{MC}(\hat{Y}_d(t)^{REF})}. \quad (11)$$

We consider the mean of each indicator over all the instants.

Results

	Sampled domains		Non-sampled domains		
Method	RB (%)	RE(%)	RB (%)	RE (%)	time (s)
Benchmark	0.28	100			0.07
PCA + LMM	0.52	20.98	5.1	200	0.53
Discretization + LMM	0.28	25.82	5.3	409	4.16
Functional regression	0.49	29	5.5	403	0.05
Courbotree	1.47	30.22	4.1	39	0.2

Table 1. Comparison of the methods, LMM stands for Linear Mixed Model

As suggested, the linear mixed model was applied on principal component scores (PCA + LMM) or directly on the discretized curves (Discretization + LMM). The benchmark is the sample mean for each domain (see Eq. (3)).

These tests showed that the integration of auxiliary information in domain mean curve estimations leads to substantial precision gains on our dataset. On sampled domains, the best results are obtained for the linear mixed models on principal components estimator (RE: 21%) or on discretized curves (RE: 26%), followed by functional linear regression (RE: 29%) and finally "courboTree"(regression trees for curves). The relative bias measures of the three methods are very small (less than one percent) on sampled domains so the proposed methods have an effect on variance reduction rather on the bias reduction. The use of the Principal Component Analysis in combination to the linear mixed model leads to a precision gain of a few percents.

On non-sampled domains, the most efficient technique is the courboTree followed by the linear mixed model on Principal Components and then the functional linear regression[6]. The relative biases of all the methods are moderate.

On our dataset, calculation times are very small: a few seconds for linear mixed models (ten times less when we use a PCA), twenty times less for regression trees and a hundred times less for functional linear regression.

5 Conclusions and perspectives

We have proposed three methods to address the problem of estimating mean load curves of small areas by sampling using auxiliary information available at the unit and area levels. These methods have been tested and compared to each other on real datasets and have been shown to lead to substantial precision gains compared to simple domain means.

[5]For the non-sampled domain, this MSE does not exist so we divide the MSE by the mean MSE over all the domains for the benchmark method.

[6]Direct estimations are impossible on non-sampled domains so the RE is the ratio of the MSE for the given estimator divided by the mean of the MSE for the benchmark estimator over sampled domains.

This work can be continued by building more efficient estimators or by robustifying the previous methods. Finally, regression trees modelling can be enhanced by replacing simple mean by functional regression on area-level auxiliary information in each leaf of the tree in order to take this information into account.

Acknowledgements: Authors wish to thank the two anonymous referees for their constructives remarks which helped us to improve much the presentation of the paper.

Bibliography

[1] Ardilly, P. (2014). *Estimation régionale de taux de pauvreté utilisant une technique de calage*, Actes du 8ème colloque francophone sur les sondages, Dijon, France.

[2] Battese, G.E., Harter, R. and Fuller, W. (1988). *An error-components model for prediction of county crop areas using survey and satellite data*, Journal of the American Statistical Association, **83**, 28–36.

[3] Breiman, L., Friedman, J., Stone, C. and Olshen, R. (1984). *Classification and regression trees*, CRC press.

[4] Cardot, H., Chaouch, M., Goga, C. and C. Labruère (2010). *Properties of Design-Based Functional Principal Components Analysis*. Journal of Statistical Planning and Inference, **140**, 75-91.

[5] Cardot, H., Dessertaine, A., Goga, C., Josserand, E. and Lardin, P. (2013). *Comparison of different sample designs and construction of confidence bands to estimate the mean of functional data: An illustration on electricity consumption*, Survey Methodology, **39**, 283–301.

[6] Cardot, H., Degras, D. and Josserand, E. (2013). *Confidence bands for Horvitz–Thompson estimators using sampled noisy functional data*, Bernoulli, **19**, 2067–2097.

[7] Cardot, H., Goga, C. and Lardin, P. (2013). *Uniform convergence and asymptotic confidence bands for model-assisted estimators of the mean of sampled functional data*, Electronic Journal of Statistics, **7**, 562–596.

[8] Deville, J.-C. and Särndal, C.-E. (1992). *Calibration estimators in survey sampling*, Journal of the American Statistical Association, **87**, 376–382.

[9] Rao, J.N.K. and Molina I. (2015). *Small area estimation*, 2nd edition, Wiley.

[10] Ramsay, J. and Silverman B. (2005). *Functional data analysis*, 2nd edition, Springer.

[11] Särndal C.-E., Swensson, B. and Wretman, J. (2003). *Model assisted survey sampling*, 2nd edition, Springer.

[12] Stephan V. and Cordogan F.(2009). *Courbotree: application des arbres de regression multivariés pour la classification de courbes*, Revue MODULAD, **33**, 129–138.

[13] Valliant, R., Dorfman, A. and Royall, R. (2000). *Finite population sampling and inference: a prediction approach*, Wiley.

Coping with level and different type of contamination by SW-estimator

Jan Ámos Víšek, *Charles University in Prague*, visek@fsv.cuni.cz

Abstract. A new estimator inheriting (hopefully) pros and removing cons of the *S-estimator* and the *least weighted squares* is proposed. It allows for both, a weight function w as well as for an objective function ρ which need not be bounded. The conditions for consistency and asymptotic representation are recalled. The combination of the weight function with the objective function allows to accommodate the estimator to the level and different type of contamination. The results of simulation studies confirm it.

Keywords. Robust regression, S-estimator, the least weighted squares, SW-estimator, consistency, asymptotic representation, simulation results.

1 Introduction

Although Siegel's *repeated median* [13] ended the pursuit for 50% breakdown point estimator of regression coefficients (which was launched by Bickel [1]), its complexity hampers implementation except for the simple regression. Rousseeuw's *least median of squares* ($\hat{\beta}^{(LMS,n,h)}$) [11] and the *least trimmed squares* ($\hat{\beta}^{(LTS,n,h)}$) [6] fortunately arrived soon after it and they fulfilled what we had expected. But the presence of order statistics of squared residuals in the definitions of both of them indicated that the study of their asymptotics could be complicated[7]. The *S-estimator* ($\hat{\beta}^{(S,n,\rho)}$) by Rousseeuw and Yohai [12] have got rid of this disadvantage and in fact it allowed, by employing results from [10], an immediate proof of its consistency. Another advantage appeared a bit later. In 1992 Hettmansperger and Sheater's study of *Engine Knock Data* [8] revealed the *switch effect* caused by zero-one objective function[8]. Then the utilization of a continuous objective function (as the S-estimator had done it) proved to be a step in the proper direction, just removing the *switch effect* (the danger of switch effect was, all after, indicated already in [9]). On the other hand, removing the order statistics of squared residuals from the definition of S-estimator brought unfortunately also an unintended negative consequence. By focusing on minimization of estimate of variance of error terms - it (potentially) loses some information of influential observations, see the results of simulations at the end of paper.

The *least weighted squares* ($\hat{\beta}^{(LWS,n,w)}$) [16] set off in the same direction as $\hat{\beta}^{(S,n,\rho)}$, employing a continuous weight function[9] but it did not remove from the definition the order statistics of squared residuals. Although it preserved the technical problems when carrying out the theoretical studies, it proved to be

[7]It confirmed the fact that we waited 20 years for the proof of consistency of LTS for multiple regression, [17].

[8]The results by Hettmansperger and Sheater were wrong (due to the use of a bad algorithm for $\hat{\beta}^{(LMS,n,h)}$, see [2]) but $\hat{\beta}^{(LTS,n,h)}$ can be computed exactly for this dataset and it exhibited correctly the *switch effect*. At the first glance it seemed that it is a consequence of the high breakdown point of $\hat{\beta}^{(LTS,n,h)}$, for a figure indicating it see [18], but as $n = 16$ and $h = 11$, the breakdown point of $\hat{\beta}^{(LTS,n,h)}$ is approximately only 30%. In fact for these data the *switch effect* was implied by zero-one objective function.

[9]The experiences from simulations hint that the weight function w should resemble the "left wing" of Tukey's function ρ for $\hat{\beta}^{(LWS,n,w)}$.

profitable from the efficiency point of view in the applications. The technical problems have been later solved by the generalization of Kolmogorov-Smirnov result [20].

It is easy to learn that $\hat{\beta}^{(LWS,n,w)}$ is not a special case of $\hat{\beta}^{(S,n,\rho)}$ and vice versa, see [21]. Then it is a straightforward step to combine these two estimators and obtain an estimator - the *S-weighted estimator* ($\hat{\beta}^{(SW,n,w,\rho)}$) (hopefully) inheriting pros and removing cons of the both parents. The present paper summarizes the up-to-now-reached theoretical results on $\hat{\beta}^{(SW,n,w,\rho)}$ and it offers the simulation results indicating that it can be adjusted not only to the level of contamination but also tailored to the different type of contamination.

2 Framework, assumptions and estimator

Let $\mathcal{N}$ denote the set of all positive integers, R the real line and R^p the p-dimensional Euclidean space. All vectors will be assumed to be column and throughout the paper we assume that all r.v.'s are defined on a basic probability space $(\Omega, \mathcal{A}, P)$, say. For a sequence of $(p+1)$-dimensional random variables (r. v.'s) $\{(X_i', e_i)\}_{i=1}^{\infty}$, for any $n \in \mathcal{N}$ and a fixed $\beta^0 \in R^p$ the linear regression model will be considered

$$Y_i = X_i'\beta^0 + e_i, \quad i = 1, 2, ..., n \quad \text{or} \quad Y = X\beta^0 + e \tag{1}$$

where $Y = (Y_1, Y_2, ..., Y_n)', X = (X_1, X_2, ..., X_n)'$ and $e = (e_1, e_2, ..., e_n)'$. Finally, we are going to make the following assumptions on the explanatory variables and the error terms (allowing generally for intercept amd discrete variables).

Conditions $\mathcal{C}1$ *The sequence $\{(V_i', e_i)'\}_{i=1}^{\infty}$ is sequence of independent p-dimensional random variables distributed according to the distribution functions (d. f.) $F_{V,e_i}(v,r) = F_V(v) \cdot F_e(\sigma_i \cdot r)$, $i \in \mathcal{N}$ where $F_{V,e}(v,r)$ is a parent d. f., $I\!EV_1 = 0$, the covariance matrix $I\!E\{V_1 V_1'\}$ is regular, $I\!Ee_i = 0$ and $\sigma_i^2 = \text{var}(e_i)$ with $0 < s < \liminf_{i\to\infty} \sigma_i \leq \limsup_{i\to\infty} \sigma_i < S < \infty$. There is ℓ, $0 \leq \ell < p$ and coordinates $V_{11}, V_{12}, ..., V_{1\ell}$ of the vector V_1 are discrete with the distribution given by $\{p_{1,v} = P(V_{11} = v_1, V_{12} = v_2, ..., V_{1\ell} = v_\ell)\}_{\{v\in\mathcal{U}\}}$ where $\mathcal{U} \subset \mathcal{T}$ and $\mathcal{T} \subset R^\ell$ is a compact. Further, the d. f. of the vector $(V_{1,\ell+1}, V_{1,\ell+2}, ..., V_{1,p-1})'$ is absolutely continuous, the density $f_{V_{1,\ell+1},V_{1,\ell+2},...,V_{1,p-1}}(v)$ is bounded by B_e. Similarly, the parent d. f. $F_e(r)$ is absolutely continuous with density $f_e(r)$ bounded by U_e. Moreover, there is $q > 1$ so that $I\!E\|V_1\|^{2q} < \infty$. Finally, consider the sequence $\{(X_i', e_i)'\}_{i=1}^{\infty}$ where $X_{i1} = 1$ and $X_{ij} = V_{i,j-1}$, $j = 2, 3, ..., p$ for all $i \in \mathcal{N}$. This sequence will be considered as the sequence of explanatory variables and of error term.*

Conditions $\mathcal{C}2$

- *$w : [0,1] \to [0,1]$ is a continuous, non-increasing weight function with $w(0) = 1$. Moreover, w is Lipschitz in absolute value, i. e. there is L such that for any pair $u_1, u_2 \in [0,1]$ we have $|w(u_1) - w(u_2)| \leq L \cdot |u_1 - u_2|$.*
- *$\rho : (0,\infty) \to (0,\infty)$, $\rho(0) = 0$, non-decreasing on $(0,\infty)$, symmetric and differentiable (denote the derivative by ψ).*
- *$\psi(r)/r$ is non-increasing for $r \geq 0$ with $\lim_{r\to 0_+} \frac{\psi(r)}{r} = 1$.*

Definition 2.1. *Let $w : [0,1] \to [0,1]$ and $\rho : [0,\infty] \to [0,\infty]$ be a weight function and an objective function, respectively. Then*

$$\hat{\beta}^{(SW,n,w,\rho)} = \underset{\beta\in R^p}{arg\,min} \left\{\sigma(\beta) \in R^+ : \frac{1}{n}\sum_{i=1}^{n} w\left(\frac{i-1}{n}\right)\rho\left(\frac{r_{(i)}(\beta)}{\sigma}\right) = b\right\} \tag{2}$$

where $b = I\!E\left\{w\left(F\left(\frac{e_i}{\sigma_0}\right)\right)\rho\left(\frac{e_1^2}{\sigma_0^2}\right)\right\}$ with F being the parent distribution function, is called the S-weighted estimator, see Víšek (2015a).

Remark. It is clear that LMS, LTS, LWS and S-estimators are special cases of the *S-weighted estimator.* □

Following Rousseeuw & Yohai [12] we can argue that the S-weighted estimator is given by $\beta \in R^p$ such that $\sigma(\beta)$ in (2) is minimal, i. e. $\hat{\sigma}(\hat{\beta}^{(SW,n,w,\rho)}) = \min_{\beta \in R^p} \sigma(\beta)$ and hence for all $\beta \in R^p$

$$\frac{1}{n}\sum_{i=1}^{n} w\left(\frac{i-1}{n}\right)\rho\left(\frac{r_{(i)}(\beta)}{\hat{\sigma}}\right) \geq b. \tag{3}$$

If the opposite sharp inequality takes place for some $\beta \in R^P$, the continuity of weight function w and of objective function ρ allows to decrease value of σ so that the required equality still holds. On the other hand, the left-hand-side in (3) attains value b for $\beta = \hat{\beta}^{(SW,n,w,\rho)}$ and it is the minimum for this expression. Hence all partial derivative have to be zero, i. e. the *normal equations*

$$\sum_{i=1}^{n} w\left(\frac{i-1}{n}\right) X_{j_i}\psi\left(\frac{Y_{j_i} - X'_{j_i}\beta}{\hat{\sigma}}\right) = 0 \tag{4}$$

have to be fulfilled (where $\psi = \rho'$) and j_i is a such index that $r^2_{j_i}(\beta) = r^2_{(i)}(\beta)$. Putting

$$\pi(\beta, j) = i \in \{1, 2, ..., n\} \qquad \Leftrightarrow \qquad r^2_j(\beta) = r^2_{(i)}(\beta) \tag{5}$$

and we arrive at

$$\sum_{j=1}^{n} w\left(\frac{\pi(\beta, j) - 1}{n}\right) X_j\psi\left(\frac{Y_j - X'_j\beta}{\hat{\sigma}}\right) = 0 \tag{6}$$

(notice that $\pi(\beta, j)$ depends also on $\omega \in \Omega$). Denoting the indicator of the set A as $I\{A\}$, the empirical distribution of the absolute values of residuals is given as

$$F^{(n)}_\beta(r) = \frac{1}{n}\sum_{i=1}^{n} I\{|r_i(\beta)| < r\} \tag{7}$$

and one easy verifies that (see [19])

$$\frac{\pi(\beta, i) - 1}{n} = F^{(n)}_\beta(|r_i(\beta)|).$$

Finally, the *normal equations* (6) can be written as[10]

$$\sum_{i=1}^{n} w\left(F^{(n)}_\beta(|r_i(\beta)|)\right) X_i\psi\left(\frac{Y_i - X'_i\beta}{\hat{\sigma}}\right) = 0. \tag{8}$$

As $\psi(0)$ is antisymmetric and $\psi(0) = 0$, the *normal equation* (8) can be handled as follows

$$\begin{aligned} &\sum_{\{i\,:\,(Y_i - X'_i\beta) \neq 0\}} w\left(F^{(n)}_\beta(|r_i(\beta)|)\right) X_i\psi\left(\frac{Y_i - X'_i\beta}{\hat{\sigma}}\right) \\ &= \sum_{\{i\,:\,(Y_i - X'_i\beta) \neq 0\}} w\left(F^{(n)}_\beta(|r_i(\beta)|)\right) X_i\left[\psi\left(\frac{Y_i - X'_i\beta}{\hat{\sigma}}\right) \cdot \frac{\hat{\sigma}}{Y_i - X'_i\beta}\right](Y_i - X'_i\beta) \\ &= \sum_{\{i\,:\,|Y_i - X'_i\beta| > 0\}} w\left(F^{(n)}_\beta(|r_i(\beta)|)\right) X_i\left[\psi\left(\frac{|Y_i - X'_i\beta|}{\hat{\sigma}}\right) \cdot \frac{\hat{\sigma}}{|Y_i - X'_i\beta|}\right](Y_i - X'_i\beta) = 0. \end{aligned} \tag{9}$$

[10] It may seem strange to speak about an *empirical distribution function* under heteroscedasticity. But it is not - see Lemma 5.1 of Appendix.

Then denoting $v(r,\sigma) = \psi(\frac{r}{\sigma})\frac{\sigma}{r}$, we finally have

$$\sum_{i=1}^{n} w\left(F_\beta^{(n)}(|r_i(\beta)|)\right) \cdot v\left(|r_i(\beta)|, \hat{\sigma}\right) X_i \left(Y_i - X_i'\beta\right) = 0. \tag{10}$$

As the function v is continuous and non-increasing on the interval $[0,\infty)$, $v(r,\sigma) = v(|r|,\sigma)$, $v(0) = 0$ and $v : [0,1] \to [0,1]$. Now, we are going to employ the idea of Frank Hampel (see [5]) that the information given by observations $z_1, z_2, ..., z_n$ (say) is the same as the information represented by the corresponding empirical distribution function. Let us define the inverse function to $F_\beta^{(n)}(r)$ as follows

$$F_{inv,\beta,\hat{\sigma}}^{(n)}(v) = \inf_{r \in R}\left\{r \in R \,:\, F_\beta^{(n)}(r) \geq v\right\}.$$

Then we have

$$F_{inv,\beta,\hat{\sigma}}^{(n)}\left(F_\beta^{(n)}(|r_i(\beta)|)\right) = |r_i(\beta)| \tag{11}$$

and putting $\tilde{\psi}_{\hat{\sigma}}(z) = \psi\left(\frac{F_{inv,\beta,\hat{\sigma}}^{(n)}(z)}{\hat{\sigma}}\right)$ we have

$$\tilde{\psi}_{\hat{\sigma}}\left(F_\beta^{(n)}(|r_i(\beta)|)\right) \cdot \frac{1}{F_{inv,\beta,\hat{\sigma}}^{(n)}\left(F_\beta^{(n)}(|r_i(\beta)|)\right)} = \psi\left(\frac{|r_i(\beta)|}{\hat{\sigma}}\right) \cdot \frac{1}{|r_i(\beta)|} = v(|r_i(\beta)|, \hat{\sigma}). \tag{12}$$

Finally, let us put $\tilde{w}(z) = w(z) \cdot v(z,\hat{\sigma}) = w(z) \cdot \tilde{\psi}_{\hat{\sigma}}(z) \cdot \frac{1}{F_{inv,\beta,\hat{\sigma}}^{(n)}(z)}$. Then, taking into account (9), (10), (11) and (12), the *normal equation* (8) attained the form

$$\sum_{i=1}^{n} \tilde{w}\left(F_\beta^{(n)}(|r_i(\beta)|)\right) X_i \left(Y_i - X_i'\beta\right) = 0. \tag{13}$$

Notice please, that $\tilde{w}$ is well defined and it fulfill $\mathcal{C}2$.

3 Consistency and asymptotic representation of the estimator

We will need the following identification condition.

Conditions $\mathcal{C}3$ *For any fixed* $\hat{\sigma} > 0$ *and any* $n \in \mathcal{N}$ *there is only one solution of*

$$(\beta - \beta^0)' \mathbb{E}\left[\sum_{i=1}^{n} \tilde{w}\left(\overline{F}_{\beta^0,\hat{\sigma}}^{(n)}(|e_i|)\right) X_i \left(e_i - X_i'(\beta - \beta^0)\right)\right] = 0$$

namely $\beta = \beta^0$ *where*

$$\overline{F}_{\beta^0,\hat{\sigma}}^{(n)}(|r|) = \frac{1}{n}\sum_{i=1}^{n} F_{\beta^0,i}(r) \quad \textit{with} \quad F_{\beta,i}(|r|) = P\left(|Y_i - X_i'\beta| < r\right) \tag{14}$$

and $\forall\,(\beta \in R^p)$ $\quad \mathbb{E}\left[\sum_{i=1}^{n} \tilde{w}\left(\overline{F}_{\beta,\hat{\sigma}}^{(n)}(|r_i(\beta)|)\right) X_i e_i\right] = 0.$

Theorem 3.1. *Let* ***Conditions*** *$\mathcal{C}1$, $\mathcal{C}2$ and $\mathcal{C}3$ be fulfilled. Then any sequence* $\left\{\hat{\beta}^{(SW,n,w,\rho)}\right\}_{n=1}^{\infty}$ *of the solutions of sequence of normal equations (8) for* $n = 1, 2, ...,$ *is weakly consistent.*

For the proof see [21] and the discussion given there.

Conditions $\mathcal{NC}1$ *The derivative* $f_e'(r)$ *exists and is bounded in absolute value by* B_e. *The derivative* $w'(\alpha)$ *exists and is Lipschitz of the first order (with the corresponding constant* J_w*). Moreover, for any* $i \in \mathcal{N}$

$$\mathbb{E}\left[w'(\overline{F}_{\beta^0,\hat{\sigma}}^{(n)}(|e_i|))\left(f_e(|e_i|) - f_e(-|e_i|)\right) \cdot e_i\right] = 0.$$

Finally, for any $j, k, \ell = 1, 2, ..., p$ $\mathbb{E}\,|X_{1j} \cdot X_{1k} \cdot X_{1\ell}| < \infty$ *(as* $F_X(x)$ *does not depend on* i*, the sequence* $\{X_i\}_{i=1}^{\infty}$ *is sequence of independent and identically distributed* p*-dimensional r.v.'s).*

Theorem 3.2. *Let Conditions $\mathcal{C}1$, $\mathcal{C}2$, $\mathcal{C}3$ and $\mathcal{NC}1$ hold. Then any sequence $\left\{\hat{\beta}^{(SW,n,w,\rho)}\right\}_{n=1}^{\infty}$ of solutions of the normal equations (13) is weakly $\sqrt{n}$-consistent, i. e.*
$\forall(\varepsilon > 0) \quad \exists(K_\varepsilon < \infty) \quad \forall(n \in \mathcal{N})$

$$P\left(\left\{\omega \in \Omega \; : \; \sqrt{n}\left\|\hat{\beta}^{(LWS,n,w)} - \beta^0)\right\| < K_\varepsilon\right\}\right) > 1 - \varepsilon.$$

For the proof see [22] and again the discussion given there.

Conditions $\mathcal{AC}1$
Let $G(v)$ be the d. f. of e^2 (for definiton of r. v. e see ***Conditions*** *$\mathcal{C}1$), i. e. $G(v) = F(\sqrt{v}) - F(-\sqrt{v})$ and $g(v)$ its density. Then for any $a \in R^+$ there is $\Delta(a) > 0$ and $L_a > 0$ so that*

$$\inf_{z \in (0, a+\Delta(a))} g(z) > L_{g,a} > 0.$$

Remark. Notice please that *Conditions* $\mathcal{AC}1$ hold for any d. f. having positive density in a neighborhood of 0. □

Theorem 3.3. *Let Conditions $\mathcal{C}1$, $\mathcal{C}2$, $\mathcal{C}3$, $\mathcal{NC}1$ and $\mathcal{AC}1$ hold and let $Q = I\!\!E\left\{w(F_{\beta^0}(|e|))X_1X_1'\right\}$ be positive definite. Then*

$$\sqrt{n}\left(\hat{\beta}^{(SW,n,w,\rho)} - \beta^0\right) = Q^{-1} \cdot \frac{1}{\sqrt{n}} \sum_{i=1}^{n} w\left(F_{\beta^0}(|e_i|)\right) \cdot X_i e_i + o_p(1). \tag{15}$$

For the proof see [23].

4 Patterns of results of simulation study

We offer here some small portion of results of simulations - for $\hat{\beta}^{(OLS,n)}$, $\hat{\beta}^{(S,n,\rho)}$, $\hat{\beta}^{(W,n,\rho)}$ (see [4][11]) and $\hat{\beta}^{(SW,n,w,\rho)}$. Data were generated by the model

$$Y_i = \mathbf{1} + \mathbf{2} \cdot X_{i2} - \mathbf{3} \cdot X_{i3} + \mathbf{4} \cdot X_{i4} - \mathbf{5} \cdot X_{i5} + e_i, \; i = 1, 2, ..., n. \tag{16}$$

The explanatory variables X_i's were generated by standard normal d. f., independent each from other, independent from error terms e_i's which were normally distributed with zero mean and heteroscedastic variances which were uniformly distributed on $[0.5, a]$ (a is specified at the captions of tables). Algorithms from [4] (employing [14] and [3]) and from [15] (which basically coinside with [7]) were used for S-, W- and SW-estimators, respectively (MATLAB codes are available on request). Tukey's function

$$\rho_c(x) = \frac{x^2}{2} - \frac{x^4}{2 \cdot c^2} + \frac{x^6}{6 \cdot c^4} \quad \text{for} \quad |x| \le c \qquad \text{and} \qquad \rho_c(x) = \frac{c^2}{6} \text{ otherwise,}$$

(see e. g. [3], the constant c is specified at the captions of tables) and the quadratic function were utilized as the objective functions ρ for S-, W- and SW-estimators, respectively. The value $b = I\!\!E\left\{w\left(F\left(\frac{e_i}{\hat{\sigma}_0}\right)\right)\rho(\frac{e_1^2}{\hat{\sigma}_0^2})\right\}$ was for given c computed as (p is dimension of model, i. e. $p = 5$)

$$b = p\frac{\chi^2_{p+2}(c^2)}{2} - p \cdot (p+2)\frac{\chi^2_{p+4}(c^2)}{2 \cdot c^2} + p \cdot (p+2) \cdot (p+4)\frac{\chi^2_{p+6}(c^2)}{6 \cdot c^4} + \frac{c^2}{6}(1 - \chi^2_p(c^2)),$$

see again [3]. Finally, the weight function $w(r) = 1$ for $r \in [0, h]$, $w(r) = 0$ for $r \in [g, 0]$ ($h < g$) and on $[h, g]$ it has the shape of the "left-wing" of Tukey's function, decreasing from 1 to 0. Due to rather

[11] W-estimator $\hat{\beta}^{(W,n,\rho)}$ is to represent an improvement of S-estimator just taking into account the Mahalanobis distances of observations from the center of gravity.

good PC[12] all the constants c, h, g and W-*level*[13] were assigned to minimize an "aggregated" (over the coordinates of $\hat{\beta}^{(estimator)}$) MSE, see below. We have generated 100 datasets[14], each contained 500 observations and we computed the estimates of regression coefficients

$$\left\{ \hat{\beta}^{(estimator,k)} = (\hat{\beta}_1^{(estimator,k)}, \hat{\beta}_2^{(estimator,k)}, \hat{\beta}_3^{(estimator,k)}, \hat{\beta}_4^{(estimator,k)}, \hat{\beta}_5^{(estimator,k)})' \right\}_{k=1}^{100}.$$

The abbreviations OLS, S, W and SW at the position of *"estimator"* indicate the method employed for the computation. Finally, we report values (for $j = 1, 2, 3, 4$ and 5)

$$\hat{\beta}_j^{(estimator)} = \frac{1}{100} \sum_{k=1}^{100} \hat{\beta}_j^{(estimator,k)} \quad \text{and} \quad \widehat{\text{MSE}} \left(\hat{\beta}_j^{(estimator)} \right) = \frac{1}{100} \sum_{k=1}^{100} \left[\hat{\beta}_j^{(estimator,k)} - \beta_j^0 \right]^2. \tag{17}$$

[12] HP Elite 7500 with Intel Core i7-3770 Processor (3.4 GHz, 8MB cache).

[13] Explanation for W-level: If the Mahalanobis distance from the center of gravity for given observation overcomes the χ^2-upper quantile for the value 1 - *W-level*, the observation is deleted from the dataset and then the S-estimator is computed.

[14] We experimented with various numbers of repetitions - smaller than 100 exhibited some instability in MSE, in the sense that repeated simulations (yielding one particular table - see below) gave (rather) different information about the dispersion of the estimates for individual datasets, - the larger gave a lower information about the preciseness of estimation by $\hat{\beta}_j^{(estimator)}$ (see (17)) just resulting in exact "true values of coefficients", see (16).

Contamination level = 1%, $h = 0.970$, $g = 0.985$, $c = 9.8$, $W\text{-}level = 0.999$					
$\hat{\beta}^{(OLS)}_{(MSE)}$	$0.978_{(0.084)}$	$1.421_{(1.568)}$	$-1.907_{(2.065)}$	$2.505_{(3.340)}$	$-3.047_{(5.185)}$
$\hat{\beta}^{(S)}_{(MSE)}$	$1.004_{(0.006)}$	$1.999_{(0.005)}$	$-3.002_{(0.004)}$	$3.992_{(0.005)}$	$-5.009_{(0.005)}$
$\hat{\beta}^{(W)}_{(MSE)}$	$1.003_{(0.006)}$	$2.001_{(0.007)}$	$-3.004_{(0.006)}$	$3.995_{(0.006)}$	$-5.005_{(0.005)}$
$\hat{\beta}^{(SW)}_{(MSE)}$	$1.003_{(0.006)}$	$1.995_{(0.006)}$	$-3.000_{(0.004)}$	$3.991_{(0.005)}$	$-5.011_{(0.005)}$

Contamination level = 2%, $h = 0.960$, $g = 0.975$, $c = 8.5$, $W\text{-}level = 0.98$					
$\hat{\beta}^{(OLS)}_{(MSE)}$	$0.950_{(0.121)}$	$0.858_{(2.571)}$	$-1.067_{(5.154)}$	$1.483_{(7.435)}$	$-1.952_{(10.970)}$
$\hat{\beta}^{(S)}_{(MSE)}$	$0.993_{(0.004)}$	$1.997_{(0.005)}$	$-3.004_{(0.005)}$	$4.011_{(0.007)}$	$-5.012_{(0.005)}$
$\hat{\beta}^{(W)}_{(MSE)}$	$0.993_{(0.004)}$	$1.996_{(0.005)}$	$-3.003_{(0.006)}$	$4.014_{(0.008)}$	$-5.012_{(0.005)}$
$\hat{\beta}^{(SW)}_{(MSE)}$	$0.992_{(0.004)}$	$1.992_{(0.005)}$	$-3.008_{(0.006)}$	$4.009_{(0.007)}$	$-5.007_{(0.005)}$

Contamination level = 3%, $h = 0.950$, $g = 0.960$, $c = 7.6$, $W\text{-}level = 0.95$					
$\hat{\beta}^{(OLS)}_{(MSE)}$	$0.987_{(0.189)}$	$0.396_{(3.391)}$	$-0.476_{(7.713)}$	$0.847_{(11.062)}$	$-0.898_{(18.278)}$
$\hat{\beta}^{(S)}_{(MSE)}$	$0.994_{(0.005)}$	$2.005_{(0.006)}$	$-2.995_{(0.006)}$	$4.015_{(0.003)}$	$-5.006_{(0.004)}$
$\hat{\beta}^{(W)}_{(MSE)}$	$0.994_{(0.005)}$	$2.011_{(0.007)}$	$-2.999_{(0.006)}$	$4.012_{(0.004)}$	$-5.014_{(0.006)}$
$\hat{\beta}^{(SW)}_{(MSE)}$	$0.992_{(0.005)}$	$2.006_{(0.006)}$	$-2.996_{(0.005)}$	$4.013_{(0.004)}$	$-5.007_{(0.005)}$

Contamination level = 5%, $h = 0.920$, $g = 0.945$, $c = 6.4$, $W\text{-}level = 0.94$					
$\hat{\beta}^{(OLS)}_{(MSE)}$	$0.921_{(0.282)}$	$-0.057_{(4.827)}$	$0.254_{(11.390)}$	$-0.268_{(19.055)}$	$0.455_{(30.643)}$
$\hat{\beta}^{(S)}_{(MSE)}$	$1.016_{(0.005)}$	$1.995_{(0.004)}$	$-2.997_{(0.006)}$	$4.009_{(0.004)}$	$-4.994_{(0.006)}$
$\hat{\beta}^{(W)}_{(MSE)}$	$1.016_{(0.005)}$	$1.997_{(0.006)}$	$-2.996_{(0.007)}$	$4.010_{(0.006)}$	$-4.995_{(0.006)}$
$\hat{\beta}^{(SW)}_{(MSE)}$	$1.013_{(0.005)}$	$1.996_{(0.006)}$	$-2.996_{(0.008)}$	$4.011_{(0.005)}$	$-4.995_{(0.006)}$

Contamination level = 10%, $h = 0.970$, $g = 0.985$, $c = 6.2$, $W\text{-}level = 0.93$					
$\hat{\beta}^{(OLS)}_{(MSE)}$	$0.933_{(0.072)}$	$1.272_{(1.635)}$	$-1.887_{(2.406)}$	$2.747_{(2.602)}$	$-3.382_{(3.513)}$
$\hat{\beta}^{(S)}_{(MSE)}$	$0.999_{(0.005)}$	$2.015_{(0.007)}$	$-3.008_{(0.006)}$	$4.012_{(0.005)}$	$-5.012_{(0.005)}$
$\hat{\beta}^{(W)}_{(MSE)}$	$0.999_{(0.005)}$	$2.016_{(0.008)}$	$-3.006_{(0.007)}$	$4.014_{(0.007)}$	$-5.012_{(0.007)}$
$\hat{\beta}^{(SW)}_{(MSE)}$	$1.001_{(0.005)}$	$2.014_{(0.006)}$	$-3.011_{(0.007)}$	$4.010_{(0.005)}$	$-5.007_{(0.006)}$

Table 1. Contamination by bad leverage points: For randomly selected observations we put $X_i = 5 * X_i^{original}$, then we put $Y_i = -X_i\beta^0 + e_i$ (for $\beta^0 = (1, 2, -3, 4, -5)'$ see (16)). Number of observations in each dataset = 500, $a = 2$.

Contamination level = 1%, $h = 0.968$, $g = 0.980$, $c = 8.3$, $W\text{-}level = 0.994$					
$\hat{\beta}^{(OLS)}_{(MSE)}$	$0.948_{(0.096)}$	$1.638_{(0.893)}$	$-2.301_{(1.132)}$	$2.915_{(1.936)}$	$-3.903_{(2.093)}$
$\hat{\beta}^{(S)}_{(MSE)}$	$0.968_{(0.023)}$	$1.994_{(0.021)}$	$-2.999_{(0.017)}$	$4.023_{(0.018)}$	$-4.979_{(0.018)}$
$\hat{\beta}^{(W)}_{(MSE)}$	$0.968_{(0.023)}$	$1.990_{(0.023)}$	$-2.992_{(0.021)}$	$4.025_{(0.022)}$	$-4.978_{(0.021)}$
$\hat{\beta}^{(SW)}_{(MSE)}$	$0.964_{(0.028)}$	$2.001_{(0.012)}$	$-3.002_{(0.014)}$	$4.003_{(0.013)}$	$-4.996_{(0.012)}$
Contamination level = 2%, $h = 0.958$, $g = 0.971$, $c = 7.2$, $W\text{-}level = 0.98$					
$\hat{\beta}^{(OLS)}_{(MSE)}$	$0.947_{(0.197)}$	$1.497_{(0.835)}$	$-1.905_{(1.855)}$	$2.753_{(2.322)}$	$-3.447_{(3.497)}$
$\hat{\beta}^{(S)}_{(MSE)}$	$1.004_{(0.019)}$	$2.008_{(0.022)}$	$-3.008_{(0.021)}$	$4.018_{(0.020)}$	$-5.016_{(0.023)}$
$\hat{\beta}^{(W)}_{(MSE)}$	$1.003_{(0.019)}$	$2.006_{(0.026)}$	$-3.011_{(0.025)}$	$4.030_{(0.027)}$	$-5.017_{(0.025)}$
$\hat{\beta}^{(SW)}_{(MSE)}$	$1.005_{(0.023)}$	$2.007_{(0.010)}$	$-3.009_{(0.013)}$	$3.999_{(0.009)}$	$-5.013_{(0.010)}$
Contamination level = 3%, $h = 0.950$, $g = 0.960$, $c = 6.8$, $W\text{-}level = 0.965$					
$\hat{\beta}^{(OLS)}_{(MSE)}$	$1.003_{(0.267)}$	$1.209_{(1.085)}$	$-1.830_{(1.773)}$	$2.671_{(2.438)}$	$-3.313_{(3.663)}$
$\hat{\beta}^{(S)}_{(MSE)}$	$1.005_{(0.021)}$	$1.988_{(0.025)}$	$-2.990_{(0.018)}$	$3.998_{(0.025)}$	$-4.993_{(0.023)}$
$\hat{\beta}^{(W)}_{(MSE)}$	$1.005_{(0.021)}$	$1.993_{(0.027)}$	$-2.993_{(0.026)}$	$3.997_{(0.028)}$	$-5.018_{(0.027)}$
$\hat{\beta}^{(SW)}_{(MSE)}$	$1.006_{(0.022)}$	$2.002_{(0.008)}$	$-3.007_{(0.007)}$	$3.998_{(0.008)}$	$-5.005_{(0.008)}$
Contamination level = 5%, $h = 0.920$, $g = 0.945$, $c = 6.2$, $W\text{-}level = 0.94$					
$\hat{\beta}^{(OLS)}_{(MSE)}$	$1.007_{(0.602)}$	$1.344_{(0.811)}$	$-1.932_{(1.601)}$	$2.642_{(2.300)}$	$-3.140_{(3.969)}$
$\hat{\beta}^{(S)}_{(MSE)}$	$1.002_{(0.026)}$	$1.983_{(0.024)}$	$-3.020_{(0.022)}$	$3.991_{(0.023)}$	$-5.030_{(0.021)}$
$\hat{\beta}^{(W)}_{(MSE)}$	$1.001_{(0.026)}$	$2.002_{(0.030)}$	$-3.021_{(0.027)}$	$3.982_{(0.026)}$	$-5.040_{(0.024)}$
$\hat{\beta}^{(SW)}_{(MSE)}$	$1.009_{(0.030)}$	$1.992_{(0.004)}$	$-3.005_{(0.004)}$	$3.997_{(0.004)}$	$-4.993_{(0.004)}$
$\hat{\beta}^{(SW)}_{(MSE)}$	$1.009_{(0.030)}$	$1.992_{(0.004)}$	$-3.005_{(0.004)}$	$3.997_{(0.004)}$	$-4.993_{(0.004)}$
Contamination level = 10%, $h = 0.820$, $g = 0.865$, $c = 6.0$, $W\text{-}level = 0.93$					
$\hat{\beta}^{(OLS)}_{(MSE)}$	$0.859_{(0.968)}$	$1.253_{(0.768)}$	$-1.817_{(1.630)}$	$2.501_{(2.477)}$	$-2.970_{(4.502)}$
$\hat{\beta}^{(S)}_{(MSE)}$	$0.981_{(0.031)}$	$1.990_{(0.023)}$	$-2.976_{(0.032)}$	$3.996_{(0.024)}$	$-4.962_{(0.028)}$
$\hat{\beta}^{(W)}_{(MSE)}$	$0.978_{(0.030)}$	$1.987_{(0.029)}$	$-2.977_{(0.032)}$	$4.006_{(0.037)}$	$-4.981_{(0.035)}$
$\hat{\beta}^{(SW)}_{(MSE)}$	$0.983_{(0.026)}$	$2.009_{(0.002)}$	$-3.000_{(0.002)}$	$3.993_{(0.003)}$	$-4.997_{(0.002)}$

Table 2. Contamination by bad leverage points as desccribed in Table 1 but data contained also good leverage points $X_i = 10 * X_i^{original}$, the same amount as of bad leverage points. Number of observations in each dataset = 500, $a = 5$.

Contamination level = 1%, $h = 0.970$, $g = 0.985$, $c = 8.9$, W-$level = 0.998$					
$\hat{\beta}^{(OLS)}_{(MSE)}$	$0.928_{(0.057)}$	$1.960_{(0.023)}$	$-2.933_{(0.022)}$	$3.908_{(0.034)}$	$-4.890_{(0.034)}$
$\hat{\beta}^{(S)}_{(MSE)}$	$0.968_{(0.021)}$	$2.003_{(0.022)}$	$-3.009_{(0.022)}$	$3.997_{(0.018)}$	$-4.988_{(0.020)}$
$\hat{\beta}^{(W)}_{(MSE)}$	$0.968_{(0.021)}$	$2.013_{(0.031)}$	$-3.011_{(0.028)}$	$4.004_{(0.024)}$	$-4.998_{(0.025)}$
$\hat{\beta}^{(SW)}_{(MSE)}$	$0.970_{(0.022)}$	$2.010_{(0.013)}$	$-2.999_{(0.010)}$	$3.993_{(0.010)}$	$-4.996_{(0.012)}$

Contamination level = 2%, $h = 0.960$, $g = 0.975$, $c = 7.7$, W-$level = 0.96$					
$\hat{\beta}^{(OLS)}_{(MSE)}$	$0.890_{(0.100)}$	$1.966_{(0.010)}$	$-2.944_{(0.010)}$	$3.929_{(0.014)}$	$-4.898_{(0.018)}$
$\hat{\beta}^{(S)}_{(MSE)}$	$0.995_{(0.026)}$	$1.979_{(0.022)}$	$-2.989_{(0.022)}$	$3.975_{(0.019)}$	$-4.988_{(0.023)}$
$\hat{\beta}^{(W)}_{(MSE)}$	$0.995_{(0.026)}$	$1.979_{(0.028)}$	$-2.984_{(0.028)}$	$3.980_{(0.023)}$	$-4.993_{(0.027)}$
$\hat{\beta}^{(SW)}_{(MSE)}$	$0.993_{(0.025)}$	$1.995_{(0.004)}$	$-2.999_{(0.004)}$	$4.000_{(0.005)}$	$-4.996_{(0.003)}$

Contamination level = 3%, $h = 0.950$, $g = 0.960$, $c = 8.2$, W-$level = 0.945$					
$\hat{\beta}^{(OLS)}_{(MSE)}$	$0.830_{(0.148)}$	$1.959_{(0.009)}$	$-2.949_{(0.008)}$	$3.925_{(0.015)}$	$-4.924_{(0.014)}$
$\hat{\beta}^{(S)}_{(MSE)}$	$1.000_{(0.020)}$	$1.992_{(0.018)}$	$-3.031_{(0.020)}$	$3.981_{(0.023)}$	$-5.004_{(0.018)}$
$\hat{\beta}^{(W)}_{(MSE)}$	$0.999_{(0.020)}$	$2.019_{(0.027)}$	$-3.028_{(0.024)}$	$3.992_{(0.027)}$	$-5.016_{(0.023)}$
$\hat{\beta}^{(SW)}_{(MSE)}$	$0.990_{(0.020)}$	$2.003_{(0.002)}$	$-2.993_{(0.003)}$	$3.993_{(0.004)}$	$-5.012_{(0.003)}$

Contamination level = 5%, $h = 0.920$, $g = 0.945$, $c = 8.35$, W-$level = 0.93$					
$\hat{\beta}^{(OLS)}_{(MSE)}$	$0.742_{(0.304)}$	$1.959_{(0.004)}$	$-2.949_{(0.006)}$	$3.934_{(0.008)}$	$-4.909_{(0.011)}$
$\hat{\beta}^{(S)}_{(MSE)}$	$1.001_{(0.024)}$	$1.984_{(0.025)}$	$-2.962_{(0.032)}$	$3.958_{(0.026)}$	$-4.956_{(0.027)}$
$\hat{\beta}^{(W)}_{(MSE)}$	$1.002_{(0.024)}$	$1.984_{(0.027)}$	$-2.983_{(0.030)}$	$3.992_{(0.030)}$	$-4.984_{(0.027)}$
$\hat{\beta}^{(SW)}_{(MSE)}$	$1.004_{(0.018)}$	$1.997_{(0.001)}$	$-2.999_{(0.001)}$	$4.002_{(0.001)}$	$-4.993_{(0.001)}$

Contamination level = 10%, $h = 0.820$, $g = 0.865$, $c = 9.1$, W-$level = 0.928$					
$\hat{\beta}^{(OLS)}_{(MSE)}$	$0.365_{(0.739)}$	$1.973_{(0.002)}$	$-2.952_{(0.003)}$	$3.948_{(0.004)}$	$-4.930_{(0.006)}$
$\hat{\beta}^{(S)}_{(MSE)}$	$0.957_{(0.037)}$	$1.910_{(0.048)}$	$-2.878_{(0.070)}$	$3.866_{(0.064)}$	$-4.751_{(0.120)}$
$\hat{\beta}^{(W)}_{(MSE)}$	$0.959_{(0.038)}$	$1.980_{(0.035)}$	$-3.000_{(0.034)}$	$4.012_{(0.029)}$	$-4.949_{(0.030)}$
$\hat{\beta}^{(SW)}_{(MSE)}$	$1.010_{(0.025)}$	$1.995_{(0.001)}$	$-2.998_{(0.001)}$	$3.999_{(0.001)}$	$-5.003_{(0.001)}$

Table 3. Contamination by outliers: For randomly selected observations we put $Y_i = 5 * Y_i^{original}$ and data contained also good leverage points $X_i = 10 * X_i^{original}$.
Number of observations in each dataset = 500, $a = 2$.

5 Conclusions

The robust estimators try to depress the influence of atypical points in data by various tools - by an appropriate shape of the objective function ρ, by the weights which are assigned utilizing an external rule or by employing the order statistics of (squared) residuals, by finding a minimal volume containing a priori given percentage of data, by minimization of an estimate of variance of error term, by minimal distance estimation, etc. In the situation which is presented in Figure 1 below, all robust estimators correctly recognize that the group of 5 points under the main cloud of data in are outliers and they try to depress their influence. The estimators which try to do it by minimizing estimated variance of error terms or by finding a minimal volume containing a priori given portion of observations or relying on minimal distance principle, very likely lose the information represented by the group of good leverage points in the right-upper corner of Figure 1. And the results exhibited in the above given tables confirm it by comparing the *mean square error* of *S-weighted estimators* with two other robust estimators, *S*- and *W-estimators*. Taking into account some previous experiences we can conjecture that the attempts to cope with contamination only by weighting down the (squared) residuals or by selecting hopefully the appropriate objective function ρ need not be sufficient. So, in a somewhat more general point of view, the results of simulations support the recommendation given already in [6] that we should employ all estimators we have at our disposal, compare the results and find the reason(s) of significant differences among them, if any.

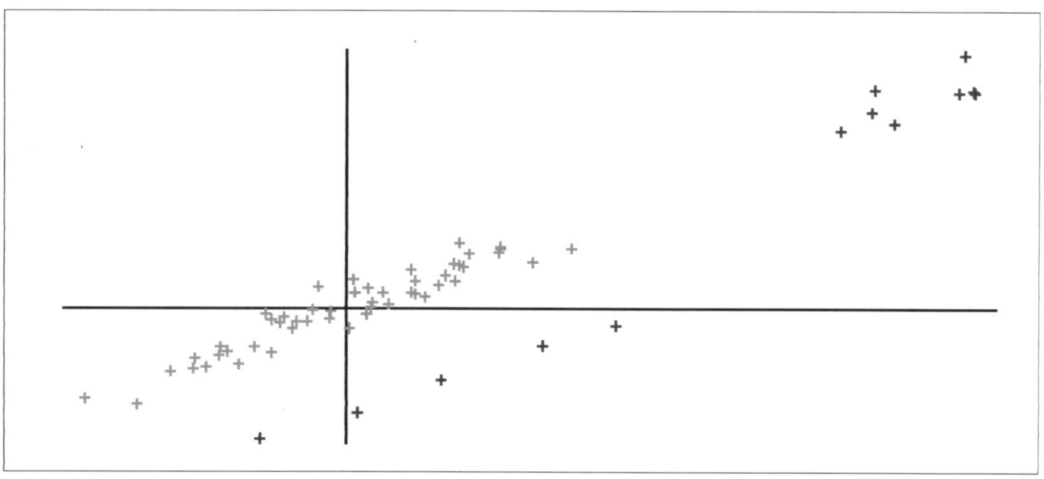

Figure 1. An example of data when outliers can cause problem if the estimator is of type of minimal volume or it is minimizing the spread of error terms - just decreasing the efficiency of estimation.

Acknowledgement

The paper was written with the support of the Czech Science Foundation project No. P402/12/G097 "DYME - Dynamic Models in Economics".

Appendix

Lemma 5.1. *Let Conditions* $\mathcal{C}1$ *hold. Recalling that* e_i*'s have different variances* $\hat{\sigma}_i^2$*, let us denote* $F_{i,\beta}(v) = P\left(|Y_i - X_i'\beta| < v\right)$ *and put*

$$\overline{F}_{n,\beta}(v) = \frac{1}{n}\sum_{i=1}^{n} F_{i,\beta}(v). \tag{18}$$

Then for any $e > 0$ *there is a constant* K_ε *and* $n_\varepsilon \in \mathcal{N}$ *so that for all* $n > n_\varepsilon$

$$P\left(\left\{\omega \in \Omega : \sup_{v \in R^+} \sup_{\beta \in \boldsymbol{R}^p} \sqrt{n}\left|F_\beta^{(n)}(v) - \overline{F}_{n,\beta}(v)\right| < K_\varepsilon\right\}\right) > 1 - \varepsilon. \tag{19}$$

For $F_\beta^{(n)}(v)$ see (7) and for the proof see [20].

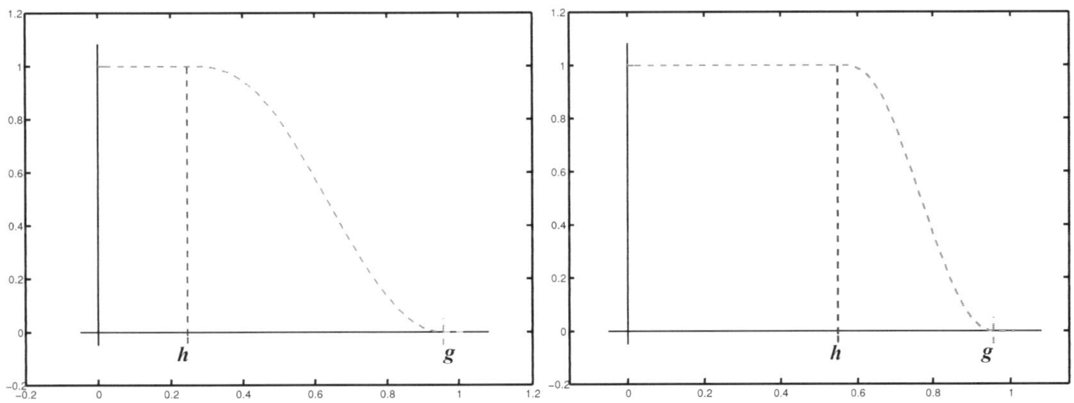

Figure 2. Examples of the weight function of Tukey's shape for SW-estimator.
There is not a theoretical result but the experiences from simulation hint that under a (serious) heteroscedasticity the left version of weight function gives better results. Noticeable results were also obtained when the heteroscedsaticity was estimated in a robust way, for the hint of classical way see [24] and robustify it.

Bibliography

[1] Bickel, P. J. (1975) *One-step Huber estimates in the linear model.* J. Amer. Statist. Assoc. 70, 428–433.

[2] Boček, P., Lachout, P. (1995) *Linear programming approach to LMS-estimation.* Memorial volume of Comput. Statist. *&* Data Analysis 19(1995), 129 - 134.

[3] Campbell, N. A., Lopuhaa, H. P., Rousseeuw, P. J. (1998) *On calculation of a robust S-estimator if a covariance matrix.* Statistics in medcine, 17, 2685 - 2695.

[4] Desborges, R., Verardi, V. (2012) *A robust instrumental-variable estimator.* The Stata Journal (2012) 12, 169 -181.

[5] Hampel, F. R. (1968) *Contributions to the theory of robust estimation.* Ph. D. thesis. University of California, Berkeley.

[6] Hampel, F. R., Ronchetti, E. M. Rousseeuw, P. J., Stahel, W. A. (1986) *Robust Statistics – The Approach Based on Influence Functions.* New York: J.Wiley *&* Sons.

[7] Hawkins, D. M. (1994) *The feasible solution algorithm for least trimmed squares regression.* Computational Statistics and Data Analysis 17, 185 - 196.

[8] Hettmansperger, T. P., Sheather, S. J. (1992) *A Cautionary Note on the Method of Least Median Squares.*The American Statistician 46, 79–83.

[9] Jurečková, J. (1984) *Regression quantiles and trimmed least squares estimator under a general design.* Kybernetika, vol. 20, pp. 345–357.

[10] Maronna, R. A., Yohai, V. J. (1981) *Asymptotic behaviour of general M–estimates for regression and scale with random carriers.* Z. Wahrscheinlichkeitstheorie verw. Gebiete 58, 7–20.

[11] Rousseeuw, P.J. (1984) *Least median of square regression.* Journal of Amer. Statist. Association 79, pp. 871-880.

[12] Rousseeuw, P. J., Yohai, V. (1984) *Robust regressiom by means of S-estimators.* In: Robust and Nonlinear Time Series Analysis. eds. Franke, J., Hărdle, W. and Martin, R. D., Lecture Notes in Statistics No. 26 Springer Verlag, New York, 256-272.

[13] Siegel, A. F. (1982) *Robust regression using repeated medians.* Biometrica, 69, 242 - 244.

[14] Verardi, V., McCathie, A. (2012) *The S-estimator of multivariate location and scatter in Stata.* The Stata Journal (2012) 12, 299 - 307.

[15] Víšek, J. Á. (1990) *Empirical study of estimators of coefficients of linear regression model.* Technical report of Institute of Information Theory and Automation, Czechoslovak Academy of Sciences (1990), number 1699.

[16] Víšek, J. Á. (2000) *Regression with high breakdown point.* Robust 2000 eds. Jaromír Antoch *&* Gejza Dohnal, (published by Union of Czech Mathematicians and Physicists), 2001, 324 - 356.

[17] Víšek, J. Á. (2006) *The least trimmed squares. Part I - Consistency.* Kybernetika (2006), vol 42, 1 - 36.

[18] Víšek, J. Á. (2010) *Robust error-term-scale estimate.* IMS Collections. Nonparametrics and Robustness in Modern Statistical Inference and Time Series Analysis: Festschrift for Jana Jurečková, Vol. 7(2010), 254 - 267.

[19] Víšek, J. Á. (2011) *Consistency of the least weighted squares under heteroscedasticity.* Kybernetika 47 , 179-206.

[20] Víšek, J. Á. (2011) *Empirical distribution function under heteroscedasticity.* Statistics 45, 497-508.

[21] Víšek, J. Á. (2015) *S-weighted estimators.* Proceedings of the 16th Conference on the Applied Stochastic Models, Data Analysis and Demographics 2015, 1031 - 1042.

[22] Víšek, J. Á. (2015) *$\sqrt{n}$-consistency of S-weighted estimators.* Submitted to the Contributions to Theoretical and Applied Statistics In honor of Corrado Gini.

[23] Víšek, J. Á. (2016) *Representatiom of SW-estimators.* Submitted to the Proceedings of 4th Stochastic Modeling Techniques and Data Analysis Conference.

[24] Wooldridge, J. M. (2006) *Introductory Econometrics. A Modern Approach.* MIT Press, Cambridge, Massachusetts, second edition 2009.

Sequential importance sampling for online Bayesian changepoint detection

Lida Mavrogonatou, *University of Glasgow*, lida.mavrogonatou@gmail.com
Vladislav Vyshemirsky, *University of Glasgow*, Vladislav.Vyshemirsky@glasgow.ac.uk

Abstract. Online detection of abrupt changes in the parameters of a generative model for a time series is useful when modelling data in areas of application such as finance, robotics, and biometrics. We present an algorithm based on Sequential Importance Sampling which allows this problem to be solved in an online setting without relying on conjugate priors. Our results are exact and unbiased as we avoid using posterior approximations, and only rely on Monte Carlo integration when computing predictive probabilities. We apply the proposed algorithm to three example data sets. In two of the examples we compare our results to previously published analyses which used conjugate priors. In the third example we demonstrate an application where conjugate priors are not available. Avoiding conjugate priors allows a wider range of models to be considered with Bayesian changepoint detection, and additionally allows the use of arbitrary informative priors to quantify the uncertainty more flexibly.

Keywords. Changepoint Detection, Bayesian Inference, Sequential Importance Sampling, Sequential Monte Carlo, Online Problems

1 Introduction

Identifying abrupt changes in the parameters of a generative model for a time series $\{x_t\}_{t=1}^T$ is a problem widely known as *changepoint detection*. A wide spectrum of changepoint detection methods has been developed with a Bayesian perspective [19, 3, 20, 8, 6, 1, 18, 21]. Some of these methods are retrospective, and require complete observation of a time series. In this paper we focus on problems where the data are obtained incrementally over time, so called *online problems*. In an online context, inferences about changepoints need to be updated each time an observation is made. An effective online Bayesian changepoint detection method was developed using conjugate priors to the exponential family of models by [1].

[21] proposed using variational approximations to expand this approach to a wider class of models. Similarly, approximations using Gaussian processes were employed by [18] to expand the utility of the online Bayesian changepoint detection algorithm. However, these two modifications are approximate, and exact inference is often desirable in critical fields. [6] developed an approach very similar to [1] which was published the same year. Although [6] extended the algorithm with direct simulation from

the posterior of the number and position of the changepoints using Sequential Monte Carlo, they are still using conjugate priors.

In this paper, we extend the method developed by [1] and [6] to a wider range of models by removing the requirement for conjugate priors, and perform inference using Sequential Importance Sampling [15]. Unlike the approach of [6], we consider a sequence of filtering distributions along posteriors of generative model parameters. This choice of filtering distributions allows us to completely avoid the conjugacy requirement, which, as aforementioned, limits model choice. Our method, in contrast to approaches of [21] and [18], performs exact inference, while sampling errors can be easily monitored and controlled. The complexity of the proposed algorithm grows linearly with new data, similarly to the methods proposed by [1] and [6].

The outline of the paper is as follows: in Section 2 we introduce the changepoint model for the proposed approach. Section 3 defines a Sequential Importance Sampling scheme for the online Bayesian changepoint detection algorithm. Experimental results from applying the proposed algorithm to a variety of changepoint detection problems are given in Section 4. The paper concludes with a discussion. The source code for the proposed algorithm and all our experiments are provided in the supplementary material.

2 Changepoint Model

We begin by adopting the changepoint model proposed by [1]. Assuming that a series of observations x_1, $x_2, \ldots x_T$ may be divided into non-overlapping product partitions [2], data within each partition p are considered i.i.d. and follow a distribution $P(x_t|\theta_p)$. A prior $\pi(\theta_p)$ is assigned to the model parameters. The parameters θ_p are considered i.i.d. between partitions. We will use the following notation for a sequence of observations from time point a to time point b:

$$x_{a:b} = \{x_t : t = a, \ldots, b\}.$$

Our goal is to estimate the posterior probability of current *run lengths* that correspond to the time since the last changepoint, given the data so far observed. The length of the current run at time point t is denoted r_t. We will use the notation x_{t,r_t} for a set of data corresponding to a run length r_t:

$$x_{t,r_t} = \begin{cases} x_{t-r_t+1:t}, & \text{if } r_t > 0, \\ \emptyset, & \text{if } r_t = 0. \end{cases}$$

As run length is unknown, the predictive density for the next coming datum can be calculated as the following:

$$P(x_{t+1}|x_{1:t}) = \sum_{r_t=0}^{t} P(x_{t+1}|x_{t,r_t})P(r_t|x_{1:t}), \tag{1}$$

where

$$P(x_{t+1}|x_{t,r_t}) = \int P(x_{t+1}|\theta_p)P(\theta_p|x_{t,r_t})d\theta_p,$$

and the posterior run length probability is defined as

$$P(r_t|x_{1:t}) = \frac{P(r_t, x_{1:t})}{P(x_{1:t})}. \tag{2}$$

The joint distribution $P(r_t, x_{1:t})$ is defined recursively

$$P(r_t, x_{1:t}) = \sum_{r_{t-1}=0}^{t-1} P(r_t|r_{t-1})P(x_t|x_{t-1,r_{t-1}})P(r_{t-1}, x_{1:t-1}), \tag{3}$$

where $P(x_t|x_{t-1,r_{t-1}})$ is the predictive probability based on the current run, and the changepoint prior $P(r_t|r_{t-1})$ is defined by a hazard function $H(r_t)$:

$$P(r_t|r_{t-1}) = \begin{cases} H(r_{t-1}+1) & \text{if } r_t = 0, \\ 1 - H(r_{t-1}+1) & \text{if } r_t = r_{t-1}+1, \\ 0 & \text{otherwise.} \end{cases} \tag{4}$$

The marginal probability $P(x_{1:t})$ in (2) is calculated as

$$P(x_{1:t}) = \sum_{r_t=0}^{t} P(r_t, x_{1:t}). \tag{5}$$

Two possible options may be considered for the current run length at the beginning of observations r_0. If it is appropriate to say that the first observation x_1 is the very first observation of the first partition of the data, we assume $P(r_0 = 0) = 1$. In a more complex scenario, when we need to consider that the process may have been running for some time before x_1, the prior for r_0 can be defined using a survival function:

$$P(r_0 = \tau) = \frac{1}{Z} F(\tau),$$

where Z is an appropriate normalisation constant, and

$$F(\tau) = \sum_{t=\tau+1}^{\infty} P(\text{run length is } t).$$

[1] as well as [6] rely on conjugate priors to calculate the predictive probability $P(x_t|x_{t-1,r_{t-1}})$ in (3). We propose estimating these probabilities with Monte-Carlo integration based on weighted samples from a generative model posterior:

$$P(x_t|x_{t-1,r_{t-1}}) = \int P(x_t|\theta_p)P(\theta_p|x_{t-1,r_{t-1}})d\theta_p \tag{6}$$

$$\approx \sum_{i=1}^{M} \omega_i P(x_t|S^{(i)}_{r_{t-1}}), \tag{7}$$

where $S^{(i)}_{r_{t-1}}$ are sampled from $P(\theta_p|x_{t-1,r_{t-1}})$ with weights ω_i, such that $\sum_{i=1}^{M} \omega_i = 1$.

This estimator is known to be unbiased with variance decreasing asymptotically to zero at the rate $1/M$ when ω_i are approximately equal [7]. At time t, this approach requires t samples $S_{r_{t-1}}$ corresponding to all possible previous run lengths from zero to t-1.

With every new datum x_t becoming available, the Online Bayesian Changepoint Detection algorithm updates a vector of probabilities $P(r_t|x_{1:t}), r_t = 0, \ldots t$ according to (2). The recursive nature of (3) allows us to evolve samples S_r from one stage of the algorithm to the next using importance sampling, establishing a Sequential Importance Sampling scheme [15] along a sequence of generative model parameter posteriors as explained in Section 3.

3 Changepoint Detection Algorithm

In Algorithm 3 we modify the Online Bayesian Changepoint Detection algorithm proposed by [1] and [6] using the Monte-Carlo estimation of the predictive probabilities (6).

Calculating the predictive probabilities in Step 3 of the algorithm requires a sample $S_{r_{t-1}}$ from the posterior of the generative model parameters $P(\theta_p|x_{t,r_{t-1}})$. We propose obtaining such a sample with importance sampling procedure. A success of such approach relies on selection of the proposal distribution in importance sampling that is relatively close to the target distribution. The structure of

Online Bayesian Changepoint Detection Algorithm based on Sequential Importance Sampling.

Step 1 *Initialise sample S_0 containing M samples from the prior of the generative model parameters with equal weights*

$$S_0^{(i)} \sim \pi(\theta_p), \quad \omega_0^{(i)} = 1/M, \quad i = 1, \ldots, M,$$

and assign

$$P(r_0 = 0) = 1, \quad \text{or} \quad P(r_0 = \tau) = \frac{1}{Z} F(\tau).$$

Step 2 *Observe new datum x_t.*

Step 3 *For every possible value of r_{t-1} from 0 to $t-1$, evaluate predictive probabilities*

$$P(x_t | x_{t-1, r_{t-1}}) = \sum_{i=1}^{M} \omega_{r_{t-1}}^{(i)} P(x_t | S_{r_{t-1}}^{(i)}).$$

Step 4 *Calculate growth probabilities for values of r_t from 1 to t*

$$P(r_t = r_{t-1} + 1, x_{1:t}) = P(r_{t-1}, x_{1:t-1}) P(x_t | x_{t-1, r_{t-1}})(1 - H(r_{t-1})).$$

Step 5 *Calculate changepoint probability*

$$P(r_t = 0, x_{1:t}) = \sum_{r_{t-1}=0}^{t-1} P(r_{t-1}, x_{1:t-1}) P(x_t | x_{t-1, r_{t-1}}) H(r_{t-1}).$$

Step 6 *Calculate marginal probability*

$$P(x_{1:t}) = \sum_{r_t=0}^{t} P(r_t, x_{1:t}).$$

Step 7 *Determine run length distribution*

$$P(r_t | x_{1:t}) = P(r_t, x_{1:t}) / P(x_{1:t}).$$

Step 8 *Update samples S_i and corresponding weights ω_i, for i from t down to 1, using importance sampling*

$$(S_i, \omega_i) = \texttt{IS}(S_{i-1}, \omega_{i-1}, x_{(t-i+1):t}).$$

The importance sampling procedure `IS` is described in Algorithm 3.

Step 9 *Sample S_0 from the prior of generative model parameters*

$$S_0^{(i)} \sim \pi(\theta_p), \quad \omega_0^{(i)} = 1/M, \quad i = 1, \ldots, M.$$

Step 10 *Go to Step 2.*

Procedure $\texttt{IS}(S_{old}, \omega_{old}, x_{t,r})$ takes a sample S_{old} weighted with ω_{old}, and a non empty subset of data $x_{t,r}$ as arguments and produces a new sample S from the generative model parameter posterior for data $x_{t,r}$ weighted with new weights ω.

Step 1 *Sample with replacement a population of M particles S^* from sample S_{old} according to weights ω_{old}.*

Step 2 *Set a new sample S to S^* perturbed with a Gaussian perturbation kernel*

$$S^{(i)} \sim \mathcal{N}(S^{*\,(i)}, \alpha \cdot Var(S_{old})),$$

where $\alpha > 0$ is a variance scaling parameter.

Step 3 *Calculate new weights*

$$\omega^{(i)} = \frac{P(x_{t,r}|S^{(i)})\pi(S^{(i)})}{\sum_{j=1}^{M} \omega_{old}^{(j)} \mathcal{N}\left(S^{(i)}; S_{old}^{(j)}, \alpha \cdot Var(S_{old})\right)}.$$

Step 4 *Calculate the Effective Sample Size of the new population according to [12]*

$$ESS = \frac{1}{\sum_{i=1}^{M} \left(\omega^{(i)}\right)^2}.$$

Step 5 *If the Effective Sample Size is smaller than $M/2$, resample S with replacement according to weights ω, and assign new particles equal weights $\omega^{(i)} = 1/M$.*

Step 6 *Return the obtained sample and corresponding weights (S, ω).*

Algorithm 3 utilises the posterior conditioned on the data $\{x_{t-r}, \ldots, x_{t-1}\}$ as the proposal distribution when sampling from the posterior conditioned on data $\{x_{t-r}, \ldots, x_{t-1}, x_t\}$. The latter data set includes only one new datum, x_t. This relationship establishes a typical Sequential Importance Sampling scheme along a sequence of generative model parameter posteriors for datasets $\{x_1\}$, $\{x_1, x_2\}$, $\{x_1, x_2, x_3\}$ and so on.

To minimise the effect of population degeneration issues, we use a Gaussian mixture approximation to the previous posterior as the proposal distribution. This mixture model prevents direct reusing of old samples from one generation to the next one. The variance scaling parameter α in Algorithm 3 controls the scale of the kernel for a smoothing approximation of the proposal distribution with a Gaussian mixture model. It is usually chosen in the range of $0.1 - 1$ and can be tuned individually to every application to obtain more effective proposal. We also measure the Effective Sample Size [12] of the obtained sample, and force resampling with replacement of the population when this metric drops below an arbitrarily selected threshold of $M/2$. This resampling allows us to drop low weight particles in the tails of the posterior, and focus more on high posterior density regions.

In practice we observed that the largest divergence between the proposal and the target distributions is frequently observed when sampling for the very first datum in the sequence using a prior sample as the proposal. In our case studies the resulting Effective Sample Size in such cases drops to about 20% of the Effective Sample Sizes observed later along the sequence of posteriors. We found it was better to use larger sample size M when the target posterior is conditioned on only one datum. In more complex cases a partial rejection control strategy [14] may be implemented to address the issues of large mismatch

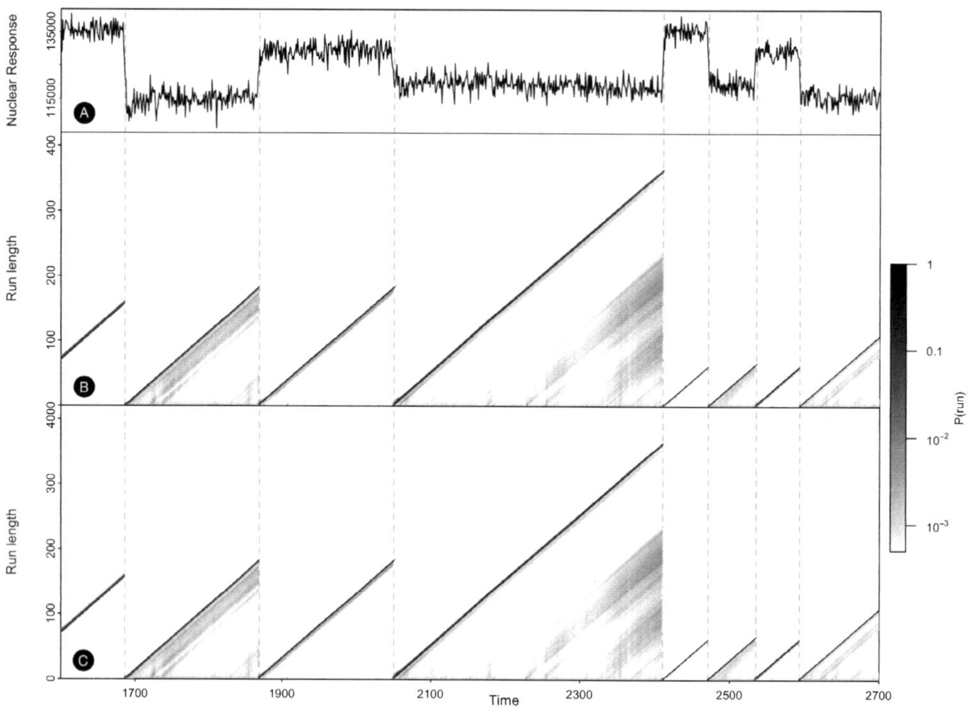

Figure 1. Changepoint detection results for the Well Log data. (A) A subset of data analysed with Online Bayesian Changepoint Detection algorithm. (B) The results obtained with our proposed method based on Sequential Importance Sampling. (C) The results obtained with [1] and [6] algorithms using conjugate priors. Both (B) and (C) depict the posterior run length over data observed so far, $P(r_t|x_{1:t})$. Darker points suggest run lengths with higher probability.

between the proposal and the target distributions.

4 Experimental Results

We apply the proposed algorithm to three data sets. In the first two examples, we replicate results of [1] and analyse the data sets with our method for comparison. In the third example, our method is applied to a new data set to demonstrate how it performs with models without conjugate priors.

Well Log Data

A sequence of measurements of nuclear magnetic response was taken during the drilling of a well. The data are used to interpret geophysical structure of the rock surrounding the well. The variations in mean reflect the stratification of the earth's crust. These data were earlier considered by [16] and [5].

A normal model with fixed variance $\sigma^2 = 4000^2$ is used as an underlying generative model for the data. The model is parametrised by single parameter μ that corresponds to the mean of the normal distribution. To compare our results to those of [1] we use the same normal prior for μ, with hyperparameters $\mu_0 = 1.15\times10^5$, and $\sigma_0^2 = 1\times10^8$. A memoryless changepoint prior was chosen using the geometric distribution and corresponding hazard function $H(r_t) = 1/\lambda$, where $\lambda = 250$.

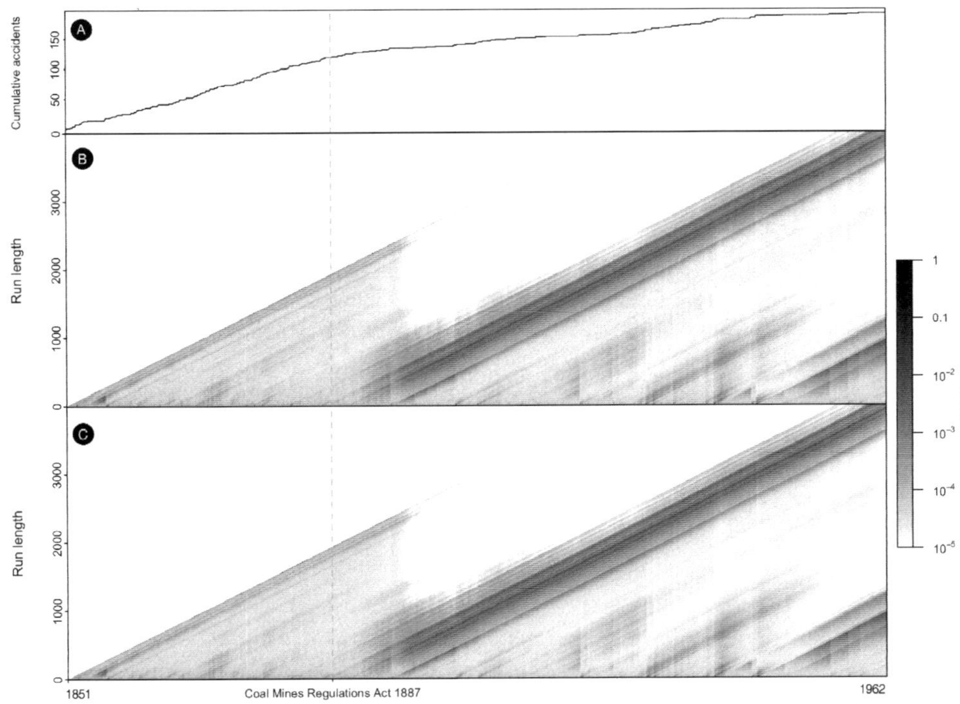

Figure 2. Changepoint detection results for the Coal Mining Disasters data. (A) The cumulative number of significant coal mining accidents between 1851 and 1962. (B) The results obtained with our proposed method based on Sequential Importance Sampling. (C) The results obtained with [1] and [6] algorithms using conjugate priors. Both (B) and (C) depict the posterior run length over data observed so far, $P(r_t|x_{1:t})$. Darker points suggest run lengths with higher probability.

A subset of the data is depicted in Figure 1. Panel A shows the original data values. Panel B shows the results obtained using the Sequential Importance Sampling approach proposed in this paper. Panel C shows the results obtained with the original Online Bayesian Changepoint Detection algorithm using conjugate priors. Notice that the drops to zero run length correspond well with the abrupt changes of the mean of the data. The differences between the results in Panel B and Panel C are very small and correspond to Monte-Carlo approximations in Sequential Importance Sampling and evaluation of the predictive distributions in (6), the mean square error between these results is 1.14×10^{-6}. Samples of 1024 particles were used in this example for larger data sets, while samples of 4096 particles were used for samples from the prior and samples for the run lengths of 1. The smallest Effective Sample Size [12] is 351, which demonstrates that there were no population degeneracy problems in the sampler. Slightly lower effective sample sizes are observed immediately after a sudden change in the mean of the data, as these cases correspond to significant updates of the parameter posteriors.

Coal Mining Disasters

To demonstrate how our method works with count data and large data sets, we applied it to a data set containing the dates of coal mining explosions that killed ten or more men between March 15, 1851 and March 22, 1962 [11]. Following [1], the data were modelled with a Poisson process by weeks, with

Gamma(1,1) prior on the rate. A geometric prior on the frequency of changepoints was selected with corresponding hazard function $H(r_t) = 1/1000$.

The results are plotted in Figure 2. The top panel shows the cumulative number of accidents. The middle panel shows the results obtained with the proposed algorithm using Sequential Importance Sampling. The bottom panel shows the results with the original Online Bayesian Changepoint Detection algorithm using conjugate priors. The results are again very similar, with only minor differences caused by Monte-Carlo estimation of predictive probabilities, the mean square error between the two results is 3.02×10^{-8}. A significant changepoint in the rate of coal mining disasters is usually attributed to the Coal Mines Regulations Act 1887 [17] that commenced as law on January 1st, 1888. This date corresponds to week 1930 in our data set and is marked in the plots with a dashed line.

As the data set contains 6000 time points, 6000 run length updates need to be performed in an online setting, and importance sampling procedure had to be performed $N(N-1)/2 = 17,997,000$ times. To keep the algorithm execution time reasonable, we were using small sample sizes of only 256 particles. The smallest effective sample size in these populations was 47, this demonstrates that we avoided population degeneracy problems [12].

Gold Prices

To demonstrate how our proposed method works with models without conjugate priors, we applied it to a new data set containing the closing prices of gold measured in USD/oz from 16th July 2014 to 16th July 2015. The data are available in the supplementary material to this paper. The data were modelled with a stochastic differential equation,

$$dG = \mu G dt + \sigma G dW,$$

where G is the price of gold, μ and σ^2 are the drift and stochastic volatility parameters respectively, and W is a Wiener process. This equation is often used in financial modelling to describe asset prices under the assumption that prices only depend on the present and not on the past states of the market. This model belongs to a class of stochastic processes known as Itô processes [10]. A significant result for such processes, known as the Itô lemma [9], allows us to derive an expression for the functions of $G(t)$. Using this lemma, logarithms of $G(t)$ are given as

$$d \log G = \left(\mu - \frac{\sigma^2}{2}\right) dt + \sigma dW.$$

Integrating this equation over the interval $[t, t+1]$ gives

$$\log G(t+1) - \log G(t) = \left(\mu - \frac{\sigma^2}{2}\right) + \sigma Z_t,$$

where $Z_t \sim \mathcal{N}(0,1)$. Using the properties of the normal distribution we can write

$$\begin{aligned} \log G(t+1) - \log G(t) &\sim \mathcal{N}(\mu - \frac{\sigma^2}{2}, \sigma^2), \\ \log \frac{G(t+1)}{G(t)} | \mu, \sigma^2 &\sim \mathcal{N}(\mu - \frac{\sigma^2}{2}, \sigma^2). \end{aligned}$$

Hence, we can model daily returns using a lognormal distribution with location $\mu - \sigma^2/2$ and scale σ.

The parameters μ and σ^2 were considered unknown random variables, and were assigned weakly informative prior distributions based on previous knowledge of gold prices. Using data for gold prices from 1968 to 2013, it was concluded that the rate of daily returns changes slightly from day to day at a maximum of $\pm 0.7\%$. The mean rate of returns is expected to have higher density closer to zero, and lower density for larger deviations. As a result, we assigned a normal prior to μ with mean $\mu_0 = 0$ and

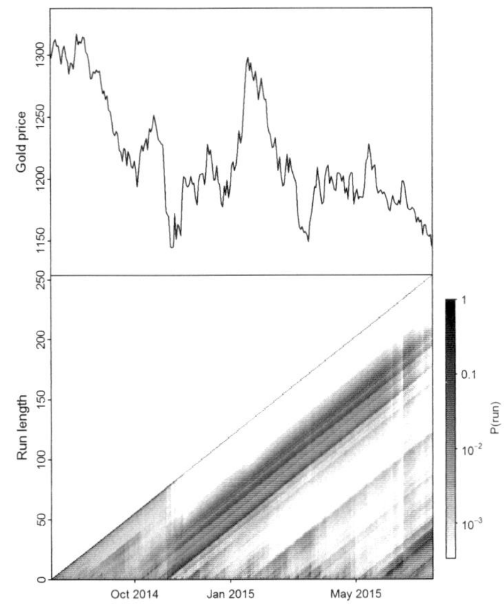

Figure 3. Changepoint analysis of the gold prices during 2014–2015. The closing market price of gold in USD/oz is plotted in the top panel. The lower panel depicts posterior run length probabilities at different dates.

variance $\sigma_0^2 = 0.005^2 = 2.5 \times 10^{-5}$. Based on the observed volatility of the historic prices, we selected an exponential prior for the volatility parameter σ^2 with mean 2.5×10^{-5}. A memoryless changepoint prior was chosen using the geometric distribution and corresponding hazard function $H(t) = 1/\lambda$, where $\lambda = 100$.

Figure 3 shows the result of changepoint analysis performed using the proposed algorithm. The most likely outcome is that the observations begin in a state with negative drift and a relatively low volatility of the prices, then some time between 8 October 2014 and 5 November 2014 the market switches to approximately zero drift with high volatility, finally, in the second half of May 2015 the market goes back to a negative drift and low volatility regime.

Significant changes in the distribution of parameter posteriors with more data becoming available required using larger populations in Sequential Importance Sampling to tackle population degeneracy problems. After a few trials with smaller populations and observing low effective sample sizes, we ended up using a population of 32768 particles for the posteriors corresponding to run length from 0 to 30, and populations of 2048 particles for posteriors corresponding to longer run lengths. The minimal effective sample size achieved with this configuration is 426, which shows no evidence of population degeneracy problems.

5 Discussion

The main structure of the proposed algorithm is similar to the one published by [1] and [6]. Sampling from the posterior of model parameters with Sequential Importance Sampling, instead of using conjugate prior updates, enables our method to perform changepoint detection with models that do not have conjugate priors. Avoiding conjugate priors also allows informative priors based on existing knowledge or observations of similar data to be used for changepoint detection in a truly Bayesian way.

[6] suggested the idea of numerical integration, and earlier gave an example of such approach using MCMC in [4]. The proposed Sequential Importance Sampling approach provides a different sampling

scheme to aid such numerical integration which does not suffer from common MCMC convergence problems and can be easily implemented in high performance computing environment.

The computational complexity of processing one more data point grows linearly as new data arrive, as with every datum one more run length needs to be considered. The requirements for data storage in computer memory also grow linearly. The computational complexity of the proposed algorithm is on the same order as for the algorithms of [1] and [6]. It must be noted that performing importance sampling is more computationally expensive in comparison to updating conjugate parametrisation. Updating conjugate parametrisation typically takes just a small constant number of arithmetic operations. Resampling the parameter posterior with SMC for a sample size M takes $O(M^2)$ operations and therefore produces large complexity scaling constants. Therefore the proposed algorithm is slower than the one that uses conjugate priors with a constant complexity proportion. For example, performing the last round of updates in the Well Log example takes the original Online Bayesian Changepoint Detection algorithm 0.000155 seconds, while our algorithm requires 5.64293 seconds. This shows that our algorithm is almost 40,000 times slower. However, Sequential Monte Carlo methods are well suited for parallel implementation using high performance computational resources, as all of the particles in the population are sampled independently and therefore can be processed at the same time. The source code provided in the supplementary material implements Sequential Importance Sampling for the three examples described in this paper using three approaches: a traditional sequential implementation, a multiprocessor parallel algorithm using OpenMP framework, and a massively parallel implementation running on a graphics processor via CUDA framework.

The examples considered in this paper use models with a small number of parameters. Unfortunately, it is well known that importance sampling is usually inefficient in high-dimensional spaces [7]. So, as the number of model parameters increases, the problem with arise in this setting. However, the number of parameters needed to observe these problems is quite high, and in many practical applications medium sized models will still be feasible.

In real world applications some heuristic simplifications can be made to limit the computational complexity of the problem. Only limited run lengths may need to be considered when monitoring some data. For example, processing the Well Log data set, we could have limited the maximal run length time to the order of a few hundred as we expect changepoints to occur on average every 250 time points. Another example would be monitoring fast changing financial markets, where the possibility of a run that goes over several years is practically zero.

select particles satisfying a desired criteria, those particles not satisfying the criteria receive zero weight. It is possible to formulate conditions under which the algorithm is guaranteed to require a finite number of attempts $M_{S_t} < \infty$ to obtain exactly M non-zero weighted particles [13].

Acknowledgement

We are grateful to Prof Dirk Husmeier and Benn MacDonald for their valuable suggestions and comments made while preparing this paper for publication.

Bibliography

[1] Adams, R. P. and MacKay, D. J. C. (2007) *Bayesian online changepoint detection.* arXiv preprint, arXiv:0710.3742.

[2] Barry, D. and Hartigan, J. A. (1992) *Product partition models for change point problems.* The Annals of Statistic, **20**, 260–279.

[3] Barry, D. and Hartigan, J. A. (1993) *A Bayesian Analysis of change point problems.* Journal of the American Statistical Association, **88**, 309–319.

[4] Fearnhead, P. (2006) *Exact and efficient Bayesian inference for multiple changepoint problems.* Statistics and Computing, **16(2)**, 203–213.

[5] Fearnhead, P. and Clifford, P. (2003) *On-line inference for hidden Markov models via particle filters.* Journal of the Royal Statistical Society: Series B (Statistical Methodology), **65(4)**, 887–899.

[6] Fearnhead, P. and Liu, Z. (2007) *On-line inference for multiple changepoint problems.* Journal of the Royal Statistical Society: Series B (Statistical Methodology), **69(4)**, 589–605.

[7] Doucet, A., de Freitas, N. and Gordon, N. (2001) *Sequential Monte Carlo Methods in Practice.* Springer, Berlin.

[8] Green, P. J. (1995) *Reversible jump Markov chain Monte Carlo computation and Bayesian model determination.* Biometrika, **82(4)**, 711–732, Biometrika Trust.

[9] Itô, K. (1944) *Stochastic Integral.* Proceedings of the Imperial Academy, **20(8)**, 519–524.

[10] Itô, K. and McKeen Jr., H. P. (1974) *Diffusion Processes and Their Sample Paths.* Springer, Berlin.

[11] Jarrett, R. G. (1979) *A note on intervals between coal-mining disasters.* Biometrika, **66(1)**, 191–193.

[12] Kong, A., Liu, J. S. and Wong, W. H. (1994) *Sequential imputations and Bayesian missing data problems.* Journal of the American Statistical Association, **89**, 278–288.

[13] Le Gland, F. and Oudjane, N. (2004) *Stability and uniform approximation of nonlinear filters using the Hilbert metric, and applications to particle filters.* The Annals of Applied Probability, **14(1)**, 144–187.

[14] Liu, J. S. (2001) *Monte Carlo Strategies in Scientific Computing.* Springer, Berlin.

[15] Liu, J. S., Chen, R. and Wong, W. H. (1998) *Rejection control and sequential importance sampling.* Journal of the American Statistical Association, **93(443)**, 1022–1031.

[16] Ó Ruanaidh, J. J. K. and Fitzgerald, W. J. (1996) *Numerical Bayesian methods applied to signal processing.* Springer, New York, 0-387-94629-2.

[17] Peace, M. W. (1888) *Coal Mines Regulation Act, 1887. 50 & 51 Victoria, Cap. 58.*, Lorimer and Gillies, London.

[18] Saatçi, Y., Turner, R. D. and Rasmussen, C. E. (2010) *Gaussian process change point models.* Proceedings of the 27th International Conference on Machine Learning (ICML-10), 927–934.

[19] Smith, A. F. M. (1975) *A Bayesian approach to inference about a change-point in a sequence of random variables.* Biometrika, **62(2)**, 407–416.

[20] Stephens, D. A. (1994) *Bayesian retrospective multiple-changepoint identification.* Applied Statistics, **43**, 159–178.

[21] Turner, R. D., Bottone, S. and Stanek, C. J. (2013) *Online Variational Approximations to non-Exponential Family Change Point Models: With Application to Radar Tracking.* Advances in Neural Information Processing Systems, 306–314.

Detection of space–time clusters for radiation data using spatial interpolation and scan statistics

Fumio Ishioka, *The Graduate School of Environmental and Life Science, Okayama University*, `fishioka@okayama-u.ac.jp`
Koji Kurihara, *The Graduate School of Environmental and Life Science, Okayama University*, `kurihara@ems.okayama-u.ac.jp`

Abstract. On March 11, 2011, a massive amount of radioactive material was released into the environment because of the Fukushima Daiichi Nuclear Power Station (NPS) accident. Surveys on the amount of radioactive materials are very important for assessing the state of the surrounding environment and planning future countermeasures. The authors attempted to detect a high-contaminant cluster accompanied by a time change for the area of evacuation in the Fukushima Prefecture from January 10–19, 2013. The data were air dose rate measured by monitoring post. As a priori analysis, the authors applied a spatial interpolation using ordinary kriging, because the observations obtained were very sparsely scattered and had extremely large dispersion and bias. The result of applying a spatial scan statistic based on echelon analysis was the detection of a significant space–time cluster that decreased with the passing of time. Moreover, the detected cluster was located in the direction of northwest from the NPS.

Keywords. Echelon analysis, Space–time cluster, Spatial scan statistic, Spatial interpolation

1 Introduction

The detection of problems such as the generation status of infective diseases or hazard maps of natural disasters is very basic and important. Some powerful and useful tools such as geographical information systems (GISs) are available, but it is very difficult to determine the location of space–time clusters for various types of spatial data in large quantities or with large time series. The aim of this study is to identify a high-contaminant cluster for the area of evacuation in the Fukushima Prefecture and to understand its temporal progress.

On March 11, 2011, a massive amount of radioactive material leaked from Tokyo Electric Power Company's Fukushima Daiichi Nuclear Power Station (NPS). This accident caused serious damage to both economic and social development, causing a wide range of problems in the environment and in food production as well. Although five years have passed since the accident, there is still intense concern about the influence of radioactive contamination, and high air dose rates and high concentrations of radionuclides are still found in some areas around the NPS [12]. Surveys on the amount of radioactive materials are very important for assessing the state of the surrounding environment and planning future countermeasures. The Nuclear Regulation Authority (NRA) of Japan inaugurated the Comprehensive Radiation Monitoring Plan on August 2, 2011 [1]. Monitoring posts are equipped with air radiation dose-rate-measuring devices at fixed locations, with data obtained at 1 m above the ground. A unit of measurement is microsieverts per hour (μSv/h). Approximately 4,400 monitors were installed in Japan as of April 1, 2016, and the Fukushima Prefecture accounts for over 85% of these. Data is logged every 10 min and stored continuously. The data are openly available at http://radioactivity.nsr.go.jp/map/ja/.

The study on cluster detection using spatial scan statistics [8] is being applied in such fields as epidemiology, astronomy, biosurveillance, and forestry, etc. Several studies on effective scan techniques for using such scan statistics have been published [2, 14, 16, 17]. However, some of them limit the shape of the detected clusters or require an unrealistic computational time if the data set is too large. To solve these problems, we proposed using an echelon spatial scan statistic [4, 5]. This method enables a cluster of an arbitrary shape to be detected even when large amounts of data are targeted. Moreover, we need to take account of both space and time because the monitoring data, measured every 10 min, provide not only the geographical location but also time series information. In this study, we applied a technique of spatial interpolation for compensating for the small number of observations of air dose rate and then detected a flexible space– time cluster by incorporating a temporal scale into an echelon spatial scan statistic. By including a temporal dimension, we could allow tracking of a time-series change of the shape of the high-contaminant radioactive cluster.

2 Materials

The study area chosen has the highest level of radiation contamination in Fukushima Prefecture. We used regularly arranged 10 km × 10 km meshes designed by the administrative organ covering the range within 37.333 degrees north latitude, 140.625 degrees east longitude, 37.667 degrees north latitude, and 141.050 degrees east longitude, containing most of the three levels of the evacuation area that the Japanese government has recognized. In addition, we set ten days as the study period, i.e., from January 10 to 19, 2013. In this period, the air dose rate remarkably decreased temporarily because of the heavy snow that fell on Japan. In such a situation, the detected cluster's size was expected to decrease with lapse of time. Figure 1 shows the study area divided into meshes. It also shows the location of the Fukushima Daiichi NPS and the monitoring posts on January 10. The total number of monitoring posts was 212 at that time.

Mean daily air dose rates with 10-min data aggregated into daily intervals were used as the analysis object in this study. The air dose rates for each monitoring post are summarized in the boxplot shown in Figure 2. We removed some false data caused by instrument anomalies, as announced on the NRA website. In Figure 2, the lines labeled from A to E were measured at the corresponding measurement points on the map of Figure 1. These points indicated remarkably higher doses than the others. Figure 2 also shows that the air dose rate decreased from January 14 to 15 as a whole. At that time, record-breaking heavy snow had fallen in the study area.

3 Spatial interpolation

Cluster detection is very difficult under these conditions, because the monitoring posts of the study area are very sparsely scattered. Accordingly, we attempted to increase the location of the analysis object by using the spatial interpolation, i.e., ordinary kriging. We describe the analysis execution for the data of January 10 here. First, we redivided the study area into smaller 500 m × 500 m meshes and assigned an air dose rate value to each of the mesh based on the corresponding post location. If two or more posts were installed in one mesh, the mesh was assigned their mean value. Figure 3 shows the division of data into the 5,440 meshes and the assigned values of air dose rate.

As is evident from the Figure 2, the dispersion and the bias of the data are extremely large. Therefore, to improve kriging precision, the observations were normalized using a Box-cox transformation ($\lambda =$

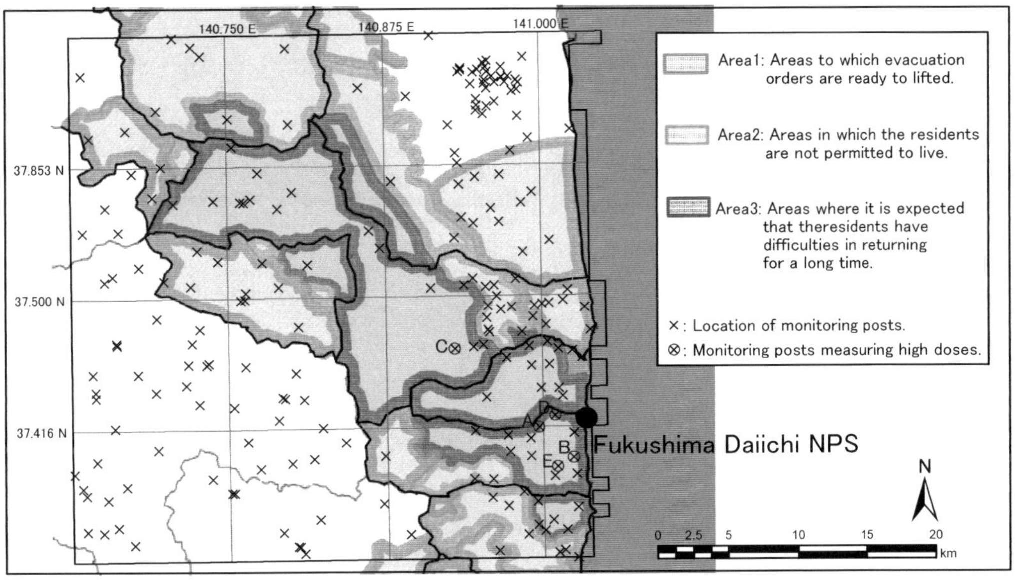

Figure 1. Study area divided most of the three levels of the evacuation area into 10 km × 10 km meshes, with the Fukushima Daiichi NPS and the location of each monitoring post on January 10.

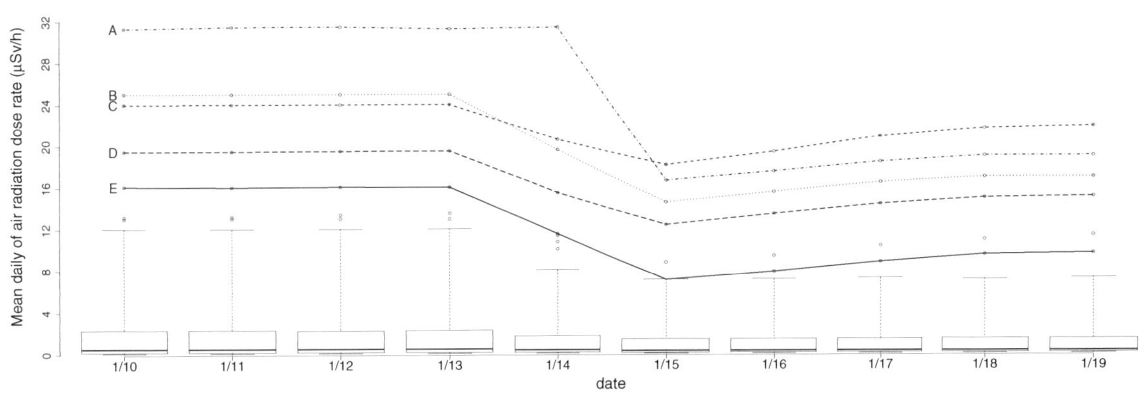

Figure 2. Time series variation of air dose rates for each monitoring post.

−0.18), and we used a model based on an anisotropic variogram. That is, by assuming a geometric anisotropy, we constructed isotropic spatial processes by performing coordinate conversion based on the rotation angle ($\theta = 150°$) of the coordinates and the anisotropy ratio ($r = 3.5$). The varigoram cloud for the transformed data is shown in Figure 4. The horizontal axis represents the intercentral distance $\|\mathbf{h}\|$ of each mesh. The vertical axis represents the dissimilarity $\gamma^*_{i,j}$ of two points $\mathbf{x}_i$ and $\mathbf{x}_j$, which can be computed as half of the squared difference between the sample data $z(\mathbf{x}_i)$ and $z(\mathbf{x}_j)$, i.e.,

$$\gamma^*_{i,j} = \frac{(z(\mathbf{x}_i) - z(\mathbf{x}_j))^2}{2}.$$

This contains information regarding the spatial structure of the sample and provides a first idea of the relationship between two points. Further, the dissimilarity γ^* depends only on the $\mathbf{h}$ value of the sample

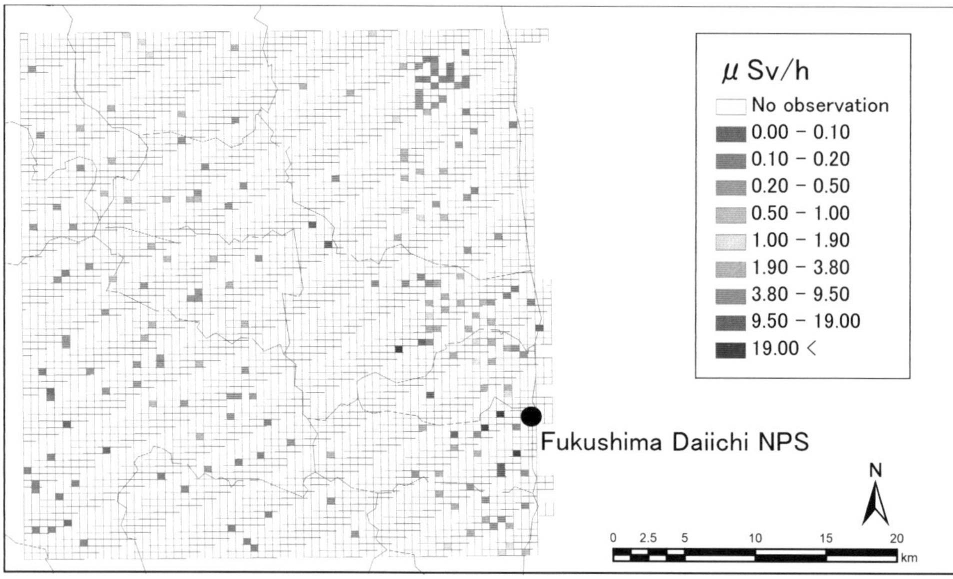

Figure 3. Study area redivided into 500 m × 500 m meshes and the assigned value of air dose rate on January 10, 2013.

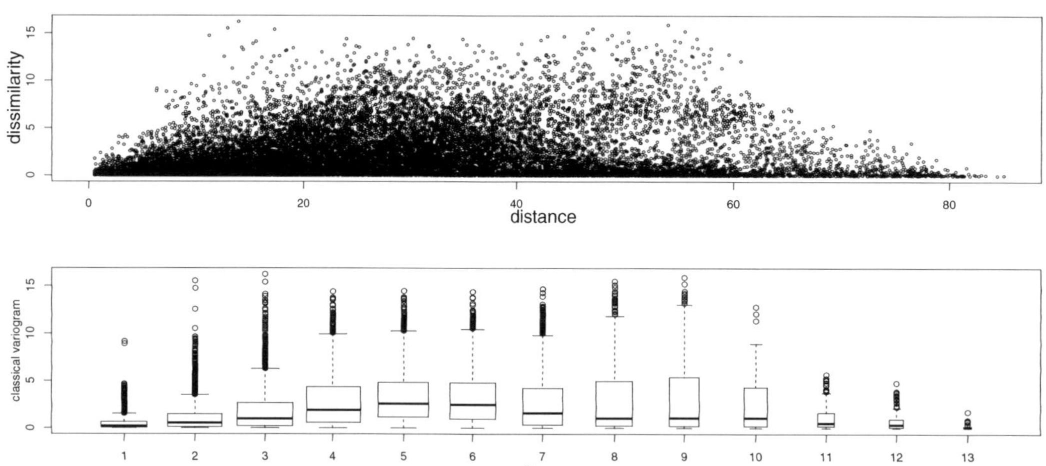

Figure 4. The variogram cloud (above) and the cloud values for each class (below) for the study area on January 10, 2013.

points $\mathbf{x}_i$ and $\mathbf{x}_i + \mathbf{h}$, hence

$$\gamma^*(\mathbf{h}) = \frac{(z(\mathbf{x}_i + \mathbf{h}) - z(\mathbf{x}_i))^2}{2}.$$

Subsequently, there could exist more than one dissimilarity value for some distance $\|\mathbf{h}\|$. On the other hand, most $\mathbf{h}$ values will be without any observation and thus still without dissimilarity value

$\gamma^*(\mathbf{h})$. To estimate a distance between two arbitrary points without directly measuring the distances between the points, the experimental variogram is assumed, and the theoretical variogram is applied to it. The experimental variogram is defined as

$$\gamma^*(\mathbf{h}) = \frac{1}{2|N(\mathbf{h})|} \sum_{N(\mathbf{h})} (z(\mathbf{x}_i) - z(\mathbf{x}_j))^2.$$

where $N(\mathbf{h}) = \{(\mathbf{x}_i, \mathbf{x}_j) : \mathbf{x}_i - \mathbf{x}_j = \mathbf{h} \text{ for } i, j = 1, 2, ..., n\}$ is the set of all pairs of points with $\mathbf{h}$ and $|N(\mathbf{h})|$ is the number of pairs in $N(\mathbf{h})$.

Fitting the variogram models

The experimental variogram $\gamma^*(\mathbf{h})$ provides a first estimate of the assumed underlying theoretical variogram $\gamma(\mathbf{h})$, which can be used to characterize the spatial structure and is needed for our future kriging methods. We must fit a variogram function to an empirical variogram, i.e., replace an empirical variogram with a theoretical variogram that is a suitable valid function. In this study, we attempted to apply the following well-known parametric variogram models. The three common parameters, θ_0, θ_1, and θ_2 relate to *nugget*, *sill* and *range*, respectively, i.e., the nugget θ_0 is defined by $\gamma(\mathbf{h})$ as $\|\mathbf{h}\| = 0$, the sill $\theta_0 + \theta_1$ is the value $\gamma(\infty) = \lim_{\|\mathbf{h}\| \to \infty} \gamma(\mathbf{h})$, and the range θ_2 is the distance at which the $\gamma(\mathbf{h})$ exceeds the sill value for the first time.

- Exponential model
$$\gamma(\mathbf{h}|\theta_0, \theta_1, \theta_2) = \theta_0 + \theta_1 \left(1 - \exp\left(-\frac{\|\mathbf{h}\|}{\theta_2}\right)\right)$$
- Gaussian model
$$\gamma(\mathbf{h}|\theta_0, \theta_1, \theta_2) = \theta_0 + \theta_1 \left(1 - \exp\left(-\frac{\|\mathbf{h}\|^2}{\theta_2^2}\right)\right)$$
- Matérn model
$$\gamma(\mathbf{h}|\theta_0, \theta_1, \theta_2) = \theta_0 + \theta_1 \left(1 - \frac{1}{2^{k-1}\Gamma(k)} \left(\frac{\|\mathbf{h}\|}{\theta_2}\right)^k K_k\left(\frac{\|\mathbf{h}\|}{\theta_2}\right)\right)$$

  with the smoothness parameter k varying from 0 to ∞, gamma function $\Gamma(\cdot)$, and modified Bessel function $K_k(\cdot)$.

Next, we need to choose the most suitable variogram model for the given empirical variogram, and the parameters θ_0, θ_1, and θ_2 of each model must be estimated. In this study, we used likelihood-based parameter estimation methods, viz., maximum likelihood (ML) and restricted maximum likelihood (REML). These methods can be used with Gaussian random fields, and we normalized our data using the Box-cox transformation. ML and REML are available as an R-package `geoR` [15] from the statistical software R. Table 1 shows the estimated parameters and AIC values for each model. Hence, taking the models with the lowest AIC yields the Matérn model, whose parameters, estimated using the REML method, have the "best" fit.

Spatial prediction

In this study, we implemented an ordinary kriging based on intrinsically stationary and isotropy assumptions, which is often used as the kriging method, to predict each air dose rate value for 500 m $\times$ 500 m meshes. The ordinary kriging predictor $Z^*(\mathbf{x}_0)$ of the value at $\mathbf{x}_0$ is given by the linear combination of $Z(\mathbf{x})$ evaluated at each sample $\mathbf{x}_i, i = 1, 2, ..., n$.

$$Z^*(\mathbf{x}_0) = \sum_{i=1}^{n} \omega_i Z(\mathbf{x}_i),$$

		θ_0	θ_1	θ_2	k	AIC
ML	Exponential	0.09	2.28	39.96	-	215.2
	Gaussian	0.21	1.82	14.49	-	222.4
	Matérn	0.13	2.09	16.75	0.85	212.3
REML	Exponential	0.09	8.44	150.80	-	207.4
	Gaussian	0.20	2.02	14.86	-	216.4
	Matérn	0.13	3.46	26.78	0.78	204.9

Table 1. Estimated parameters using ML or REML, and their respective AICs.

where $\omega_i, i = 1, 2, ..., n$ provides the unknown weights corresponding to the influence of the variable $Z(\mathbf{x})$. We can obtain the dissimilarity value between the point $\mathbf{x}_0$ and the ith observed point by using the estimated variogram model. Under the restriction conditions of $\sum_{i=1}^{n} \omega_i = 1$, the computation of $Z^*(\mathbf{x}_0)$ is conducted using the Lagrange multiplier. The predicted map is shown in the Figure 5.

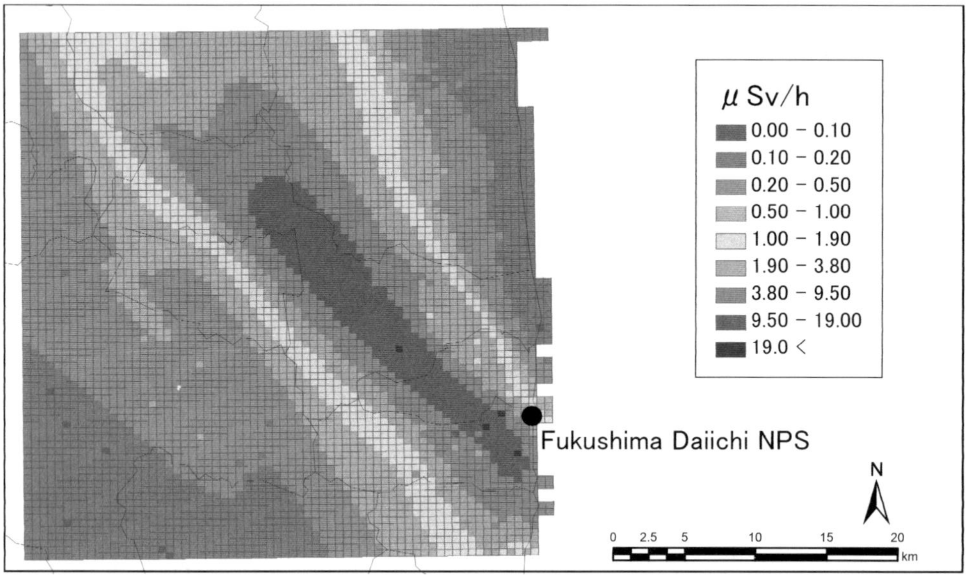

Figure 5. The predicted map of air dose rate obtained by ordinary kriging at January 10, 2013.

Similarly, we applied ordinary kriging to the data from January 11 to 19, 2013. In all cases, Mattern model of REML was chosen as the suitable variogram function by the AIC criterion. The estimated parameters of each day are shown in Table 2. In addition, Figure 6 shows the predicted maps of air dose rate for each day, computed using ordinary kriging.

4 Echelon spatial scan statistic

Spatial scan statistic based on normal model

The spatial scan statistic is a popular method used in disease surveillance for the detection of disease clusters [8]. This statistical approach can detect clusters of any size located anywhere. It is commonly used to evaluate the statistical significance of temporal and geographical clusters without requiring any

	θ_0	θ_1	θ_2	k
Jan. 11	0.13	3.47	26.72	0.79
Jan. 12	0.13	3.39	25.36	0.81
Jan. 13	0.12	3.39	26.13	0.79
Jan. 14	0.14	3.64	21.68	0.90
Jan. 15	0.14	3.72	20.74	0.89
Jan. 16	0.14	3.57	21.07	0.88
Jan. 17	0.14	3.47	20.56	0.89
Jan. 18	0.15	3.73	20.41	0.90
Jan. 19	0.15	3.55	19.66	0.90

Table 2. Estimated parameters for Matérn model based on REML each day.

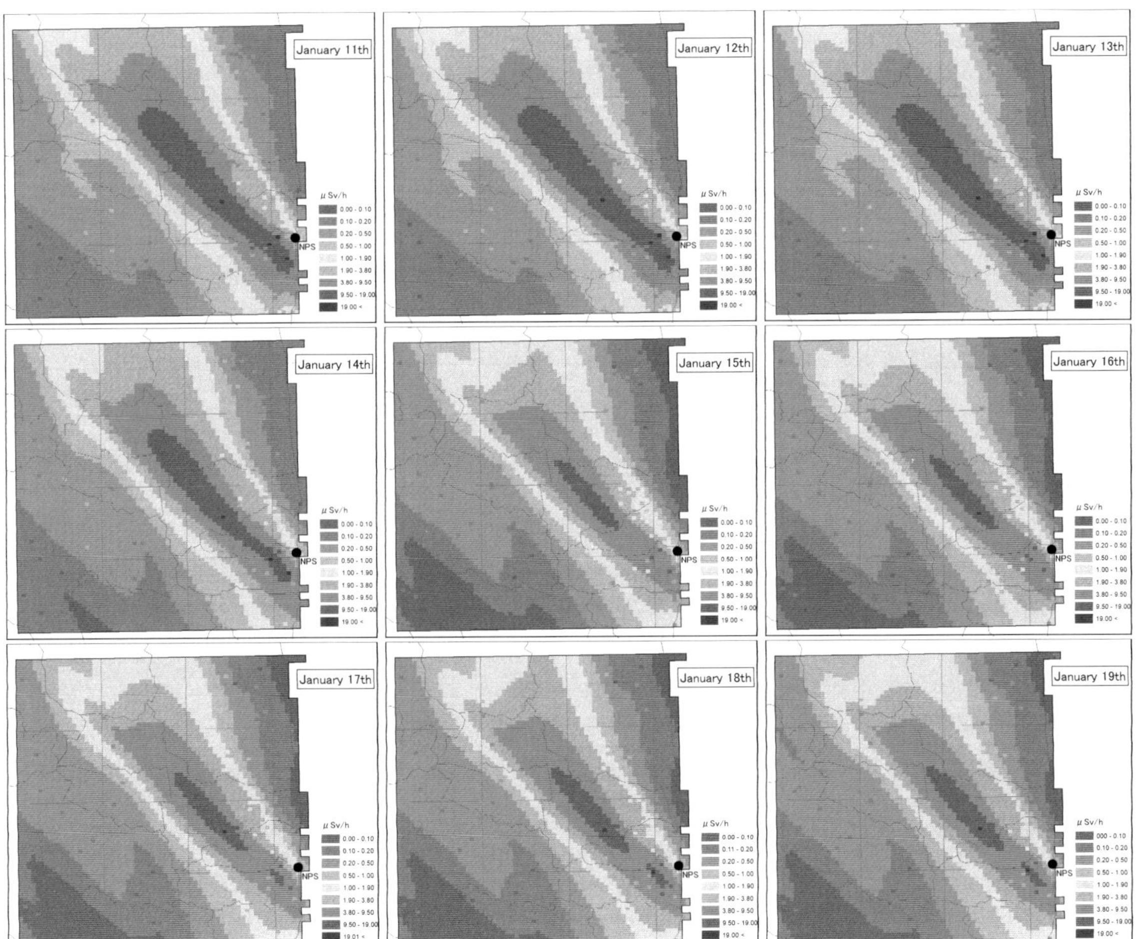

Figure 6. Predicted map of air dose rate from January 11 to 19, 2013.

prior assumptions about their location, time period, or size. The precise model to be used depends on the nature and the probability distribution of data [3, 6, 7, 9, 10]. In this study, we selected a scan statistic appropriate for continuous data with a normal distribution.

The spatial scan statistic is defined through a large number of overlapping windows called Z. For each window Z, a log likelihood ratio (LLR) is calculated. Then, the test statistic is defined as the maximum LLR over all windows. Here, $x_i, i = 1, 2, ..., N$ is an observation value for the locations i at the fixed latitude and longitude. The notation n_z is used to denote the number of observations within window Z. $x_z = \sum_{i \in Z} x_i$ denotes the sum of observed values within window Z. Under the null hypothesis, all observations come from the same distribution, with mean μ and variance σ^2. The maximum likelihood estimates of mean and variance are then estimated as $\hat{\mu}_0 = \sum_i x_i/N$ and $\hat{\sigma_0^2} = \sum_i (x_i - \hat{\mu}_0)^2/N$, respectively. We can therefore write the log likelihood function as

$$\ln L_0 = -N \ln \sqrt{2\pi} - N \ln \sqrt{\hat{\sigma_0^2}} - \frac{N}{2}$$

Under the alternative hypothesis, there is one cluster that has a larger mean than areas outside the cluster. We need to obtain the maximum likelihood estimators that are specific to each window Z in this case. Means inside and outside the window are obtained by setting $\hat{\mu}_z = x_z/n_z$ and $\hat{\mu}_{z^c} = x_{z^c}/n_{z^c} = (\sum_i x_i - x_z)/(N - n_z)$, respectively. The maximum likelihood estimate for the common variance is defined by $\hat{\sigma}_z^2 = (\sum_{i \in Z}(x_i - \hat{\mu}_z) + \sum_{i \notin Z}(x_i - \hat{\mu}_{z^c}))/N$. We can therefore express the log likelihood for window Z by

$$\begin{aligned} \ln L(Z) &= -N \ln \sqrt{2\pi} - N \ln \sqrt{\hat{\sigma}_z^2} - \frac{1}{2\hat{\sigma}_z^2}\left(\sum_{i \in Z}(x_i - \hat{\mu}_z) + \sum_{i \notin Z}(x_i - \hat{\mu}_{z^c})\right) \\ &= -N \ln \sqrt{2\pi} - N \ln \sqrt{\hat{\sigma}_z^2} - \frac{N}{2} \end{aligned}$$

As the test statistic, we use the maximum log likelihood ratio.

$$\begin{aligned} LLR(Z) &= \ln L(Z)/\ln L_0 \\ &= N \ln \sqrt{\hat{\sigma}^2} - N \ln \sqrt{\hat{\sigma}_z^2} \end{aligned}$$

The window with the maximum likelihood ratio constitutes the most likely cluster (MLC), or the cluster least likely to have occurred by chance. A P-value is estimated using Monte Carlo hypothesis testing by generating a large set of random data by randomly permuting the observed value. With randomization conducted in this way, the correct α level will be maintained even if the observations do not come from a truly normal distribution. We next must find a window Z whose $LLR(Z)$ is high. An important problem is how to scan the study area effectively and efficiently, because we cannot realistically calculate $LLR(Z)$ for all patterns of window Z consisting of spatially linked subsets of i.

Echelon scanning method

Echelon analysis [11, 13] provides an objective description of regions using spatial structures based on vertical intervals in each region. Regional data have real referenced values h_i within a spatial region $i \in G, i = 1, 2, ..., m$ for an entire area G consisting of m regions. Then, the data are expressed in the form of (i, h). Figure 7 (left) shows an example of nine regions labeled A to I and their values h. This regional data are divided into the same structured area, as shown in Figure 7 (center). These parts are called echelons. The first, second, third, and fourth echelons are peaks; the fifth echelon is a foundation of peaks; and the sixth and seventh echelons are foundations of peaks and foundations, respectively. Each region belongs to a specific echelon. For example, the first peak consists of region {A} and the third peak consists of the regions {H, E}. Finally, the spatial structure of this regional data is provided by the echelon dendrogram shown in Figure 7 (right).

Each mesh assigned the predicted air dose rate has spatial and temporal information. To apply the echelon technique to our mesh data, we needed to define two kinds of spatial neighboring information. The first involved geographically adjacent relationships. Here, we defined a mesh to have contiguity relationships with all upper and lower and right and left adjacent meshes: a so-called rook-type neighborhood.

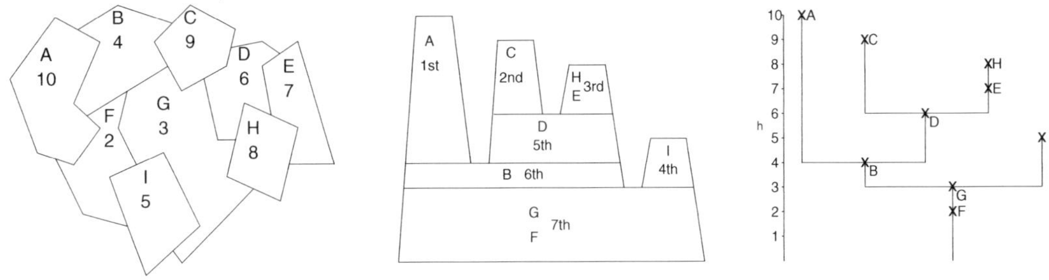

Figure 7. Regional data (left), division to the same echelon (center), and echelon dendrogram (right).

The second challenge was in defining the time series variation. Here, we assumed a three-dimensional structure obtained by adding the axis of time on geographic two-dimensional data, which then defined a continuous relationship over three days, in which each mesh connected the same mesh in the preceding and following days. This is based on the idea that the state of one place is most affected by the preceding or the following state of the same place. The echelon scanning process was performed using the following steps.

1. Represent a topological hierarchy for spatial data by an echelon dendrogram.
2. Scan a region and add it to the window Z, from the upper to the bottom echelons.
3. Consider the window Z with $\max_Z LLR(Z)$ as a cluster.

5 Results

Figure 8 shows the echelon dendrogram that describes the hierarchical structure of the predicted daily mean air dose rate from January 10 to 19, 2013 in the study area. The MLC under the constraint of presetting a maximum cluster size at 5,000 meshes is drawn on the dendrogram. There was a single significant space–time cluster with $LLR(Z)$ of $34,235.01$. To test for statistical significance, 99 replications of Monte Carlo hypothesis testing were performed. This yielded a value of $p < 0.01$. The detected area has geographical and time information, and we can describe the maps of the MLC as in Figure 9. The cluster detected by our study can be considered from two perspectives. From the geographical perspective, the locations identified were not in around the NPS, but were in the direction of northwest from the NPS. From the temporal perspective, the cluster had a decreasing number of cluster locations as time advanced. In particular, it was greatly reduced after January 14. These changes might be due to the influence of weather conditions.

6 Conclusions

This study identified a significant space-time cluster of air radiation dose rates in the area of the highest level of radiation contamination in the Fukushima Prefecture during the period from January 10 to 19, 2013. Before detecting the cluster, we attempted to use the spatial interpolation technique to increase the amount of information that could be analyzed. In this study, we assumed the intrinsically stationary, and performed coordinate transformation for the isotropic spatial processes. However, the around of extremely high rate mesh might be estimated lower than actual rate as a possible influence. About a strict validation of the stationary in our situation needs more discussion.

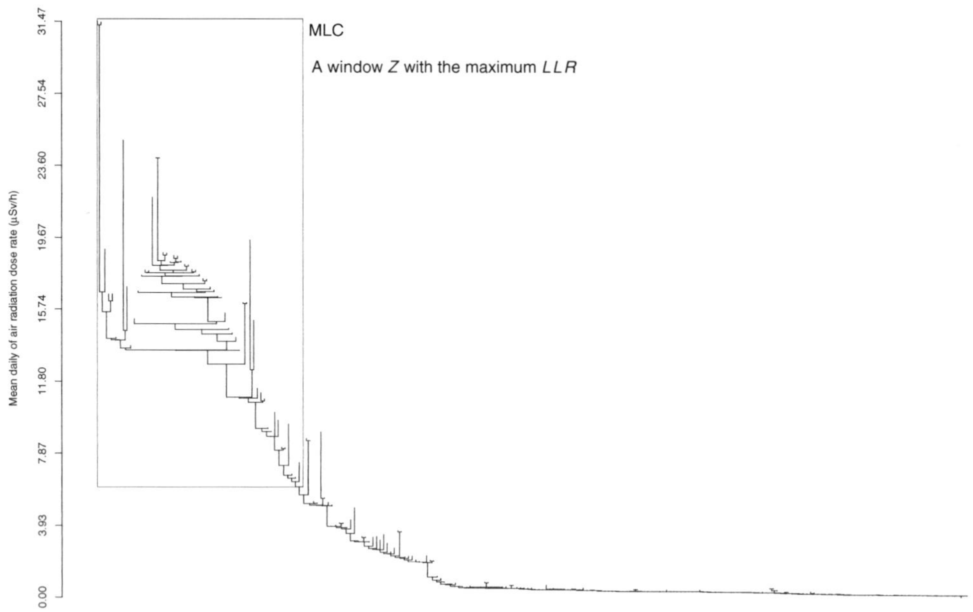

Figure 8. Echelon dendrogram for the daily mean of air dose rates from January 10 to 19, 2013 in the study area.

The echelon scan statistic enables detection of clusters having various shapes and high likelihood ratios, because the scanning process is based on the core spatial structure of the data. Our method considerably reduces the number of scanned windows Z by converting the data to a simple tree structure. It is therefore superior to other scanning methods when large amounts of data are to be handled. In addition, with the appropriate adjacency information provided, we could also detect the time–space cluster. In the future, we aim to apply this method to various environments or large amount data.

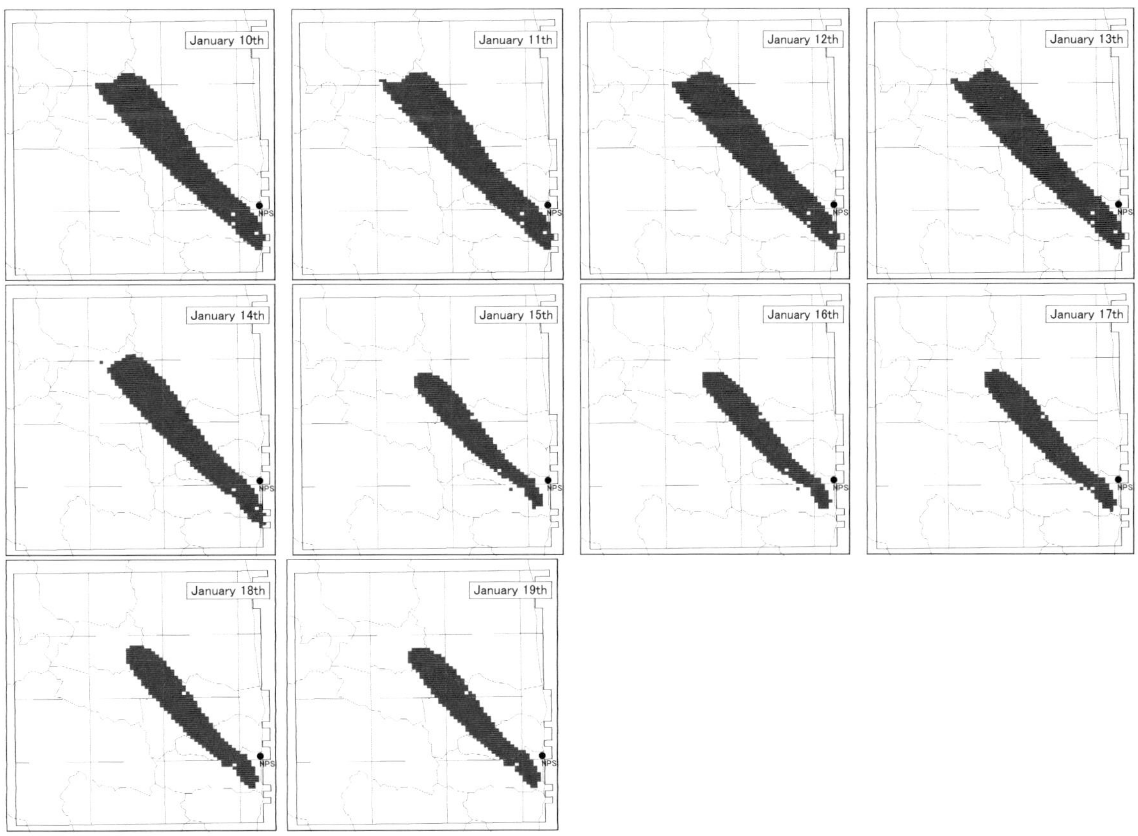

Figure 9. Geographical location of simultaneous space–time cluster from January 10 to 19, 2013.

Bibliography

[1] Nuclear Regulation Authority, Japan. Comprehensive Radiation Monitoring Plan. (2015) http://radioactivity.nsr.go.jp/en/

[2] Duczmal, L. and Assunção, R.A. (2004) *A simulated annealing strategy for the detection of arbitrarily shaped spatial clusters.* Computational Statistics and Data Analysis, **45**, 269–286.

[3] Huang, L., Tiwari, R., Zuo, J., Kulldorff, M. and Feuer, E. (2009) *Weighted normal spatial scan statistic for heterogeneous population data.* Journal of the American Statistical Association, **104**, 886–898.

[4] Ishioka, F., Kurihara, K., Suito, H., Horikawa, Y. and Ono, Y. (2007) *Detection of hotspots for 3-dimensional spatial data and its application to environmental pollution data.* Journal of Environmental Science for Sustainable Society, **1**, 15–24.

[5] Ishioka, F. and Kurihara, K. (2012) *Detection of spatial clusters using echelon scanning method.* Proceedings of COMPSTAT2012. 20th International Conference on Computational Statistics (Edited by Colubi A et al.), Heidelberg: Physica-Verlag, 341–352.

[6] Jung, I., Kulldorff, M. and Klassen, A. (2007) *A spatial scan statistic for ordinal data.* Statistics in Medicine, **26**, 1594–1607.

[7] Jung, I., Kulldorff, M. and Richard, O.J. (2010) *A spatial scan statistic for multinomial data.* Statistics in Medicine, epub.

[8] Kulldorff, M. (1997) *A spatial scan statistic.* Communications in Statistics: Theory and Methods, **26**, 1481–1496.

[9] Kulldorff, M., Mostashari, F., Duczmal, L., Yih, K., Kleinman, K. and Platt, R. (2007) *Multivariate spatial scan statistics for disease surveillance.* Statistics in Medicine, **26**, 1824–1833.

[10] Kulldorff, M., Huang, L. and Konty, K. (2009) *A scan statistic for continuous data based on the normal probability model.* International Journal of Health Geographics, **8**, 58.

[11] Kurihara, K. (2004) *Classification of geospatial lattice data and their graphical representation.* Classification, Clustering, and Data Mining Applications (Edited by Banks D et al.), Springer, 251–258.

[12] METI Measures and Requests in response to the Great East Japan Earthquake. Ministry of Economy, Trade and industry. http://www.meti.go.jp/english/

[13] Myers, W.L, Patil, G.P. and Joly, K. (1997) *Echelon approach to areas of concern in synoptic regional monitoring.* Environmental and Ecological Statistics, **4**, 131–152.

[14] Patil, G.P. and Taillie, C. (2004) *Upper level set scan statistic for detecting arbitrarily shaped hotspots.* Environmental and Ecological Statistics, **11**, 183–197.

[15] Ribeiro Jr. P.J. and Diggle, P.J. (2015) *geoR: A package for geostatistical data analysis using the R software.* http://www.leg.ufpr.br/geoR/

[16] Tango, T. and Takahashi, K. (2005) *A flexible spatial scan statistic for detecting clusters.* International Journal of Health Geographics, **4**, 11.

[17] Tango, T. and Takahashi, K. (2012) *A flexible spatial scan statistic with a restricted likelihood ratio for detecting disease clusters.* Statistics in Medicine, **31**, 4207–4218.

A Bayesian approach for the transformed gamma degradation process

Massimiliano Giorgio, *Second University of Naples*, massimiliano.giorgio@unina2.it
Maurizio Guida, *University of Salerno & National Research Council (CNR)*, mguida@unisa.it
Fabio Postiglione, *University of Salerno*, fpostiglione@unisa.it
Gianpaolo Pulcini, *National Research Council (CNR)*, g.pulcini@im.cnr.it

Abstract. Very recently, a new degradation process, namely the transformed gamma (TG) process, has been proposed in the literature to describe Markovian degradation processes whose increments over disjoint intervals are not independent, so that the degradation growth over a future time interval can depend both on the current age and the current state (degradation level) of the unit. This paper proposes a Bayesian estimation approach for such a process, that is based on prior information relative to the sign (positive or negative) of the correlation between the degradation increment and the current state or age of the unit. Several different prior distributions are then proposed, reflecting the knowledge of the analyst. A Markov Chain Monte Carlo technique, based on the adaptive Metropolis algorithm, is used for estimating the TG parameters and some functions thereof, such as the residual reliability of a unit, as well as for predicting future degradation growth. Finally, the proposed approach is applied to a real dataset consisting of wear measures of the liners of the 8-cylinder engine which equips a cargo ship.

Keywords. Degradation process, Transformed gamma process, Bayesian estimation, degradation growth prediction, Markov Chain Monte Carlo.

1 Introduction

A large body of literature has addressed the problem of developing stochastic process models able to provide an effective description of real degradation phenomena. Very recently, a number of stochastic processes have been proposed to describe degradation phenomena where the degradation increment over a future time interval is no longer independent of the observed history, but can depend on the current state of the unit (as well as on its current age), so that the degradation increments over disjoint time intervals are not independent random variables [1], [2], [3] and [4].

Within these (Markovian) state-dependent degradation process models, the transformed gamma (TG) process [4] seems to be very attractive due to its mathematical tractability. For example, unlike the other Markovian state-dependent processes proposed in the literature, the conditional distribution of the degradation growth under the TG process is available in closed form. In addition, since the TG process can be viewed as a non-linear transformation of the gamma process [5], it constitutes a natural choice for modelling degradation phenomena when degradation growth takes place gradually over time in a sequence of tiny increments. Thus, the TG process seems to be suitable to describe degradation phenomena caused by continuous use, such as wear, chemical corrosion, fatigue, and so on.

Estimation procedures of the TG process parameters based on the maximum likelihood method have been discussed in [4], while the Bayesian approach has been not yet considered. In this paper, in order to fill in this gap, a Bayesian procedure is proposed that allows prior information based on knowledge of the physics of the observed degradation phenomenon to be introduced in the inferential procedure. In this way, more accurate estimates of the model parameters and functions thereof can be achieved. Moreover, the Bayesian approach allows interval estimates and predictions to be easily obtained, whereas classical approaches, such as the maximum likelihood one, generally involve asymptotic approximation of the distribution of the estimators. In particular, the prior information is formulated in terms of the sign (positive or negative) of the correlation between the degradation growth in a future time interval and the degradation level reached by the unit at the current age or the time required to reach the current degradation level.

Posterior inference is made on the process parameters and on several functions thereof, such as the residual reliability, by using Markov Chain Monte Carlo (MCMC) techniques. Prediction of the degradation increment over a future time interval is also provided. Finally, the proposed procedure is applied to a real dataset given in [6], that consists of the wear measures of the liners of the eight-cylinder engine equipping a cargo ship of the Grimaldi Lines.

2 The transformed gamma process

Let $\eta(t)$ be a non-negative, monotone increasing function of time t, hereinafter called "age function", with $\eta(0) = 0$, and let $g(w)$ be a non-negative, monotone increasing and differentiable function of the degradation level w, hereinafter called "state function", with $g(0) = 0$. An increasing degradation process $\{W(t); t \geq 0\}$ is said to be a TG process with age function $\eta(t)$ and state function $g(w)$ if it possesses the following properties:

1. the degradation increments over disjoint time intervals are (possibly) not independent;
2. the degradation increment $\Delta W(t, t+\Delta t) \equiv W(t+\Delta t) - W(t)$ over the time interval $(t, t+\Delta t)$ depends on the process history up to t through the current time t and the current state (degradation level) $w_t = W(t)$, only, being independent on the past;
3. the (conditional) distribution of $\Delta W(t, t+\Delta t)$ is given by:

$$f_{\Delta W(t,t+\Delta t)}(\delta|w_t) = g'(w_t+\delta) \frac{g(w_t, w_t+\delta)^{\eta(t,t+\Delta t)-1}}{\Gamma[\eta(t,t+\Delta t)]} \exp[-g(w_t, w_t+\delta)], \quad \delta > 0, \tag{1}$$

   where $g'(w_t+\delta)$ is the derivative of the state function $g(w)$ evaluated at $w_t+\delta$, $g(w_t, w_t+\delta) = g(w_t+\delta) - g(w_t)$, $\eta(t, t+\Delta t) = \eta(t+\Delta t) - \eta(t)$, and $\Gamma(\cdot)$ is the complete gamma function.

If $\eta(t)$ is linear with t, the (conditional) distribution of $\Delta W(t, t+\Delta t)$ depends on the interval width Δt and not on the current age t, so that the TG process is said to be age-independent. On the other side, if $g(w)$ is linear with w, the distribution of $\Delta W(t, t+\Delta t)$ does not depend on the current degradation level w_t, and the TG process reduces to a (state-independent) gamma process.

From (1), the probability density function (pdf) and the cumulative distribution function (Cdf) of the degradation level $W(t)$ at the time t of a new (unused) unit are given, respectively, by

$$f_{W(t)}(w) = g'(w) \frac{[g(w)]^{\eta(t)-1}}{\Gamma[\eta(t)]} \exp[-g(w)], \tag{2}$$

$$F_{W(t)}(w) = \frac{\mathrm{IG}[g(w); \eta(t)]}{\Gamma[\eta(t)]}, \tag{3}$$

where IG $(y; s)$ is the (lower) incomplete gamma function.

Several functional forms for the age and state functions can be chosen, such as the power-law and the exponential function suggested in [4] and [6]. Following [6], in this paper a power-law function is used both for $\eta(t)$ and for $g(w)$:

$$\eta(t) = (t/a)^b \quad \text{and} \quad g(w) = (w/\alpha)^\beta. \tag{4}$$

Under such formulation, the TG process becomes age-independent when $b = 1$, and is state-independent when $\beta = 1$. The mean and variance of the degradation level $W(t)$ are in closed form, and given by:

$$E\{W(t)\} = \alpha \frac{\Gamma\left[(t/a)^b + 1/\beta\right]}{\Gamma\left[(t/a)^b\right]} \quad \text{and} \quad V\{W(t)\} = \alpha^2 \left(\frac{\Gamma\left[(t/a)^b + 2/\beta\right]}{\Gamma\left[(t/a)^b\right]} - \frac{\Gamma^2\left[(t/a)^b + 1/\beta\right]}{\Gamma^2\left[(t/a)^b\right]} \right). \tag{5}$$

The (conditional) residual reliability $R_t(\tau|w_t)$, that is, the probability that, given the current degradation level $W(t) = w_t$, the level $W(t+\tau)$ reached at the future time $t+\tau$ does not exceed a given threshold level w_{max}, is given by:

$$R_t(\tau|w_t) = \Pr\{W(t+\tau) \le w_{max}|w_t\} = \Pr\{\Delta W(t, t+\tau) \le w_{max} - w_t|w_t\}$$

$$= \frac{\text{IG}\left\{(w_{max}/\alpha)^\beta - (w_t/\alpha)^\beta ; \left[(t+\tau)/a\right]^b - (t/a)^b\right\}}{\Gamma\left\{\left[(t+\tau)/a\right]^b - (t/a)^b\right\}}. \tag{6}$$

From (3), since $\Pr\{W(t) \le w\} = \Pr\{T(w) \ge t\}$, the distribution of the age $T(w)$ at which a given degradation level w is reached is given by:

$$f_{T(w)}(t) = -\frac{\text{d}}{\text{d}t} \frac{\text{IG}\left[g(w); \eta(t)\right]}{\Gamma\left[\eta(t)\right]} = -\frac{\text{d}}{\text{d}t} \frac{\text{IG}\left[(w/\alpha)^\beta ; (t/a)^b\right]}{\Gamma\left[(t/a)^b\right]}. \tag{7}$$

By using arguments in [4], the pdf in (7) can be given in a more tractable form that does not involve a numerical derivation:

$$f_{T(w)}(t) = \frac{b}{a}\left(\frac{t}{a}\right)^{b-1} \frac{1}{\Gamma\left[(t/a)^b\right]} \left\{ \text{IG}\left[(w/\alpha)^\beta ; (t/a)^b\right] \left(\psi[(t/a)^b] - \ln[(w/\alpha)^\beta]\right) \right.$$

$$\left. + \sum_{k=0}^{\infty} \frac{(-1)^k (w/\alpha)^{\beta\left[(t/a)^b + k\right]}}{\left[(t/a)^b + k\right]^2 k!} \right\}, \tag{8}$$

where $\psi(\cdot)$ denotes the digamma function. If w is set equal to the threshold value w_{max}, then the pdf (8) provides the distribution of the lifetime $T = T(w_{max})$ of the unit.

It can be empirically showed that the behavior of the age and state function affects the correlation between the degradation growth $\Delta W(t, t+\Delta t)$ during the future time interval $(t, t+\Delta t)$ and the degradation level reached at the current age or the time required to reach the current degradation level. In particular, if the state function is concave downwards, as occurs when the shape parameter β in (4) is less than 1, the degradation increment is positively correlated to the degradation level $W(t)$ reached at the current age t, given Δt. It means that the larger the degradation $W(t)$ reached at the age t, the more rapidly the degradation grows in the future. On the contrary, if $g(w)$ is convex downwards, $\Delta W(t, t+\Delta t)$ and $W(t)$, given t and Δt, are negatively correlated. As mentioned before, if $g(w) \propto w$, the process is state-independent and $\Delta W(t, t+\Delta t)$ and $W(t)$ are uncorrelated.

Likewise, if the age function is concave downwards, as occurs when the shape parameter b in (4) is less than 1, the variables $\Delta W\left[T(w), T(w)+\Delta t\right]$ and $T(w)$, given w and Δt, are negatively correlated, where $T(w)$ is the age at which the degradation level w is reached. On the contrary, if $\eta(t)$ is convex downwards, $\Delta W\left[T(w), T(w)+\Delta t\right]$ and $T(w)$, given w and Δt, are positively correlated. Finally, if

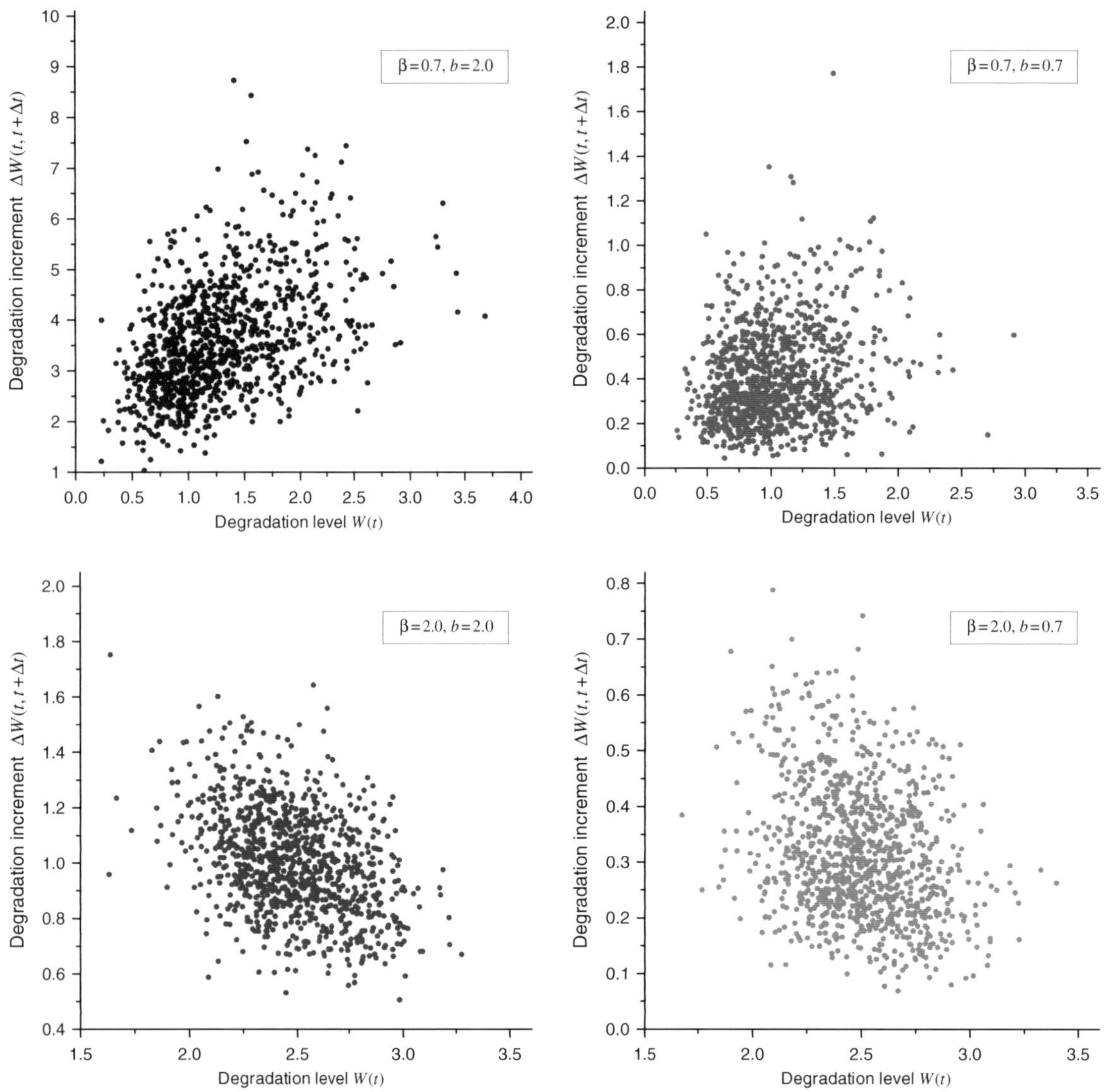

Figure 1. Plots of pseudo-random samples of $W(t)$ and $\Delta W(t, t+\Delta t)$, given $t=5$, with $\Delta t=2$, $\beta=0.7$ or 2.0, and $b=0.7$ or 2.0.

$\eta(t) \propto t$, $\Delta W\left[T(w), T(w)+\Delta t\right]$ and $T(w)$ are uncorrelated. In this latter case, as previously noted, the process is age-independent, because the conditional distribution of the increment $\Delta W(t, t+\Delta t)$, given the current state, does not depend functionally on the current age. Note that the sign of these correlations is independent on the value of t, Δt, w, and Δw, respectively, as well as on the value of the scale parameters a and α.

For illustrative purpose, Figure 1 shows some pseudo-random samples (of size 1000) drawn from the joint distribution of $W(t)$ and $\Delta W(t, t+\Delta t)$, with $t=5$, $\Delta t=2.0$, $\beta=0.7$ or 2.0, and $b=0.7$ or 2. The effect of the parameter β on the correlation sign is evident, as well as we can see that the shape parameter b of the age function has no effect on the sign of the correlation.

Similar plots are obtained when sampling from the joint distribution of $\Delta W\left[T(w), T(w)+\Delta t\right]$ and $T(w)$. In this case the correlation is negative when $b<1$ and positive when $b>1$.

3 The Bayesian inferential procedure

Let us suppose that m units operate under identical conditions, and that each unit is inspected n_i times at possibly not equal ages $t_{i,k}$ $(k=1,\ldots,n_i)$. Let $w_{i,k}=W\left(t_{i,k}\right)$ denote the degradation level of the unit i measured at the k-th inspection time $t_{i,k}$. Then, under the assumption that the degradation processes are TG with $\eta(t)=(t/a)^b$ and $g(w)=(w/\alpha)^\beta$, the likelihood function relative to the observed data $\boldsymbol{w}=(w_{1,1},\ldots,w_{1,n_1},\ldots,w_{m,1},\ldots,w_{m,n_m})$ is given by:

$$L\left(\boldsymbol{w}|\boldsymbol{\theta}\right)=\left(\frac{\beta}{\alpha}\right)^N\prod_{i=1}^{m}\left(\prod_{k=1}^{n_i}\left(\frac{w_{i,k}}{\alpha}\right)^{\beta-1}\frac{\left[(w_{i,k}/\alpha)^\beta-(w_{i,k-1}/\alpha)^\beta\right]^{(t_{i,k}/a)^b-(t_{i,k-1}/a)^b-1}}{\Gamma\left[(t_{i,k}/a)^b-(t_{i,k-1}/a)^b\right]}\right)\times\exp\left[-\left(\frac{w_{i,n_i}}{\alpha}\right)^\beta\right], \qquad (9)$$

where $N=\sum_{i=1}^{m}n_i$ is the total number of observations, $t_{i,0}=0$ and $w_{i,0}=0$ for all i, and $\boldsymbol{\theta}=(a,b,\alpha,\beta)$ denotes the vector of the TG parameters.

We now assume that the analyst possesses prior information on the sign of the (possible) correlation between $\Delta W\left[T(w),T(w)+\Delta t\right]$ and $T(w)$, given generic values of w and Δt, and on the sign of the correlation between $\Delta W(t,t+\Delta t)$ and $W(t)$, given generic values of t and Δt. Thus, taking into account the relationship between the correlation sign and the behavior of the age and state functions, the analyst is able to formulate a prior information on the shape parameters b and β. In particular, referring to β, the analyst can pick one of the following assumptions:

1. no information is available on β, and then the vague prior $g(\beta)\propto 1/\beta$ is used;
2. the correlation between $\Delta W(t,t+\Delta t)$ and $W(t)$ is known to be positive, and then the following Beta prior is used:

$$g(\beta)=\frac{\beta^{p-1}\left(1-\beta\right)^{q-1}}{B(p,q)}, \quad 0\leq\beta\leq 1;\ \ p,q>0, \qquad (10)$$

   whose parameters can be obtained once a prior mean and variance on β, say $E\{\beta\}=p/(p+q)$ and $V\{\beta\}=pq/\left[(p+q)^2(p+q+1)\right]$, are formulated. If the only prior information is that $\beta<1$, then the hyper-parameters p and q can be both set equal to 0.5;
3. the correlation is assumed to be null or weak, and hence the following gamma prior with mean equal to 1 is used:

$$g(\beta)=\frac{p^p\beta^{p-1}}{\Gamma(p)}\exp\left(-p\beta\right), \quad \beta>0;\ \ p>0, \qquad (11)$$

   whose parameter p is determined on the basis of the prior variance $V\{\beta\}=1/p$;
4. the correlation is known to be negative, and then the following 3-parameter gamma distribution, with unit location parameter, is used:

$$g(\beta)=\frac{q^p(\beta-1)^{p-1}}{\Gamma(p)}\exp\left[-q(\beta-1)\right], \quad \beta>1;\ \ p,q>0, \qquad (12)$$

   whose parameters can be obtained once a prior mean and variance on β, say $E\{\beta\}=p/q+1$ and $V\{\beta\}=p/q^2$, are formulated. If the only prior information is that $\beta>1$, then the hyper-parameters of (12) can be set equal to $p=q=0$, so that $g(\beta)\propto 1/(\beta-1)$.

The same distributions can be used as prior pdf on b, by denoting the prior hyper-parameters of $g(b)$ by r and s, in place of p and q, respectively, taking however into account that $b < 1$ implies a negative correlation between $\Delta W\left[T(w), T(w) + \Delta t\right]$ and $T(w)$, and $b > 1$ implies a positive correlation. Thus, the Beta prior (10) has to be used when the correlation between $\Delta W\left[T(w), T(w) + \Delta t\right]$ and $T(w)$ is known to be negative, and the 3-parameter gamma prior (12) has to be used when the above correlation is positive.

Finally, we assume that no prior information is available on the scale parameters, so that the vague priors $g(a) \propto 1/a$ and $g(\alpha) \propto 1/\alpha$ are used. Thus, the joint posterior pdf of the TG parameters is then given by:

$$\pi\left(a, b, \alpha, \beta | \boldsymbol{w}\right) \propto L\left(\boldsymbol{w} | \boldsymbol{\theta}\right) g(b) g(\beta)/(a \cdot \alpha), \tag{13}$$

from which the posterior pdf of any process parameter or function thereof can be derived. Note that the posterior pdf (13) is always proper even if some of the prior pdf are improper.

The posterior predictive distribution of the degradation increment $\Delta W_i = W(t_{i,n_i} + \Delta t) - W(t_{i,n_i})$ of unit i during the future time interval $(t_{i,n_i}, t_{i,n_i} + \Delta t)$ given $W(t_{i,n_i}) = w_{i,n_i}$, can be formulated as:

$$f_{\Delta W_i}(\delta | \boldsymbol{w}) = \int_\beta \int_\alpha \int_b \int_a \pi\left(a, b, \alpha, \beta | \boldsymbol{w}\right) f_{\Delta W_i}(\delta | w_{i,n_i}) \, \mathrm{d}a \, \mathrm{d}b \, \mathrm{d}\alpha \, \mathrm{d}\beta, \tag{14}$$

yielding from (1):

$$f_{\Delta W_i}(\delta | w_{i,n_i}) = \frac{\beta}{\alpha} \left(\frac{w_{i,n_i} + \delta}{\alpha}\right)^{\beta - 1} \frac{\left\{\left[(w_{i,n_i} + \delta)/\alpha\right]^\beta + (w_{i,n_i}/\alpha)^\beta\right\}^{\eta\left(t_{i,n_i}, t_{i,n_i} + \Delta t\right) - 1}}{\Gamma\left[\eta\left(t_{i,n_i}, t_{i,n_i} + \Delta t\right)\right]} \times \exp\left\{-\left(\frac{w_{i,n_i} + \delta}{\alpha}\right)^\beta + \left(\frac{w_{i,n_i}}{\alpha}\right)^\beta\right\}. \tag{15}$$

Similarly, by using (8) with $w = w_{max}$ instead of $f_{\Delta W_i}(\delta | w_{i,n_i})$ in (14), we can obtain the posterior predictive distribution of the lifetime T of new units, say $f_T(t|\boldsymbol{w})$.

4 The Markov Chain Monte Carlo procedure

The Bayesian inferential procedure presented in Section 3 could be implemented, in line of principle, by adopting numerical multivariate integration that, however, is often unfeasible or highly time consuming in the practice. Thus, in this paper we adopt an MCMC technique in order to reduce the computational burden and thus the execution time of the computer code.

In particular, we use the software package OpenBUGS [7], that implements different families of MCMC algorithms, such as Gibbs, Metropolis and slice sampling. The adaptive Metropolis algorithm [8] is here adopted because it typically provides good convergence characteristics in OpenBUGS also in the presence of distributions not included in the software like the distribution of $\Delta W(t, t + \Delta t)$ in (1). We draw a four-dimensional vector sample of size M, say $\boldsymbol{\theta}_j = (a_j, b_j, \alpha_j, \beta_j)$, $j = 1, \ldots, M$, from the posterior pdf $\pi(a, b, \alpha, \beta|\boldsymbol{w})$ in (13), generated after a sufficiently large burn-in period performed to make negligible the influence of the starting point of the numerical procedure. Convergence to the stationary (target) distribution $\pi(a, b, \alpha, \beta|\boldsymbol{w})$ of the Markov Chain is also monitored and assessed.

From the vector sample $\boldsymbol{\theta}_j$, the posterior mean and the $(1-\gamma)$ highest posterior density (HPD) interval of each parameter can be estimated: the former is given by the mean of the corresponding elements of the posterior sample, for instance $E\{\alpha|\boldsymbol{w}\} = \int_0^\infty \alpha \cdot \pi(\alpha|\boldsymbol{w}) \, \mathrm{d}\alpha \cong \sum_{j=1}^M \alpha_j/M$, while the latter is obtained by ordering the posterior sample and selecting the shortest interval containing the fraction $(1 - \gamma)$ of the sample.

The posterior sample of any function $h(\boldsymbol{\theta})$ of the TG parameters, such as the residual reliability $R_t(\tau|w_t)$ in (6) or the mean degradation $E\{W(t)\}$ in (5), is simply given by $h_j = h(\boldsymbol{\theta}_j)$, $j = 1, \ldots, M$, from which the posterior pdf, the mean and the HPD interval of such a quantity are easily obtained [9].

By using the conditional distribution (1) of the degradation increment $\Delta W(t, t+\Delta t)$, given the current degradation level w_t, a posterior sample of the degradation increment is obtained. In particular, by applying the method of composition (see, e.g., [10]), the conditional increment $\delta_j|w_t$, $j = 1, \ldots, M$, given w_t, is obtained by firstly generating a sample of size M, say $z_j|w_t$, $j = 1, \ldots, M$, from a gamma distribution with unit scale parameter and shape parameter $\eta_j(t, t+\Delta t) = [(t+\Delta t)/a_j]^{b_j} - (t/a_j)^{b_j}$, and then by transforming each element $z_j|w_t$ of the pseudo-random sample by

$$\delta_j|w_t = \alpha_j \left[z_j|w_t + (w_t/\alpha_j)^{\beta_j}\right]^{1/\beta_j} - w_t. \tag{16}$$

From the posterior sample obtained from (16), the posterior predictive pdf $f_{\Delta W(t,t+\Delta t)}(\delta_j|w_t)$, the posterior mean, and the $(1-\gamma)$ HPD interval of the conditional increment are easily derived.

Likewise, a posterior sample of the lifetime T, say t_j, $j = 1, \ldots, M$, is obtained by first generating a sample of size M, say u_j, $j = 1, \ldots, M$, from a Uniform standard distribution, and then by searching the value of t_j such that

$$\frac{\text{IG}\left[(w_{max}/\alpha_j)^{\beta_j}; (t_j/a_j)^{b_j}\right]}{\Gamma\left[(t_j/a_j)^{b_j}\right]} - u_j = 0. \tag{17}$$

5 Numerical application

Let now consider the wear measures, given in Table 1, of the liners of the 8-cylinder engine which equips a cargo ship of the Grimaldi Lines. A total of 23 inspections were carried out during a total operating time of 185,000 hours. Due to the caliper sensitivity, all of the wear measures are rounded up to the nearest multiple of 0.05 mm. In Figure 2, the observed paths are depicted, where the measured points are linearly connected for graphical display.

This wear dataset was initially analyzed in [11] under a pure age-dependent Markov degradation model, thus assuming that the degradation growth during a future time interval depends only on the current age of the unit, and not on the current state (the wear level). In Figure 3 (b) the empirical point estimates of the process variance, for ten different values of t, are plotted. This plot clearly shows that the variance of the observed process does not monotonically increase with the age t, as it should occur in case of any pure age-dependent degradation process whose variance increases monotonically with the age. Thus, a degradation model, which is not purely age-dependent, is required to adequately describe the observed process. Note that, since the inspection times can vary from unit to unit, and hence the wear measures generally refer to different operating time of the liners, the empirical estimate of the variance is obtained by using an interpolation procedure at selected equispaced times as suggested in [1]. Afterwards, this wear dataset was analyzed within the TG process in [6] in order to illustrate

i	$t_{i,1}$	$w_{i,1}$	$t_{i,2}$	$w_{i,2}$	$t_{i,3}$	$w_{i,3}$	$t_{i,4}$	$w_{i,4}$
1	11,300	0.90	14,680	1.30	31,270	2.85		
2	11,300	1.50	21,970	2.00				
3	12,300	1.00	16,300	1.35				
4	14,810	1.90	18,700	2.25	28,000	2.75		
5	10,000	1.20	30,450	2.75	37,310	3.05		
6	6,860	0.50	17,200	1.45	24,710	2.15		
7	2,040	0.40	12,580	2.00	16,620	2.35		
8	7,540	0.50	8,840	1.10	9,770	1.15	16,300	2.10

Table 1. Wear $w_{i,k}$ [mm] accumulated by liner i up to the inspection time $t_{i,k}$ [hours].

a condition-based maintenance policy for deteriorating units. Maximum likelihood estimates were then obtained.

In order to perform the Bayesian estimation and prediction procedure, we have to formulate the prior information.

We then assume that the analyst, on the basis of similar wear processes that he has previously observed, knows that the wear increment $\Delta W(t, t+\Delta t)$ during a future time interval is strongly negatively correlated to the current level $W(t)$, and that the wear process is age-independent (i.e., $\Delta W\,[T(w), T(w)$ $+\Delta t$ and $T(w)$ are uncorrelated). Thus, he chooses the 3-parameter gamma pdf (12) with prior mean $E\{\beta\} = 3.0$ and prior variance $V\{\beta\} = 1.0$ as prior distribution on β, so that $p = 4$ and $q = 2$, and the gamma prior (11) with unit mean and variance equal to 0.5 as prior information on b, so that the hyper-parameter of the gamma prior is $r = 2.0$.

The inferential procedure is based on $N = 5 \cdot 10^5$ samples obtained by the adaptive Metropolis algorithm implemented in the OpenBUGS software, with a burn-in period of 10^5 iterations and a thinning interval equal to 100. The execution time of the OpenBUGS routine is about 1 hour and 15 minutes on a notebook based on an Intel® Core™ i7 CPU@2.60GHz, showing the feasibility of the proposed Bayesian MCMC procedure.

The posterior means of the process parameters are $E\{a|\boldsymbol{w}\} = 5092$ hours, $E\{b|\boldsymbol{w}\} = 1.611$, $E\{\alpha|\boldsymbol{w}\} = 0.764$ mm and $E\{\beta|\boldsymbol{w}\} = 2.265$, while the corresponding 90% HPD intervals are (1819 h, 8243 h), (1.110, 2.090), (0.330 mm, 1.182 mm), and (1.425, 3.053). For a comparative purpose, the ML estimates given in [6] are $\hat{a} = 5107$ hours, $\hat{b} = 1.701$, $\hat{\alpha} = 0.750$ mm, and $\hat{\beta} = 2.31$, whereas the approximate 90% confidence intervals based on the log-normal approximation for the distribution of the ML estimators of the (positive) parameter are, respectively (1985 h, 13138 h), (1.077, 2.686), (0.322 mm, 1.746 mm), and (1.305, 4.099). It should be noted that the HPD intervals of both b and β do not include the value 1, and the lower limits are both greater than 1. This implies that the wear increment is negatively correlated to the current wear level, but positively correlated to the current age. In addition, we note that all the HPD intervals are narrower than the corresponding confidence intervals, thus showing how the Bayes procedure based on informative priors provide more accurate estimates.

In Figure 3 the posterior mean and the 90% HPD interval of the mean and variance of the wear level $W(t)$ are depicted, and compared to the empirical estimates. We have that the posterior mean of the mean wear $E\{W(t)\}$ is very close to the empirical estimate, and the 90% HPD interval is very narrow and includes all the empirical estimates. The posterior mean of the wear variance is able to describe the

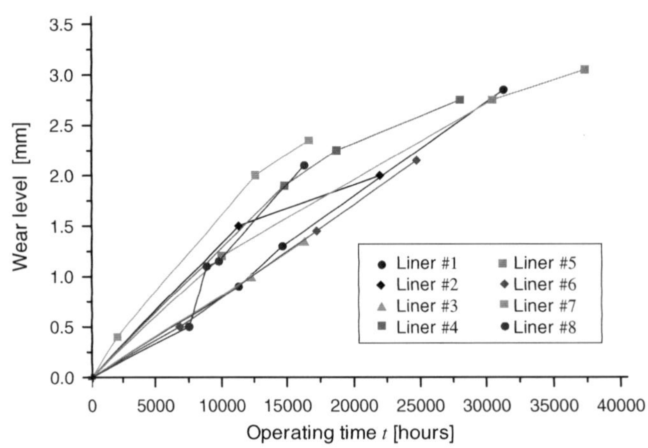

Figure 2. Observed paths of the liner wear (the measured points are linearly connected for graphical opportunity).

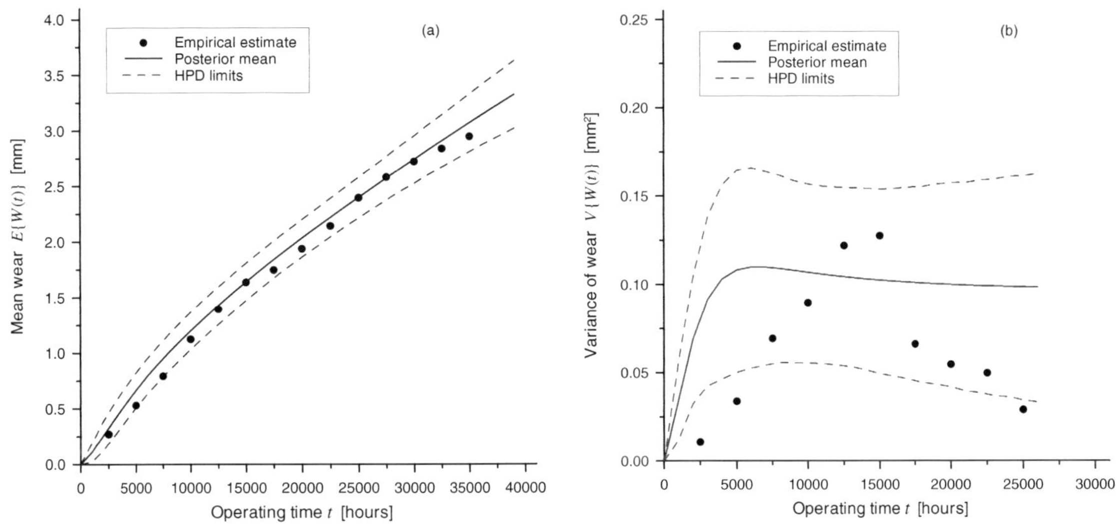

Figure 3. Empirical and Bayesian estimates of mean (a) and variance (b) of the wear process.

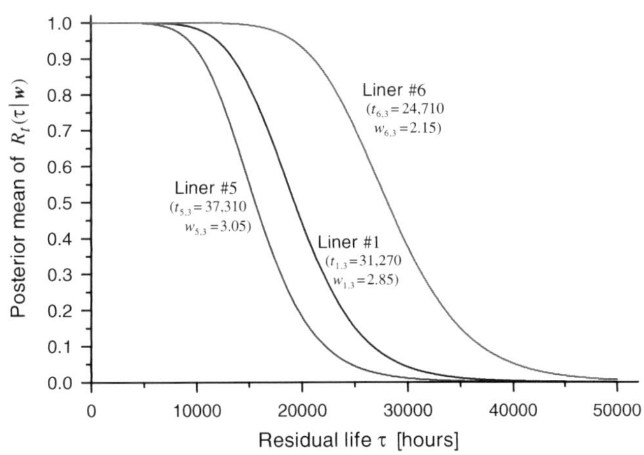

Figure 4. Posterior mean of the residual reliability of liners #1, 5, and 6.

non-monotone behavior of the empirical estimates, but increases initially more quickly than the empirical estimates and decreases quite more slowly. However, it should be considered that the empirical estimates of the variance are based on a few number of "points", and that the "points" are not observed but obtained through linear interpolation of the observations.

Figure 4 gives the posterior mean of the residual reliability (6) of the liners #1, 5, and 6, by setting $w_{max} = 4$ mm. As it might be expected, the larger the current degradation level of a liner, the lower the residual reliability of that liner.

In Figure 5 the posterior predictive distribution (14) of the wear increment $\Delta W(t_{i,n_i}, t_{i,n_i} + \Delta t)$ during the future time interval of width $\Delta t = 20{,}000$ hours, relative to the liners #3, 6, and 8, is depicted. The dependence of the current age and current wear level on the growth of the wear process is there highlighted. In particular, the wear increment relative to the liner #3 is much larger than the increment of liner #8 because, although their current age is the same, the current wear level of liner

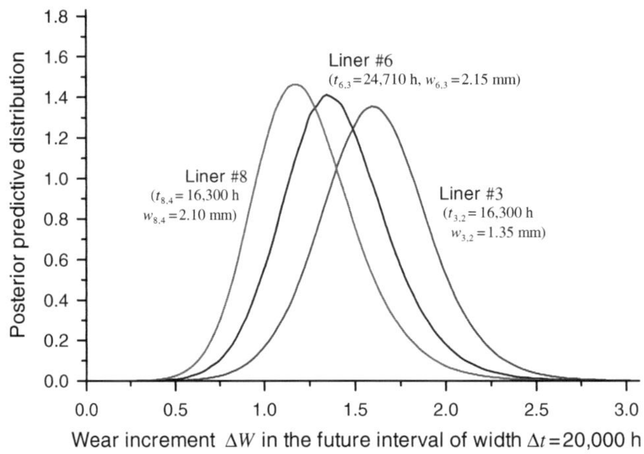

Figure 5. Posterior predictive distribution of the wear increment during the future interval of 20,000 hours process of liners #3, 6, and 8.

#3, say $w_{3,2} = 1.35$ mm, is smaller than the current wear level of liner #8, say $w_{8,4} = 2.10$ mm, and the wear increment is negatively correlated to the current wear level (the shape parameter β is larger than 1). Likewise, the wear increment relative to the liner #6 is much larger than the increment of liner #8 because, although their current wear level is about the same, the current age of liner #6, say $t_{6,3} =$ 24,710 hours, is larger than the current age of liner #8, say $t_{8,4} =$ 16,300 hours, and the wear increment is positively correlated to the current age (the shape parameter b is larger than 1).

We have also predicted the lifetime of a new liner; the posterior mean is equal to 51,132 hours and the 90% HPD interval is (39,189 h, 63,028 h).

6 Conclusions

In this paper, a Bayesian estimation procedure of the parameters of the transformed gamma (TG) degradation process has been proposed, when physical/technological prior information on the correlation between the future degradation increment and the current state or age is available. Different types of prior information, reflecting different information of the degradation process under study, have been considered and briefly discussed. The posterior distribution of the parameters of the TG process, as well as of other quantities of interest such as the residual reliability, has been derived based on a Markov Chain Monte Carlo technique implemented in OpenBUGS. From these posterior distributions, point and interval estimates have been obtained. Prediction on the liner lifetime and on the degradation increment over a future time interval has been also discussed. The proposed Bayesian approach and estimation procedure have been applied to a real case study consisting of the wear process of the liners of the 8- cylinder engine which equip a cargo ship of the Grimaldi Lines, thus showing the feasibility of the suggested Bayesian MCMC procedure.

Bibliography

[1] Giorgio, M., Guida, M. and Pulcini, G. (2010) *A state-dependent wear model with an application to marine engine cylinder liners.* Technometrics, **52**, 172–187.

[2] Giorgio, M., Guida, M. and Pulcini, G. (2011) *An age- and state-dependent Markov model for degradation processes.* IIE Transactions, **43**, 621–632.

[3] Guida, M. and Pulcini, G. (2011) *A continuous-state Markov model for age- and state-dependent degradation processes.* Structural Safety, **33**, 354–366.

[4] Giorgio, M., Guida, M. and Pulcini, G. (2015) *A new class of Markovian processes for deteriorating units with state dependent increments and covariates.* IEEE Transactions on Reliability, **64**, 562–578.

[5] Abdel-Hameed, M. A. (1975) *A gamma wear process.* IEEE Transactions on Reliability, **24**, 152–154.

[6] Giorgio, M., Guida, M. and Pulcini, G. (2015) *A condition-based maintenance policy for deteriorating units. An application to the cylinder liners of marine engine.* Applied Stochastic Models in Business and Industry, **31**, 339–348.

[7] Lunn, D., Spiegelhalter, D., Thomas, A. and Best, N. (2009) *The BUGS project: Evolution, critique and future directions (with discussion).* Statistics in Medicine, **28**, 3049–3082.

[8] Haario, H., Saksman, E. and Tamminen, J. (2001) *An adaptive Metropolis algorithm.* Bernoulli, **7**, 223–242.

[9] Robert, C. P. and Casella, G. (2010) *Introducing Monte Carlo Methods with R.* Springer, New York.

[10] Tanner, M.A. (1996) *Tools for statistical inference: methods for the exploration of posterior distribution and likelihood function.* Third Edition. Springer, New York.

[11] Giorgio, M., Guida, M. and Pulcini, G. (2007) *A wear model for assessing the reliability of cylinder liners in marine Diesel engines.* IEEE Transactions on Reliability, **56**, 158–166.

Time series forecasting with a learning algorithm: an approximate dynamic programming approach

Ricardo A. Collado, *Stevens Institute of Technology*, `rcollado@stevens.edu`
Germán G. Creamer, *Stevens Institute of Technology*, `gcreamer@stevens.edu`

Abstract. We consider the basic problem of re-fitting a time series over a finite period of time and formulate it as a stochastic dynamic program. By changing the underlying Markov decision process we are able to obtain a model that at optimality considers historical data as well as forecasts of future outcomes. We design lookahead dynamic methods for the solution of our Markov decision process. By recursively applying this idea over a range of future time periods, look-ahead dynamic programming methods effectively react to changes in the data and consider the stream of future outcomes obtained from our model decisions. These techniques give rise to models calibrated to historical data which at any point in time would be optimally positioned to react to possible future data stream.

Keywords. Computational finance, machine learning, dynamic programming, time series

1 Introduction

Energy markets and their design are central to a broad social issue: providing efficient energy to sustain economic growth around the world. Similarly, financial markets support economic growth of our expanding global market economy. Debates on costs and remedies can improve if we could better predict pricing of such markets.

From a different perspective, energy and financial market data present measurable outputs of complex systems, and are valuable datasets to test algorithms that seek to predict social behavior. Market prices respond to changes in the availability of raw materials and downstream refinement processes. They also respond to geopolitical events and speculative behavior in markets. Thus, predicting such markets calls for a variety of different techniques. In this paper we propose techniques similar to ensemble methods to predict energy and financial market values. Such methods were used to solve the Netflix challenge and achieved an unprecedented prediction rate of customer movie preferences on a massive dataset. Ensemble methods, however, often involve combining and switching algorithms at random, or based on past performance. We strive to find an informed way to choose methods, or combinations of methods, at a particular point in time by understanding more about context. This may be accomplished by looking at multiple intersecting, high frequency time series and better understanding the text surrounding a particular moment, as text reflects geopolitical events, overall sentiment, and broader trends in populations. It may also involve making predictions about the future using a model, letting a view of the future inform the choice of method in the present.

Time series analysis and external forces.

Historical data is used in time series analysis to model their stochastic mechanism and predict their future value [12]. The complex and chaotic nature of the forces acting on energy and financial markets tend to defeat the predictive time series systems at critical or extreme events. For example, during the past 120 years we have faced a steady stream of financial and banking crises, stock market bubble bursts, and credit crunches that seem to happen ever more often [16, 26]. Likewise, electricity markets exhibit random spikes in spot prices with potential disastrous results to those involved who are not properly protected [19, 10]. The effects of these catastrophic events are exacerbated by the fact that our time series analysis tools often fail to anticipate sudden changes in data, leaving participants exposed to huge losses. The ramifications of these losses have enormous economic and social impact [18, 7].

On the one hand, current theories for financial crises are based on behavioral considerations such as herd behavior [13, 8], positive feedback in multiple Nash equilibria from coordination games [22], and on models that analyze the economy as a complex adaptive system with a network structure governing the interaction of its players [20, 2, 3] On the other hand, energy market prices are influenced not only by seasonal fluctuations but by local weather, temperature, wind, local cost and availability of fuel, war, competition, regulation, and other socio-economic factors [9, 15, 14]. These developments give strong indication that in order to obtain better forecasts for energy and financial markets we should consider alternative data sources with social and text based information.

Our goal is to develop context-based, dynamic stochastic systems capable of fitting time series to data by combining time series methods with forecasting data obtained from social and text based datasets. In this paper we focus on the dynamics and develop stochastic dynamic programming algorithms that fit time series model (or combination of models) to an energy dataset by relying on the use of an external forecast for future values. This external forecast function relies on exogenous information capable of giving a good forecast with some accuracy. In future research we plan to consider external forecast functions based on sparse data obtained from text and sentiment analysis of energy-related (more generally, time series-related) datasets.

2 Time series approximation as a stochastic dynamic program.

Time series methods commonly used to fit financial prices and electricity futures regress structural models on historical data with the goal of obtaining functions capable of forecasting future values with some performance guarantees [1, 5, 21]. By relying solely on historical data these techniques fail to consider possible future outcomes generated by the model and its implications on the model itself. From the point of view of dynamic programming, these time series model selection processes form myopic policies that lack the forecasting power attained by looking ahead into future outcomes [24, 6]. In general, these myopic policies do not easily adapt to extreme changes in the new data, requiring constant calibration in order to fit the model properly.

In the following we describe a controlled Markov decision process and show how the common method of fitting time series is a suboptimal policy solution for this.

A controlled Markov decision process.

We define a controlled Markov decision process and introduce our notation based on the presentation of [17, 27]. Let $(\mathcal{S}, \mathcal{B}_\mathcal{S})$ and $(\mathcal{A}, \mathcal{B}_\mathcal{A})$ be Borel spaces. We call $\mathcal{S}$ the *state space* and $\mathcal{A}$ the *action space*. To each state $s \in \mathcal{S}$ we associate a set of admissible actions $A(s) \subseteq \mathcal{A}$ in such a way that the map $s \mapsto A(s)$ defines a measurable multifunction. We call the multifunction $A(\cdot)$ an *action set* and define its graph as

$$\text{graph}(A) = \{(s,a) \in \mathcal{S} \times \mathcal{A} \,|\, a \in A(s)\}\,.$$

Let $\mathcal{P}$ denote the set of probability measures on $(\mathcal{S}, \mathcal{B}_{\mathcal{S}})$ endowed with the usual weak topology. A *controlled kernel* is a measurable function $Q : \text{graph}(A) \to \mathcal{P}$. So, for every state $s \in \mathcal{S}$ and action $a \in A(s)$ the value of $Q(s,a)$ is a probability measure on the state space $(\mathcal{S}, \mathcal{B}_{\mathcal{A}})$. This can be interpreted as the probability of reaching a state given that we are in state s and take action $a \in A(s)$, which for a Borel set $B \subseteq \mathcal{S}$ is denoted by $Q(B \mid s, a)$. A *cost function* is a measurable function $c : \text{graph}(A) \to \mathbb{R}$.

Our *controlled Markov* model has a state pace $\mathcal{S}$, an action space $\mathcal{A}$ and sequences of action sets A_t, controlled kernels Q_t, and cost functions c_t, $t = 0, 1, 2, \ldots$. A *Markov policy* (or simply a *policy*) is a sequence of measurable functions $\pi_t : \mathcal{S} \to \mathcal{A}$, such that $\pi_t(s) \in \mathcal{A}_t(s)$, for all $s \in \mathcal{S}_t$ and all $t = 0, 1, 2 \ldots$. A Markov policy is *stationary* if $\pi_t = \pi_0$, for all $t = 1, 2, \ldots$.

In this paper we assume that the initial state s_0 is fixed.

Let Π be the set of all policies. Each policy results in a cost sequence that we could optimize. The classical *finite horizon expected value problem* looks for the policy $\pi^* = \{\pi_0^*, \ldots, \pi_{T-1}^*\}$ that minimizes the expected sums of cost:

$$\min_{\pi \in \Pi} \mathbb{E} \left[\sum_{t=0}^{T-1} c_t\left(s_t, \pi_t(s_t)\right) + c_T(s_T) \right], \tag{1}$$

where $\pi = (\pi_0, \ldots, \pi_{T-1})$ and $c_T : \mathcal{S} \to \mathbb{R}$ is a measurable function of *final cost.*

Under some reasonable assumptions problem (1) has an optimal Markov policy that can be described by the famous Bellman's dynamic programming equations, see [25].

Traditional time series approximation.

In this section we state the traditional method of time series forecasting as a controlled Markov decision process.

Let $(\Omega, \mathcal{F}, P)$ be a probability space with σ-algebra $\mathcal{F}$ and probability measure P and consider a real-valued stochastic process $\{X_t \mid t = 0, \ldots, T\}$ of which we know its initial value, i.e. $X_0 = \{\phi_0\}$, where ϕ_0 denotes the constant function with value ϕ_0. Our objective is to forecast $\{X_t\}$ with a time series model (Ω, P), where $\mathrm{P} = \{\mathrm{P}_\theta : \theta \in \Theta\}$ is parameterized by $\Theta \subseteq \mathbb{R}^d$. For example, P could be the set of all autoregressive models of order p given by $r_t = \phi_0 + \phi_1 r_{t-1} + \cdots + \phi_p r_{t-p} + \epsilon_t$, where $\{\epsilon_t\}$ is a Gaussian noise series of mean $\mu = 0$ and variance σ^2. In this case $\Theta = \{(\phi_1, \ldots, \phi_p) \mid \phi_1, \ldots, \phi_p \in \mathbb{R}\} = \mathbb{R}^p$, since ϕ_0, μ, and σ^2 are fixed.

We define a dynamic programming problem based on the natural sample filtration $\mathcal{F}_0 \subset \cdots \mathcal{F}_T \subset \mathcal{F}$ on Ω. This standard construction is detailed below.

Define the set of admissible states at time $k = 0, \ldots, T$ as the set $\mathcal{S}_k$ of all sample sequences $x_0, \ldots, x_k$ of the stochastic process $\{X_t\}$. Let $\mathcal{S} = \bigcup_{t=0}^{T} \mathcal{S}_t$ be the state space and $A_t(s) = \Theta, \forall s \in \mathcal{S}, t = 0, \ldots, T-1$. Our intention is that at any state and time our action selects a time series model from Θ to approximate the observations. For $t = 0, \ldots, T-1$ and a state-action pair $(s,a) \in \text{graph}(\mathcal{A}_t)$ we define the controlled kernel $Q_t(s,a)$ as the probability measure induced on $\mathcal{S}$ by random variable X_{t+1}. Notice that kernel $Q_t(s,a)$ does not depend on the selection of model $a = (\phi_1, \ldots, \phi_p)$. In other words, our current approximation does not have any influence on the observed random process.

At time $t = 0, \ldots, T-1$, for $(s,a) \in \text{graph}\,\mathcal{A}_t$ we define the cost function $c_t(s,a)$ as the result of a (predetermined) goodness of fit test for the observations $s = (x_0, \ldots, x_t)$ and the model selection $a = (\phi_1, \ldots, \phi_p)$. At time T the cost function does not depend on actions so it is denoted as a function of S, $c_T(s)$, Our method prefer smaller values and this should be reflected in the choice of cost function, so smaller values of $c_t(s,a)$ should reflect a "better" fitting model. For example $c_t(s,a)$ could be the result of the Akaike information criterion on (s,a). Alternatively, we could use the Kolmogorov-Smirnov statistic against the proposed model a or a truncated mean squared error of (s,a) (truncation is necessary to satisfy the optimality requirements for Bellman's equations). The choice of cost function depends on the application at hand.

Conditions for optimality and Bellman's equation.

With these definitions we consider the corresponding finite horizon expected value problem described by eq. (1). In order to guarantee a solution we require the following conditions to be satisfied for every $t = 0, \dots, T-1$:
(i) For every $s \in \mathcal{S}$, the set $A_t(s)$ is compact;
(ii) For all $s \in \mathcal{S}$, the cost function $c_t(s, \cdot)$ is lower semicontinuous;
(iii) Let P_0 be a fixed probability measure on $(\mathcal{S}, \mathcal{B}_\mathcal{S})$, where $\mathcal{B}_\mathcal{S}$ denotes the usual Borel σ-algebra on $\mathcal{S}$. For every measurable selection $a_t(\cdot) \in A_t(\cdot)$, the functions $s \mapsto c_t(s, a_t(s))$ and $c_T(\cdot)$ are elements of $\mathcal{L}_1(\mathcal{S}, \mathcal{B}_\mathcal{S}, P_0)$;
(iv) The stochastic kernel function $Q_t(s, \cdot)$ is continuous.

Under these conditions eq. (1) has an optimal solution in the form of a Markov policy π^* [25, 4]. Moreover, the *Principle of Optimality* states that there exists a policy that is optimal for every state [25, p. 87]. Such an optimal policy π^* for (1) minimizes the time aggregated error (as measured by our goodness of fit test) over the time horizon $[0, T]$ but also stays optimal at every time t. This means that at time t and state s, the optimal policy π^* will pick the model given by the parameters $a = (\phi_1, \dots, \phi_p)$ that fits the observed samples better (according to the criterion stated by the cost function) but also fits optimally the future stream of possible observations. We can see this by analyzing the optimal policy π^* obtained from the set of Bellman's equations stated below.

Consider the value functions $v_t : \mathcal{S} \to \mathbb{R}$, $t = 1, \dots, T$, given recursively by:

$$v_T(s) = c_T(s), \quad s \in \mathcal{S}, \tag{2}$$

$$v_t(s) = \min_{a \in A_t(s)} \left\{ c_t(s, a) + \mathbb{E}\left[v_{t+1} \,|\, s, a \right] \right\}, \; s \in \mathcal{S}, \; t = T-1, \dots, 0, \tag{3}$$

where $\mathbb{E}[\,\cdot\,|\, s, a]$ is the expected value subject to the probability measure given by the kernel $Q(s, a)$. Then an optimal Markov policy $\pi^* = \{\pi_0^*, \dots, \pi_{T-1}^*\}$ exists and satisfies the equations:

$$\pi_t^*(s) \in \arg\min_{a \in A_t(s)} \{ c_t(s, a) + \mathbb{E}\left[v_{t+1} \,|\, s, a \right] \}, \; s \in \mathcal{S}, \; t = T-1, \dots, 0. \tag{4}$$

Conversely, any measurable solution of eqs. (2) to (4) is an optimal Markov policy π^*.

Looking closely at the expectation terms in eqs. (4) and (3), we realize that $\mathbb{E}[v \,|\, s, a] = \mathbb{E}[v \,|\, s, a']$, for any value function v and any $(s, a), (s, a') \in \operatorname{graph}(\mathcal{A})$. This means that eq. (4) can be rewritten as

$$\pi_t^*(s) \in \arg\min_{a \in A_t(s)} \{ c_t(s, a) \}, \; s \in \mathcal{S}, \; t = T-1, \dots, 0. \tag{5}$$

This implies that optimal policy π^* is purely myopic in the sense that it completely disregards any information about future outcomes and bases its decision entirely on historical data. This policy is the customary method of fitting time series models and it leads to the loss of any meaningful information about the future time series. In the next section we propose a new time series fitting scheme that eliminates this problem.

3 Lookahead policies for time series approximation.

To overcome limitations imposed by myopic policies, we propose look-ahead dynamic programming methods for time series model selection and updating process. Under this paradigm we leverage dynamic programming methods that rely on historical and future outcomes to obtain model selection policies. Unlike traditional time series methods, solutions obtained from dynamic programs are not simple model parameter solutions but policies that describe each action to take (or in our case which time series model to consider) under possible data outcomes. By recursively applying this idea over a range of future time periods, look-ahead methods effectively react to changes in the data and consider the stream of future outcomes obtained from our model decisions. These techniques give rise models calibrated to historical data which at any point in time would be optimally positioned to react to possible future data stream.

In order to design our look-ahead dynamic programs, we need to describe the Markov decision process we intent to follow as well as describe the means to evaluate and measure future contributions of current model decisions. We do this in the following subsections.

A controlled Markov decision process for lookahead time series approximation.

We consider an extension to the Markov decision process described in Section 2. Consider the same $(\Omega, \mathcal{F}, P)$ probability space, real-valued stochastic process $\{X_t \mid t = 0, \dots, T\}$, and natural sample filtration $\mathcal{F}_0 \subset \cdots \mathcal{F}_T \subset \mathcal{F}$ on Ω. As before, we aim to forecast $\{X_t\}$ with a time series model (Ω, P), where $\mathrm{P} = \{\mathrm{P}_\theta : \theta \in \Theta\}$ is parameterized by $\Theta \subseteq \mathbb{R}^d$.

Let $\mathcal{S}_0 = \{(x_0, \emptyset, \emptyset)\}$ and define the set of states admissible at time $t = 1, \dots, T$ by

$$\mathcal{S}_t = \left\{ (x_t, h_{t-1}, \theta_{t-1}) \;\middle|\; \begin{array}{l} x_t \text{ is an observation from } X_t \text{ obtained at time } t, \\ h_t = x_0, \dots, x_{t-1} \text{ is a sample sequence of } \{X_k\}_{k=0}^{t-1}, \\ \theta_{t-1} = (\phi_1, \dots, \phi_p) \text{ is a parameter from } \Theta \end{array} \right\}.$$

At time $0 \le t \le T$, we interpret the state $s_t = (x_t, h_{t-1}, \theta_{t-1}) \in \mathcal{S}_t$ as being the current observation x_t at time t together with the history h_{t-1} of previous observations. Coordinate θ_{t-1} describes an approximation to X_t from parameterized space P. As before, let $\mathcal{S} = \bigcup_{t=0}^{T} \mathcal{S}_t$ be the state space and let $A_t(s) = \Theta, \forall s \in \mathcal{S}, t = 0, \dots, T-1$. This means that at any state and time we are allowed to select a time series model from Θ to approximate our current and past observations.

For $t = 0, \dots, T-1$ and a state-action pair $(s, a) \in \operatorname{graph}(\mathcal{A}_t)$ we define a controlled kernel $Q_t(s, a)$ as the probability measure induced on $\mathcal{S}|_a := \{(x, h, \theta) \in \mathcal{S} \mid \theta = a\}$ by random variable X_{t+1} and extended this the whole space $\mathcal{S}$. Notice that kernel $Q_t(s, a)$, where $s = (x_t, h_{t-1})$ only "adds weight" for the measure to elements of the form $(x_{t+1}, h_t, \theta_t) \in \mathcal{S}$ such that $h_t = h_{t-1}, x_t$ and $\theta_t = a$.

Let $s_t = (x_t, h_{t-1}, \theta_{t-1})$ and $h_{t-1} = x_0, \dots, x_{t-1}$. At time $t = 0, \dots, T-1$, for $(s_t, \theta_t) \in \operatorname{graph} \mathcal{A}_t$ we have a cost function $c_t(s, \theta_t) = \gamma(s_t, \theta_t) + r\,\delta(s_t, \theta_{t-1}, \theta_t)$, where $\gamma(s_t, \theta_t)$ is the result of a goodness of fit test for the observations $x_0, \dots, x_{t-1}, x_t$ and the model selection given by parameter $\theta_t \in \Theta$. We define

$$\delta(s_t, \theta_{t-1}, \theta_t) := 1 - \exp\left\{-\lambda \left|\mathbb{E}\left[\mathrm{P}_{\theta_t} \mid x_0, \dots, x_{t-1}, x_t\right] - \mathbb{E}\left[\mathrm{P}_{\theta_{t-1}} \;\middle|\; x_0, \dots, x_{t-1}, x_t\right]\right|\right\}, \tag{6}$$

where P_{θ_k} denotes the random variable given at time k by the time series model obtained from parameter θ_k. The constants $r, \lambda \ge 0$ are scaling factors used to balance the penalty imposed by δ on the cost function. Basically, δ adds a penalty for changing "too much" our previous model selection. An example of a cost function is given by:

$$c_t[(x_t, h_{t-1}, \theta_{t-1}),\, \theta_t] = \mathrm{AIC}\left(\theta_t \mid h_{t-1}, x_t\right) + r\left(1 - \exp\left\{-\lambda \left|\mathbb{E}\left[\mathrm{P}_{\theta_t} \mid h_{t-1}, x_t\right] - \mathbb{E}\left[\mathrm{P}_{\theta_{t-1}} \;\middle|\; h_{t-1}, x_t\right]\right|\right\}\right), \tag{7}$$

where $r, \lambda \ge 0$.

Bellman's equations and the need to look ahead.

Under the conditions stated in section 2 the finite horizon expected value problem from eq. (1) is guaranteed to have a solution via the corresponding the Bellman's optimality equations. Of these, the continuity and boundness properties of the cost functions depend on the time series model parametrized space. In general, we can say that these conditions are not to stringent and should be met by a wide range of applications.

Applying Bellman's optimality equations eqs. (2) to (4) gives an optimal policy π^*. Unfortunately, the dependance on future streams of data imposed by the changes in the cost function renders it impossible to be able to obtain the value functions necessary for the definition of the optimal policy. For this reason we resource to look-ahead policies that "capture the effect of decisions now by explicitly optimizing in the future using some approximation of information that is not known now" [24, p. 222]. In order to design our look-ahead dynamic programs, we need to describe the means to evaluate and measure future contributions of current model decisions. We do this by defining contribution functions that evaluate time-marginal contributions of each decision against its impact on some data forecast.

Selecting the data forecast is a crucial step to our methods. It is clear that our forecast must consider historical data; but relying solely on this would give rise to dynamic "look-ahead" methods that effectively fail to look into the future. In order for our methods to succeed, we need forecasts that rely not only on historical data but on alternative data sources with observed predictive qualities to our selected data. The purpose of this paper is the initial exploration of these ideas and as such we rely on a simple forecast obtained by adding white noise to the actual data. Future contributions will be tailored to develop the necessary theory and methods to use forecasts obtained from external sources such as text and sentiment analysis of related data. This is particularly appealing when dealing with applications to financial and energy markets due to the large volume of data easily available and ready to mine for sentiment.

The lookahead method.

Assume that at time t we have access to forecast random variables $\widehat{X}^t_{t+1}, \ldots, \widehat{X}^t_{t+h}$, which are discrete approximations to $X_{t+1}, \ldots, X_{t+h}$, respectively. Each of the forecasts induces a discrete probability measure on $\mathcal{S}$. In practice we expect the forecasts to be finite random variables with very few atoms. This has the effect of significantly reducing our calculations, but at the expense of having coarse results. The idea of the lookahead policy is to make a decision at time t by solving an approximation (given by the forecast functions) of the whole problem over the time window $t, t+1, \ldots, t+h$. At time t after solving the approximation problem, we obtain a value function approximation $\hat{v}_t : \mathcal{S} \to \mathbb{R}$ which we use at the end of the algorithm to define a suboptimal policy.

Suppose we are at time $0 \leq t \leq T-1$ and let $\hat{v}_t(s)$ denote the approximation to the value function at state s. We obtain $\hat{v}_t(s)$ by solving

$$\hat{v}_t(s_t) = \min_{a_t \in A_t(s_t)} \left\{ c_t(s_t, a_t) + \mathbb{E}\left[\bar{v}_{t+1} \,|\, s_t, a_t\right] \right\}, \tag{8}$$

where $\bar{v}_{t+1}$ is the random variable of the total value accrued over the next h stages as given by the forecast random variables $\widehat{X}^t_{t+1}, \ldots, \widehat{X}^t_{t+h}$. That is,

$$\begin{aligned} &\mathbb{E}\left[\bar{v}_{t+1} \,|\, s_t, a_t\right] = \\ &\mathbb{E}_{\widehat{X}_{t+1}}\left[\min_{a_{t+1} \in A_{t+1}(s_{t+1})} \left\{ c_{t+1}(s_{t+1}, a_{t+1}) + \mathbb{E}_{\widehat{X}_{t+2}}\left[\min_{a_{t+2} \in A_{t+2}(s_{t+2})} \left\{ c_{t+2}(s_{t+2}, a_{t+2}) + \cdots \right.\right.\right.\right. \\ &\left.\left.\left.\left. \cdots + \mathbb{E}_{\widehat{X}_{t+h}}\left[\min_{a_{t+h} \in A_{t+h}(s_{t+h})} \left\{ c_{t+h}(s_{t+h}, a_{t+h}) \right\} \,\middle|\, s_{t+h-1}, a_{t+h-1} \right] \right\} \cdots \,\middle|\, s_{t+1}, a_{t+1} \right] \right\} \,\middle|\, s_t, a_t \right], \end{aligned} \tag{9}$$

where $\mathbb{E}_{\widehat{X}_{t+i}}$ is the expectation taken with respect the random variable approximation $\widehat{X}_{t+2}$. This is a multistage stochastic optimization problem and its solution depends on the properties of the random variables, cost functions, and approximations. In the best case we could obtain a convex optimization problem which can be solved solve via specialized methods.

We could further simplify our method by keeping the model selection (action $a_t \in A(s_t)$) over the whole approximation time window. This would remove the penalty for changing models $\delta_t + i, i = 1, \ldots, h$ during the approximation, but the penalty remains in effect at the time of selecting a next model for the calculation of $\hat{v}_{t+1}$. Doing this simplification seems natural since we would be considering the stream of cost attained in future times in case we do not change the model selected at time t. To do this we replace eq. (9) by

$$\mathbb{E}\left[\bar{v}_{t+1} \mid s_t, a_t\right] = \mathbb{E}_{\widehat{X}_{t+1}}\Big[c_{t+1}(s_{t+1}, a_t) + \mathbb{E}_{\widehat{X}_{t+2}}\left[c_{t+1}(s_{t+2}, a_t) + \cdots \right.$$
$$\left. \cdots + \mathbb{E}_{\widehat{X}_{t+h}}\left[c_{t+h}(s_{t+h}, a_t) \mid s_{t+h-1}\right] \cdots \,\Big|\, s_{t+1}\right] \,\Big|\, s_t\Big]. \quad (10)$$

This allow us to obtain a scenario formulation of eq. (8) via the use of nonanticipativity constraints:

$$\begin{aligned} \min_{a_t \in A_t(s_t)} \quad & \mathbb{E}_{\widehat{X}_{t+1},\ldots,\widehat{X}_{t+h}}\left[\sum_{i=0}^{h} c_{t+i}(s_{t+i}, a_t)\right] \\ \text{s.t. } \quad & s_{t+i} = \mathbb{E}\left[s_{t+i} \,\middle|\, \widehat{X}_{t+i}\right], \quad i = 1, \ldots, h. \end{aligned} \quad (11)$$

There are many specialized methods that we can leverage to solve eq. (11), particularly if this is a convex optimization problem.

In this way for every $t = 0, \ldots, T-1$ we obtain $\hat{v}_t(s)$, $\forall s \in \mathcal{S}_t$. This can easily be done if the state space is finite or if we are willing to consider a discretization of the state space. In case the state space is infinite, we need to use more sophisticated methods to obtain our value function approximations. These methods include the use of basis functions, kernel regression, radial basis, and Dirichlet clouds (see [24] for a in-depth treatment of many of these techniques).

Monte Carlo method.

The Monte Carlo algorithm runs iteratively a prefixed N number of times and on each iteration it cycles over time periods $t = 0, \ldots, T$. The output of the Monte Carlo algorithm are functions $\hat{V}_t : \mathcal{S} \to \mathbb{R}$, each approximating its corresponding value function v_t. At each iteration we obtain a new value function approximation $\hat{v}_t^n, t = 1 \ldots, T$ where we use a sample path of observations and the old value approximation $\hat{v}_{t-1}^n$. We do this by selecting a convex combination of the old value approximation and the new one.

The convergence of algorithm 3 depends on the properties of forecasts $\widehat{X}_{t+1}, \ldots, \widehat{X}_{t+h}$ and on stepsizes $\{\alpha_n\}_{n=0}^{N}$. In general it's extremely difficult to obtain bounds or guarantees of convergence for a small number of Monte Carlo samples. If we allow infinitely many iterations of algorithm 3, then we consider nonnegative stepsizes $\{\alpha_n\}_{n=0}^{\infty}$, such that $\sum_{n=0}^{\infty} \alpha_n = \infty$ and $\sum_{n=0}^{\infty} (\alpha_n)^2 < \infty$ [24, p. 437]. These are basic requirements for the convergence of many different Monte Carlo algorithms with step sizes and it is only natural to require it.

Other methods.

The core of our method lies in selecting suitable value function approximations and forecast functions capable of capturing time series "intent" in our applications.

The selection of approximation method is critical due to the large volume and high frequency of our data. As is common in other applications with big amounts of data, any approximation method requiring more than linear time and space might prove to be infeasible for very large problems. For this reason it makes sense to focus on approximation schemes of linear nature such as finite look-ahead, linear basis function, and piecewise linear approximations.

The methods described so far are tailored to combine data from alternative data sources with financial and economic historical data-driven time series models. We expect these methods to give rise to historical data fitted models capable of following trends and reacting to changes outlined by the alternative data.

This leads to the creation of new time series methods obtained through the combination of multiple time series models via dynamic programming. Another way to combine time series is using an expert weighting algorithm which selects the most appropriate combined algorithm in every time period based on the performance of each method. This algorithm has its roots in the weighted majority algorithm introduced by Littlestone and Warmuth [23], and applied to the financial domain by Creamer and Freund [11].

Monte Carlo Algorithm

Step 0 Initialize $\overline{v}_t^0(s), \forall t, s$, select an initial state s_0^1 and set $n = 1$.

Step 1 For $t = 0, \ldots, T$ do:

1. Update the state variable: Observe a value x_t^n of the stochastic process, let s_t^n be obtained from x_n^t and model selection at $t-1$, and get forecasts $\widehat{X}_{t+1}^n, \ldots, \widehat{X}_{t+h}^n$.
2. Solve
$$\hat{v}_t^n = \min_{a_t \in A_t(s_t^n)} \left\{ c_t(s_t^n, a_t) + \mathbb{E}\left[\bar{v}_{t+1} \,|\, s_t^n, a_t\right] \right\},$$
as described by eq. (11). Let a_t^n be the obtained optimal solution to the minimization problem.
3. Update the value function approximation $\bar{v}_t^{n-1}$:
$$\bar{v}_t^n(s) = \begin{cases} (1-\alpha_{n-1})\bar{v}^{n-1}(s) + \alpha_{n-1}\hat{v}_t^n, & \text{if } s = s_t^n, \\ \bar{v}_t^{n-1}(s), & \text{otherwise.} \end{cases}$$

Step 2 Let $n = n + 1$. If $n < N$, go to ***Step 1***.

Step 3 Return the value functions $\left\{\bar{v}_t^N \,\middle|\, t = 0, 1, \ldots, T\right\}$.

---

4 Numerical results

We applied our methods to a dataset of natural gas prices. On this dataset we ran a series of experiments with different time series models. The models considered are ARIMA, suppor vector machine (SVM), and random forest (RF). We compared these against a logistic regression benchmark. We also considered an application of the expert weighting algorithm applied to the financial market by [11].

In order to evaluate our experiments we made a number of simplifications. The forecast functions were obtained by adding white noise to the actual "future" values. In this way we are able to test our models in the case where the forecast is consistent with the forecasted data. The other simplification we made was to consider only the path generated by the data we had (so we didn't need to generate policies for all the state space).

Figure 1-a shows comparisons between benchmark traditional (static) time series methods and our dynamic programming approach for RF, SVM, and ARIMA. Figure 1-a also includes a logistic regression benchmark (traditional time series) and the expert weighting algorithm where the "experts" are the dynamic versions of RF, SVM, and ARIMA. The expert weighted forecast shows an improvement in relation to ARIMA, and have similar results to SVM. However, RF outperforms it. The dynamic version of ARIMA is only slightly better than its benchmark. Figure 1-b shows the evolution of the weights obtained from the expert weighting algorithm at every time step.

In our experiments we significantly beat the non-dynamic benchmarks consistently. These results show that dynamically combining historical methods with data forecasts have a huge potential of giving improved methods of forecast and approximation of time series. Based on these encouraging results we plan to continue further research in this direction. Future work focus on studying the statistical properties of our dynamic methods. We also plan to develop better approximation methods to solve the time-consuming dynamic optimization process that we obtain from our dynamic modeling.

[A. Mean square error (MSE) of dynamic and static (benchmark) methods]

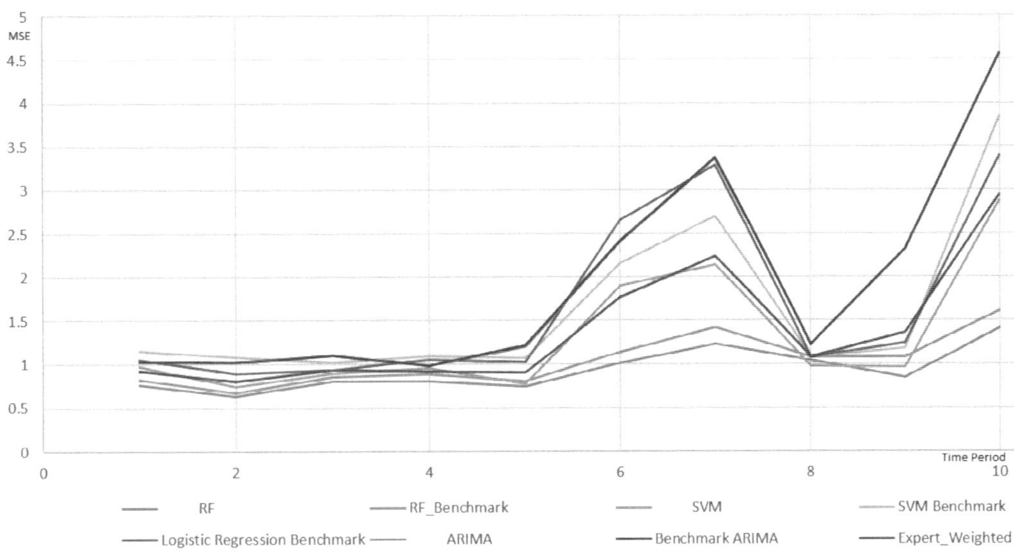

[B. Weights from the expert weighting algorithm]

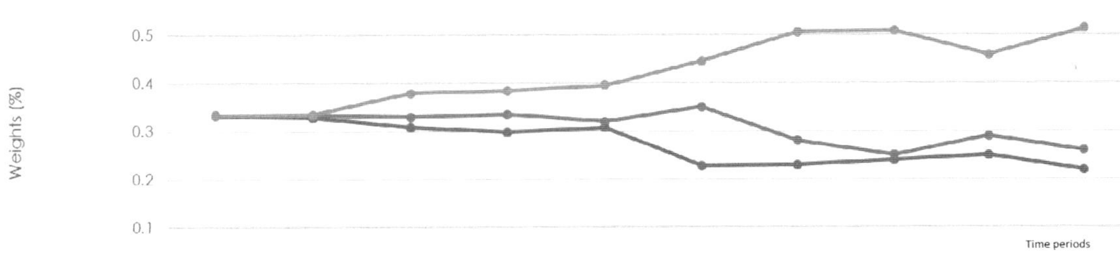

	1	2	3	4	5	6	7	8	9	10
ARIMA	0.33253515	0.328158	0.308158	0.298158	0.307158	0.227158	0.229158	0.239158	0.249158	0.219158
SVM	0.333159691	0.333483	0.32987	0.33487	0.31987	0.34987	0.27987	0.24987	0.28987	0.25987
Random Forest	0.334305159	0.335	0.37893	0.38393	0.39493	0.44493	0.50493	0.50793	0.45793	0.51293

Figure 1. Mean square error (MSE) of the dynamic and static (benchmark) methods (A) and weights from the expert weighting algorithm of the dynamic methods (B). RF and SVM stand for random forest and support vector machine respectively.

Bibliography

[1] Ahmed, N. K., Atiya, A. F., Gayar, N. E., and El-Shishiny, H. (2010) *An Empirical Comparison of Machine Learning Models for Time Series Forecasting.* Econometric Reviews, **29** (5-6), 594–621.

[2] Anand, K., Kirman, A., and Marsili, M. (2010) *Epidemics of rules, information aggregation failure and market crashes.* Working Papers halshs-00545144, HAL.

[3] Bacchetta, P. and van Wincoop, E. (2003) *Can Information Heterogeneity Explain the Exchange Rate Determination Puzzle?* Technical Report 3808, C.E.P.R. Discussion Papers.

[4] Bertsekas, D. P. (1995) *Dynamic Programming and Optimal Control, Two Volume Set.* Athena Scientific, 2nd edition.

[5] Bontempi, G., Ben Taieb, S., and Borgne, Y.-A. (2013) *Machine Learning Strategies for Time Series Forecasting.* In Aufaure, M.-A. and Zimnyi, E., editors, *Business Intelligence*, **138** of *Lecture Notes in Business Information Processing*, Springer Berlin Heidelberg, 62–77.

[6] Busoniu, L., Babuska, R., Schutter, D., and Ernst, D. (2010) *Reinforcement Learning and Dynamic Programming Using Function Approximators.* Automation and Control Engineering. Taylor & Francis.

[7] Butrica, B. A., Johnson, R. W., and Smith, K. E. (2011) *Potential Impacts of the Great Recession on Future Retirement Incomes.* Pension Research Council.

[8] Cipriani, M. and Guarino, A. (2014) *Estimating a Structural Model of Herd Behavior in Financial Markets.* American Economic Review, **104** (1), 224–251.

[9] Conejo, A., Carrión, M., and Morales, J. (2010) *Decision Making Under Uncertainty in Electricity Markets.* International Series in Operations Research & Management Science. Springer.

[10] Coulon, M., Powell, W. B., and Sircar, R. (2013) *A model for hedging load and price risk in the Texas electricity market.* Energy Economics, **40** 0, 976 – 988.

[11] Creamer, G. and Freund, Y. (2010) *Automated trading with boosting and expert weighting.* Quantitative Finance, **10** (4), 401–420.

[12] Cryer, J. and Chan, K. (2008) *Time Series Analysis: With Applications in R.* Springer Texts in Statistics. Springer.

[13] Devenow, A. and Welch, I. (1996) *Rational herding in financial economics.* European Economic Review, Elsevier, **40** (3-5), 603–615.

[14] Glachant, J. (2009) *Electricity Reform in Europe: Towards a Single Energy Market.* Edward Elgar.

[15] Gomez-Exposito, A., Conejo, A., and Canizares, C. (2008) *Electric Energy Systems: Analysis and Operation.* Electric Power Engineering Series. Taylor & Francis.

[16] Gorton, G. B. (2012) *Misunderstanding Financial Crises: Why We Don't See Them Coming.* Oxford University Press, USA.

[17] Hernández-Lerma, O. and Lasserre, J. B. (1996) *Discrete-time Markov control processes: basic optimality criteria.* Springer.

[18] Hurd, M. D. and Rohwedder, S. (2010) *Effects of the financial crisis and great recession on American households.* Technical report, National Bureau of Economic Research.

[19] Kim, J. H. and Powell, W. B. (2011) *An hour-ahead prediction model for heavy-tailed spot prices.* Energy Economics, **33** (6), 1252 – 1266.

[20] Kirman, A. (2010) *The Economic Crisis is a Crisis for Economic Theory.* CESifo Economic Studies, **56** (4), 498–535.

[21] Krollner, B., Vanstone, B., and Finnie, G. (Apr. 2010) *Financial time series forecasting with machine learning techniques: A survey.* Paper presented at the European symposium on artificial neural networks: Computational and machine learning, Bruges, Belgium.

[22] Krugman, P. (1979) *A model of balance-of-payments crises.* Journal of Money, Credit, and Banking, **11**, 311–325.

[23] Littlestone, N. and Warmuth, M. K. (1994) *The weighted majority algorithm.* Information and Computation, **108**, 212–261.

[24] Powell, W. (2011) *Approximate Dynamic Programming: Solving the Curses of Dimensionality.* Wiley Series in Probability and Statistics. John Wiley & Sons, 2 edition.

[25] Puterman, M. L. (1994) *Markov decision processes: discrete stochastic dynamic programming.* Wiley series in probability and statistics. Wiley-Interscience.

[26] Reinhart, C. M. and Rogoff, K. S. (2009) *This Time is Different: Eight Centuries of Financial Folly.* Princeton University Press, Princeton, NJ.

[27] Ruszczyński, A. (2010) *Risk-averse dynamic programming for Markov decision processes.* Math. Programming, **125**, 235–261.

Trees Garrote for Regression Analysis

Masatoshi Nakamura, *Oita University*, masatoshi-nakamura@ds-pharma.co.jp
Yoshimichi Ochi, *Oita University*, ochi@oita-u.ac.jp
Masashi Goto, *Biostatistical Research Association, NPO*, bra-goto@ybb.ne.jp

Abstract. In regression analysis, stochastic models are often constructed to model relationships between outcomes and explanatory variables, and we derive statistical interpretations for data based on these models. However, if we use only linear regression models, constructing a true model reflecting actual characteristics can be difficult. A tree-structured approach is recommended, such as classification and regression trees (CART), which develops a tree and provides an interpretation of the data based on the fundamental model derived from the tree. Random Forest (RF) involves an ensemble learning method based on the trees and can predict outcomes more precisely. However, RF cannot provide a tree-structured model for interpreting the data. Here, a nonnegative garrote (NNG), a shrinkage estimator, is examined, and trees garrote (TG) is proposed as an adjustment of RF based on NNG. Some shrinkage estimators for ensemble learning are reported to yield better predictive performance. In addition, TG can lead to tree-structured models that are useful for interpretation of data. Simulation studies show that the proposed method is highly accurate predictively. Finally, two case studies of diabetes and prostate cancer data illustrate descriptive features of tree-structured models based on TG.

Keywords. Tree-structured approaches, Ensemble learning, CART, Random Forest, Nonnegative garrote

1 Introduction

In a regression analysis, a stochastic model is often constructed to the relationship between an outcome and explanatory variables, and we derive statistical interpretations for data based on this model, but when there is one response variable with multiple explanatory variables, it is rarely possible to describe real phenomena with linear regression models. If we use only such models, then formulating an accurate model reflecting particular characteristics can be difficult. In general, mechanisms of events occurring behind real data are often non-linear. To say that the non-linear so that there is a variety of non-linear model, considering this circumstance, we must depart from the traditional paradigm of linearity in creating models. In addition, an approach that is easy to interpret and flexible to construct a model based on the data is required not only to predict a response variable but also to identify factors that is useful for interpretations.

As alternative methods for fulfilling these requirements, we recommend tree-structured approaches that develop a tree structure and derive interpretations of the data based on the fundamental model derived from the tree. A typical useful method for the tree-structured model is classification and regression trees (CART) [1]. A tree-structured model was constructed using the subspace of outcomes divided by some cut-off value of explanatory variables within the sample space together with estimates of corresponding parameters on each subspace. Moreover, this model can be expressed as a tree containing nodes with child nodes determined by values of selected explanatory variable and terminal nodes that express the subspace of outcomes. Based on the graphical interpretation of the potential relationship between outcomes and explanatory variables for given data, we can derive insights related to the factors and their interactions affecting the outcome.

Noteworthy, ensemble learning methods, studied primarily in connection with machine learning, have been incorporated into the tree-structured model. Random Forest (RF) [2] involves an ensemble learning method based on trees and can predict outcomes precisely. On the other hand, RF cannot suggest one or a few trees for exploratory interpretation of data. Especially, in terms of the interpretability viewpoint, RF is often criticized on the grounds that hundreds of trees with many nodes are involved in the regression model, resulting in regression relationships that are difficult to visualize [3].

From that viewpoint, rule ensembles that use shrinkage estimators have been proposed [4]. This method uses a least absolute shrinkage and selection operator (LASSO)-type penalty [5] on the coefficients of each rule in RF, rendering RF more interpretable by reducing the very large number of rules in the hundreds of trees that RF requires. However, we cannot obtain more detailed information concerning the relationship between an outcome and explanatory variables based on trees such as CART. Hence, we want to select one or a few representative trees from the hundreds of trees in RF for interpreting the data.

To attain both predictive accuracy and interpretability, we propose trees garrote (TG) as a new tree-structured model with high predictive accuracy and that is interpretable through one or a few representative trees. We examine nonnegative garrote (NNG) [6], a shrinkage estimator, and formulate TG as an adjustment of RF based on NNG. In our technique, entire trees are removed or weighted by penalty, and a prediction is represented as a weighted average of predicted outcomes of remaining trees. An adjustment of RF based on LASSO is reported to yield better predictive performance [7]. In addition, for interpreting data, TG can lead to useful trees selected from RF by the NNG-type penalty. We consider a few trees with high-order weights as representative trees and can visualize these trees for interpretation.

We describe the notation and algorithm of TG in Section 2. We evaluate the predictive accuracy of our proposed method through simulation in Section 3, and we examine case studies using our proposed method in Section 4. We summarize our conclusions and futher discussion in Section 5.

2 Trees Garrote method

In this section, we examine RF, NNG, TG notation, and the procedure for creating trees.

Random Forest method

[2] formalized the concept of an RF with hundreds of trees as a predictor consisting of an ensemble of classifiers $\{h(\boldsymbol{x};\theta_b), b = 1,\ldots,B\}$, where $\boldsymbol{x}$ is the P -dimensional explanatory variable vector and $\boldsymbol{\theta} = \theta_1,\ldots,\theta_B$ are independent and identically distributed (i.i.d.) probability vectors with information involving split-rules of each classifier. $h(\boldsymbol{x};\theta_b)$'s are CART-like structures. Denote the data as $(y,\boldsymbol{x})$, and assume n samples are obtained as $(y_1,\boldsymbol{x}_1),\ldots,(y_n,\boldsymbol{x}_n)$, where y is a one-dimensional outcome (real number).

The tree-structured model of RF ensembles the classifiers as the average of $\widehat{h}(\boldsymbol{x};\theta_b)$ constructed by B times bootstrapping from the given data, described as follows:

$$f_{RF} = \frac{1}{B}\sum_{b=1}^{B}\widehat{h}(\boldsymbol{x};\theta_b). \quad (1)$$

For decreasing the prediction error (i.e., improving prediction accuracy), the RF procedure has two important aspects, as follows:

1. To reduce prediction error of each classifier, each classifier is larger than original CART, and these classifiers are not pruning of some nodes because of constructing them more adaptively for each bootstrap sample.
2. To obtain the lower correlation between classifiers, the classifiers are constructed using different procedures by adding two random selections as follows:
   a) Each classifier is constructed from a bootstrap sample from the given data
   b) Each node of the classifier is divided by a best split value among the explanatory variables selected at random.

Non Negative Garrote method

The NNG estimator was introduced by [6] as a shrinkage estimator that both shrinks and zeroes coefficients estimated with regression analysis. Ridge regression is a traditional shrinkage estimator in terms of reducing variance and/or resolving multicollinearity [8].

NNG estimators are intended to obtain a more robust estimator. When we get the original estimated coefficients $\{\beta_p\}_0^P$ with explanatory variables $\boldsymbol{x} = x_1, \ldots, x_P$ based on a regression model, these are optimized by the NNG estimator $\{c_p\}_0^P$, estimated under the following conditions:

$$\{\widehat{c}_p\}_0^P = \arg\min_{\{c_p\}_1^P} \sum_{n=1}^{N} \left(y_n - \widehat{\beta_0} - \sum_{p=1}^{P} c_p \widehat{\beta_p} x_{np} \right)^2 \text{ subjet to } \sum_{p=1}^{P} c_p \leq s \text{ and } c_p \geq 0, p = 1, ..., P, \tag{2}$$

where s is the shrinkage parameter. Comparing NNG with ridge regression, the differences are $c_p \geq 0, p = 1, ..., P$ and replacing the sum of the absolute value of parameter distance (l_1 norm) penalty with the sum of the square value of parameter distance (l_2 norm) penalty. However, the transition of each coefficient dependent on the shrinkage parameter is decidedly different; in other words, many coefficients are estimated as zero, and many variables are removed from the model. Generalized cross-validation or k-fold cross-validation is used to estimate s.

There are other methods used as shrinkage estimators. [5] proposed the LASSO, which puts no limit on the values of parameters, and [10] proposed the elastic net, which uses a linear combination of the l_1 and l_2 norm as the penalty. [7] proposed an adjustment of RF based on the LASSO.

Notation of Trees Garrote

TG makes an adjustment by adding NNG to ensemble trees in RF. The explanatory variables x_{np} in (2) are replaced with the tree-structured classifiers $h(\boldsymbol{x}; \boldsymbol{\theta})$ created by RF. The coefficients of each classifier are estimated using the following equation:

$$\{\widehat{c}_b\}_1^B = \arg\min_{\{c_b\}_1^B} \sum_{n=1}^{N} \left(y_n - \sum_{b=1}^{B} c_b h(\mathbf{x}, \theta_b) \right)^2 \text{ subjet to } \sum_{b=1}^{B} c_b = 1 \text{ and } c_b \geq 0, b = 1, ..., B.$$

Here, let the coefficients $\{\widehat{\beta}_b\}_0^B$ in (2) to be defined as $\widehat{\beta}_0 = 0, \widehat{\beta}_b = 1, b = 1, \ldots, B$, because the ensemble procedure in RF is constructed based on the equally weighted mean value of all classifiers outcomes. As an additional assumption, in the case of the ensemble procedure in TG, the shrinkage parameter s in (2) is fixed at one; hence, the optimized value of s is not estimated because sufficient accuracy for prediction can be maintained even with $s = 1$ and there are disadvantages in implementing this estimation in terms of the computer load. [3] offers the same recommendation from a different viewpoint.

The tree-structured TG model is constructed as follows:

$$f_{\mathrm{TG}} = \sum_{b=1}^{B} \widehat{c}_b \widehat{h}(\mathbf{x}, \theta_b) \tag{3}$$

Figure 1 shows a flow chart for constructing a TG model.

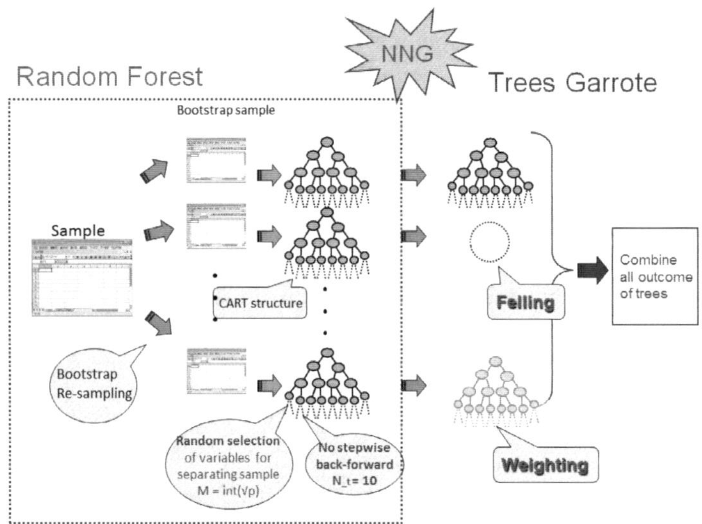

Figure 1. A flow chart of Trees Garrote

Algorithm of Trees Garrote. The details of the TG algorithm are as follows:

1. According to random sampling with replacement from all sample $\mathbf{L} = \{(y_n, \boldsymbol{x}_n); n = 1, ..., N\}$ (i.e., bootstrap-resampling), bootstrap-samples $\{\mathbf{L}_{boot}^{(b)}, b = 1, \ldots, B\}$ of sample size N are created.

2. Based on the B bootstrap samples, B trees constructed as CART-like structures $\{h(\boldsymbol{x}, \theta_b), b = 1, \ldots, B\}$ are created based on RF methods. Here, we set up the RF parameters as follows:

   a) Let the number of randomly selected variables M from explanatory variables $\{\boldsymbol{x}_p, p = 1, \ldots, P\}$ intended to search for the best split value for branching off the nodes (separating sample) be defined as the maximum integer that does not exceed $\sqrt{P}$. A best split variable is decided from M explanatory variables and this value is selected from observed values of M explanatory variables.

   b) Set the maximum sample size in each terminal node N_t to ten, that is, if the sample size in a node is less than ten, this node is treated as a terminal node and not be further divided.

TG3. Finally, the ensemble trees (3) are constructed using TG such that all classifiers $\{h(\boldsymbol{x}, \theta_b), b = 1, \ldots, B\}$ are weighted (i.e., $\widehat{c}_b > 0$) or felled (i.e. $\widehat{c}_b = 0$) by NNG.

Discription Procedure of Tree-Creating.

TG can lead to useful trees for data interpretation. We regard a few trees from TG as representative trees and visualize these trees for interpretation.

Selection of the representative trees. We can easily assume that a few trees with the largest weights $\widehat{c}_b$ in (3) contribute most to the prediction. Therefore, in TG, we regard such trees with high-order weights as representative trees and can visualize the trees as CART-like structures. This can be seen as a proposal to rank the trees in RF by their contribution to the predictive performance. If we use other shrinkage estimators in place of NNG, for example, the LASSO or the elastic net, then negative c_b's are included, and ranking the trees becomes difficult. Actually, [7] can not show the representative trees based on LASSO-RF because it is difficult to use LASSO's coefficients including negative value for selecting the representative trees.

Pruning a tree for interpretation. When a tree is selected from RF, it is usually too big for data interpretation. The tree requires pruning of some nodes and reconstruction to a smaller tree with improved data interpretability. This pruning procedure is performed after selection of the representative trees. we also consider to select the representative trees after all trees is pruned, however don't do that because the trees is not represent from TG model, that is, all pruned trees is not contained in the TG.

Let a measure of the non-uniformity in a node be defined as the standard deviation (SD) and standard error (SE) of the sample included in the node. The SD and SE in a node are estimated using out-of-bag samples as follow:

$$SD = \sqrt{\sum_{n=1}^{n_{oob}} (y_n - \bar{y})^2}, \qquad SE = \frac{SD}{\sqrt{n_{oob}}},$$

where n_{oob} is the number of data included in the node. A bootstrap-sample is used in RF to construct a tree. On the other hand, in the original sample, there are leftover samples, called out-of-bag samples, from the bootstrap sample. Such out-of-bag samples are not used for creating the tree; hence, we can use them to estimate SD and SE values in each tree node.

Let the pooled SD of two child nodes as terminal nodes are defined as SD_{pooled}, and let the SD and SE of a parent node of these two child nodes be SD_P and SE_P, respectively. Two child nodes are pruned and the parent node is designated a terminal node, when SD_P, SE_P and SD_{pooled} satisfy the condition

$$SD_P - SE_P < SD_{pooled}.$$

When there are no out-of-bag in two terminal node, then they are pruned. However, the out-of-bag sample contain around 30% of all sample, therefore the estimation of SD and SE by using out-of-bag sample is more evaluable than using 5 or 10 fold cross-validation (when 5 or 10 cross-validation is used to estimate SD and SE, 20% or 10% of all sample are evaluated, respectively).

Number of representative trees. It is important to decide how many representative trees we need to interpret the data. In this paper, we determine those trees by comparison with the CART model, because CART is the most popular method for constructing a tree for interpretation. The one standard error (1 SE) rule are used for selecting the optimal CART model. The SE is estimated by ten-fold cross-validation. We take the mean squared error (MSE) between the original and predicted outcomes as a measure of the predictive accuracy of the model. Let the predicted value of the representative trees be the mean of the predicted values from all representative trees, and when the trees are increased one by one according to the order $\widehat{c}_b$, the number of representative trees is defined when the following condition regarding MSE is satisfied for the first time:

$$MSE(CART) \geq MSE(mean(RepresentativeTrees)).$$

This inequality has an aim that Representative Trees has predictive accuracy which is not inferior or at least the same accuracy than CART, and therefore we can compared representative trees with CART about interpretability for data.

3 Simulations for Predictive Accuracy

We evaluate the performance of TG and RF in terms of predictive accuracy via a numerical experiment from three models as follows:

Model 1: Simple interaction model (See [11]).

$$\begin{aligned} y &= 3x_1x_2 + 3x_3 + 1.5x_4 + 2x_4x_5 + \epsilon. \\ \epsilon &: N(0,1) \\ x_1, ..., x_6 &: Uni(0,1) \end{aligned}$$

Model 2: Non-linear model I (See [12]).

$$\begin{aligned} y &= 10sin(\Pi x_1x_2) + 20(x_3 - 0.5)^2 + 10x_4 + 5x_5 + \epsilon. \\ \epsilon &: N(0,1) \\ x_1, ..., x_5 &: Uni(-\sqrt{3}, \sqrt{3}) \end{aligned}$$

Model 3: Non-linear model II (See [12]).

$$\begin{aligned} y &= 9\prod_{j=1}^{3} \exp\{-3(1-x_j)^2\} - 0.8\exp\{-2(x_4 - x_5)\} + 2\sin^2(\pi x_6) - 2.5(x_7 - x_8) + \epsilon. \\ \epsilon &: N(0,\sigma^2)[\sigma^2/\mathrm{Var}[f(x)] = 0.5] \\ x_1, ..., x_8 &: Uni(0,1) \end{aligned}$$

A training sample (N = 500) generated from each simulation model is used to create RF and TG models. The number of bootstrap samples is 300 (B = 300). A test sample (N = 100) is used for calculating MSEs between observed values and estimated values, as follows:

$$MSE = \frac{1}{100}\sum_{i=1}^{100}(y_i - \widehat{f}(\mathbf{x}_i))^2.$$

Results. To compare the predictive accuracy of TG and RF, we simulate 1,000 training samples for each model and show box plots of MSE(TG)/MSE(RF) for all simulated datasets for each simulation model (Figure 2).

In nearly all of the simulated datasets, the ratio of MSE is less than one regardless of the model, and it is shown that TG predicts the outcomes much more accurately than RF. the one of the reason for getting more accurately is that TG implicate RF, that is, when all TG's coefficient are the same, RF equal TG.

Let the removal rate be defined as the number of weights estimated to be zero divided by the number of bootstrap samples B. The mean removal rates for each model are 79.9% (Model 1), 75.4% (Model 2), and 76.1% (Model 3). Although the remaining trees ranged from approximately 20% to 25% for NNG, , TG still provides an effective adjustment of RF.

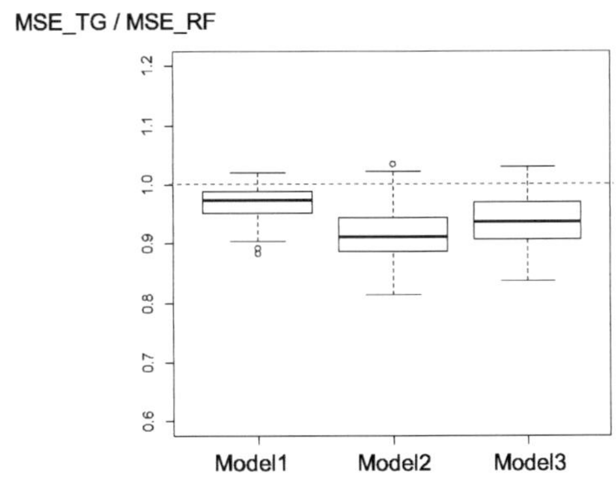

Figure 2. Box plots of MSE(TG)/MSE(RF)

4 Case Studies for tree creation

Diabetes Study (See [13])

Ten baseline variables, age, sex, body mass index (BMI), average blood pressure (BP), and six blood serum measurements (TC: total cholesterol, LDL: LDL cholesterol, HDL: HDL cholesterol, US: urinary sugar, Hb: HbA1c, GLU: glucose) were obtained for each diabetes patients (n = 442), along with the response of interest, a quantitative measure of disease progression one year after baseline.

All samples were used in creating the TG model. The number of bootstrap samples was 300 (B = 300). Tree-structured models were created using both CART and TG methods. CART and TG trees with the highest weight cb are shown in the left and right panels, respectively, of Figure 3. Both trees have nodes divided by Hb and BMI, but the TG tree has a leading node dividing at Age = 57. This information provides a new insight regarding the relationship between outcome and age.

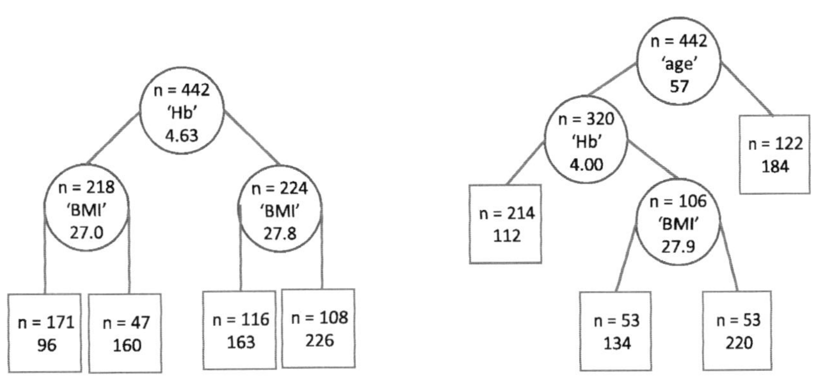

Figure 3. Trees created by CART (left) and trees garrote (right) methods

Selection of multiple representative trees. We consider selecting multiple trees based on TG. TG involves constructing many trees, and we can select multiple trees for data interpretation. In this situation, we expect that a few trees will be selected that have higher values of coefficients C_b's.

We compared the predictive accuracy between multiple trees and CART. A training sample (N = 398 [90% of all samples]) was used to create TG and CART models. The training sample was selected randomly from all samples. The number of bootstrap samples for TG was 300 (B = 300). The remaining test sample (N = 44 [10% of all samples]) was used for calculating MSE values between observed and estimated values. The MSE values for one to ten trees of highest weights were calculated and compared with the MSE of the CART tree. This process was repeated 100 times, and the mean MSE was calculated.

The mean MSE values for one to ten trees of highest weight were calculated and compared with the mean MSE of the CART tree (Figure 4).

The mean MSE values of representative trees decreases as the number of trees increases, and when the number of multiple trees is three, the mean MSE of multiple trees became smaller than the mean MSE of CART. Hence, we selected three trees from those with the highest-valued coefficients c_b for data interpretation.

The three representative trees are shown in Figure 5. The values of the coefficients c_b of each tree are 0.092, 0.084, and 0.076, respectively. The second tree is nearly the same as the CART tree, but the third tree dividing LDL and SEX, instead of Hb, is different from the CART tree. This information provides an additional insight.

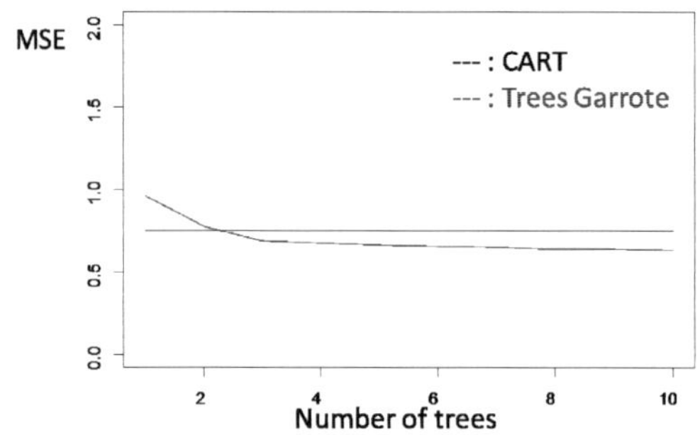

Figure 4. Comparison of mean MSE values of CART and representative trees of trees garrote

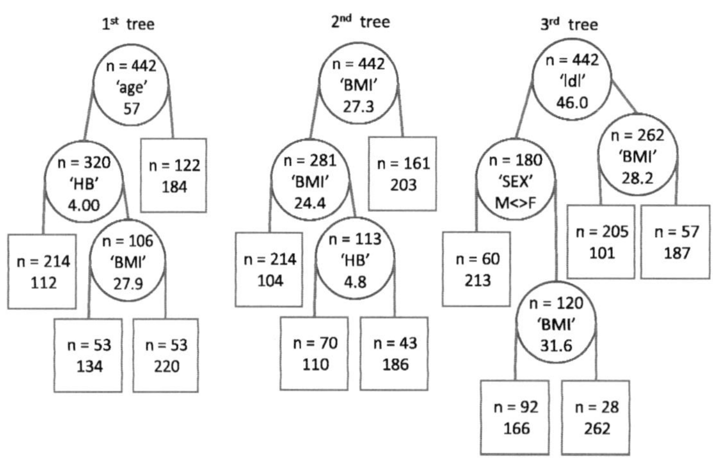

Figure 5. Multiple representative trees for diabetes study

Prostate Cancer Data (See [5])

The prostate cancer data (n = 97) come from a study that examined the correlation between the level of prostate specific antigen (PSA) and a number of clinical measures in men who were about to receive a radical prostatectomy. We fit a tree-structure model to log(PSA). The eight baseline variables were cancer volume (VOL), prostate weight (WT), age, benign prostatic hyperplasia amount (LBPH), seminal vesicle invasion (SVI), capsular penetration (LCP), Gleason score (GS), and percentage Gleason scores four or five (GS45).

All samples were used for creating the TG model. The number of bootstrap samples was 80 (B = 80). A tree-structured model was created using both methods, CART and TG. CART and TG trees with the highest weight cb are shown in the left and right panels, respectively, of Figure 6. The CART tree is divided only by VOL, but the TG tree is divided additionally by GS and GS45.

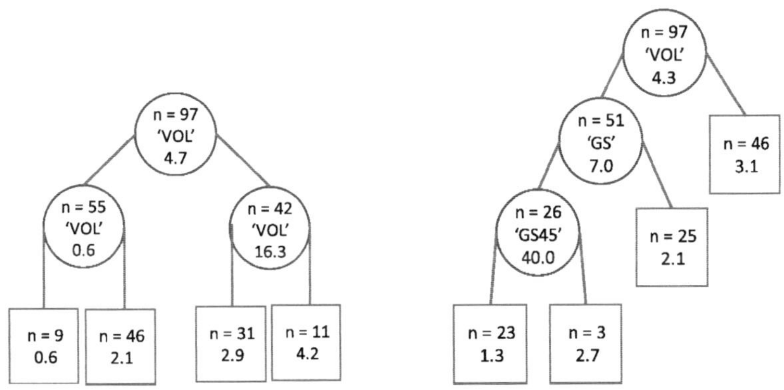

Figure 6. CART (left) and trees garrote(right) trees

Select multiple representative trees. We compared the predictive accuracy of multiple trees and CART. A training sample (N = 88 [90% of all samples]) was used to create TG and CART models. The number of TG bootstrap samples was 80 (B = 80). The remaining test samples (N = 9 [10% of all samples]) were used to calculate MSE values between observed values and estimated values. The MSE values for one to ten trees of highest weight were calculated and compared with the MSE values of the CART tree. This process was repeated 100 times and the mean MSE was calculated.

The mean MSE values for one to ten trees of highest weight were calculated and compared with the mean MSE of the CART tree (Figure 7). When the number of multiple trees is four, the mean MSE of multiple trees becomes smaller than the mean MSE of CART. Hence, we selected four trees from the highest values of the coefficients c_b for data interpretation.

The four representative trees are shown in Figure 8. The values of the coefficients cb of individual trees are 0.112, 0.097, 0.097, and 0.095, respectively. The first, second, and third trees are divided by GS and have the same split value (eight) as GS. The third and fourth trees are divided by SVI, rather than VOL. This information cannot be obtained from the CART tree.

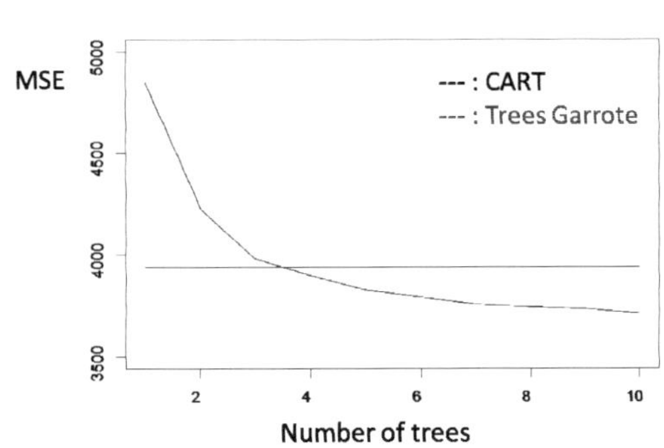

Figure 7. Comparison of mean MSE values of CART and representative trees of trees garrote

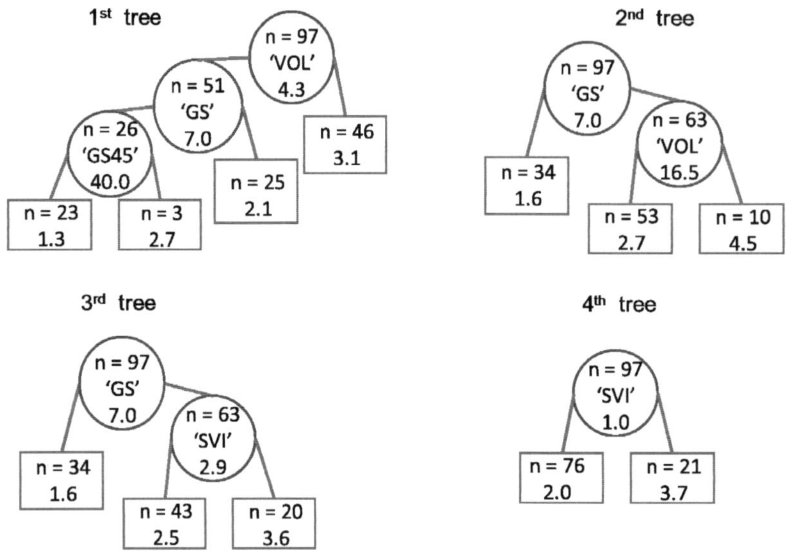

Figure 8. Multiple representative trees for Prostate Cancer Data

5 Conclusions

In regression analysis, we often face a tradeoff between predictive accuracy and interpretability. Many ensemble techniques have strong predictive power, but they are often criticized for difficulty with regard to interpretation of the specific regression relationship, since hundreds of trees with many nodes are involved in the regression model. In contrast, linear regression models are more interpretable, but suffer from unimpressive predictive accuracy.

To retain both predictive accuracy and interpretability, we proposed TG, which has high predictive accuracy and can be interpreted through representative trees. Some trees are removed by zero coefficients of NNG, and the prediction is represented as a weighted average of the remaining trees. We can surmise that trees having the highest weight contribute most to the prediction. Therefore, in TG, we regard a few trees that have high-order weights as representative trees, and can visualize the data using the representative CART-like structure trees. Furthermore, a simulation and case studies elucidated the merits of this technique, finding that TG has better performance with regard to prediction error and can visualize one or a few trees, potentially providing new meaningful insights regarding the data.

Discussion

TG make an ensemble of classifiers based on RF, that is, the ensemble are made by using bootstrap sampling and random selection of variables dividing nodes. Boosting and Importance sample learning ensemble (ISLE) [4] adopt other procedures to make an ensemble, that is, the ensemble are made by fitting residual of the given ensemble. For the TG ensemble we chose RF based approach because we considered that selection of representative trees should be made by the same criteria from the data. On the other hands, in terms of the structure of models, the methods of integrating the rules which indicate how to move sample from top node to each node are suggested, such as Rule Ensemble [4] and Forest Garrote [3], unlike ours of integrating weighted trees. These methods can provide the interesting interpretation of data, but cannot show the trees for the data. We focus on the trees type representation that can describe the procedure how all sample reach each subgroup divided by values of explanatory variables. Meanwhile, regarding predictive accuracy, some papers have been reported the superiorlity of predictive accuracy of the methods of making an ensemble by fitting residual or integrating rules for the model. In this paper, the predictive accuracy of TG is evaluated by comparing with RF, and the predictive accuracy of RF has already been compared with various methods mentioned above including, LASSO-RF [7], original CART [1], and some linear or non-linear regression models in several papers. It may be possible to compare the predictive accuracy of TG indirectly with other methods via those results, however, we understand the importance of directly compare with many methods. Hence, we prepare the data analysis and simulations. In addition, for evaluation of TG, three small simulations and two case studies presented here, and we are currently undertaking simulation studies to evaluate the extent that TG can recapture true structure via adaptation of representative trees.

Bibliography

[1] Breiman, L., Friedman, J. H., Olshen, R. A. and Stone, C. J. (1984). *Classification and Regression Trees.* Belmont, CA: Wadsworth.

[2] Breiman, L. (2001). *Random forests.* Machine Learning, **45**, 5-32.

[3] Meinshausen, N. (2009). *Forest Garrote.* Electronic Journal of Statistics, **3**, 1288-1304.

[4] Friedman, J. H. and Popescu, B. E. (2008). *Predictive learning via rule ensembles.* Annals of Applied Statistics. **2**(3), 916-954.

[5] Tibshirani, R. (1996). *Regression shrinkage and selection via the Lasso.* journal of the royal statistical society, **58**(1), 267-288.

[6] Breiman, L. (1995). *Better subset regression using the nonnegative garrote.* Technometrics, **37**,373-384.

[7] Nakamura, M., Shimokawa, T., Sakamoto, W. and Goto, M. (2013). *Regression Analysis Using Lasso Random Forest* Computational Statistics of Japan, **26**, 1-15.

[8] Hoerl, E. and Kennard, R. W. (1970) *Ridge Regression: Biased Estimation for Nonorthogonal Problems.* Technometrics, **12**(1), 55-67.

[9] Craven, P. and Wahba, G. (1979). *Smoothing Noisy Data with Spline Functions.* Numerische Mathematik. **31**, 377-403.

[10] Zou, H. and Trevor H. (2005). *Regularization and Variable Selection via the Elastic Net.* Journal of the Royal Statistical Society. **67**(2), 301-320.

[11] Sugimoto, T., Simokawa, T. and Goto, M. (2005). *Tree-structured approaches and recent advances.* Computational Statistics of Japan, **18**, 123-164.

[12] Friedman, J. H. (1991). *Multivariate adaptive regression splines.* The annals of statistics, **19**(1), 1-67.

[13] Efron, B., Hastie, T., Johnstone, I. and Tibshirani, R. (2004). *Least angle regression.* The Annals of Statistis, **32**, 407-499.

On mixture modelling with multivariate skew distributions

Sharon X. Lee, *Department of Mathematics, University of Queensland, Brisbane QLD 4072, Australia*, s.lee11@uq.edu.au
Geoffrey J. McLachlan, *Department of Mathematics, University of Queensland, Brisbane QLD 4072, Australia*, g.mclachlan@uq.edu.au

Abstract. In recent years, there has been increasing use of non-normal distributions in the modelling and analysis of heterogeneous data. Attention here is focussed on the use of skew symmetric distributions with multivariate skewing functions that allow for the modelling of skewness in p arbitrary directions in the feature space, where p is the number of variables. In particular, various multivariate skew normal and skew t-distributions are considered corresponding to some commonly used characterizations in the literature. Parameter estimation for these distributions and mixtures of them can be obtained via the Expectation-Maximization (EM) algorithm. However, the E-step for such models typically involves the calculation of multidimensional integrals that are computationally expensive to evaluate. Some approaches are therefore considered to reduce the computation time required for the fitting of these models. In addition to methods that are directly applicable to single-threaded implementation, an approach is developed to utilize the processing resources available from machines with multiple cores. An example on a real dataset will be given to illustrate the approach.

Keywords. mixture models, skew distributions, EM algorithm, model-based clustering

1 Introduction

Non-normal distributions have attracted increasing attention in recent times due to their usefulness in modelling and analyzing data that exhibit non-normal features. Following the seminal paper on the classical skew normal distribution [6] and its subsequent generalization to the multivariate case [11], a substantial body of research has focussed on the development of flexible distributions that can take asymmetric distributional shapes. Among the many non-normal distributions proposed in the literature, the skew normal and skew t-distributions are gaining popularity. They have been studied extensively in the model-based clustering literature (see, for example, [8, 26, 31, 32, 39, 49] and the references therein). Their usefulness have also been demonstrated in many applications from a range of related fields including astrophysics [45], financial risk analysis and modelling [1, 12, 28, 33, 48], fisheries science [14], flow cytometry [16, 21, 22, 34, 35, 37, 38, 42, 44, 43, 46], image segmentation [27], pharmaceutical science [47], and the social sciences [5, 40]. Among other non-normal distributions considered in the literature, there are mixtures of normal-inverse-Gaussian distributions [24] and mixtures of multiple-scaled distributions [15, 50], but the focus of this paper is on skew normal and skew t-distributions.

There exists various characterizations of the skew normal and skew t-distributions that produce different versions with varying degrees of flexibility. This is related to how the latent skewing variable is formulated in the stochastic representations of these distributions. They also have implications on the type of skewness that can be modelled by the density. We discuss these issues in Section 2. A survey of some of the more commonly used characterizations can be found in [2, 3, 7] and the recent monograph [10]. For finite mixtures of these distributions, see [26, 27, 29, 32].

Parameter estimation for these mixture models via the EM algorithm can be difficult and time consuming. In Section 4, we consider some strategies to reduce the computation time required for the fitting of these models. In particular, we outline some procedures for generating starting values for the EM algorithm and the incorporation of trimming into each iteration. Furthermore, to take advantage of the computing resources of modern multicore machines, we describe in Section 4 an implementation developed for such operating environment. These approaches are illustrated on a flow cytometric dataset in Section 5.

2 Skew distributions

In this paper, we focus on a fairly general characterization of a skew symmetric distribution known as the canonical fundamental skew distribution [4]. This characterization has an arbitrary matrix of skewness parameters, thus allowing for the modelling of skewness in different directions simultaneously. Some of the more commonly used skew normal and skew t-distributions correspond to special cases of this characterization.

We begin with the definition of the canonical fundamental skew normal (CFUSN) distribution. Let $\boldsymbol{U}_1$ and $\boldsymbol{U}_0$ be independent random vectors that follow the central normal distribution with covariance matrix $\boldsymbol{\Sigma}$ and $\boldsymbol{I}_q$, respectively. Here $\boldsymbol{\Sigma}$ is a $p \times p$ positive definite matrix and $\boldsymbol{I}_q$ denotes the q-dimensional identity matrix. Then the p-dimensional random vector $\boldsymbol{Y}$ defined as a convolution of $\boldsymbol{U}_0$ and $\boldsymbol{U}_1$,

$$\boldsymbol{Y} = \boldsymbol{\mu} + \boldsymbol{\Delta} \left|\boldsymbol{U}_0\right| + \boldsymbol{U}_1, \tag{1}$$

follows the CFUSN distribution. In (1), $\boldsymbol{\mu}$ is a p-dimensional location vector and $\boldsymbol{\Delta}$ is a general $p \times q$ matrix. It can be shown that the density of $\boldsymbol{Y}$ is given by

$$f_{\text{CFUSN}_{p,q}}(\boldsymbol{y}; \boldsymbol{\mu}, \Sigma, \boldsymbol{\delta}) = 2^q \phi_p(\boldsymbol{y}; \boldsymbol{\mu}, \boldsymbol{\Omega})\, \Phi_q\left(\boldsymbol{\Delta}^T \boldsymbol{\Omega}^{-1}(\boldsymbol{y} - \boldsymbol{\mu}); \boldsymbol{0}, \boldsymbol{I}_q - \boldsymbol{\Delta}^T \boldsymbol{\Omega}^{-1} \boldsymbol{\Omega}\right), \tag{2}$$

where $\boldsymbol{\Omega} = \boldsymbol{\Sigma} + \boldsymbol{\Delta}\boldsymbol{\Delta}^T$, $\phi_p(.; \boldsymbol{\mu}, \boldsymbol{\Omega})$ is the density of the p-variate normal distribution with mean vector $\boldsymbol{\mu}$ and covariance matrix $\boldsymbol{\Omega}$, and $\Phi_p(.; \boldsymbol{\mu}, \boldsymbol{\Omega})$ is the corresponding distribution function. The matrix $\boldsymbol{\Delta}$ in (2) contains the skewness parameters, which regulate the skewness in the CFUSN density. If $\boldsymbol{Y}$ follows the CFUSN distribution (2), we write $\boldsymbol{Y} \sim CFUSN_{p,q}(\boldsymbol{\mu}, \boldsymbol{\Sigma}, \boldsymbol{\Delta})$. It can be observed that when the skewness matrix is zero, that is, $\boldsymbol{\Delta} = \boldsymbol{0}$, the CFUSN density (2) reduces to the multivariate normal distribution.

An equivalent generalization of the t-distribution can be formulated using the stochastic representation in (1) with the joint distribution of $\boldsymbol{U}$ and $\boldsymbol{U}_0$ replaced by a multivariate t-distribution. The resulting density is known as the canonical fundamental skew t (CFUST) distribution, and can be expressed as

$$f_{\text{CFUST}_{p,q}}(\boldsymbol{y}; \boldsymbol{\mu}, \Sigma, \boldsymbol{\Delta}, \nu) = 2^q\, t_p(\boldsymbol{y}; \boldsymbol{\mu}, \boldsymbol{\Omega}, \nu)\; T_q\left(\boldsymbol{c}(\boldsymbol{y})\sqrt{\frac{\nu + p}{\nu + d(\boldsymbol{y})}}; \boldsymbol{0}, \boldsymbol{\Lambda}, \nu + p\right), \tag{3}$$

where

$$\begin{aligned}
\boldsymbol{\Omega} &= \boldsymbol{\Sigma} + \boldsymbol{\Delta}\boldsymbol{\Delta}^\top, \\
\boldsymbol{c}(\boldsymbol{y}) &= \boldsymbol{\Delta}^\top \boldsymbol{\Omega}^{-1}(\boldsymbol{y} - \boldsymbol{\mu}), \\
\boldsymbol{\Lambda} &= \boldsymbol{I}_q - \boldsymbol{\Delta}^\top \boldsymbol{\Omega}^{-1} \boldsymbol{\Delta}, \\
d(\boldsymbol{y}) &= (\boldsymbol{y} - \boldsymbol{\mu})^\top \boldsymbol{\Omega}^{-1} (\boldsymbol{y} - \boldsymbol{\mu}).
\end{aligned}$$

In (3), the scalar parameter ν is a scalar degrees of freedom that regulate the tails of the distribution. Here we let $t_p\left(\boldsymbol{y}; \boldsymbol{\mu}, \boldsymbol{\Omega}, \nu\right)$ denote the p-dimensional t-distribution with location parameter $\boldsymbol{\mu}$, scale matrix $\boldsymbol{\Omega}$, and degrees of freedom ν, and $T_q(.)$ is the q-dimensional (cumulative) t-distribution function. We write $\boldsymbol{Y} \sim CFUST_{p,q}(\boldsymbol{\mu}, \boldsymbol{\Sigma}, \boldsymbol{\Delta}, \nu)$ if $\boldsymbol{Y}$ follows the CFUST distribution (3). If $\boldsymbol{\Delta} = \mathbf{0}$, the CFUST density reduces to the multivariate t-distribution. Some properties of the CFUST distribution are described in [4]. In addition, it should be noted that the CFUSN and CFUST distributions suffer an identifiability issue [32]. Their densities are invariant under permutations of the columns of the skewness matrix, which implies the number of free parameters in $\boldsymbol{\Delta}$ reduces to $pq - (q-1)^2$.

Imposing constraints on the skewness matrix

As mentioned previously, the CFUST distribution encompasses as special and/or limiting cases some of the more commonly used skew normal and skew t-distributions. This includes the classical skew normal distribution [11] and the skew normal and skew t distributions [42]. They were referred to as the restricted and unrestricted skew distributions in the terminology in [26, 29], respectively. The unrestricted skew normal and skew t-distributions are obtained by constraining $\boldsymbol{\Delta}$ to be a diagonal matrix in (2) and (3), respectively. For the classical skew normal and skew t-distributions [9], they are equivalent to the CFUSN and the CFUST distributions by letting $q = 1$ or taking $\boldsymbol{\Delta}$ to be a matrix of zeros except for one column [32]. This formulation of the skew normal and skew t-distribution can be shown to be equivalent to that considered in [13, 20, 25, 42]; see [29] for details.

Some of the implications of the constraints on $\boldsymbol{\Delta}$ as imposed by the restricted and unrestricted characterizations are discussed in the recent paper [39]. Briefly, the dimension of latent skewing variable $\boldsymbol{U}_0$, denoted by q, is related to the number of directions in which the skewness of the density is concentrated. On the other hand, the matrix of skewness parameters $\boldsymbol{\Delta}$ regulates the orientation or direction of skewness.

For the restricted skew distribution, the constraint $q = 1$ implies that $\boldsymbol{U}_0$ in (1) is a scalar random variable. Thus the realizations of this latent skewing variable U_0 are confined to lie on a line in the p-dimensional feature space. This essentially means that skewness is concentrated in a single direction regardless of the dimension of the feature space. Hence the restricted skew distributions are limited to modelling skewing that is concentrated in a single direction.

In the case of the unrestricted skew distribution, $\boldsymbol{U}_0$ is a p-dimensional random vector but $\boldsymbol{\Delta}$ is a $p \times p$ diagonal matrix. Thus it allows for the modelling of skewness along p directions that are uncorrelated. This effectively means that skewness is concentrated along the directions that are parallel to the axes of the feature space. Hence the unrestricted skew distributions are best suited to modelling data with skewness along the the directions of the axes of the feature space.

The CFUSN and CFUST distributions have an arbitrary $p \times q$ matrix of skewness of parameters and the latent skewing variable has dimension q. Note that q is not necessary smaller than p. Thus these two distributions are fairly flexible and can model skewness along q different directions. There is no restriction on the correlation between these directions of skewness. Some examples demonstrating the difference between these three characterizations of the skew t-distribution on simulated and real datasets have been presented in [32]. For further discussion on this topic, the reader is to referred to [39].

3 Parameter estimation for finite mixtures of skew normal and t-mixture models

For model-based clustering applications, we are interested in the fitting of finite mixtures of CFUSN and CFUST distributions. The density of a g-component finite mixture model is defined as a convex linear combination of g densities, given by

$$f(\boldsymbol{y}; \boldsymbol{\Psi}) = \sum_{h=1}^{g} \pi_h f(\boldsymbol{y}; \boldsymbol{\theta}_h), \tag{4}$$

where $\boldsymbol{\theta}_h$ contains the unknown parameters of the hth component density which includes the elements of $\boldsymbol{\mu}_h$ and $\boldsymbol{\Delta}_h$ (and ν_i for the CFUST case), and the distinct elements of $\boldsymbol{\Sigma}_h$. $\boldsymbol{\Psi} = (\pi_1, \ldots, \pi_{g-1}, \boldsymbol{\theta}_1^T, \ldots, \boldsymbol{\theta}_g^T)^T$ is the vector of all unknown parameters of the mixture model. The mixing proportions π_i are non-negative and sum to one. When component densities $f(\boldsymbol{y}; \boldsymbol{\theta}_i)$ in (4) take the form of (2) and (3), we obtain a finite mixture of CFUSN (FM-CFUSN) distributions and of CFUST (FM-CFUST) distributions, respectively. Their densities are given by

$$f(\boldsymbol{y};\boldsymbol{\Psi}) = \sum_{h=1}^{g} \pi_h f_{\text{CFUSN}_{p,q}}(\boldsymbol{y}; \boldsymbol{\mu}_h, \boldsymbol{\Sigma}_h, \boldsymbol{\Delta}_h) \quad \text{and} \quad f(\boldsymbol{y};\boldsymbol{\Psi}) = \sum_{h=1}^{g} \pi_h f_{\text{CFUST}_{p,q}}(\boldsymbol{y}; \boldsymbol{\mu}_h, \boldsymbol{\Sigma}_h, \boldsymbol{\Delta}_h, \nu_h), \tag{5}$$

respectively. We write $\boldsymbol{Y} \sim \text{FM-CFUSN}_{p,q}(\boldsymbol{\Psi})$ and $\boldsymbol{Y} \sim \text{FM-CFUST}_{p,q}(\boldsymbol{\Psi})$ to denote that $\boldsymbol{Y}$ has a FM-CFUSN and FM-CFUST distribution, respectively.

EM algorithm

The EM algorithm for the fitting of the FM-CFUSN and FM-CFUST models can be implemented by noting the hierarchical representation of the CFUSN and CFUST distributions. For brevity, we focus here on the FM-CFUST model, but the corresponding expressions for the FM-CFUSN model can be obtained in a similar manner (see [36] for technical details). The derivations of the E and M-steps are given in [32] and are not repeated here. It was shown that on the $(k+1)$th iteration, the E-step requires the following five conditional expectations to be calculated,

$$z_{hj}^{(k)} = \frac{\pi_h f(\boldsymbol{y}_j; \boldsymbol{\mu}_h^{(k)}, \boldsymbol{\Sigma}_h^{(k)}, \boldsymbol{\delta}_h^{(k)}, \nu_h^{(k)})}{f(\boldsymbol{y}_j; \boldsymbol{\Psi}^{(k)})}, \tag{6}$$

$$w_{hj}^{(k)} = \left(\frac{\nu_h^{(k)} + p}{\nu_h^{(k)} + d_h^{(k)}(\boldsymbol{y}_j)}\right) \frac{T_q\left(\boldsymbol{c}_h^{(k)}(\boldsymbol{y}_j)\sqrt{\frac{\nu_h^{(k)}+p+2}{\nu_h^{(k)} d_h^{(k)}(\boldsymbol{y}_j)}}; \boldsymbol{0}, \boldsymbol{\Lambda}_h^{(k)}, \nu_h^{(k)} + p + 2\right)}{T_q\left(\boldsymbol{c}_h^{(k)}(\boldsymbol{y}_j)\sqrt{\frac{\nu_h^{(k)}+p}{\nu_h^{(k)} d_h^{(k)}(\boldsymbol{y}_j)}}; \boldsymbol{0}, \boldsymbol{\Lambda}_h^{(k)}, \nu_h^{(k)} + p\right)}, \tag{7}$$

$$e_{1hj}^{(k)} = w_{hj}^{(k)} - \log\left(\frac{\nu_h^{(k)} + d_h^{(k)}(\boldsymbol{y}_j)}{2}\right) - \left(\frac{\nu_h^{(k)} + p}{\nu_h^{(k)} + d_h^{(k)}(\boldsymbol{y}_j)}\right) + \psi\left(\frac{\nu_h^{(k)} + p}{2}\right), \tag{8}$$

$$\boldsymbol{e}_{2,hj}^{(k)} = w_{hj}^{(k)} E_{\boldsymbol{\Psi}^{(k)}}[\boldsymbol{u}_{hj} \mid \boldsymbol{y}_j], \tag{9}$$

$$\boldsymbol{e}_{3hj}^{(k)} = w_{hj}^{(k)} E_{\boldsymbol{\Psi}^{(k)}}[\boldsymbol{u}_{hj}\boldsymbol{u}_{hj}^T \mid \boldsymbol{y}_j], \tag{10}$$

where $\boldsymbol{U}_{hj} \mid \boldsymbol{y}_j$ has a q-dimensional truncated t-distribution given by

$$\boldsymbol{U}_{hj} \mid \boldsymbol{y}_j \sim tt_q\left(\boldsymbol{c}_{hj}^{(k)}, \left(\frac{\nu_h^{(k)} + d_h(\boldsymbol{y}_j)}{\nu_h^{(k)} + p + 2}\right)\boldsymbol{\Lambda}_h^{(k)}, \nu_h^{(k)} + p + 2; \mathbb{R}^+\right).$$

It should be noted that (7), (9), and (10) need to be evaluated numerically. Expressions and routines for calculating the truncated moments (9) and (10) are discussed in [26]. Also, it should be noted that $e_{1hj}^{(k)}$ can be evaluated using different approaches, two of which are described in the above reference. The use of the approximate OSL approach to calculate $e_{1hj}^{(k)}$ can result in the incomplete-data likelihood not increasing monotonically. This conditional expectation can be calculated more accurately by a power series derived in [30, 32] for which monotonicity of the likelihood is preserved.

On the $(k+1)$th iteration of the the M-step, the estimates of the parameters of the FM-CFUST model are updated by the following expressions:

$$\pi_h^{(k+1)} = \frac{1}{n}\sum_{j=1}^{n} z_{hj}^{(k)}, \tag{11}$$

$$\boldsymbol{\mu}_h^{(k+1)} = \frac{\sum_{j=1}^{n} z_{hj} w_{hj}^{(k)} \boldsymbol{y}_j - \boldsymbol{\Delta}_h^{(k)} \sum_{j=1}^{n} z_{hj}^{(k)} \boldsymbol{e}_{2hj}^{(k)}}{\sum_{j=1^n}^{n} z_{hj}^{(k)} w_{hj}^{(k)}}, \tag{12}$$

$$\boldsymbol{\Delta}_h^{(k+1)} = \left[\sum_{j=1}^{n} z_{hj}^{(k)} \left(\boldsymbol{y}_j - \boldsymbol{\mu}_h^{(k+1)}\right) \boldsymbol{e}_{2hj}^{(k)\top}\right] \left[\sum_{j=1}^{n} z_{hj}^{(k)} \boldsymbol{e}_{3hj}^{(k)}\right]^{-1}, \tag{13}$$

$$\begin{aligned}\boldsymbol{\Sigma}_h^{(k+1)} = & \left\{\sum_{j=1}^{n} z_{hj}^{(k)} \left[w_{hj}^{(k)} \left(\boldsymbol{y}_j - \boldsymbol{\mu}_h^{(k+1)}\right)\left(\boldsymbol{y}_j - \boldsymbol{\mu}_h^{(k+1)}\right)^T - \boldsymbol{\Delta}_h^{(k+1)} \boldsymbol{e}_{2hj}^{(k)} \left(\boldsymbol{y}_j - \boldsymbol{\mu}_h^{(k+1)}\right)^\top\right.\right. \\ & \left.\left. - \left(\boldsymbol{y}_j - \boldsymbol{\mu}_h^{(k+1)}\right) \boldsymbol{e}_{2hj}^{(k)\top} \boldsymbol{\Delta}_h^{(k+1)^\top} + \boldsymbol{\Delta}_h^{(k+1)} \boldsymbol{e}_{3hj}^{(k)\top} \boldsymbol{\Delta}_h^{(k+1)^\top}\right]\right\} \left[\sum_{j=1}^{n} z_{hj}^{(k)}\right]^{-1}.\end{aligned} \tag{14}$$

An update of the degrees of freedom ν_h is obtained by solving the following equation for $\nu_h^{(k+1)}$,

$$0 = \left(\sum_{h=1}^{n} z_{hj}^{(k)}\right)\left[\log\left(\frac{\nu_h^{(k+1)}}{2}\right) - \psi\left(\frac{\nu_h^{(k+1)}}{2}\right) + 1\right] - \sum_{j=1}^{n} z_{hj}^{(k)} \left(e_{1hj}^{(k)} - w_{hj}^{(k)}\right), \tag{15}$$

where $\psi(\cdot)$ denotes the digamma function.

4 Speeding up the fitting of skew mixture models

The EM algorithm described in the previous section is quite computationally intensive. In particular, the evaluation of (9) and (10) relies on routines that involve multiple calculations of the multivariate t-distribution function. The latter is an intractable multidimensional integral that is computationally expensive to evaluate using numerical methods. To reduce the computation time required to perform the EM algorithm in Section 3, we may consider strategies such as finding good starting values, incorporating the trimming approach, and scaling the algorithm to run on multiple core machines.

Generating good starting values

As the log likelihood function for mixture distributions may exhibit a complicated profile with many local maxima and the EM algorithm is sensitive to its initial values, it is important to choose good starting values. Three strategies for generating valid initial values for the EM algorithm for the FM-CFUST model are described in [31]. The first approach is to start the EM algorithm with the solution given by one of the nested models of a CFUST distribution, for example, the results from fitting a normal or t-mixture model. The second approach is based on the moments of an unrestricted multivariate skew normal (uMSN) distribution. It has mean and covariance matrix given by

$$E(\boldsymbol{Y}_j) = \boldsymbol{\mu} + \sqrt{\frac{2}{\pi}}\boldsymbol{\delta} \text{ and } \text{cov}(\boldsymbol{Y}_j) = \boldsymbol{\Sigma} + (1 - \frac{2}{\pi})\boldsymbol{\Delta}^2, \tag{16}$$

respectively. On rearranging the above expressions, an expression for $\boldsymbol{\mu}$ and $\boldsymbol{\Sigma}$ in terms of $\boldsymbol{\delta}$ (note that by definition $\boldsymbol{\Delta} = \text{diag}(\boldsymbol{\delta})$ for the uMSN distribution). To obtain an initial value for $\boldsymbol{\delta}^{(0)}$, we scale the diagonal elements of $\boldsymbol{\Sigma}^{(0)}$ by an arbitrary proportion $(1-a)$ where $a \in [0,1]$. This leads a set of expressions given by

$$\boldsymbol{\delta}^{(0)} = \pm\sqrt{\frac{\pi(1-a)}{\pi-2}}\boldsymbol{s}^*, \quad \boldsymbol{\Sigma}^{(0)} = \boldsymbol{S} + (a-1)\,\mathrm{diag}(\boldsymbol{s}^*), \quad \text{and} \quad \boldsymbol{\mu}^{(0)} = \bar{\boldsymbol{y}} - \sqrt{\frac{2}{\pi}}\boldsymbol{\delta}^{(0)}, \tag{17}$$

where the sign of each element of $\boldsymbol{\delta}^{(0)}$ is given by the sign of the corresponding element of the sample skewness measure. In (17), $\boldsymbol{s}^*$ is a p-dimensional vector containing the diagonal elements of the sample covariance matrix $\boldsymbol{S}$, and $\bar{\boldsymbol{y}}$ denotes the sample mean. The initial degrees of freedom is set (initially) to a large number to reflect a uMSN distribution.

A third approach is based on the transformation $\boldsymbol{X}_j = \boldsymbol{C}\boldsymbol{Y}_j$, where $\boldsymbol{C}$ is an orthogonal matrix such that the covariance matrix of $\boldsymbol{X}_j$, $\mathrm{cov}(\boldsymbol{X}_j)$, is diagonal. A uMST distribution is then fitted to the transformed vectors $\boldsymbol{X}_j$. It follows that an initial value for $\boldsymbol{\mu}$ and for $\boldsymbol{\Delta}$ can be given by $C^\top\hat{\boldsymbol{\mu}}$ and $C^\top\hat{\boldsymbol{\Delta}}$, respectively, where $\hat{\boldsymbol{\mu}}$ and $\hat{\boldsymbol{\Delta}}$ are the estimates of $\boldsymbol{\mu}$ and $\boldsymbol{\Delta}$ obtained by fitting the unrestricted skew t-distribution to the $\boldsymbol{X}_j$.

The above strategies are described for a single component distribution. In the case where a mixture of CFUST distributions is to be fitted, we first cluster the $\boldsymbol{Y}_j$ into g clusters and apply the methods described above separately within each cluster.

Trimming and constrained estimation

Recently, García-Escudero et al. [19] applied the trimming approach [17, 18, 41] and constrained estimation [23] to the FM-CFUSN model. The concept of trimming is to tentatively discard a small proportion of observations in the data that are deemed as least plausible to occur under the (current) estimated model. Note that the observations that are 'trimmed' can be different in each iteration. Although not specifically designed to speed up the EM algorithm, incorporating trimming into the EM iterations is effectively skipping the calculation of the conditional expectations in the E-step for the specified proportion of observations in the data. Hence one can expect that the overall computation time can potentially be reduced by approximately the same proportion.

Implementation for machines with multiple processing units

With the widespread availability of multicore machines, building multithreaded implementations of software to utilize available resources is gaining popularity. Algorithms such as the model fitting procedure described in Section 3 can benefit substantially if they can be reformulated to support machines with multiple cores and processors. We briefly describe one possible approach. It can be observed that the expressions in the E- and M-steps of the EM algorithm have the same form for each component. With the exception of (6), the evaluation of the remaining four conditional expectations in the E-step for component h does not require knowledge about the other $g-1$ components. Similarly, for the M-step, evaluation of (11) to (15) can be performed independently for each component. Hence a straightforward implementation is to schedule these components to run in parallel, that is, each component of the E- and M-steps is to be executed on a separate thread. At the end of each iteration, the results from these individual threads are collected and combined for the calculation of the log likelihood and other measures. Concerning (6), as the denominator is the sum of the numerators across all g components, the latter can be evaluated separately for each component as part of the E-step and then combined later to obtain $z_{hj}^{(k)}$. Technical details of the approach will appear in a forthcoming paper.

5 Clustering of stem cells

We consider a dataset collected by the British Columbia Cancer Agency consisting of measurements of single cells in a blood sample obtained from a patient of a hematopoietic stem cell transplant (HSCT)

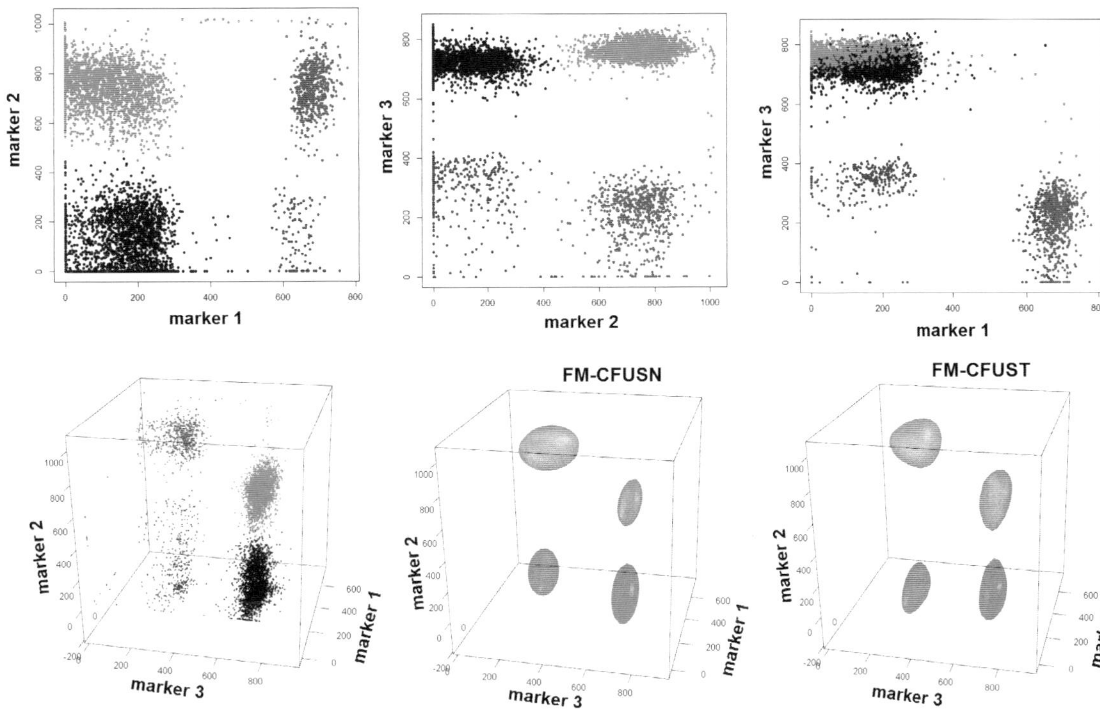

Figure 1. Bivariate contour plots of the fitted FM-CFUST model to the stem cells dataset. Top row: scatter plots of the data, where the colours corresponds to different cell populations. Bottom row: a scatter plot of the data in 3D is shown in the left panel and the density contours of each component of the fitted FM-CFUSN and FM-CFUST models are shown in the middle and right panels, respectively.

experiment. The four variables correspond to four different protein markers used by experts to discriminate between four cell populations in the sample. There were 6,143 cells in the sample, each belonging to one of four distinct cell types. Pairwise plots of the markers for this dataset are displayed in the top row of Figure 1, where the colours of the cells correspond to the labels given by the expert analyst.

The FM-CFUST model was fitted to this dataset using the algorithm described in Section 3. The same model was fitted again using the multicore implementation described in Section 4. For comparison, the FM-CFUSN model was also fitted to the dataset, using the same initialization strategy as for the FM-CFUST model. The contours of the densities of the fitted FM-CFUSN and the FM-CFUST models are depicted in the second row of Figure 1. The FM-CFUSN model attained a misclassification rate (MCR) of 0.00407 for this dataset, whereas the FM-CFUST model reduced it by around 32% (MCR = 0.00277). Concerning the computation time, it was observed that by running the multithreaded implementation of the EM algorithm we achieved a reduction in the total computation time of approximately three hours (equivalent to approximately 60%) on a typical quad-core machine. Note that it is expected that the actual percentage of reduction in computation time would be less than the theoretical speed up of 75%, since the latter ignores the overhead costs associated with the setting up of the multi-threaded implementation. In addition, we consider also the fitting of the FM-CFUSN model using the trimming approach. In this case, it was observed that incorporating a trimming of 10% for the FM-CFUSN model reduces the computation time by around five minutes, which is equivalent to approximately 12% of the total computation time.

6 Conclusions

We have discussed some of the commonly used characterizations of the various versions of the multivariate skew t (MST) distribution in the model-based literature. These different versions of the MST distribution correspond to different constraints placed on the latent skewing variables in the stochastic representation of the MST distribution. In practice, these constraints can affect the flexibility of these skew densities in modelling skewness in arbitrary directions in the feature space. Here the focus is on the CFUST distribution that contains the other restricted versions as special cases and provides a very flexible basis for modelling skewness and long-tailedness.

We have also discussed an EM algorithm for the fitting of the CFUST and CFUSN mixture models, including different strategies for generating starting values and the incorporation of the trimming approach. In addition, we have outlined a multithreaded implementation of the algorithm for use on multicore machines. For the real dataset considered in this paper (with $n = 6,143$, $p = 3$, and $g = 4$), the latter implementation reduced the computation time by approximately 60% compared to the traditional implementation.

Bibliography

[1] Abanto-Valle, C.A., Lachos, V.H. and Dey, D.K. (2015) *Bayesian estimation of a skew-student-t stochastic volatility model.* Methodology and Computing in Applied Probability, **17**, 721–738.

[2] Arellano-Valle, R.B. and Azzalini, A. (2006) *On the unification of families of skew-normal distributions.* Scandinavian Journal of Statistics, **33**, 561–574.

[3] Arellano-Valle, R.B., Branco, M.D. and Genton, M.G. (2006) *A unified view on skewed distributions arising from selections.* The Canadian Journal of Statistics, **34**, 581–601.

[4] Arellano-Valle, R.B. and Genton, M.G. (2005) *On fundamental skew distributions.* Journal of Multivariate Analysis, **96**, 93–116.

[5] Asparouhov, T. and Muthén, B. (2016) *Structural equation models and mixture models with continuous non-normal skewed distributions.* Structural Equation Modeling: A Multidisciplinary Journal, **23**, 1–19.

[6] Azzalini, A. (1985) *A class of distributions which includes the normal ones.* Scandinavian Journal of Statistics, **12**, 171–178.

[7] Azzalini, A. (2005) *The skew-normal distribution and related multivariate families.* Scandinavian Journal of Statistics, **32**, 159–188.

[8] Azzalini, A., Browne, R.P., Genton, M.G. and McNicholas, P. (2016) *On nomenclature for, and the relative merits of, two formulations of skew distributions.* Statistics & Probability Letters, **110**, 201–206.

[9] Azzalini, A. and Capitanio, A. (2003) *Distributions generated by perturbation of symmetry with emphasis on a multivariate skew t distribution.* Journal of the Royal Statistical Society B, **65**, 367–389.

[10] Azzalini, A. and Capitanio, A. (2014) *The Skew-Normal and Related Families*, Institute of Mathematical Statistics Monographs, UK: Cambridge University Press.

[11] Azzalini, A. and Dalla Valle, A. (1996) *The multivariate skew-normal distribution.* Biometrika, **83**, 4, 715–726.

[12] Bernardi, M. (2013) *Risk measures for skew normal mixtures.* Statistics & Probability Letters, **83**, 1819–1824.

[13] Branco, M.D. and Dey, D.K. (2001) *A general class of multivariate skew-elliptical distributions.* Journal of Multivariate Analysis, **79**, 99–113.

[14] Contreras-Reyes, J.E. and Arellano-Valle, R.B. (2013) *Growth estimates of cardinalfish (Epigonus Crassicaudus) based on scale mixtures of skew-normal distributions.* Fisheries Research, **147**, 137–144.

[15] Forbes, F. and Wraith, D. (2014) *A new family of multivariate heavy-tailed distributions with variable marginal amounts of tailweight: application to robust clustering.* Statistics and Computing, **24**, 971–984.

[16] Frühwirth-Schnatter, S. and Pyne, S. (2010) *Bayesian inference for finite mixtures of univariate and multivariate skew-normal and skew-t distributions.* Biostatistics, **11**, 317–336.

[17] Gallegos, M.T. and Ritter, G. (2009) *Trimmed ML estimation of contaminated mixtures.* Sankhya A, 164–220.

[18] García-Escudero, L.A., Gordaliza, A. and Mayo-Iscar, A. (2014) *A constrained robust proposal for mixture modeling avoiding spurious solutions.* Advances in Data Analysis and Classification, **8**, 27–43.

[19] García-Escudero, L.A., Greselin, F. and Mayo-Iscar, A. (2015) *Robust clustering for hetereogenous skew data.* Book of Abstracts of CLADAG 2015.

[20] Gupta, A.K. (2003) *Multivariate skew-t distribution.* Statistics, **37**, 359–363.

[21] Ho, H.J., Lin, T.I., Chang, H.H., Haase, H.B., Huang, S. and Pyne, S. (2012) *Parametric modeling of cellular state transitions as measured with flow cytometry different tissues.* BMC Bioinformatics, **13**, (Suppl 5): S5.

[22] Hu, X., Kim, H., Brennan, P.J., Han, B., Baecher-Allan, C.M., De Jager, P.L., Brenner, M.B. and Raychaudhuri, S. (2013) *Application of user-guided automated cytometric data analysis to large-scale immunoprofiling of invariant natural killer t cells.* Proceedings of the National Academy of Sciences USA, **110**, 19030–19035.

[23] Ingrassia, S. and Rocci, R. (2007) *Constrained monotone EM algorithms for finite mixture of multivariate Gaussians.* Computational Statistics & Data Analysis, **51**, 5339–5351.

[24] Karlis, D. and Santourian, A. (2009) *Model-based clustering with non-elliptically contoured distributions.* Statistics and Computing, **19**, 73–83.

[25] Lachos, V.H., Ghosh, P. and Arellano-Valle, R.B. (2010) *Likelihood based inference for skew normal independent linear mixed models.* Statistica Sinica, **20**, 303–322.

[26] Lee, S. and McLachlan, G.J. (2014) *Finite mixtures of multivariate skew t-distributions: Some recent and new results.* Statistics and Computing, **24**, 181–202.

[27] Lee, S.X. and McLachlan, G.J. (2013) *Model-based clustering and classification with non-normal mixture distributions.* Statistical Methods and Applications, **22**, 427–454.

[28] Lee, S.X. and McLachlan, G.J. (2013) *Modelling asset return using multivariate asymmetric mixture nodels with applications to estimation of Value-at-Risk*, In: *Proceedgins of the 20th International Congress on Modelling and Simulation*, J. Piantadosi, R.S. Anderssen and J. Boland (Eds.). Melbourne: Modelling and Simulation Society of Australia and New Zealand, pp. 1228–1234.

[29] Lee, S.X. and McLachlan, G.J. (2013) *On mixtures of skew-normal and skew t-distributions.* Advances in Data Analysis and Classification, **7**, 241–266.

[30] Lee, S.X. and McLachlan, G.J. (2014) *Maximum likelihood estimation for finite mixtures of canonical fundamental skew t-distributions: the unification of the unrestricted and restricted skew t-mixture models.* arXiv:1401.8182.

[31] Lee, S.X. and McLachlan, G.J. (2015) *EMMIXcskew: an R package for the fitting of a mixture of canonical fundamental skew t-distributions.* arXiv:1509.02069.

[32] Lee, S.X. and McLachlan, G.J. (2016) *Finite mixtures of canonical fundamental skew t-distributions: The unification of the restricted and unrestricted skew t-mixture models.* Statistics and Computing, **26**, 573–589.

[33] Lee, S.X. and McLachlan, G.J. (2016) *Risk measures based on multivariate skew normal and skew t-mixture models.* In *Asymmetric Dependence in Finance*, J. Alcock and S. Satchell (Eds.). Hokoben, New Jersey: Wiley. To appear.

[34] Lee, S.X., McLachlan, G.J. and Pyne, S. (2014) *Supervised classification of flow cytometric samples via the Joint Clustering and Matching (JCM) procedure.* arXiv:1411.2820.

[35] Lee, S.X., McLachlan, G.J. and Pyne, S. (2016) *Modelling of inter-sample variation in flow cytometric data with the Joint Clustering and Matching (JCM) procedure.* Cytometry A, **89**, 30–43.

[36] Leemaqz, S.X. (2014) *Finite Mixture Modelling using Multivariate Skew Distributions*, Ph.D. thesis, School of Mathematics and Physics, University of Queensland, Australia.

[37] Lin, T.I., McLachlan, G.J. and Lee, S.X. (2016) *Extending mixtures of factor models using the restricted multivariate skew-normal distribution.* Journal of Multivariate Analysis, **143**, 398–413.

[38] Lin, T.I., Wu, P.H., McLachlan, G.J. and Lee, S.X. (2015) *A robust factor analysis model using the restricted skew t-distribution.* TEST, **24**, 510–531.

[39] McLachlan, G.J. and Lee, S.X. (2016) *Comment on "On nomenclature for, and the relative merits of, two formulations of skew distributions" by A. Azzalini, R. Browne, M. Genton, and P. McNicholas.* Statistics & Probability Letters, **116**, 1–5.

[40] Muthén, B. and Asparouhov, T. (2014) *Growth mixture modeling with non-normal distributions.* Statistics and Medicine, **34**, 1041–1058.

[41] Neykov, N., Filzmoser, P., Dimova, R. and Neytchev, P. (2007) *Robust fitting of mixtures using the trimmed likelihood estimator.* Computational Statistics & Data Analysis, **52**, 299–308.

[42] Pyne, S., Hu, X., Wang, K., Rossin, E., Lin, T.I., Maier, L.M., Baecher-Allan, C., McLachlan, G.J., Tamayo, P., Hafler, D.A., De Jager, P.L. and Mesirow, J.P. (2009) *Automated high-dimensional flow cytometric data analysis.* Proceedings of the National Academy of Sciences USA, **106**, 8519–8524.

[43] Pyne, S., Lee, S. and McLachlan, G. (2015) *Nature and man: The goal of bio-security in the course of rapid and inevitable human development.* Journal of the Indian Society of Agricultural Statistics, **69**, 117–125.

[44] Pyne, S., Lee, S.X., Wang, K., Irish, J., Tamayo, P., Nazaire, M.D., Duong, .T., Ng, S.K., Hafler, D., Levy, R., Nolan, G.P., Mesirov, J. and McLachlan, G. (2014) *Joint modeling and registration of cell populations in cohorts of high-dimensional flow cytometric data.* PLOS ONE, **9**, 10.1371/journal.pone.0100334.

[45] Riggi, S. and Ingrassia, S. (2013) *A model-based clustering approach for mass composition analysis of high energy cosmic rays.* Astroparticle Physics, **48**, 86–96.

[46] Rossin, E., Lin, T.I., Ho, H.J., Mentzer, S.J. and Pyne, S. (2011) *A framework for analytical characterization of monoclonal antibodies based on reactivity profiles in different tissues.* Bioinformatics, **27**, 2746–2753.

[47] Schaarschmidt, F., Hofmann, M., Jaki, T., Grün, B. and Hothorn, L.A. (2015) *Statistical approaches for the determination of cut points in anti-drug antibody bioassays.* Journal of Immunological Methods, **25**, 295–306.

[48] Soltyk, S. and Gupta, R. (2011) *Application of the multivariate skew normal mixture model with the EM algorithm to Value-at-Risk.* In: *Proceedings of the 19th International Congress on Modelling and Simulation*, F. Chan, D. Marinova and R.S. Anderssen (Eds.). Perth: Modelling and Simulation Society of Australia and New Zealand, pp. 1638–1644.

[49] Vrbik, I. and McNicholas, P.D. (2014) *Parsimonious skew mixture models for model-based clustering and classification.* Computational Statistics & Data Analysis, **71**, 196–210.

[50] Wraith, D. and Forbes, F. (2015) *Location and scale mixtures of gaussians with flexible tail behaviour: Properties, inference and application to multivariate clustering.* Computational Statistics & Data Analysis, **90**, 61–73.

A modified control chart for monitoring the multihead weighing process

Alexander Pulido-Rojano, *Industrial Engineering Department, Universidad Simon Bolivar, Av. 59 No. 59-92. A.A. 50595, Barranquilla, Colombia*, apulido3@unisimonbolivar.edu.co
J. Carlos García-Díaz, *Centre for Quality and Change Management, Universitat Politecnica de Valencia,Camino de Vera, s/n. 46022, Valencia, Spain*, juagardi@eio.upv.es

Abstract. The modified control charts are used for monitoring and control of the manufacturing processes which are considered as six-sigma process, ensuring a probability of out-of-specification product acceptably small. The use of these charts is based on the idea that the cost of identifying and correcting special causes is much higher than the cost of off-target products. Therefore, the process mean is essentially acceptable as long as it is anywhere within the specification limits. These concepts have been applied to the packaging process in multihead weighers. The weight of the packed product, seen as the quality characteristic to be monitored, must be as close to a specified target weight and comply with applicable regulations. In order to design the modified control chart and comply with requirements for its implementation, the packaging process has been previously optimised and improved through a packaging strategy. The strategy seeks to reduce the variability in the selection of the total weight of the package and it is evaluated through a proposed packing algorithm. In this way, a set of numerical experiments were conducted to examine the solutions generated and which are subsequently monitored.

Keywords. Quality control, Modified control chart, Reduction of variability, Combinatorial optimization, Packaging process, Multihead weighing process.

1 Introduction

The usual control charts use the three-sigma control limits to monitor the quality characteristics of a process. These charts are designed to distinguish between common and special causes of variation. However, when the process capability (C_p) is equal to or greater than 2 and the cost of identifying and correcting special causes is very large the three-sigma control limits are uneconomical. In these cases, the use of modified control charts are a good option for the monitoring of process. The modified control limits of a modified control chart allow to the process mean to vary over an interval, as far as 1.5 times its standard deviation (σ_p) from the desired target while ensuring that no out-of-specification product is produced [2, 9].

The interval at which the process mean can vary usually is represented by μ_L and μ_U, which are the smallest and largest permissible values of the mean, respectively. This ensures that the fraction nonconforming will be less than a threshold known as δ. Therefore, to establish this interval, the δ value must be established in advance, normally the lowest possible.

In this paper, we have designed a modified $\overline{X}$ chart to the multihead weighing process. This process is a packaging process in which machines of high technology, known as multihead weighers, are used. Prior to the application of the modified control limits, the packing process has been optimised and improved through a packaging strategy seeking to reduce the variability in the weight of the packed product and at the same time, the increase of the process capability (see later sections 3 and 4).

To specify the modified control limits of a modified $\overline{X}$ chart, it will be assumed that the packing process output is normally distributed (as we will see in the section 2). The modified limits are established taking as input the μ_L and μ_U values. Furthermore, due to their conceptual basis, these limits are wider than the usual three-sigma limits. For the calculation of μ_L and μ_U both the specification limits and the standard deviation of the process are considered. As shown below:

$$\mu_L = LSL + Z_\delta \sigma_p \tag{1}$$

$$\mu_U = USL - Z_\delta \sigma_p \tag{2}$$

Where LSL and USL are the lower and upper specification limits and Z_δ is the upper 100(1 - δ) percentage point of the standard normal distribution [4, 9]. In this way, the lower and upper control limits can be calculated by using the following equations:

$$LCL = \mu_L - \frac{Z_\alpha \sigma_p}{\sqrt{N}} = LSL + \left(Z_\delta - \frac{Z_\alpha}{\sqrt{N}} \right) \sigma_p \tag{3}$$

$$UCL = \mu_U + \frac{Z_\alpha \sigma_p}{\sqrt{N}} = USL - \left(Z_\delta - \frac{Z_\alpha}{\sqrt{N}} \right) \sigma_p \tag{4}$$

Notice that the modified control limits are designed to monitor whether the process mean is between μ_L and μ_U.

The present document is structured in the following way: In section 2 the multihead weighing process is introduced. In Section 3 the proposed approach of optimization is presented. Section 4 shows the results and analysis of the numerical experiments. Section 5 offers the conclusions of this work.

2 Multihead weighing process

To illustrate the packing process, in this section we will present both the components of multihead weigher and a brief explanation of the multihead packing process.

The multihead weigher

Multihead weighers or combinational weighers are used to provide accurate weights at high packing speed and are currently the most used dosing method for many kinds of products, also including those with heterogeneous characteristics [8]. The control of the actual content of the packaged is regulated by the directive 76/211/EC and must be implemented by factories, plants of packaging and importers. The regulations state that consumers be informed about the quantity and be protected against short measures, while allowing businesses the flexibility to control quantity on the production line within specific tolerances [10].

The combinational weighers, designed in the mid 1980s, uses combination weighing techniques to achieve dispensed weighments that are closer to the desired target weight than with conventional weighing techniques. They have a number of weighing heads that statically weigh the product; these weight data are fed to a computer, which calculates all of the possible combinations of product weights in order to dispense the best combination (closest match to target weight) to a packaging machine.

The weighing system consists of three elements, namely: A system to automate product feed to the weighing stations (depending on the layout of the machine, the feed system is configured either in a radial or in line construction), a system to collect product and feed it into a weighing hopper (this system consists of a set of hoppers which commonly known as feed hoppers) and a set of weighing hoppers. As stated above, the weight data contained in the weighing hoppers is fed to an electronic system to combine the data from the other weighing hoppers on the machine to determine which hoppers should be discharged to the downstream process (weight hoppers combination) [7, 11]. Figure 1 presents the schematic of the basic combinational weigher components.

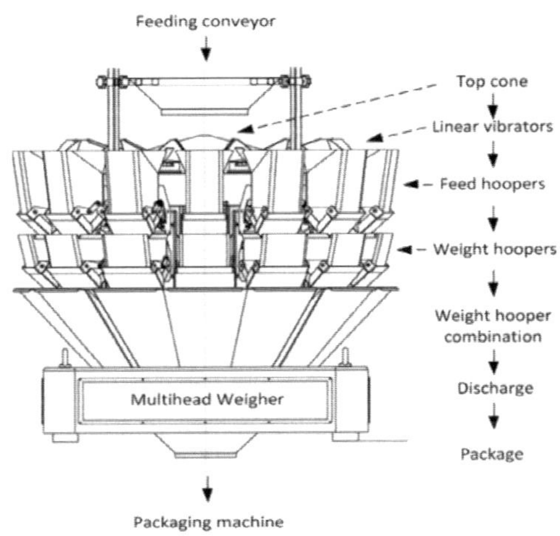

Figure 1. Arrangement of feeders and hoppers of a radial multihead weigher.

Packing process

The multihead packing process starts when a quantity of food is placed into each weighing hopper i, ($i = 1, 2, \ldots, n$) [6] and the weight signal is transmitted to the built-in computer. The computer calculates the combinations of weights that come closest to the desired weight T, and the combination of the closest weights is ejected from the corresponding hoppers. The resulting empty hoppers are supplied with new quantities of food. The computer continuously repeats this process until it obtains the number of packages (Q) needed one by one. The goal is to choose a subset H' from the set H of the current n weighing hoppers to produce a goods package. It needs to be mentioned that the number of possible different hopper subsets H' depends on the number k of hoppers to be combined each packing operation. This causes the optimization problem that focuses on minimizing the difference between the actual and the target package weight is seen as a NP-complete subset-sum combinatorial problem [3] when k is neither previously fixed nor constant [5]. This paper deals with the case in which the number of hoppers k to be combined at each packing operation is constant and fixed in advance. If we assume that the weights X_i in the hoppers follow a normal probability distribution and all the hoppers were independently filled according to the same distribution N (μ,σ) and the k hoppers were randomly selected in each packing operation, then the weight of packages would be known to follow a normal distribution N ($k\mu$, $\sqrt{k}\sigma$) where the average package mean weight $k\mu$ is expected to equal the target T. The value of $\sqrt{k}\sigma$ (the standard deviation if hoppers were selected at random) is considered to be an index of quality in the packaging process. However, the subset of hoppers to be discharged H' is actually not selected at random but in a driven way, usually such that the total weight $W = \sum_{i \epsilon H'} X_i$ is as close as T as possible. Therefore,

$\sigma^2_{package} = VAR\left(\sum_{i \epsilon H'} X_i\right)$ [1, 12].

3 The approach of optimization

The proposed approach of optimization include both the packaging strategy and algorithm. The strategy is based on study the filling setting of weighing hoppers while the algorithm seeks to evaluates the strategy through of a procedure to carry out the packing operation.

Packaging strategy

As has been mentioned in section 2, in this work is assumed that the number of hoppers k to be combined at each packing operation is constant and fixed in advance. This cause that the supply of product to the weighing hoppers must be set at $\mu = T/k$. However, the packaging strategy proposed considers the general case in which each hopper i is expected to be filled with a different average quantity of food (instead of a common value μ). In this paper we will explore the case in which several hoppers weights are set in such a way that share the same average quantity of food, as this has been proven to be an efficient strategy to reduce package variability [1, 8].
In our particular case, we propose to divide the n weighing hoppers into five subgroups (n_1, n_2, n_3, n_4 and n_5 with $n = \sum_{j=1}^{5} n_j$) and provide with an unequal amount of product to each subgroup, on average (μ_1, μ_2, μ_3, μ_4 and μ_5, respectively). During the packing operation the filling setting is establish in the following way: $\mu_1 = \mu - 1.5\sigma$, $\mu_2 = \mu - 1\sigma$, $\mu_3 = \mu$, $\mu_4 = \mu + 1\sigma$ and $\mu_5 = \mu + 1.5\sigma$. As an example in the calculation of the μ_j values, suppose T = 1000 gr., k = 2, σ = 7.07 gr. In these conditions the μ_j values would be: μ_1 = 500 - 1.5(7.07) = 489.39 gr., μ_2 = 500 - 1(7.07) = 492.93 gr., μ_3 = 1000/2 = 500 gr., μ_4 = 500 + 1(7.07) = 507.07 gr. and μ_5 = 500 + 1.5(7.07) = 510.61 gr. In this way some hoppers would share the same value for μ_j depending on which subgroup (n_1, n_2, n_3, n_4 and n_5) the hopper belongs to.

Packing algorithm

The procedure proposed to carry out the packing operation is explained in this section. This enumerative procedure is made for each packed product and can be implemented in software systems installed in control unit of the multihead weigher. The packing algorithm consists in 4 steps, namely:

- Step 1. Feed the n weighing hoppers according their respective setting (μ_1, μ_2, μ_3, μ_4 and μ_5).
- Step 2. The weights in hoppers are used to calculate the k weight combinations say, $C_t = \frac{n!}{k!(n-k)!}$ combinations. The closest one to target weight (T) is chosen if is within a confidence level (1 - α) of 99.73%, i.e., $T \pm Z_{\alpha/2}\sqrt{k}\sigma$. Where $Z_{\alpha/2}$ represents the critical value of the standard normal probability distribution N (0, 1) for a significance level α. Then, the optimal combination is packed and we go to step 4.
- Step 3. If all the total weights (as result of all the combinations C_t of k hoppers) are outside the confidence level all hoppers are discharged. The hoppers are supplied with new weights according their respective setting and we go back to step 2.
- Step 4. If the required total number of packages (Q) is not completed, then the empty hoppers are supplied with the next new weights according their respective setting and we go back to step 2. Otherwise, the packing process ends.

The step 3 of the algorithm describes a situation in which all hoppers should be discharged in order to avoid producing packages that would not meet the quality requirements for the final product in terms of weight. When that happens, all this discharged product could be taken and reused in the process again, for instance. Figure 2 shows the flowchart for the proposed packing procedure.

The packing algorithm was implemented in Pascal and run on a personal computer with Windows 7 Home Premium (64bit), Intel Core i5-3317U CPU (1.7 GHz) and 4 GB memory. Figure 3 shows the user interface of this prototype for the case in which the number of hoppers k to be combined is equal

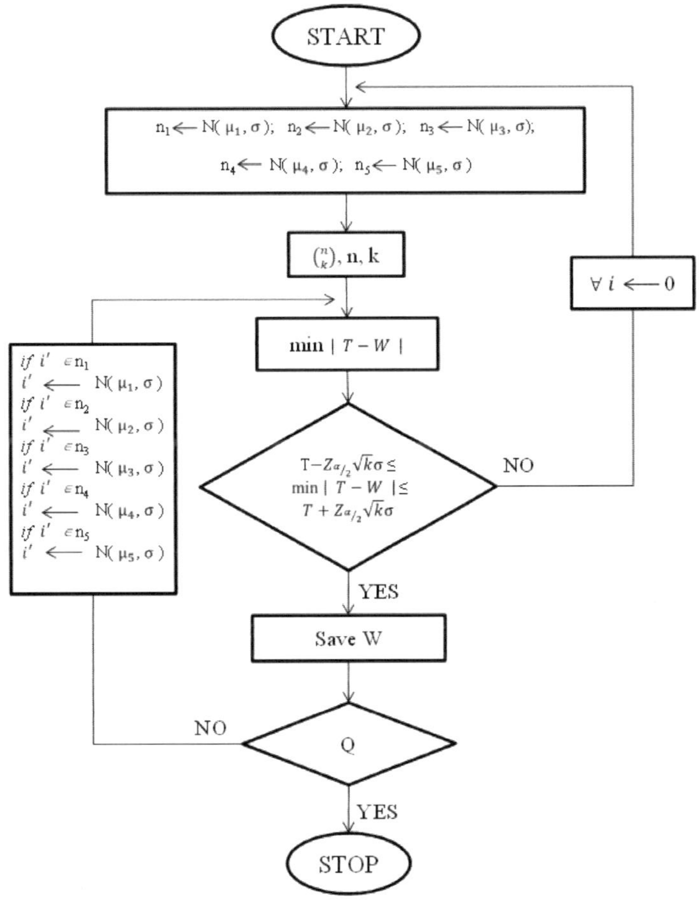

Figure 2. Flowchart for the packing procedure

to 5. The software outputs are: The proportion of iterations each hopper has been used, the average weight of the number total of packages produced ($\mu_{package}$), the standard deviation of the number total of packages produced ($\sigma_{package}$) and percentage of discharge for confidence level (DCL). The results are shown in Table 1.

4 Results and analysis

The packing strategy has been evaluated through the packing algorithm for k values between 2 and 7. The numerical experiments were realized for a target weight T = 1000 gr and a number total of hoppers $n = 8$, with $n_1 = 1$, $n_2 = 2$, $n_3 = 2$, $n_4 = 2$ and $n_5 = 1$. Besides, we use the expected coefficient of variation (CV) of the final package – say, if the hoppers were selected at random – for calculation of the standard deviation of weights in every hopper (σ) as an input in the packaging process. e.g., if $CV = (\sqrt{k}\sigma)/T \cdot 100 = 1\%$, $T = 1000$ gr and $k = 2$. Theoretically we have to $\sqrt{k}\sigma = 10$ gr and therefore $\sigma = 7.07$ gr. However, as explained in subsection 2, does not mean that $\sqrt{k}\sigma$ will be the actual variability obtained in the package produced through our packing strategy. During the simulation 10000 packages were produced in the packing operations for each k value. The directive 76/211/EEC states that the

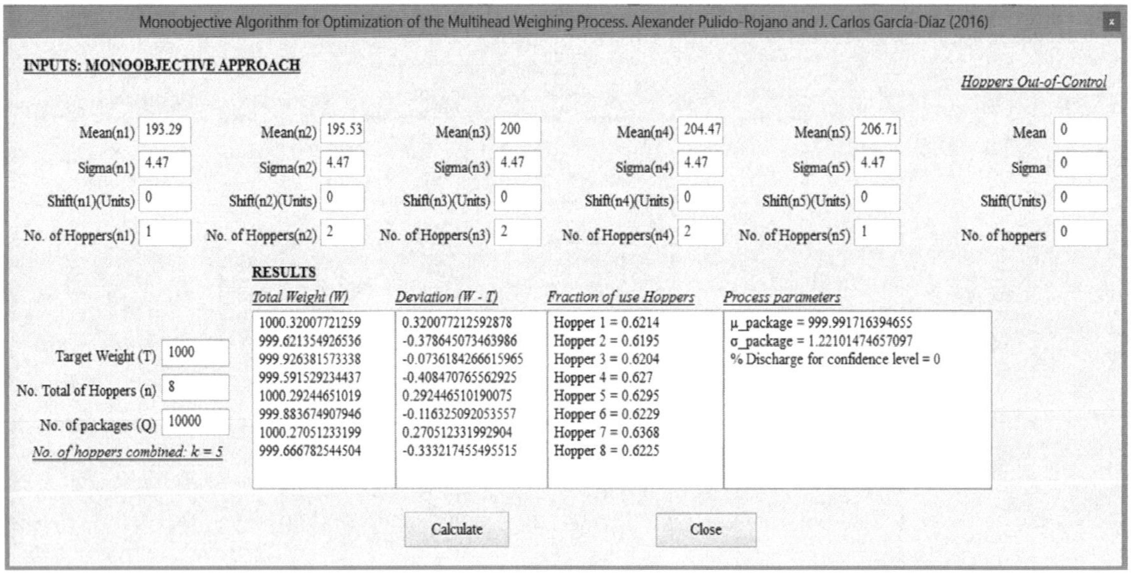

Figure 3. User interface of the prototype software developed

maximum permissible error to a $T = 1000$ gr is 15 gr. Therefore, from the point of view of the products and the consumers, the lower and upper specification limits would be 985 gr and 1015 gr, respectively. Table 1 presents the results of the average weight of the number total of packages produced ($\mu_{package}$), the standard deviation of the number total of packages produced ($\sigma_{package}$), the process capability index (C_p) and percentage of discharge for confidence level (DCL) for CV values of 1%, 2.5% and 5%.

The results show that with the use of the packaging strategy the mean process is not affected by an increase in the CV values. Note that no discharges of product from hoppers were presented; ensuring that at least one of the total weights, as a result of all combinations, was within the confidence level for each packing operation. Also it is observed that the $\sigma_{package}$ values increase when the expected coefficient of variation of the final weight (CV) also increases, as expected. The numerical experiments show that the lowest variability and the highest values of C_p are achieved when three weighing hoppers are combined for a CV value of 1%. Furthermore, it is interesting to see that combinations of three and four hoppers don't have the same effect on $\sigma_{package}$ when CV = 1%, even though combinations of four hoppers result in the highest total number of combinations to choose from. The results confirm that the packing process can be considered as a process with six-sigma capacity if three or four hoppers are combined and the CV takes values of 1% and 2.5%. Based on the above analysis, it is proceed to calculate the modified control limits for a value of $k = 3$ and a CV = 1%. This is the optimum operating condition which allows to minimize the variability in the weight of the package. We used eqs. (1), (2), (3) and (4) with Z_δ and Z_α replaced by 3.7 and 3.0, respectively. In this particular case, the Z_δ value corresponds to a fraction nonconforming of at most $\delta = 0.0001$. The values of μ_L and μ_U have been obtained assuming that the mean may drift as much as $1.5\sigma_{package}$ from target weight, as already mentioned. Furthermore, it is important to note that σ_p is replaced by $\sigma_{package}$. All these values are presented in Table 2 for a sample size of $N = 1$.

Figure 4 is the modified control chart in the monitoring of the total weight (W) for 200, 500, 1000 and 5000 packed products. Graphs show the process behaviour after establishing the modified control limits through the standard deviation ($\sigma_{package}$) estimated by simulation of our packing algorithm. In this regard, the process does not present an untypical behaviour and therefore the modified control chart is established for monitoring the packing process.

Inputs					Outputs			
$\sqrt{k}\sigma$	k	C_t	μ	σ	$\mu_{package}$	$\sigma_{package}$	DCL	C_p
10	2	28	500.00	7.07	999.99	2.34	0.00	2.13
	3	56	333.33	5.77	999.99	0.74	0.00	6.79
	4	70	250.00	5.00	999.99	0.77	0.00	6.52
	5	56	200.00	4.47	999.95	1.24	0.00	4.05
	6	28	166.67	4.08	1000.00	2.41	0.00	2.08
	7	8	142.86	3.78	999.92	4.71	0.00	1.06
25	2	28	500.00	17.68	1000.08	6.08	0.00	0.82
	3	56	333.33	14.43	1000.02	1.88	0.00	2.66
	4	70	250.00	12.50	1000.03	1.65	0.00	3.02
	5	56	200.00	11.18	1000.03	3.06	0.00	1.64
	6	28	166.67	10.21	1000.06	5.86	0.00	0.85
	7	8	142.86	9.45	999.73	11.76	0.00	0.43
50	2	28	500.00	35.36	999.97	11.96	0.00	0.42
	3	56	333.33	28.87	1000.05	3.76	0.00	1.33
	4	70	250.00	25.00	1000.01	3.60	0.00	1.39
	5	56	200.00	22.36	999.93	6.53	0.00	0.77
	6	28	167.67	20.41	999.90	12.38	0.00	0.40
	7	8	142.86	18.90	999.87	23.70	0.00	0.21

Table 1. Simulation results from the packing algorithm for different values of the number of hoppers to be combined and the expected coefficient of variation of the final package.

CV(%)	$\sqrt{k}\sigma$	k	μ_L	μ_U	LSL	USL	LCL	UCL
1	10	3	998.90	1001.10	985.00	1015.00	985.53	1014.47

Table 2. Parameters of the modified $\overline{X}$ chart for the multihead weighing process

5 Conclusions

We have designed a modified $\overline{X}$ chart for monitoring and control of the multihead weighing process. The modified control limits have been established to ensure a fraction nonconforming of at most 0.0001. Prior to this, the process was optimised and improved through a packing strategy which seeks to reduce the variability in the selection of the total weight of the package with respect to a desired target weight. The strategy has been evaluated through a proposed packing algorithm which simulated the packing process for different values of the number of hoppers combined in a multihead weigher with eight weighing hoppers. In this sense, exact algorithms were developed to evaluate the several values of the number of combined hoppers. The results indicate that both the packaging strategy and algorithm can resolve the packing problem in an efficient way, to the point where the process can be considered as a process with six-sigma capacity. It was concluded that combinations of three weighing hoppers reduce the variability in the package when the expected coefficient of variation of the final weight is minimal, even though combinations of three hoppers not result in the highest total number of combinations. We recommended, for future research, make deeper studies for determine the relationship between the number of combined hoppers and the total number of hoppers in the multihead weigher.

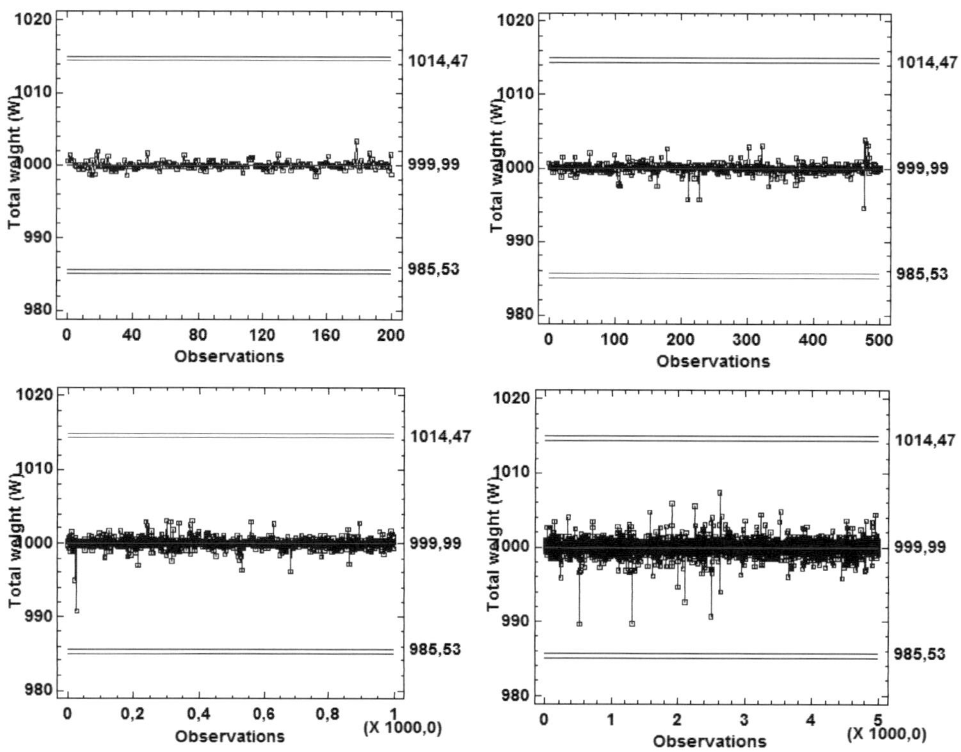

Figure 4. Modified $\overline{X}$ chart for the multihead weighing process simulating different values for the number of needed packages.

Bibliography

[1] Barreiro, J.J., González, C. and Salicrú, M. (1998) Optimization of Multiweighing Packing Proceeding. Top, 6(1), 37-44.

[2] Duncan, A.J. (1986) Quality control and industrial statistics. Illinois: Irwin.

[3] Garey, M.R. and Johnson, D.S. (1979) Computers and Intractability: A guide to the Theory of NP-Completeness. New York: WH Freeman and Company.

[4] Grant, E.L. and Leavenworth, R. (1996) Statistical Quality Control. New York: McGraw-Hill.

[5] Imahori, S., Karuno, Y., Nagamochi, H. and Wang, X. (2011) Kansei engineering humans and computers: Efficient dynamic programming algorithms for combinatorial food packing problems. International Journal of Biometrics, 3(3), 228-245.

[6] Karuno, Y., Nagamochi, H. and Wang, X. (2007) Bi-criteria food packing by dynamic programming. Journal of the Operations Research Society of Japan, 50(8), 376-389.

[7] Keraita, J.N. and Kim, K.H. (2006) A Study on the optimum scheme for Determination of Operation time of Line Feeders in Automatic Combination Weighers. Journal of Mechanical Science and Technology, 20(10), 1567-1575.

[8] Keraita, J.N. and Kim, K.H. (2007) A Weighing Algorithm for Multihead Weighers. International Journal of Precision Engineering and Manufacturing, 8(1), 21-26.

[9] Montgomery, D.C. (2009) Statistical Quality Control. New York: John Wiley & Sons.

[10] Pulido-Rojano, A. and García-Díaz, J.C. (2014) Optimization of multihead weighing process using the Taguchi loss function. Proceedings of IIE International 8th International Conference on Industrial Engineering and Industrial Management and XX International Conference on Industrial Engineering and Operations Management, Málaga, Spain, 305-312.

[11] Pulido-Rojano, A., García-Díaz, J.C. and Giner-Bosch, V. (2015) A multiobjective approach for optimization of the multihead weighing process. Proceedings of the International Conference on Industrial Engineering and Systems Management, Seville, Spain, 426-434.

[12] Salicrú, M., González, C. and Barreiro, J.J. (1996) Variability Reduction with Multiweighing Proceedings. Top, 4(2), 319-329.

On multivariate extensions of the Mixed Tempered Stable distribution

Asmerilda Hitaj, *University of Milano-Bicocca*, asmerilda.hitaj1@unimib.it
Friedrich Hubalek, *Vienna University of Technology*, fhubalek@fam.tuwien.ac.at
Lorenzo Mercuri, *University of Milan*, lorenzo.mercuri@unimi.it
Edit Rroji, *University of Trieste*, erroji@units.it.

Abstract. We consider a generalization of Normal Variance Mean Mixtures and name it multivariate Mixed Tempered Stable distribution. Properties of this distribution and its capacity in capturing fat tails are discussed supported by simulation analysis. We point out that this distribution is suitable in reproducing stylized facts and different dependence structures of asset returns.

Keywords. Mixed Tempered Stable, Infinitely divisible, Heavy tails, Dependence structure

1 Introduction

The Mixed Tempered Stable (MixedTS from now on) distribution has been introduced in [9] and used for portfolio selection in [4]. It is a generalization of the Normal Variance Mean Mixtures [1] since the structure is similar but its definition generates a dependence of higher moments on the parameters of the Tempered Stable [7] that replaces the Normal distribution.
In these notes we present the multivariate MixedTS distribution and discuss its main features. The dependence between components in the mixing random variable controls the dependence structure in the new distribution. In literature, a similar approach has been used by Semeraro in [11] for the construction of the multivariate Variance Gamma distribution. However, the Semeraro model that considers as mixing random variable a Gamma distribution, i.e. semi-heavy tailed, seems to be too restrictive for describing the joint distribution of asset returns as observed in [3]. In particular, the signs of the skewness of the marginal distributions determine the sign of the covariance. We show that we overcome these limitations due to the presence of the additional parameters coming from the Tempered Stable random variable in the multivariate definition of the MixedTS.
We review the main results of the univariate MixedTS and extend them in the multivariate context. Analytical formulas for higher moments and a simulation study are the backbones of the comparison of our approach with the Semeraro model.

2 Univariate Mixed Tempered Stable

Let us recall the definition of the univariate Mixed Tempered Stable distribution.

Definition 2.1. *A random variable (r.v.) Y is Mixed Tempered Stable distributed if:*

$$Y = \mu_0 + \mu V + \sqrt{V} X \tag{1}$$

where, conditioned on the positive r.v. V, the r.v. X follows a classical Tempered Stable distribution with parameters $\left(\alpha, \lambda_+\sqrt{V}, \lambda_-\sqrt{V}\right)$ i.e.:

$$X|V \sim stdCTS(\alpha, \lambda_+\sqrt{V}, \lambda_-\sqrt{V}) \tag{2}$$

with parameters μ_0, $\mu \in \Re$, $\alpha \in (0,2]$ and $\lambda_+, \lambda_- \in \Re^+$.

It is possible to calculate the first four moments of this distribution, which are reported in the following Proposition.

Proposition 2.1.
Under the assumption that $E\left[V^4\right] < \infty$, the first four moments of the MixedTS are:

$$\begin{cases} E[Y] = \mu_0 + \mu E[V] \\ Var[Y] = \mu^2 Var(V) + E[V] \\ m_3(Y) = \mu^3 m_3(V) + 3\mu Var(V) + (2-\alpha)\frac{\left(\lambda_+^{\alpha-3} - \lambda_-^{\alpha-3}\right)}{\left(\lambda_+^{\alpha-2} + \lambda_-^{\alpha-2}\right)} E[V] \\ m_4(Y) = \mu^4 m_4(V) + 6\mu^2 E\left[(V - E(V))^2 V\right] + 4\mu(2-\alpha)\frac{\lambda_+^{\alpha-3} - \lambda_-^{\alpha-3}}{\lambda_+^{\alpha-2} + \lambda_-^{\alpha-2}} Var(V) \\ +(3-\alpha)(2-\alpha)\frac{\left(\lambda_+^{\alpha-4} + \lambda_-^{\alpha-4}\right)}{\left(\lambda_+^{\alpha-2} + \lambda_-^{\alpha-2}\right)} E[V], \end{cases} \tag{3}$$

where m_3 and m_4 are respectively the third and fourth central moments.

We observe that m_3 and m_4 depend on the mixing r.v. V and the tempering parameters λ_- and λ_+. Indeed we can have asymmetric distributions even if $\mu = 0$. It is worth to note that parameters μ_0 and μ may have an economic interpretation. In particular, μ_0 can be thought to represent the risk free rate and μ the risk premium for each unit of the variance process V. Through Normal Variance Mean Mixtures is not possible to have negatively skewed distributions with $\mu > 0$. From an economic point of view this means that it is not possible to have a positive risk premium for each unit of variance when the asset distribution is negatively skewed as usually observed in the market.

Proposition 2.2.
The characteristic function of the MixedTS, obtained by applying the law of iterated expectation, is:

$$\begin{aligned} E\left[e^{iuY}\right] &= E\left[E\left[e^{iu\left(\mu_0 + \mu V + \sqrt{V}X\right)}\middle| V\right]\right] \\ &= e^{iu\mu_0 + \Phi_V(iu\mu + L_{stdCTS}(u;\ \alpha,\ \lambda_+,\ \lambda_-))}, \end{aligned} \tag{4}$$

where the $L_{stdCTS}(u; \alpha, \lambda_+, \lambda_-)$ is the characteristic exponent of a Classical Tempered Stable r.v. defined as:

$$L_{stdCTS}(u;\ \alpha,\ \lambda_+,\ \lambda_-) = \frac{(\lambda_+ - iu)^\alpha - \lambda_+^\alpha + (\lambda_- + iu)^\alpha - \lambda_-^\alpha}{\alpha(\alpha-1)\left(\lambda_+^{\alpha-2} + \lambda_-^{\alpha-2}\right)} + \frac{iu\left(\lambda_+^{\alpha-1} - \lambda_-^{\alpha-1}\right)}{(\alpha-1)\left(\lambda_+^{\alpha-2} + \lambda_-^{\alpha-2}\right)}.$$

The MixedTS has as special cases some well known distributions used in literature for modeling financial return time series. In particular, if we assume that $V \sim \Gamma(a, \sigma^2)$ we get the Variance Gamma introduced in [8] for $\alpha = 2$ and the Standardized Classical Tempered Stable (see [5]) if we add the constraint $\sigma = \frac{1}{\sqrt{a}}$ and let a go to infinity. By choosing:

$$\begin{aligned} \lambda_+ &= \lambda_- = \lambda \\ a &= 1 \\ \sigma &= \lambda^{\frac{\alpha-2}{2}} \gamma^{\frac{\alpha}{2}} \sqrt{\left| \frac{\alpha(\alpha-1)}{\cos\left(\alpha\frac{\pi}{2}\right)} \right|} \end{aligned} \tag{5}$$

and computing the limit for $\lambda \to 0^+$ we obtain the Geometric Stable distribution (see [6]).

It is worth noting that, if V is gamma distributed, the characteristic function in (4) identifies the distribution associated to a time-changed Lévy process [2]. The time-changed Lévy process gives rise to an infinitely divisible distribution (see, theorems 7.10 and 30.1 in [10]). Using the infinite divisibility property we have that:

$$\{\Phi(Y)\}^t = e^{iu\mu_0 t + \phi_{V_t}(iu\mu + L_{stdCTS}(u,\alpha,\lambda_+,\lambda_-))} = \Phi(Y_t) \tag{6}$$

where $V_t \sim \Gamma(at, \sigma^2)$ and Y_t is a MixedTS$(\mu_0 t, \mu, \sigma^2, at, \alpha, \lambda_+, \lambda_-)$ from where we are able to introduce a MixedTS Lévy process defined as follows:

Definition 2.2. *Let* $\left(\Omega, \mathcal{F}, \{\mathcal{F}_t\}_{t\geq}, P\right)$ *be a filtered probability space, we define a MixedTS Lévy process* $(Y_t)_{t\geq 0}$ *such that:*

- $Y_0 = 0$.
- *The increments are independent and stationary.*
- $Y_{t-s} := Y_t - Y_s$ *is MixedTS distributed with parameters* $\left(\mu_0(t-s), \mu, \sigma^2, a(t-s), \lambda_+, \lambda_-\right)$.

In Figures 1 and 2 we plot the sample paths of a MixedTS process for fixed values of parameter α and different combinations of tempering parameters λ_+ and λ_-. We observe from Figure 2 that λ_+ and λ_- control the asymmetry in the distribution of the increments. For $\lambda_+ > \lambda_-$ the distribution of the increments is negatively skewed while, for $\lambda_+ < \lambda_-$ is positively skewed.

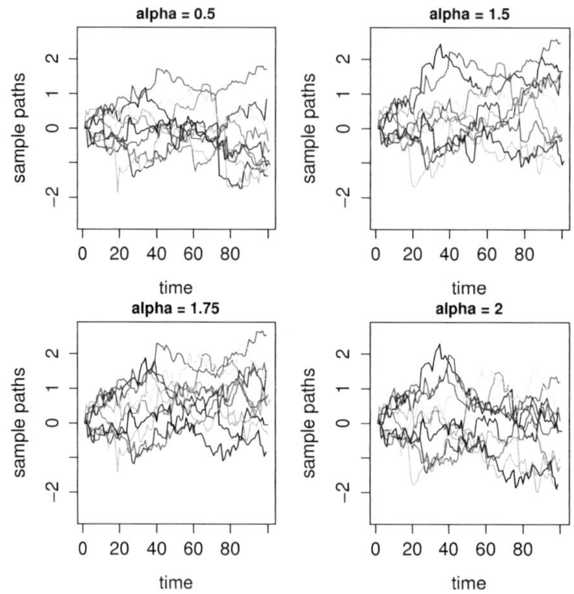

Figure 1. Sample paths of MixedTS process with parameters $\mu_0 = 0$, $\mu = 0$, $\sigma = 0.2$, $\lambda_+ = 1$, $\lambda_- = 1$ and varying α

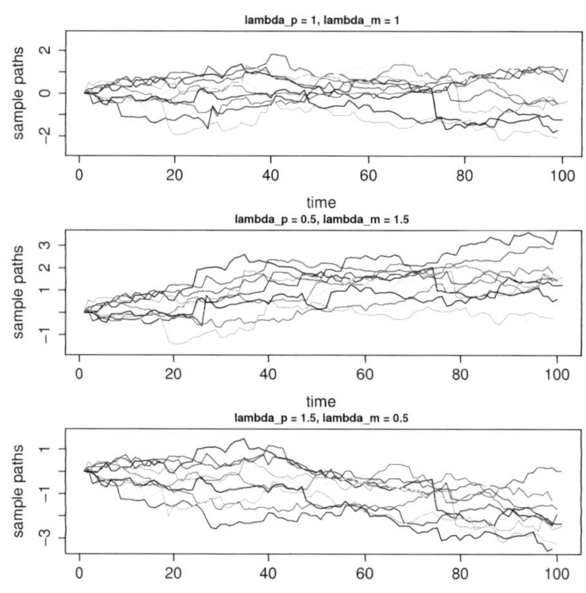

Figure 2. Sample paths of MixedTS process with parameters $\mu_0 = 0$, $\mu = 0$, $\sigma = 0.2$, $\alpha = 1.5$ and varying λ_+ and λ_-

3 Multivariate Mixed Tempered Stable distribution; definition and properties

In this section we define the multivariate MixedTS and discuss its properties.

Definition 3.1. *A random vector $Y \in \mathcal{R}^N$ follows a multivariate MixedTS distribution if its ith component is defined as:*

$$Y_i = \mu_{0,i} + \mu_i V_i + \sqrt{V_i} X_i, \tag{7}$$

where $X_i|V_i \sim stdCTS(\alpha_i, \lambda_{+,i}\sqrt{V_i}, \lambda_{-,i}\sqrt{V_i})$ and V_i is the ith component of a multivariate random vector V, defined as:

$$V_i = G_i + a_i\ Z,\ \ a_i \geq 0. \tag{8}$$

G_i and Z are infinitely divisible random variables with positive support, where $\{G_i\}_{i=1}^N$ and Z independent.

From (8) we have that the distribution V_i is infinitely divisible and the components in (7) are MixedTS distributed. In this short note, Y_i is assumed to be MixedTS distributed with Gamma mixing density. For this reason we need either $G_i \sim \Gamma(l_i, m_i)$ and $a_i = 0$ or $G_i \sim \Gamma(l_i, m_i)$, $Z \sim \Gamma(n, k)$ and

$$a_i = \frac{k}{m_i} \quad \Rightarrow\ a_i Z \sim\ \Gamma(n, m_i). \tag{9}$$

The first requirement implies that the ith component is independent from the others. In the remaining part of this work, we study the case where the condition (9) holds, i.e. we consider dependent components. From Proposition 2.1 it follows that:

- The mean of the general i^{th} element Y_i is given by:

$$E\left[Y_i\right] = \mu_{0,i} + \mu_i \frac{l_i + n}{m_i}. \tag{10}$$

- The variance (σ_i^2) of the ith element Y_i is given by:

$$\sigma_i^2 = \left(1 + \frac{\mu_i^2}{m_i}\right)\frac{(l_i + n)}{m_i}. \tag{11}$$

- The skewness of the ith component is given by:

$$\begin{aligned} s_{iii} &= E\left[\left(Y_i - E\left(Y_i\right)\right)^3\right] \\ &= E\left[\left[\mu_i\left(V_i - E\left(V_i\right)\right) + \sqrt{V_i} X_i\right]^3\right] \\ &= \left[(2-\alpha_i)\frac{\lambda_{+,i}^{\alpha_i-3} - \lambda_{-,i}^{\alpha_i-3}}{\lambda_{+,i}^{\alpha_i-2} + \lambda_{-,i}^{\alpha_i-2}} + \left(3 + 2\frac{\mu_i^2}{m_i}\right)\frac{\mu_i}{m_i}\right]\frac{(l_i + n)}{m_i}. \end{aligned} \tag{12}$$

After straightforward calculations, the covariance (σ_{ij}) between the ith (Y_i) and the jth (Y_j) element is:

$$\sigma_{i,j} = \frac{\mu_i \mu_j}{m_i m_j} n. \tag{13}$$

Level curves and comparison with Multivariate Variance Gamma

The multivariate MixedTS inherits from its univariate version the same level of flexibility. For instance choosing all $\alpha_i = 2$ for $i = 1, .., N$ we obtain as a special case the multivariate Variance Gamma introduced by Semeraro in [11]. The Semeraro model has the same structure as in (7) but instead of each X_i we have W_i where $W_1, .., W_N$ are independent Standard Normals. We recall the definition of the multivariate Variance Gamma introduced in [11].

Definition 3.2. *A random vector $Y^S \in \mathcal{R}^N$ is a Multivariate Variance Gamma if its components are defined as:*

$$Y_i^S = \mu_{0,i} + \mu_i V_i + \sigma_i \sqrt{V_i} W_i.$$

for $i = 1, ..., N$; the W_i's are independent standard normals and the V_i's are random variables each defined as:

$$V_i = G_i + a_i Z \text{ for } i = 1, ..., N \text{ and } a_i \geq 0,$$

with $G_i \sim \Gamma(l_i, m_i)$, $Z \sim \Gamma(n, k)$; for fixed i we have that G_i, Z and W_i are independent.

The moments of the multivariate Variance Gamma can be easily calculated. In particular, the ith component of the mean vector is:

$$E(Y_i^S) = \mu_{0,i} + \frac{\mu_i}{m_i}(l_i + n).$$

The components of the covariance matrix are given by

$$\begin{aligned} Var(Y_i^S) &= \left(\tfrac{\mu_i^2}{m_i^2} + \tfrac{\sigma_i^2}{m_i}\right)(l_i + n) && \text{for } i = j \\ Cov(Y_i^S, Y_j^S) &= \tfrac{\mu_i}{m_i}\tfrac{\mu_j}{m_j} n && \text{for } i \neq j. \end{aligned} \tag{14}$$

The skewness of each component is:

$$s_{iii} = \left(2\frac{\mu_i^3}{m_i^3} + 3\frac{\sigma_i^2 \mu_i}{m_i^2}\right)(l_i + n). \tag{15}$$

As remarked in [3], this model is not able to capture some aspects often observed in the behavior of financial time series. For instance, it is not able to reproduce negative correlation between assets with negatively skewed marginal distributions. Let us consider two components i and j and assume that their skewnesses have discordant signs. From equation (15) we can observe that μ_i and μ_j must also have discordant signs, which leads to a negative covariance in (14), i.e. we can not have two components with discordant skewness signs and positive correlation. Following the same reasoning we deduce that we can not have two components with the same skewness sign and negative correlation. These limitations become problematic in portfolio selection where time series of asset returns are assumed to be multivariate Variance Gamma distributed.
In the multivariate MixedTS, the sign of the skewness of the marginals may differ and have positive covariance. This is obvious from equation (12), since the skewness sign does not depend only on the sign of μ_i but also on the signs of $\lambda_{+,i}$ and $\lambda_{-,i}$.
In Figures 3 and 4 we plot the level curves of joint densities of bivariate MixedTS which can not be reproduced with the multivariate Variance Gamma proposed in [11]. In Figure 3 we have the case where the marginal distributions have discordant skewness signs and positive correlation. In Figure 4 the marginals are negatively skewed distributions with negative correlation.

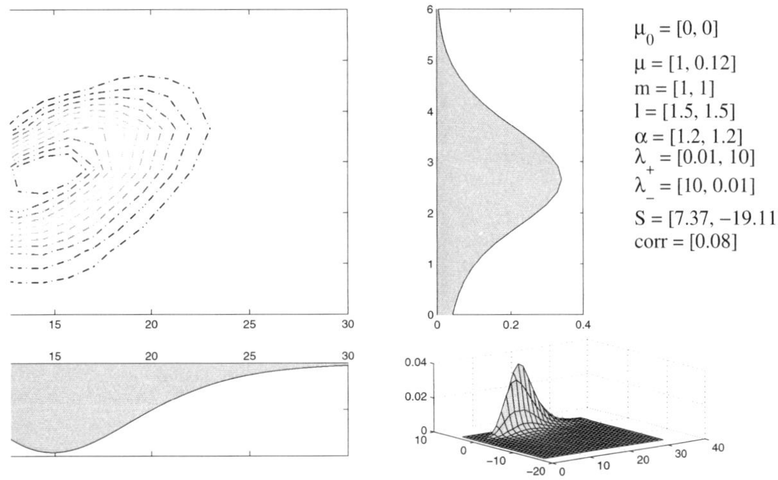

Figure 3. Level curves, marginal distributions and joint density of a bivariate MixedTS. In this case $skew(Y_1) = 7.37$, $skew(Y_2) = -19.11$ and correlation is positive.

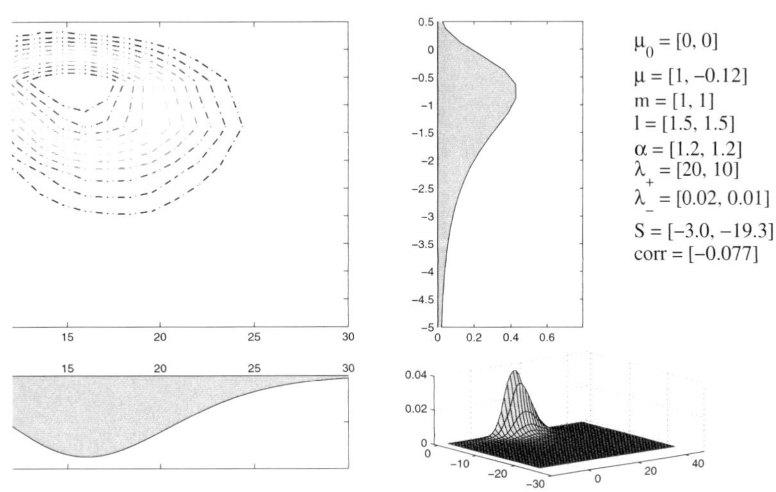

Figure 4. Level curves, marginal distributions and joint density of a bivariate MixedTS. In this case $skew(Y_1) = -3$, $skew(Y_2) = -19.3$ and correlation is negative.

4 Conclusion

In these notes we presented an infinitely divisible multivariate distribution that is a generalization of the Normal Variance Mean Mixtures. This new multivariate model is characterized by a high flexibility in terms of ability in reproducing different sign combinations of higher moments. These features are quite important for financial time series as the multivariate Variance Gamma distribution is able to capture different dependence structures. The results here reported are discussed through a direct comparison of the proposed model with the Multivariate Variance Gamma constructed in a similar way.

Bibliography

[1] Barndorff-Nielsen, O., Kent, J. and Sørensen, M. (1982). *Normal variance-mean mixtures and z distributions.* International Statistical Review, **50**, (2), 145–159.

[2] Carr, P. and Wu, L. (2004). *Time-changed Lévy processes and option pricing.* Journal of Financial Economics, **71**, (1), 113 – 141

[3] Hitaj, A. and Mercuri, L. (2013). *Hedge fund portfolio allocation with higher moments and mvg models.* Chapter **14** on Advances in Financial Risk Management: Corporates, Intermediaries and Portfolios, Palgrave Macmillan UK, 331–346.

[4] Hitaj, A., Mercuri, L. and Rroji, E. (2015). *Portfolio selection with independent component analysis.* Finance Research Letters, **15**, 146–159

[5] Kim, Y.S., Rachev, S. T., Bianchi, M. L. and Fabozzi, F.J. (2008). *Financial market models with Lévy processes and time-varying volatility.* Journal of Banking and Finance, **32** (7), 1363–1378

[6] Kozubowski, T. J. , Podgórski, K. and Samorodnitsky, G. (1999). *Tails Of Lévy Measure Of Geometric Stable Random Variables.* Extremes, **1** (3), 367–378

[7] Küchler, U. and Tappe, S. (2013). *Tempered stable distributions and processes.* Stochastic Processes and their Applications, **123** (12), 4256–4293.

[8] Madan, D. and Seneta, E. (1990). *The variance gamma (v.g.) model for share market returns.* Journal of Business, **63** (4), 511–524.

[9] Rroji, E. and Mercuri, L. (2015). *Mixed tempered stable distribution.* Quantitative Finance, **15** (9), 1559–1569.

[10] Sato, K. I. (1999). *Lévy processes and infinitely divisible distributions.* Cambridge University Press, 2nd edition.

[11] Semeraro, P. (2008). *A multivariate variance gamma model for financial applications.* International Journal of Theoretical and Applied Finance, **11** (1), 1–18.

Extension to mixed models of the Supervised Component-based Generalised Linear Regression

Jocelyn Chauvet, *University of Montpellier*, jocelyn.chauvet@umontpellier.fr
Catherine Trottier, *University of Montpellier 3*, catherine.trottier@univ-montp3.fr
Xavier Bry, *University of Montpellier*, xavier.bry@univ-montp2.fr
Frédéric Mortier, *Cirad UR BSEF*, frederic.mortier@cirad.fr

Abstract. We address the component-based regularisation of a multivariate Generalized Linear Mixed Model (GLMM). A set of random responses Y is modelled by a GLMM, using a set X of explanatory variables, a set T of additional covariates, and random effects used to introduce the dependence between statistical units. Variables in X are assumed many and redundant, so that regression demands regularisation. By contrast, variables in T are assumed few and selected so as to require no regularisation. Regularisation is performed building an appropriate number of orthogonal components that both contribute to model Y and capture relevant structural information in X. To estimate the model, we propose to maximise a criterion specific to the Supervised Component-based Generalised Linear Regression (SCGLR) within an adaptation of Schall's algorithm. This extension of SCGLR is tested on both simulated and real data, and compared to Ridge- and Lasso-based regularisations.

Keywords. Component-model, Multivariate GLMM, Random effect, Structural Relevance, Regularisation, SCGLR.

1 Data, Model and Problem

A set of q random responses $Y = \{y^1, \dots, y^q\}$ is assumed explained by two different sets of covariates $X = \{x^1, \dots, x^p\}$ and $T = \{t^1, \dots, t^r\}$, and a random effect $\xi = \{\xi^1, \dots, \xi^q\}$. Explanatory variables in X are assumed many and redundant while additional covariates in T are assumed selected so as to preclude redundancy. Explanatory variables in T are thus kept as such in the model. By contrast, X may contain several unknow structurally relevant dimensions $K < p$ important to model and predict Y, how many we do not know. X is thus to be searched for an appropriate number of orthogonal components that both capture relevant structural information in X and contribute to model Y.

Each y^k is modelled through a Generalised Linear Mixed Model (GLMM) [7] assuming conditional distributions from the exponential family. More specifically in this work, the n statistical units are not considered independent but partitioned into N groups. The random effects included in the GLMM aim at modelling the dependence of units within each group.

Over the last decades, component-based regularisation methods for Generalised Linear Models (GLM) have been developped. In the univariate framework, i.e. when $Y = \{y\}$, Bastien et al. [1] proposed an extension to GLM of the classical Partial Least Square (PLS) regression, combining generalised linear regressions of the dependent variable on each of the regressors considered separately. However, doing so, this method does not take into account the variance structure of the overall model when building a component. Still in the univariate framework, Marx [6] proposed a more appropriate Iteratively Reweighted Partial Least Squares (IRPLS) estimation that builds PLS components using the weighting matrix derived from the GLM. More recently, Bry et al. [2] extended the work by Marx [6] to the multivariate framework with a technique named Supervised Component-based Generalised Linear Regression (SCGLR). The basic principle of SCGLR is to build optimal components common to all the dependent variables. To achieve it, SCGLR introduces a new criterion which is maximised at each step of the Fisher Scoring Algorithm (FSA).

Besides, regularisation methods have already been developped for GLMM, in which the random effects allow to model complex dependence structure. Eliot et al. [3] proposed to extend the classical ridge regression to Linear Mixed Models (LMM). The Expectation-Maximisation algorithm they suggest includes a new step to find the best shrinkage parameter - in the Generalised Cross-Validation (GCV) sense - at each iteration. More recently, Groll and Tutz [4] proposed an L_1-penalised algorithm for fitting a high-dimensional GLMM, using Laplace approximation and efficient coordinate gradient descent.

Instead of using a penalty on the norm of the coefficient vector, we propose to base the regularisation of the GLMM estimation on SCGLR-type components.

2 Reminder on SCGLR with additional covariates

In this section, we consider the simplified situation where each y^k is modelled through a GLM (without random effect) and only one component is calculated ($K = 1$). Moreover, let us use the following notations:

Π_E^M : orthogonal projector on space E, with respect to some metric M.
$\langle X \rangle$: space spanned by the column-vectors of X.
M' : transpose of any matrix (or vector) M.

The first conceptual basis of SCGLR consists in searching for an optimal component $f = Xu$ common to all the y's. Therefore, SCGLR adapts the classical FSA to predictors having colinear X-parts. To be precise, for each $k \in \{1, \dots, q\}$, the linear predictor writes:

$$\eta^k = (Xu)\gamma_k + T\delta_k$$

where γ_k and δ_k are the parameters associated with component $f = Xu$ and covariates T respectively. For identification, we impose $u'Au = 1$, where A may be any symmetric definite positive matrix. Assuming that both the y's and the n statistical units are independent, the likelihood function L can be written:

$$L(y|\eta) = \prod_{i=1}^{n} \prod_{k=1}^{q} L_k(y_i^k|\eta_i^k)$$

where L_k is the likelihood function relative to y^k. Owing to the product $\gamma_k u$, the "linearised model" (LM) on each step of the associated FSA for the GLM estimation is not indeed linear: an alternated least squares step was designed. Denoting z^k the classical working variables on each FSA's step and W_k^{-1} their variance-covariance matrix, the least squares on the LM consists in the following optimisation (see [2]):

$$\min_{u'Au=1} \sum_{k=1}^{q} \left\| z^k - \Pi_{\langle Xu,T\rangle}^{W_k} z^k \right\|_{W_k}^2 \quad \Longleftrightarrow \quad \max_{u'Au=1} \sum_{k=1}^{q} \left\| \Pi_{\langle Xu,T\rangle}^{W_k} z^k \right\|_{W_k}^2$$

which is also equivalent to :

$$\max_{u'Au=1} \psi_T(u), \quad \text{with} \quad \psi_T(u) = \sum_{k=1}^{q} \left\|z^k\right\|^2_{W_k} \cos^2_{W_k}\left(z^k, \langle Xu, T\rangle\right) \tag{1}$$

The second conceptual basis of SCGLR consists in introducing a closeness measure of the component $f = Xu$ to the strongest structures in X. Indeed, ψ_T is a mere goodness-of-fit measure, and must be combined with a structural relevance measure to get regularisation. Consider a given weight-matrix W - e.g. $W = \frac{1}{n}I_n$ - reflecting the a priori relative importance of units, the most structurally relevant component would be the solution of:

$$\max_{u'Au=1} \phi(u), \quad \text{with} \quad \phi(u) = \left(\sum_{j=1}^{p} \langle Xu|x^j\rangle^{2l}_W\right)^{\frac{1}{l}} = \left(\sum_{j=1}^{p} (u'X'Wx^j x^{j\prime} WXu)^l\right)^{\frac{1}{l}} \tag{2}$$

Tuning parameter l allows to draw components towards more (greater l) or less (smaller l) local variable bundles as depicted on Figure 1.

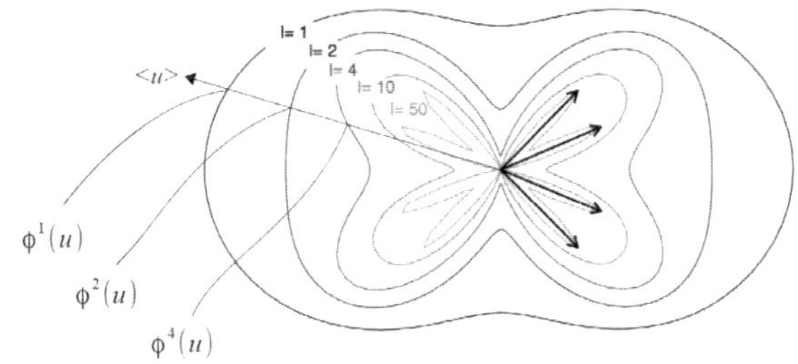

Figure 1. Polar representation of $\phi^l(u)$ according to the value of l, in the elementary case of four coplanar variables.

To sum things up, s being a parameter that tunes the importance of the structural relevance relative to the goodness-of-fit, SCGLR attempts a trade-off between (1) and (2), solving:

$$\max_{u'Au=1} \left[\phi(u)\right]^s \left[\psi_T(u)\right]^{1-s} \tag{3}$$

3 Adapting SCGLR to grouped data

We propose to adapt SCGLR to grouped data, for which the independence assumption of the statistical units is no longer valid (e.g. longitudinal or spatial correlated data). The whithin-group dependence is modelled by a random effect. Consequently, each of the y^k is ultimately modelled through a GLMM. We call this adaptation "Mixed-SCGLR".

Single component model estimation

We first present the method's principle. Then we give a Bayesian justification of the involved Henderson systems. Finally, we present our algorithm's steps.

Principle We maintain predictors colinear in their X-parts. Introducing a random effect in each predictor and still imposing $u'Au = 1$, the linear predictors now write:

$$\forall k \in \{1, \ldots, q\}\,, \ \eta^k_\xi = (Xu)\gamma_k + T\delta_k + U\xi^k \tag{4}$$

The random group effect is assumed different accross responses, yielding q random effects $\xi^1, \ldots, \xi^q$. These are assumed independent and normally distributed: $\mathcal{N}_N(0, D_k = \sigma^2_k A_k)$, where N is the number of groups and A_k a known matrix ($A_k = I_N$ in general).

Owing to the GLMM dependence structure, the FSA was adapted by Schall [9]. We, in turn, adapt Schall's algorithm to our component-based predictor (4), by introducing the following alternated procedure at each step:

- Given γ_k, δ_k, ξ^k and σ^2_k, we build the component $f = Xu$ by solving a (3)-type program, which attempts a compromise between goodness-of-fit and structural relevance criterions.
- Given u, z^k_ξ being the classical working variables of the Schall's algorithm and $W^{-1}_{\xi,k}$ their conditional variance-covariance matrix, parameters γ_k, δ_k and ξ^k are estimated by solving the following Henderson system, which, subsequently, allows us to estimate σ^2_k :

$$\begin{pmatrix} (Xu)'W_{\xi,k}(Xu) & (Xu)'W_{\xi,k}T & (Xu)'W_{\xi,k}U \\ T'W_{\xi,k}(Xu) & T'W_{\xi,k}T & T'W_{\xi,k}U \\ U'W_{\xi,k}(Xu) & U'W_{\xi,k}T & U'W_{\xi,k}U + D_k^{-1} \end{pmatrix} \begin{pmatrix} \gamma_k \\ \delta_k \\ \xi^k \end{pmatrix} = \begin{pmatrix} (Xu)'W_{\xi,k}\, z^k_\xi \\ T'W_{\xi,k}\, z^k_\xi \\ U'W_{\xi,k}\, z^k_\xi \end{pmatrix} \tag{5}$$

We chose Henderson's method [5] since it is quicker than EM, for instance.

A Bayesian justification of the Henderson systems The conditional distribution of the data, given the random effects, is supposed to belong to the exponential family, i.e. for each $k \in \{1, \ldots, q\}$, the conditional density of $Y^k_i \,|\, \xi^k$ may be written:

$$f_{Y^k_i | \xi^k}\left(y^k_i, \theta^k_i\right) = \exp\left\{ \frac{y^k_i \theta^k_i - b_k\left(\theta^k_i\right)}{a_{k,i}(\phi_k)} + c_k\left(y^k_i, \phi_k\right) \right\}$$

Linearisation step : Denoting G_k the link function of variable y^k, g_k its first derivative and μ_k the conditional expectation (i.e. $\mu^k := \mathbb{E}(Y^k \,|\, \xi^k)$), the working variables are obtained as:

$$z^k_\xi = \eta^k_\xi + e^k, \quad \text{where} \quad e^k_i = (y^k_i - \mu^k_i) g_k(\mu^k_i)$$

Their conditional variance-covariance matrices are:

$$W^{-1}_{\xi,k} := \mathrm{Var}\left(z^k_\xi \,|\, \xi^k\right) = \mathrm{Diag}\left(\left[g_k\left(\mu^k_i\right)\right]^2 \mathrm{Var}\left(Y^k_i \,|\, \xi^k\right)\right)_{i=1,\ldots,n}$$

Estimation step : Our estimation step is based on the following lemma about normal hierarchy:

Lemma 3.1. *Given*

$$y|\theta \sim \mathcal{N}(M\theta, R)$$
$$\theta \sim \mathcal{N}(\alpha, \Omega)$$

the posterior distribution is $\theta|y \sim \mathcal{N}(\hat\theta, C)$, *where* $C = \left(M'R^{-1}M + \Omega^{-1}\right)^{-1}$ *and* $\hat\theta$ *satisfies:*

$$C^{-1}\hat\theta = M'R^{-1}y + \Omega^{-1}\alpha. \tag{6}$$

Given u, we just apply Schall's method [9] with our regularised linear predictors (4), which is equivalent to consider the following modelling:

$$z_\xi^k | \gamma_k, \delta_k, \xi^k \sim \mathcal{N}\left((Xu)\gamma_k + T\delta_k + U\xi^k,\ W_{\xi,k}^{-1} \right) \tag{7}$$

$$(\gamma_k, \delta_k, \xi^k) \sim \mathcal{N}\left(\begin{pmatrix} \gamma_k^{(0)} \\ \delta_k^{(0)} \\ 0 \end{pmatrix}, \begin{pmatrix} B_k^\gamma & 0 & 0 \\ 0 & B_k^\delta & 0 \\ 0 & 0 & D_k \end{pmatrix} \right)$$

We suggest to choose a noninformative prior distribution for parameters γ_k et δ_k (as inspired by Stiratelli et al. [10]) imposing $B_k^{\gamma^{-1}} = B_k^{\delta^{-1}} = 0$. Current estimates of parameters γ_k, δ_k and ξ^k are thus obtained solving (6), which is equivalent to the Henderson system (5).

Finally, as mentioned in [9], given estimate $\widehat{\xi^k}$ for ξ^k, we have the following updates for the maximum likelihood estimation of the variance parameters $\sigma_k^2,\ k \in \{1, \dots, q\}$:

$$\sigma_k^2 \longleftarrow \frac{\widehat{\xi^k}' A_k^{-1} \widehat{\xi^k}}{N - \frac{1}{\sigma_k^2} tr\left(A_k^{-1} C_k\right)} \quad \text{where} \quad C_k = \left(U' W_{\xi,k} U + D_k^{-1} \right)^{-1}$$

The algorithm Component $f = Xu$ is still found solving a (3)-type program, adapting the expression of ψ_T. Indeed, conditional on the random effects ξ^k, the working variables z_ξ^k are assumed normally distributed according to (7). We thus modify the previous goodness-of-fit measure, taking into account the variance of z_ξ^k conditional on ξ^k. In case of grouped data,

$$\psi_T(u) = \sum_{k=1}^{q} \left\| z_\xi^k \right\|_{W_{\xi,k}}^2 \cos^2_{W_{\xi,k}} \left(z_\xi^k, \langle Xu, T \rangle \right) \tag{8}$$

Extracting higher rank components

Let $F^h = \{f^1, \dots, f^h\}$ be the set of the first h components. An extra component f^{h+1} must best complement the existing ones plus T, i.e. $T^h := F^h \cup T$. So f^{h+1} must be calculated using T^h as additional covariates. Moreover, we must impose that f^{h+1} be orthogonal to F^h, i.e.:

$$F^{h'} W f^{h+1} = 0$$

Component $f^{h+1} := Xu^{h+1}$ is thus obtained solving:

$$\begin{cases} \max \quad [\phi(u)]^s \ [\psi_{T^h}(u)]^{1-s} \\ \text{subject to:} \quad u'Au = 1 \text{ and } D^{h'} u = 0 \end{cases} \tag{9}$$

where $\psi_{T^h}(u) = \sum_{k=1}^{q} \left\| z_\xi^k \right\|_{W_{\xi,k}}^2 \cos^2_{W_{\xi,k}} \left(z_\xi^k, \langle Xu, T^h \rangle \right)$ and $D^h = X'WF^h$.

In Appendix, we give an algorithm to maximise, at least locally, any criterion on the unit sphere: the Projected Iterated Normed Gradient (PING) algorithm. Varying the initialisation allows us to increase confidence that the maximum reached is global. It allows us to build components of rank $h > 1$ by solving programs (9) and also component of rank $h = 1$ if we impose $T^h = T$ and $D^h = 0$ in the aforementioned program.

Current iteration of the single component Mixed-SCGLR

Step 1 Computation of the component
Set:

$$u^{[t]} = \underset{u'Au=1}{\arg\max}\ [\phi(u)]^s\ [\psi_T(u)]^{1-s} \quad \text{where } \psi_T \text{ is defined by (8)}$$
$$f^{[t]} = Xu^{[t]}$$

Step 2 Henderson systems
For each $k \in \{1, \ldots, q\}$, solve the following system:

$$\begin{pmatrix} f^{[t]\prime}W^{[t]}_{\xi,k}f^{[t]} & f^{[t]\prime}W^{[t]}_{\xi,k}T & f^{[t]\prime}W^{[t]}_{\xi,k}U \\ T'W^{[t]}_{\xi,k}f^{[t]} & T'W^{[t]}_{\xi,k}T & T'W^{[t]}_{\xi,k}U \\ U'W^{[t]}_{\xi,k}f^{[t]} & U'W^{[t]}_{\xi,k}T & U'W^{[t]}_{\xi,k}U + D^{[t]^{-1}}_k \end{pmatrix} \begin{pmatrix} \gamma_k \\ \delta_k \\ \xi^k \end{pmatrix} = \begin{pmatrix} f^{[t]\prime}W^{[t]}_{\xi,k}\, z^{k\,[t]}_\xi \\ T'W^{[t]}_{\xi,k}\, z^{k\,[t]}_\xi \\ U'W^{[t]}_{\xi,k}\, z^{k\,[t]}_\xi \end{pmatrix}$$

Call $\gamma^{[t]}_k$, $\delta^{[t]}_k$ and $\xi^{k\,[t]}$ the solutions.

Step 3 Updating variance parameters
For each $k \in \{1, \ldots, q\}$, compute:

$$\sigma^{2\,[t+1]}_k = \frac{\xi^{k\,[t]^{\prime}} A^{-1}_k \xi^{k\,[t]}}{N - \frac{1}{\sigma^{2\,[t]}_k} tr\left(A^{-1}_k C^{[t]}_k\right)} \quad \text{and} \quad D^{[t+1]}_k = \sigma^{2\,[t+1]}_k A_k$$

Step 4 Updating working variables and weighting matrices
For each $k \in \{1, \ldots, q\}$, compute:

$$\eta^{k\,[t]} = f^{[t]}\gamma^{[t]}_k + T\delta^{[t]}_k + U\xi^{k\,[t]}$$
$$\mu^{[t]}_{k,i} = G^{-1}_k\left(\eta^{k\,[t]}_i\right),\ i = 1, \ldots, n$$
$$z^{k\,[t+1]}_i = \eta^{k\,[t]} + \left(y^k_i - \mu^{[t]}_{k,i}\right) g_k\left(\mu^{[t]}_{k,i}\right),\ i = 1, \ldots, n$$
$$W^{[t+1]}_{\xi,k} = Diag\left(\left(\left[g_k\left(\mu^{[t]}_{k,i}\right)\right]^2 \mathrm{Var}\left(Y^k_i \mid \xi^k\right)^{[t]}\right)^{-1}\right)_{i=1,\ldots,n}$$

Step 1–4 are repeated until stability of u and parameters γ_k, δ_k and σ^2_k is reached.

4 Simulation study in the canonical Gaussian case

A simple simulation study is conducted to characterise the relative performances of Mixed-SCGLR and Ridge- and Lasso-based regularisations in the LMM framework [3, 4]. We focus on the multivariate case, i.e. several $y's$, with many redundant explanatory variables. To do so, two random responses $Y = \{y^1, y^2\}$ are generated, and explanatory variables X are simulated so as to contain three independent bundles of variables: X_1, X_2 and X_3. Each explanatory variable is assumed normally distributed with mean 0 and variance 1. The level of redundancy within each bundle is tuned with parameter $\tau \in \{0.1, 0.3, 0.5, 0.7, 0.9\}$. To be precise, correlations among explanatory variables whithin bundle X_j are:

$$corr(X_j) = \tau \mathbf{1}\mathbf{1}' + (1-\tau)I$$

Besides, bundle X_1 (15 variables) models and predicts only y^1, bundle X_2 (10 variables) only y^2, while bundle X_3 (5 variables) is designed to be a bundle of noise. Considering no additional covariates ($T = 0$), we thus simulate Y as :

$$\begin{cases} y^1 = X\beta_1 + U\xi^1 + \varepsilon^1 \\ y^2 = X\beta_2 + U\xi^2 + \varepsilon^2 \end{cases} \tag{10}$$

We consider the case of $N = 10$ groups and $R = 10$ units per group. Consequently, the random-effect design matrix can be written: $U = I_N \otimes \mathbf{1}_R$. All variables whithin each bundle are assumed to contribute homogeneously to predict Y. Then our choice of the fixed parameters are:

$$\beta_1 = (\underbrace{0.3,\ldots,0.3}_{5\text{ times}}, \underbrace{0.4,\ldots,0.4}_{5\text{ times}}, \underbrace{0.5,\ldots,0.5}_{5\text{ times}}, \underbrace{0,\ldots\ldots\ldots\ldots,0}_{15\text{ times}})'$$

$$\beta_2 = (\underbrace{0,\ldots\ldots\ldots\ldots,0}_{15\text{ times}}, \underbrace{0.3,..,0.3}_{3\text{ times}}, \underbrace{0.4,..,0.4}_{4\text{ times}}, \underbrace{0.5,..,0.5}_{3\text{ times}}, \underbrace{0,\ldots,0}_{5\text{ times}})'$$

Finally, residual variability and within groups variability are fixed to $\sigma_k^2 = 1$. We thus simulate random effects and noise respectively as: $\xi^k \sim \mathcal{N}_N(0, \sigma_k^2 I_N)$ and $\varepsilon^k \sim \mathcal{N}_n(0, \sigma_k^2 I_n)$, where $k \in \{1, 2\}$ and $n = NR$.

On the whole, $M = 500$ simulations are conducted for each value of τ, according to model (10). Simulation m provides two fixed-effects estimations: $\hat{\beta}_1^{(m)}$ and $\hat{\beta}_2^{(m)}$. Unlike Mixed-SCGLR, both LMM-Ridge and (G)LMM-Lasso are not designed for multivariate responses: estimations are computed separately in these cases. Consequently, for each method, we decide to retain only the one which provides the lower relative error. Their mean over M simulations (MLRE) is defined as:

$$\text{MLRE} = \frac{1}{M}\sum_{m=1}^{M} \min\left(\frac{\left\|\hat{\beta}_1^{(m)} - \beta_1\right\|^2}{\|\beta_1\|^2}, \frac{\left\|\hat{\beta}_2^{(m)} - \beta_2\right\|^2}{\|\beta_2\|^2} \right)$$

In table 1, we summarise the optimal regularisation parameters selected via cross-validation. Corresponding MLRE's are presented in Table 2 to which we added the results provided without regularisation. As expected, in both Ridge and Lasso regularisations, the shrinkage parameter value increases with τ. On the other hand, the greater τ, the more Mixed-SCGLR (with $l = 4$ as recommended in [8]) focuses on the main structures in X that contribute to model Y. The average value of s is approximatively 0.5, which means that there is no significant preference between goodness-of-fit and structural relevance. Except for $\tau = 0.1$, Mixed-SCGLR provides the most precise fixed effect estimates despite the sophistication of the dependence structure and the high level of correlation among explanatory variables. Indeed, if there are no actual bundles in X ($\tau \simeq 0$), looking for structures in X may lead Mixed-SCGLR to be slightly less accurate. Conversely, the stronger the structures (high τ), the more efficient our method.

	(G)LMM-Lasso	LMM-Ridge	Mixed-SC(G)LR	
	Optimal shrinkage parameter	Optimal shrinkage parameter	Optimal number of components K	Optimal tuning parameter s
$\tau = 0.1$	63.4	24.1	25	0.53
$\tau = 0.3$	111.2	53.7	5	0.53
$\tau = 0.5$	171.3	73.2	3	0.51
$\tau = 0.7$	220.6	78.2	2	0.51
$\tau = 0.9$	254.9	84.9	2	0.52

Table 1. Optimal regularisation parameter values obtained by cross-validation over 500 simulations

	LMM	(G)LMM-Lasso	LMM-Ridge	Mixed-SC(G)LR
$\tau = 0.1$	0.141	0.083	0.090	0.094
$\tau = 0.3$	0.340	0.180	0.124	0.105
$\tau = 0.5$	0.686	0.413	0.150	0.059
$\tau = 0.7$	1.571	0.913	0.189	0.061
$\tau = 0.9$	5.022	2.431	0.261	0.050

Table 2. Mean Lower Relative Errors (MLRE's) associated with the optimal parameter values

In order to highlight the power of Mixed-SCGLR for model interpretation, we represent on Figure 2 the correlation scatterplots obtained for $\tau = 0.5$, $l = 4$, $s = 0.51$, and $K = 3$. It clearly appears that y^1 is explained by the first bundle and y^2 by the second. The third component calculated catches the third bundle, which appears to play no explanatory role.

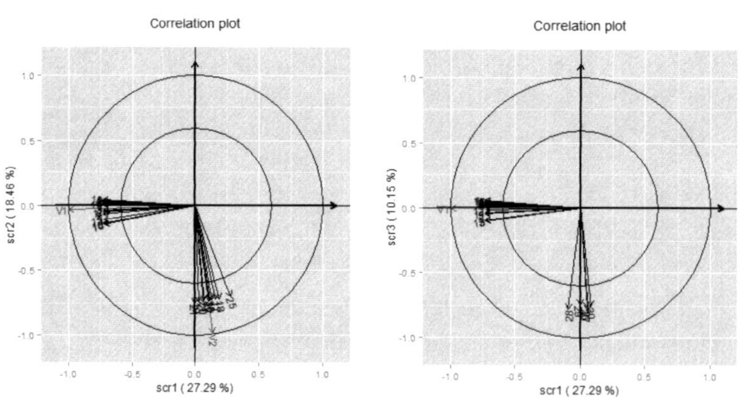

Figure 2. Correlation scatterplots given by Mixed-SCGLR method on the simulated data. The left hand side and the right hand side plots respectively show component planes $(1, 2)$ and $(1, 3)$. Both X-part linear predictors related to y^1 and y^2 are considered supplementary variables.

5 Real data example in the canonical Poisson case

Genus is a dataset built from the "CoForChange" study, which was conducted on $n = 2600$ developped plots divided into $N = 22$ forest concessions. It gives the abundance of $q = 94$ common tree genera in the tropical moist forest of the Congo-Basin, $p = 56$ geo-referenced environmental variables, and $r = 2$ other covariates which describe geology and anthropogenic interference. Geo-referenced environmental variables are used to represent:

- 29 physical factors linked to topography, rainfall, or soil moisture,
- 25 photosynthesis activity indicators obtained by remote sensing: EVI (Enhanced Vegetation Index), NIR (Near InfraRed channel index) and MIR (Mid-InfraRed channel index)),
- 2 indicators which describe stand of trees height.

Physical factors are many and redundant (monthly rainfalls are highly correlated, for instance, and related to the geographic location). So are all photosynthesis activity indicators. Therefore, all these variables are put in X. By contrast, as geology and anthropogenic interference are weakly correlated and interesting per se, we put them in the set T of additional covariates.

It must be noted that the abundances of species given in *Genus* are count data. For each random response $y^1, \ldots, y^q$ we thus choose a Poisson regression with log link:

$$\forall k \in \{1, \ldots, q\}, \quad y^k \sim \mathcal{P}\left(\exp\left[\sum_{j=1}^{K} \left(Xu^j\right)\gamma_j + T\delta_k + U\xi^k\right]\right)$$

Among a series of parameter choices, values $l = 4$ and $s = 0.5$ prove to yield components very close to interpretable variable bundles. We therefore keep these parameter values in order to find - through a cross-validation procedure - the number of components K which minimises the Average Normalised Root Mean Square Error (AveNRMSE) defined as:

$$\text{AveNRMSE} = \frac{1}{q}\sum_{k=1}^{q}\sqrt{\frac{1}{n}\sum_{i=1}^{n}\left(\frac{y_i^k - \hat{y}_i^k}{\bar{y}^k}\right)^2}$$

On Figure 3, we plot the AveNRMSE's for $K \in \{0, 1, \ldots, 25\}$. As one can see, the best models are the ones with 12, 14 and 16 components. We retain the most parcimonious of them, i.e the one with 12 components. Two examples of correlation scatterplots we obtain are given on Figure 4, in which the X-parts of linear predictors are considered supplementary variables.

6 Conclusion

Mixed-SCGLR is a powerful trade-off between multivariate GLMM estimation (which cannot afford many and redundant explanatory variables) and PCA-like methods (which take no account of responses in building components). While preserving the qualities of the plain version of SCGLR, the mixed one performs well on grouped data, and provides robust predictive models based on interpretable components. Compared to penalty-based approaches as Ridge or Lasso, the orthogonal components built by Mixed-SCGLR reveal the multidimensional explanatory and predictive structures, and greatly facilitate the interpretation of the model.

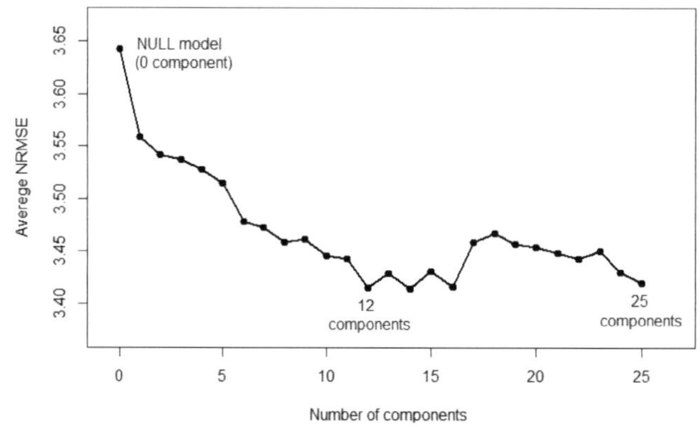

Figure 3. AveNRMSE's as a function of the number of components, obtained by a cross-validation procedure. The "null model" does not include any explanatory variables in X, but only additional covariates in T.

Acknowledgement

The extended data *Genus* required the arrangement and the inventory of 140.000 developped plots across four countries : Central African Republic, Gabon, Cameroon and Democratic Republic of Congo. The authors thank the members of the CoForTips project for the use of this data.

Appendix

The Projected Iterated Normed Gradient (PING) is an extension of the iterated power algorithm, solving any program which has the form:

$$\max_{\substack{u'Au=1\\ D'u=0}} h(u) \tag{11}$$

Note that putting $v := A^{1/2}u$, $g(v) := h(A^{-1/2}v)$ and $C := A^{-1/2}D$, program (11) is strictly equivalent to program (12):

$$\max_{\substack{v'v=1\\ C'v=0}} g(v) \tag{12}$$

In our framework, the particular case $C = 0$ (no extra-orthogonality contrain) allows us to find the first rank component. Denoting $\Pi_{C^\perp} := I - (C'C)^{-1}C'$ and $\Gamma(v) := \nabla_v g(v)$, a Lagrange multiplier- based reasoning gives the basic iteration of the PING algorithm:

$$v^{[t+1]} = \frac{\Pi_{C^\perp}\Gamma\left(v^{[t]}\right)}{\left\|\Pi_{C^\perp}\Gamma\left(v^{[t]}\right)\right\|} \tag{13}$$

Despite the fact that iteration (13) follows a direction of ascent, it does not guarantee that g actually increases on every step. Algorithm PING therefore repeats the following steps until convergence of v is reached:

Step 1 Set: $\kappa^{[t]} = \dfrac{\Pi_{C^\perp}\Gamma\left(v^{[t]}\right)}{\left\|\Pi_{C^\perp}\Gamma\left(v^{[t]}\right)\right\|}$

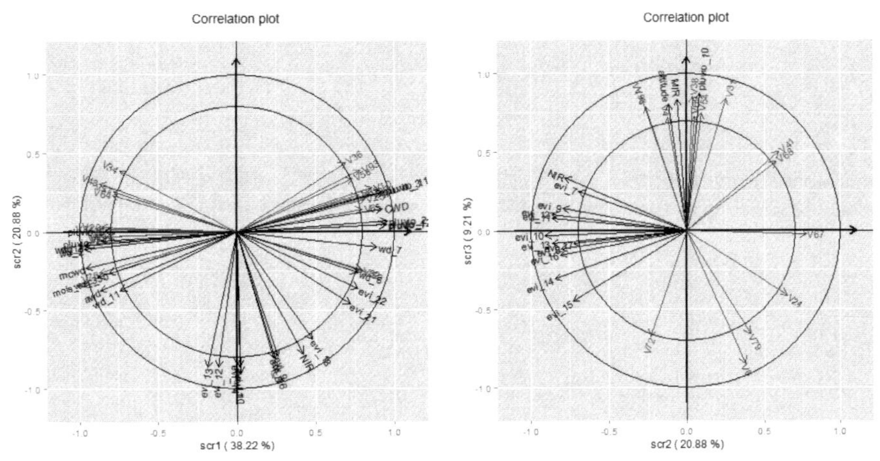

Figure 4. Two examples of correlations scatterplots on data *Genus*. The left hand side plot displays only variables having cosine greater than 0.8 with component plane $(1, 2)$. It reveal two patterns: a "rain-wind"-pattern driven by the Pluvio's and Wd's variables and a photosynthesis-pattern driven by the Evi's. On the right hand side, we plot variables having cosine greater than 0.7 with component plane $(2, 3)$. Component 3 reveals a bundle driven by variables Altitude, MIR and Pluvio10, which prove important to model and predict several y's.

Step 2 A Newton-Raphson unidimensional maximisation procedure is used to find the maximum of $g(v)$ on the arc $\left(v^{[t]}, \kappa^{[t]}\right)$ and take it as $v^{[t+1]}$.

Bibliography

[1] Bastien, P., Esposito Vinzi, V. and Tenenhaus, M. (2004) *PLS generalized linear regression.* Computational Statistics & Data Analysis, **48**, 17–46.

[2] Bry, X., Trottier, C., Verron, T. and Mortier, F. (2013) *Supervised component generalized linear regression using a PLS-extension of the Fisher scoring algorithm.* Journal of Multivariate Analysis, **119**, 47-60.

[3] Eliot, M., Ferguson, J., Reilly, M.P. and Foulkes, A.S. (2011) *Ridge Regression for Longitudinal Biomarker Data.* The International Journal of Biostatistics, **7**, 1–11.

[4] Groll, A. and Tutz, G. (2014) *Variable Selection for Generalized Linear Mixed Models by L1-Penalized Estimation.* Statistics and Computing, **24**, 137–154.

[5] Henderson, C.R. (1975) *Best linear unbiaised estimators and prediction under a selection model.* Biometrics, **31**, 423-447.

[6] Marx, B.D. (1996) *Iteratively reweighted partial least squares estimation for generalized linear regression.* Technometrics, **38**, 374-381.

[7] McCulloch, C.E. and Searle, S.R (2001) *Generalized, Linear, and Mixed Models.* John Wiley & Sons.

[8] Mortier, F., Trottier, C., Cornu, G., Bry, X. (2015) *SCGLR - An R Package for Supervised Component Generalized Linear Regression.*

[9] Schall, R. (1991) *Estimation in generalized linear models with random effects.* Biometrika, **78**, 719-727.

[10] Stiratelli, R., Laird, N., and Ware, J.H. (1984) *Random-Effects Models for Serial Observations with Binary Response.* Biometrics, **40**, 961–971.

Evaluation of robust PCA for supervised audio outlier detection

Sarka Brodinova, *Vienna University of Technology*, sarka.brodinova@tuwien.ac.at
Thomas Ortner, *Vienna University of Technology*, thomas.ortner@tuwien.ac.at
Peter Filzmoser, *Vienna University of Technology*, p.filzmoser@tuwien.ac.at
Maia Zaharieva, *Vienna University of Technology*, maia.zaharieva@tuwien.ac.at
Christian Breiteneder, *Vienna University of Technology*, christian.breiteneder@tuwien.ac.at

Abstract. Outliers often reveal crucial information about the underlying data such as the presence of unusual observations that require for in-depth analysis. The detection of outliers is especially challenging in real-world application scenarios dealing with high-dimensional and flat data bearing different subpopulations of potentially varying data distributions. In the context of high-dimensional data, PCA-based methods are commonly applied to reduce dimensionality and to reveal outliers. Thus, a thorough empirical evaluation of various PCA-based methods for the detection of outliers in a challenging audio data set is provided. The various experimental data settings are motivated by the requirements of real-world scenarios, such as varying number of outliers, available training data, and data characteristics in terms of potential subpopulations.

Keywords. Outlier detection, Robust PCA, Audio data, Experiments

1 Introduction

Outlier identification is an essential data mining task. Outliers do not only contaminate distributions and, thus, estimations based on the distributions, moreover, they often are the prime focus of attention. In many fields outliers carry significant, even crucial information for applications such as fraud detection, surveillance, and medical imaging. In this paper, we employ outlier detection in an automated highlight detection application for audio data. This is a first step towards the identification of key scenes in videos, where the audio is a fundamental component.

Outlier detection gets considerably more difficult in a high-dimensional space or when there are less observations than variables available (flat data). In a high-dimensional space, data becomes sparse and distances between observations differ very little. To justify the application of distance-based similarity measures in such a situation, the reduction of dimensionality is an inevitable course of action. A well-established approach for this purpose is the use of principal component analysis (PCA), which transforms the original variables to a smaller set of uncorrelated variables keeping as much of the total variance as possible [8]. This step removes the curse of high dimensionality for this subspace. Nevertheless, it has been shown, that even though in theory distance functions loose their meaningfulness in high dimensionality, the orthogonal complement of the principal component (PC) space might still hold crucial differences in the distance and, thus, important information for outlier detection [20].

The focus of this paper is the thorough empirical comparison of PCA-based methods for high-dimensional and flat data, that are suitable for outlier detection in audio data. We compare classical PCA with its robust versions in terms of sensitivity regarding changes in the setup such as the percentage of outliers and the size or the distribution of the data sets. A crucial aspect in this context is the proper choice of number of components used for the construction of the PC space. We propose to manually label a small number of observations and to use those labels to estimate the best possible number of PCs without any prior knowledge of the data structure. This concept creates a reasonable situation for real-world applications. Thus, an estimation for the optimal number of components is performed throughout all the experiments including an analysis regarding the number of pre-labeled observations itself. Furthermore, we outline an approach for the optimization of critical values used for outlier detection by employing the additional information from the labeled observations, which can greatly increase the robustness of the outlier detection towards the number of chosen components.

2 Related work

Several authors perform simulation studies to explore the performance of the classical and various robust PCA-based methods in different scenarios in the context of outlier detection, such as varying degree of data contamination, data dimensionality, and missing data, e.g. [12][15][16][19]. For example, Pascoal et al. [12] compare the classical PCA approach [8] with five robust methods: spherical PCA [10], two projections pursuit techniques [1][2], and the ROBPCA approach [6] in different contamination schemes. The results show that ROBPCA outperforms the compared methods in terms of estimated recall. Similarly, Sapra [15] shows that a robust PCA approach based on projection pursuit [4] outperforms the classical PCA even for data sets with more variables than observations. In a recent simulation study, Xu et al. [19] show that for the generated data settings the performance of ROBPCA and techniques based on projection pursuit degrades substantially in terms of expressed variance as the dimensionality of the data increases. However, the authors only consider the first few principal components and focus on a data setting where the observations and the variables are of the same magnitude. Usually, simulation studies are performed for very specific data settings, e.g. all observations/variables follow a predefined distribution. However, real data have more complex data structures than synthetic data and, thus, outlier detection on real data is even more challenging. Current evaluations on real data sets are often limited by the number of available data. As a result, a thorough investigation of different outlier detection methods for various data settings is barely feasible. For example, Sapra [15] performs an evaluation on a small set of financial data with 120 observations. Hubert et al. [6] report evaluations on three low-sampled real data sets with varying dimensionality. While evaluations on multiple data sets provide an estimation of the robustness of the investigated approaches, no general conclusions about the sensitivity to specific data aspects can be made. Experiments with larger real data sets are commonly tailored to the evaluation of the performance of outlier detection methods for a particular data without any variation of the experimental settings, e.g. [3][17]. In contrast, we employ a large real data set in the simulation of different experimental settings and perform a thorough evaluation of the sensitivity of the explored approaches with respect to varying data aspects.

3 Evaluation setup

Compared approaches

In general, algorithms for estimating the PC space are based on either eigenvector decomposition of the empirical covariance matrix, singular value decomposition (SVD) of the (mean-centered) data matrix, or on projection-pursuit (PP) technique. We compare several approaches including both classically and robustly estimated PCs which are suitable for high-dimensional flat data. PCA-based outlier detection can be employed using two different distances for each observation derived from the PC space [6]: score distance, SD, and orthogonal distance, OD:

$$SD_i^{(k)} = \sqrt{\sum_{j=1}^{k} \frac{t_{ij}^2}{l_j}}, \quad OD_i^{(k)} = ||\mathbf{x}_i - \mathbf{P}\mathbf{t}_i||, \quad i = 1, \ldots, n, \tag{1}$$

where $\mathbf{t}_i = (t_{i1}, \ldots, t_{ik})^\top$ are the score vectors in the PC space, $\mathbf{x}_i$ is the ith observation of the data matrix $\mathbf{X}$, and the index k refers to the number of PCs. While SD represents the distance of observations in the estimated subspace to the center of data, OD measures the distance of the observations to the subspace. Two thresholds are used to detect outliers. For the SD, the 97.5% quantile of the χ^2 distribution with k degrees of freedom, i.e. $c_{SD}^{(k)} = (\chi^2_{k,0.975})^{1/2}$, and for the OD, 97.5% quantile of the standard normal distribution, $c_{OD}^{(k)} = (\hat{\mu} + \hat{\sigma} z_{0.975})^{3/2}$, can be taken as the critical values. The estimation of $\hat{\mu}$ (resp. $\hat{\sigma}$) can be obtained using the median (resp. MAD) of the values of $OD_i^{2/3}$ (see [6] for more details). If either threshold is exceeded, the respective observation is classified as an outlier.

clPCA: Classical (non-robust) PCA [8] for flat data is performed by means of SVD which is directly related to eigenvalue decomposition of the classical empirical covariance [18]. The columns of the loading matrix $\mathbf{P}$ are the right singular vectors and the variance l_j corresponding to the j-th singular value. However, the classical covariance is sensitive to outliers [6] and the resulting PCs do not describe the true data structure.

OGK PCA [11] is a PCA-based approach using robust covariance matrix estimation. The method starts by robustly scaling the data, $\mathbf{Y} = \mathbf{X}\mathbf{D}^{-1}$, where $\mathbf{D} = diag\{\hat{\sigma}(X_1) \ldots \hat{\sigma}(X_p)\}$ is the robustly estimated univariate dispersion of each column X_j of the data matrix $\mathbf{X}$, and $\hat{\sigma}$ is computed by using τ-estimation of univariate dispersion. Next, the Gnanadesikan-Kettenring estimator [5] is computed for all variable pairs of $\mathbf{Y}$ resulting in a robust correlation matrix, $\mathbf{U}$, where $U_{jk} = cov(Y_j, Y_k)$, $j, k = 1, \ldots, p$. The eigenvector decomposition of the correlation matrix $\mathbf{U} = \mathbf{E}\mathbf{\Lambda}\mathbf{E}^\top$ allows for the projection of the data onto the directions of the eigenvectors, $\mathbf{Z} = \mathbf{Y}\mathbf{E}$. Finally, the covariance matrix is transformed back to the original space, $\mathbf{S_X} = \mathbf{DELE}^\top\mathbf{D}^\top$, where $\mathbf{L} = diag\{\hat{\sigma}(Z_1) \ldots \hat{\sigma}(Z_p)\}$ and $\mathbf{DE}$ is the loading matrix of p orthogonal eigenvectors of dimension k and corresponds to the direction of the principal components.

GRID PCA [1] is a robust PCA approach using the GRID search algorithm. It employs the PP method to project the data on a direction which maximizes the robust variance of the projected data [9]. GRID first sorts the variables in decreasing order according to the robust dispersion. The first projection direction is found in the plane spanned by the first two sorted variables and it passes through the robust center and a grid point. The remaining variables successively enter the search plane to obtain the first optimal direction. The algorithm searches the subsequent directions in a similar way by imposing orthogonality until there is no improvement in maximizing the robust variance.

ROBPCA [6] combines robust PP techniques [9] with robust covariance estimation. First, the data space is reduced to an affine subspace using a SVD [7]. In the next step the least outlying observations are identified using the univariate Minimum Covariance Determinant (MCD) location and scale estimator [13]. The covariance matrix, $\mathbf{S}_0$, of the least outlying points is subsequently used to select a number of components k and to project the data on the subspace determined by the first k eigenvectors of $\mathbf{S}_0$. The FAST-MCD algorithm [14] is employed to obtain a robust scatter matrix, $\mathbf{S} = \mathbf{PLP}^\top$, where $\mathbf{P}$ is the loading matrix of p orthogonal eigenvectors of dimension k and $\mathbf{L}$ the diagonal matrix of k eigenvalues.

PCOut [3] is a method already comprising an outlier detection algorithm, in contrast to the previously described approaches. First, the observations being far away from the center of the main body of the data are identified, i.e. *location* outliers. Then, the detection of *scatter* outliers generated from a model with the same location as the main data but with a different covariance structure is conducted. Outlier detection is performed in the subspace using the robustly scaled PCs which contribute to about 99% of the total variance.

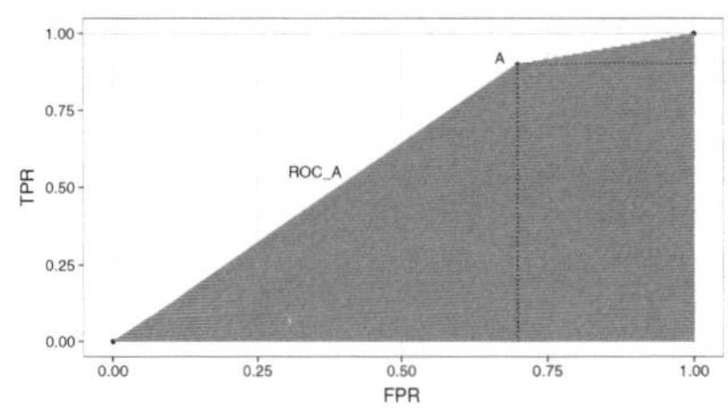

Figure 1. ROC curve construction.

Performance measures

We evaluate the performance of the compared approaches in terms of true positive rate, TPR, and false positive rate, FPR: $TPR = TP/(TP + FN), FPR = FP/(FP + TN)$, where TP refers to the *true positives* (correctly identified outliers), FN to the *false negatives* (outliers declared as normal observations), FP to the *false positives* (normal observations declared as outliers), and TN to the *true negatives* (correctly identified normal observations). Additionally, we calculate the area under the Receiver Operating Characteristics (ROC) curve (AUC) representing the trade-off between TPR and FPR by a single value. Figure 1 illustrates the construction of a ROC curve for an example evaluation A. The estimation of the corresponding AUC of A is obtained in such way that the area is divided into regular shapes and summed up which results in $AUC = 1/2\ (1 + TPR - FPR)$. Note that when the algorithm does not detect any outlier (i.e. both TPR and FPR are zero) the AUC according to the above formula is equal to 0.5. Although the two extreme scenarios (no outlier detected and random prediction) are not identical, they are both not desired output in terms of effectiveness of outlier detection approaches. It should also be noted that the number of regular observations is much higher than the number of outliers. Thus, the defined AUC measure is much more sensitive towards changes in the total number of positively identified than towards negatively identified outliers. While this looks disproportional at first, the focus of the performed evaluations is the successful detection of outliers. Therefore, in this concept the high sensitivity towards single changes in TP is a welcome side effect.

Data set

We employ a high-dimensional, real-world audio data set of approximately 8,700 observations to construct different challenging experimental settings, i.e. flat data, varying number of outliers and available training data, etc. The data set covers the three fundamental audio types: music, speech, and environmental sounds. Each observation is represented by a set of 50 (partially) multi-dimensional features, i.e. each feature consists of one or more variables, resulting in a feature vector of 679 dimensions in total. Features were selected in order to capture a wide range of audio properties and to represent the particular qualities of the three audio types equally well. The feature set comprises features that operate in the temporal and frequency domains, e.g. features for zero crossings, amplitude or brightness, features from the MPEG7 standard, perceptional features, and various cepstral coefficients.

The observations are approximately equally distributed across the three audio types. However, the underlying data structures are strongly varying due to present subpopulations of different sizes, e.g. different genres in the music samples and different voices in the speech samples. When constructing the data sets for the experiments and for the performance evaluation of the investigated approaches for outlier detection, we exploit the available labels, e.g. we define TV speech data, the largest subpopulation, as main group and select observations from environmental sounds as "outliers". This is a very challenging approach: While, usually, speech and music recordings can be easily separated by the employment of suitable features, this does not hold for environmental sounds. Environmental sounds cover a wide range of noises that sometimes have great similarities with speech data, sometimes with music and often they are just different.

The outlier detection approaches based on the two distance measures (OD and SD) employ three data sets: training, validation, and test set. The PC space spanned by k components is constructed with the observations coming from the training set. Additionally, we calculate the two critical values for the orthogonal distance, c^k_{OD}, and for the score distance, c^k_{SD}. These measures are exclusively derived from loadings and scores of the training data. Next, the observations from the validation set are projected onto the constructed PC space spanned by k PCs. An observation having an orthogonal or score distance larger than the respective critical value is declared as an outlier. This procedure is conducted with varying number of components k to select the optimal number of components, k_{opt}, in terms of maximizing AUC. The use of validation set in this context prevents potential overfitting of the estimated parameter, k_{opt}, to the characteristics of the training data. Finally, we perform an evaluation on the test data with respect to the optimal number of components k_{opt} from the validation set and the PC space spanned by k_{opt} determined by observations from the training set. Finally, we perform the evaluation on the test set using the parameters estimated on the training set.

We rescale the data to make variables comparable using the mean and standard deviation of the variables in the training set. The reason for applying a non-robust scaling is the presence of many variables which are almost constant but a small proportion of values has huge deviations. The robust MAD for such variables would be very small and this would artificially increase the whole data range during the scaling. As a consequence, many of the regular observations would be made indistinguishable from real outliers. The assignment of the observations to training, validation, and test sets is done randomly and all evaluations are based on 100 replications. Since we have a larger pool of available data, independent training and validation data were constructed repeatedly. We think this is preferable over cross-validation, which would typically be used in situations where independent validation data are not available.

4 Experimental results

In this section we present the performed experiments which focus on the sensitivity of the investigated approaches with respect to the percentage of outliers, size of training and validation sets, and data characteristics. We report results in terms of AUC, TPR, FPR, number of PCs, and the corresponding standard errors (SE) over the 100 randomly initialized replications for each experiment.

Sensitivity to the percentage of outliers

For this evaluation we consider TV recordings (the biggest speech subgroup) as regular observations and we randomly select observations from both environmental and music samples as outliers. We split the data equally into training, validation, and test sets, corresponding to approximately 360 regular observations per set.

In a first experiment, we calculate the PC space using only the regular observations from the training set and we consider different percentage of outliers for the validation and test sets: 2%, 5%, and 10% of the main observations (see Table 1). The results show that clPCA performs similar or better than the robust PCA methods, while PCOut is capable of finding only approximately half of the outliers (indicated by the low TPR). Although the performance of clPCA and its robust counterparts degrades slightly by decreasing the percentage of outliers, ROBPCA does not indicate such dependency. SE remains at a very low level during the experiments for all methods.

In a second experiment, we consider that the training set is not free of outliers in order to explore their impact on the constructed PC space. The results show that the robust PCA methods clearly outperform clPCA. PCOut performs as poorly as in the first experiment. While the number of outliers does not show any clear dependency on the resulting AUC, this is not the case for the number of PCs. GRID PCA reduces the number of selected PCs with decreasing contamination in contrast to the remaining methods.

%	Method	Pure training set				Training set with outliers			
		AUC (SE_{AUC})	k (SE_k)	TPR (SE_{TPR})	FPR (SE_{FPR})	AUC (SE_{AUC})	k (SE_k)	TPR (SE_{TPR})	FPR (SE_{FPR})
10	clPCA	0.948 (0.002)	122 (1)	0.953 (0.005)	0.058 (0.003)	0.531 (0.005)	152 (16)	0.067 (0.011)	0.004 (0.001)
	GRID PCA	0.943 (0.002)	144 (2)	0.943 (0.004)	0.056 (0.002)	0.921 (0.003)	140 (1)	0.896 (0.006)	0.053 (0.001)
	ROBPCA	0.907 (0.002)	57 (2)	0.918 (0.007)	0.103 (0.004)	0.891 (0.004)	75 (3)	0.887 (0.007)	0.106 (0.005)
	OGK PCA	0.936 (0.002)	191 (4)	0.946 (0.005)	0.074 (0.003)	0.929 (0.003)	281 (4)	0.926 (0.006)	0.068 (0.002)
	PCOut	0.602 (0.006)	- (-)	0.516 (0.060)	0.311 (0.002)	0.624 (0.004)	- (-)	0.351 (0.007)	0.103 (0.002)
5	clPCA	0.946 (0.003)	136 (2)	0.949 (0.007)	0.057 (0.003)	0.557 (0.007)	205 (15)	0.124 (0.015)	0.009 (0.002)
	GRID PCA	0.942 (0.003)	163 (2)	0.937 (0.007)	0.054 (0.002)	0.933 (0.003)	152 (2)	0.923 (0.007)	0.057 (0.002)
	ROBPCA	0.916 (0.003)	51 (2)	0.929 (0.006)	0.097 (0.004)	0.918 (0.004)	54 (3)	0.928 (0.008)	0.092 (0.004)
	OGK PCA	0.931 (0.003)	168 (5)	0.931 (0.008)	0.070 (0.003)	0.925 (0.003)	203 (6)	0.918 (0.008)	0.067 (0.004)
	PCOut	0.609 (0.007)	- (-)	0.538 (0.016)	0.320 (0.006)	0.621 (0.006)	- (-)	0.359 (0.012)	0.118 (0.005)
2	clPCA	0.912 (0.007)	164 (4)	0.865 (0.016)	0.040 (0.003)	0.565 (0.009)	173 (15)	0.141 (0.019)	0.011 (0.002)
	GRID PCA	0.937 (0.007)	190 (4)	0.860 (0.015)	0.039 (0.003)	0.914 (0.006)	174 (4)	0.872 (0.013)	0.044 (0.002)
	ROBPCA	0.913 (0.005)	39 (3)	0.911 (0.010)	0.085 (0.005)	0.918 (0.005)	39 (3)	0.916 (0.011)	0.081 (0.005)
	OGK PCA	0.905 (0.007)	129 (6)	0.862 (0.015)	0.052 (0.004)	0.908 (0.007)	139 (7)	0.865 (0.016)	0.050 (0.004)
	PCOut	0.604 (0.009)	- (-)	0.531 (0.021)	0.323 (0.006)	0.615 (0.009)	- (-)	0.420 (0.022)	0.191 (0.011)

Table 1. Evaluation results for different percentage of outliers (%).
px

ROBPCA tends to select a considerably lower number of PCs than its counterparts. The achieved results in terms of AUC suggest that the use of robust PCA methods is recommended when there is no guarantee that the training set is free of outliers. In a real-world scenario this can not always be satisfied. Therefore, we take this into account and all further experiments consider training set containing outliers.

Sensitivity to the size of training, validation, and test sets

In this experiment, we investigate whether varying the size of training, validation, and test sets considerably influences the performance of the compared approaches. Again, we consider the biggest speech subgroup as main observations and we add 5% from the instances from music and environmental sounds as outliers. We divide the data into training, validation, and test sets according to different partitions ranging from 0.33/0.33/0.33 to 0.05/0.05/0.90 corresponding to the size of sets from 378/378/380 to 57/57/1022 observations. Note that the percentage of outliers is the same in each data set.

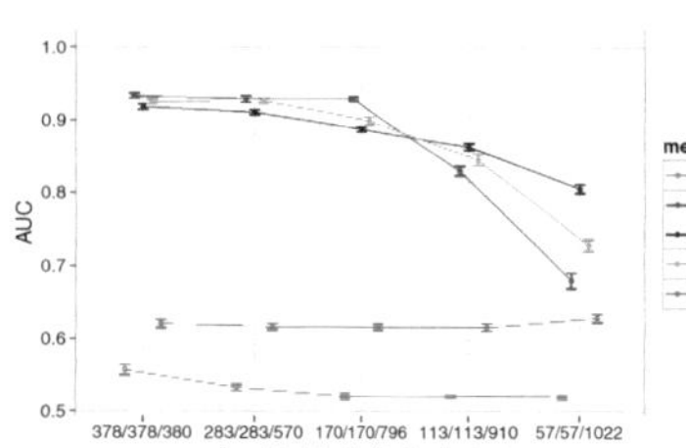

Figure 2. AUC

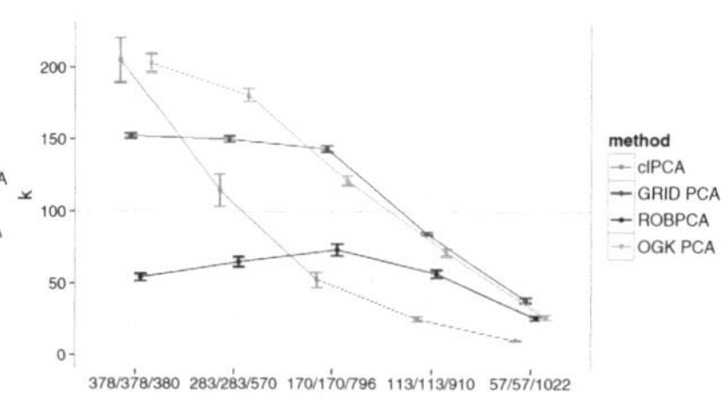

Figure 3. Number of PCs

Figure 4. Evaluation results for varying size of training, validation, and test sets.
px

Figure 4 shows the results of the evaluation in terms of AUC and number of PCs necessary to distinguish outliers from main observations. clPCA fails since the training set contains outliers. The performance of the robust PCA methods decreases with the reduction of the size of training and validation sets. GRID PCA achieves a high AUC and outperforms the remaining methods even if the size of the available training set is reduced to 170 instances. AUC falls rapidly when considering smaller data size. In contrast, ROBPCA yields still a reasonable AUC in the most extreme setting (57 observations). For a

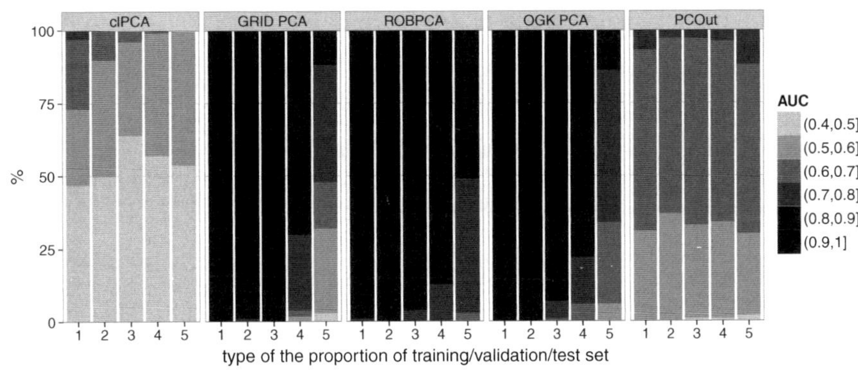

Figure 5. Detailed investigation of the resulting AUC during the replications for different partitions of training, validation, and test set (training/validation/test set) corresponding to the following size of sets: **1**: 378/378/380, **2**: 283/283/570, **3**: 170/170/796, **4**: 113/113/910, and **5**: 57/57/1022 observations.

px

more detailed investigation, we visualize the distribution of the resulting AUC during the 100 replications for each method. Figure 5 illustrates that PCOut and clPCA perform similar in each situation since the distribution of observed intervals is almost identical. This does not hold for the other three methods. The proportion of AUC ranging from 0.9 to 1 representing the results of ROBPCA decreases with the size reduction of training and validation sets. Considering the performance of OGK PCA, we observe that the largest proportion of AUC between 0.9 and 1 is attained when the sample size of training set is 283, and subsequently reduced size to 57 instances causes that the majority of AUC achieves the values between 0.5 and 0.8. The results of GRID PCA reveal very large proportion of AUC from the interval $(0.9, 1]$ in the first three situations. However, when the size of training and validation sets is reduced from 113 to 57, almost half of the AUC values are in the interval $(0.7, 0.4]$. Figure 3 shows that the number of PCs selected by ROBPCA is independent from the size of the sets and it tends to choose fewer PCs while the number of components in case of the other methods is affected by decreasing the number of observations in the training and validation sets. This is given by the method itself but also by the size of the employed training set. Moreover, the number of PCs selected by clPCA deviates considerably during the replications in the first three situations. In contrast, GRID PCA indicates small SE of the selected numbers of components.

Sensitivity to the size of the validation set

Our last experiment employed a training set containing outliers to construct the PC space and calculate two critical values. That means, the available information about labels is required only for the validation set to select the optimal number of PCs. Additionally, the results from the experiment indicated that some of the compared approaches perform well even if the size of validation set is reduced to 170 or 57 instances. These findings motivated us to explore how many observations in the validation set need to be labeled to achieve satisfying results. We fix training and test sets to the same size, 378 observations, and vary the number of observation in the validation set from 21 up to 378 instances. We simulate the biggest speech subgroup as the main observations and we add 5% from the other two audio groups as outliers. PCOut is not included to this experiment since it does not use a validation set.

Figure 6 shows that both GRID PCA and OGK PCA are sensitive to the size of the validation set. Additionally, the AUC deviates considerably with decreasing size of the validation set. In contrast, ROBPCA performs well independently from the number of instances in the validation set and achieves a high AUC even if the size of the validation set is small in comparison to the training and test set. In general, the number of PCs (see Figure 7) decreases with reducing the size of validation set and deviates during the replications. ROBPCA indicates both small SE and slight decline in the selected number of PCs. To stress our conclusion that available labeled validation data set can be small to achieve reasonable results, we change the main group to music and perform the same experiment. The size of training and test set is fixed to 168 instances. Figure 8 and Figure 9 indicate very similar performance and ROBPCA outperforms the remaining methods in all investigated situations.

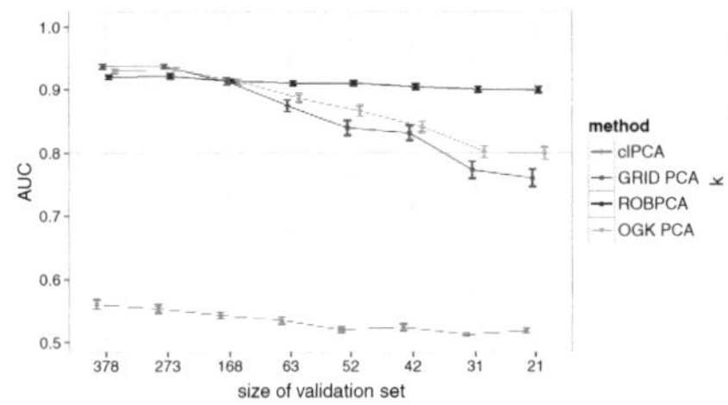

Figure 6. AUC

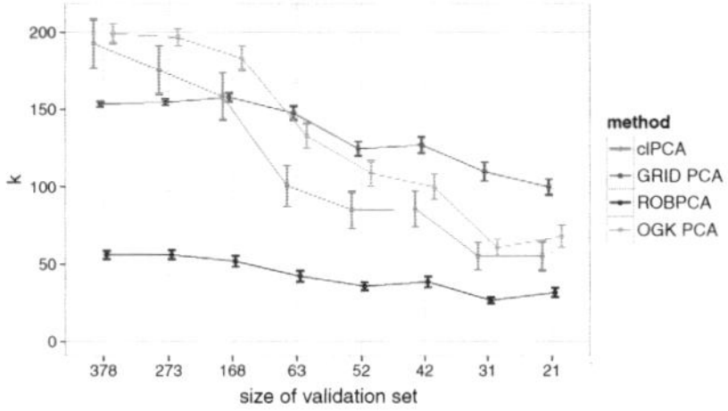

Figure 7. Number of PCs

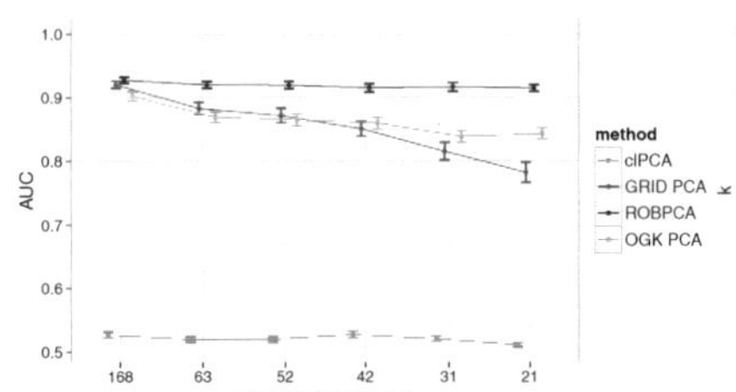

Figure 8. AUC

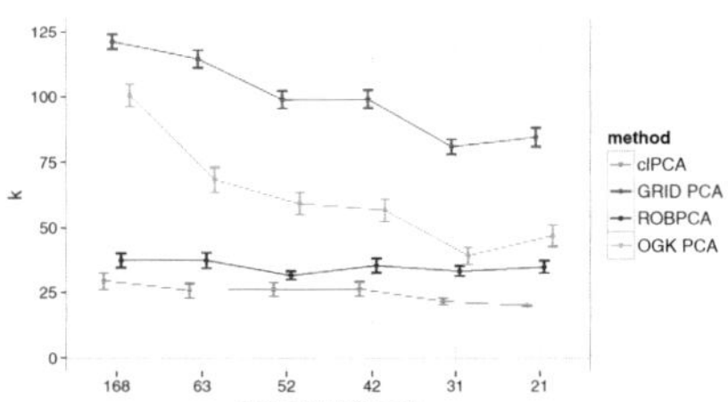

Figure 9. Number of PCs

Figure 10. Evaluation results for varying sizes of the validation set. Top row: main observations from speech data. Bottom row: music.

px

Sensitivity to the data characteristics

In this experiment we explore the sensitivity of the compared approaches to the underlying data characteristics with respect to varying data structures given by the different subpopulations in the audio dataset. We simulate the main observations consisting of three randomly selected audio subgroups with different sample size and the percentage of outliers is fixed to 5% of the corresponding main observations. We investigate the case of one majority subgroup present in the main observations and, in a next step, several subgroups. Figure 13 shows that the performance is slightly better when a single majority group is considered. Although ROBPCA and GRID PCA achieve a higher AUC, the results indicate that these two methods face difficulties in coping with multi-group data structures. clPca completely fails with AUC of 0.5. Additionally, the values for the the SE of AUC are considerably higher than in the previous experiments. Overall, there is no clear dependency between the number of PCs and the different multi-group data structure.

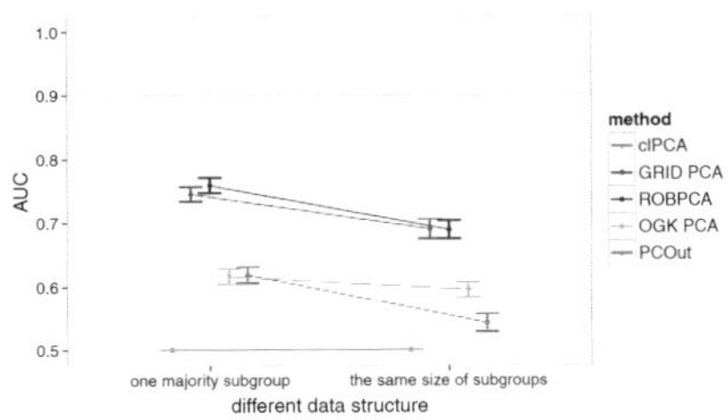

Figure 11. AUC

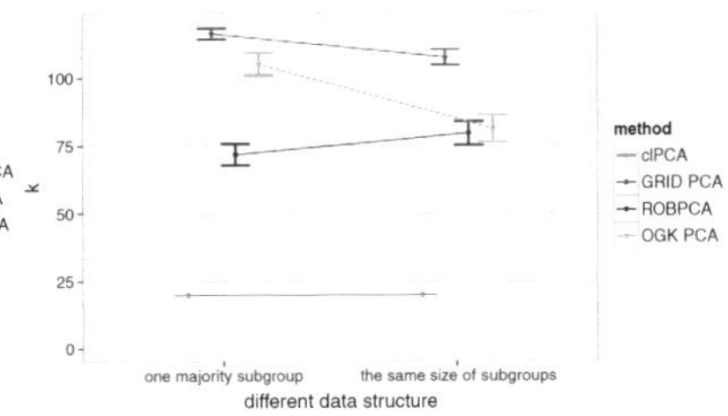

Figure 12. Number of PCs

Figure 13. Evaluation results for different data structures.
px

5 Discussion on critical values

The critical values are given by the quantiles of a χ^2 distribution for the SDs and the quantiles of the unknown distribution for the ODs which can be estimated by a robust Wilson-Hilferty approximation. Both critical values are based on the assumption of multivariate, normally distributed main observations. Those critical values are always an approximation since the distribution itself is estimated from the given observations. The central χ^2 distribution of the SDs and the non-central χ^2 distribution of ODs get distorted if the assumptions of normality are violated. In our experiments, we clearly observed data structures, which do not follow a normal distribution. We partly absorb this effect by using robust estimations. Therefore, the majority of observations can be properly modeled based on a normal distribution. To cope with the distorted distributions of the distances in addition to using robust estimations, we suggest to take advantage of the availability of validation data and to adjust the critical values. For this purpose, we can maximize the AUC performance for each fixed number of components, varying the critical values for SDs and ODs. Note, that the only meaningful critical values are the distances given by pre-labeled outliers. All other possible values will increase the FPR, without affecting the TPR. Thus, the necessary computational effort is very acceptable.

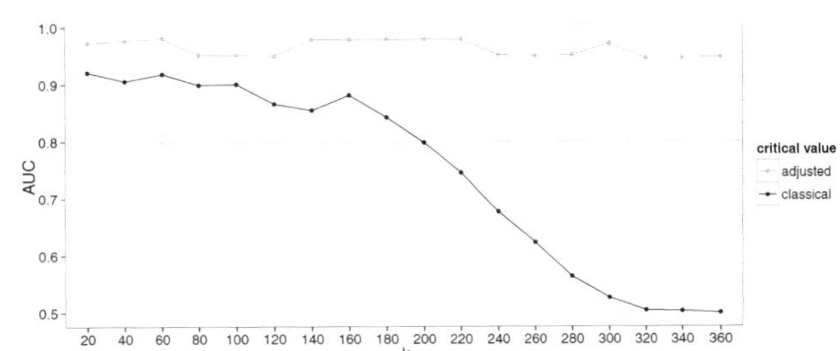

Figure 14. Comparison of AUC values depending on the number of components. While the quality of the classification for the classical critical values is highly depending on the chosen number of components, the adjusted critical values remains at an almost constant level.
px

The main benefit of this procedure is the resulting robustness towards the number of chosen PCs. Figure 14 shows this effect for ROBPCA for one example of speech main observations with 5% outliers. The experiment indicates that performing the outlier detection for a low number of PCs is sufficient. Thus, even though the adjustment needs computation time, the total computational effort decreases, since it is no longer necessary to calculate a whole range of different numbers of PCs. At the same time, the risk of choosing an inappropriate number of PCs vanishes with increasing number of observations. It can be easily shown that the adjustment will asymptotically always perform at least as good as the theoretical critical values with increasing numbers of validation observations. If the theoretical assumptions of multivariate normal distribution holds where observations with large Mahalanobis distance are classified as outliers, the adjustment converges to the provided theoretical critical value due to the law of large numbers. For any non-normal distribution it converges to the respective true critical value and, therefore, it outperforms the theoretical critical values, derived from false assumptions. However, for large number of observations, especially outlying observations, the adjustment converges. Thus, the method should only be used to analyze setups where enough outlying observations allow for a proper estimation of the ROC curve.

6 Conclusion

In this paper we compared different PCA-based algorithms for outlier detection in the context of a high-dimensional audio data set. Since the classical PCA [8] is sensitive to the presence of outliers in the training data, we employed several, well-established robust PCA methods, such as GRID PCA [1], ROBPCA [6], OGK [11], and PCOUT [3], to better reveal outlying samples. We performed a thorough investigation of the sensitivity of the employed approaches with respect to different data properties, percentage of outliers, and size of the available training data. In all of those settings, ROBPCA performed at the same level as the GRID and OGK algorithms. However, ROBPCA showed much lower sensitivity towards changes in the number of available training and validation observations. The reason for this property is the fewer necessary number of PCs to properly model the data structure. If the number of available observations is too low to create the necessary PCA space or to properly evaluate the used PCA space, the quality of the outcome decreases. We therefore recommend the usage of ROBPCA for outlier detection in similar setups where few pre-labeled observations allow for the individual estimation of a proper number of PCs. Further utilization of pre-labeled observations is possible by adjusting the critical values for outlier detection if the observations do not follow a normal distribution. In such a situation, if the number of observations, especially outliers, is big enough, the adjustment can significantly improve the quality of the proposed procedures, providing a more robust set of critical values, which are able to cope with skewed distributions.

Acknowledgments

This work has been partly funded by the Vienna Science and Technology Fund (WWTF) through project ICT12-010 and by the K-project DEXHELPP through COMET - Competence Centers for Excellent Technologies, supported by BMVIT, BMWFW and the province Vienna. The COMET program is administrated by FFG.

Bibliography

[1] Croux, C., Filzmoser, P. and Oliveira, M. (2007) *Algorithms for Projection-Pursuit Robust Principal Component Analysis.* Chemometr. and Intell. Lab. Sys., **41**, 15:1–15:58.

[2] Croux, C. and Ruiz-Gazen, A. (2005) *High breakdown estimators for principal components: the projection-pursuit approach revisited.* Journal of Multivariate Analysis, **95(1)**, 206–226.

[3] Filzmoser, P., Maronna, R. and Werner, M. (2008) *Outlier Identification in High Dimensions.* Computational Statistics & Data Analysis, **52**, 1694–1711.

[4] Filzmoser, P., Serneels, S., Croux, C. and Van Espen, P. (2006) *Robust Multivariate Methods: The Projection Pursuit Approach* From Data and Information Analysis to Knowledge Engineering, 81–124.

[5] Gnanadesikan, R. and Kattenring, J. R. (1972) *Robust Estimates, Residuals, and Outlier Detection with Multiresponce Data.* Biometrics, **28**, 81–124.

[6] Hubert, M., Rousseeuw, P. and Vanden Branden, K. (2005) *ROBPCA: A New Approach to Robust Principal Component Analysis.* Technometrics, **47**, 64–79.

[7] Hubert, M., Rousseeuw, P. and Verboven, S. (2002) *A fast method for robust principal components with applications to chemometrics.* Chemometr. Intell. Lab., **60**, 101–111.

[8] Jolliffe, I.T. (2002) *Principal Component Analysis.* Springer.

[9] Li, G. and Chen, Z. (1985) *Projection-Pursuit Approach to Robust Dispersion Matrices and Principal Components: Primary Theory and Monte Carlo.* Journal of the American Statistical Association, **80**, 759–766.

[10] Locantore, N. et al. (1999) *Robust principal component analysis for functional data.* Test, **8(1)**, 1–73.

[11] Maronna, R. and Zamar, R. (2002) *Robust estimates of location and dispersion for high-dimensional data sets.* Technometrics, **43**, 307–317.

[12] Pascoal, C., Oliveira, M., Pacheco, A. and Valadas, R. (2010) *Detection of outliers using robust principal component analysis: A simulation study.* Combining Soft Computing and Statistical Methods in Data Analysis, 499–507.

[13] Rousseeuw, P. J. (1984) *Least Median of Squares Regression.* Journal of the American Statistical Association, **79**, 871–880.

[14] Rousseew, P. and van Driessen, K. (1999) *A fast algorithm for the minimum covariance determinant estimator.* Journal of the American Statistical Association, **41**, 212–223.

[15] Sapra, K. S. (2010) *Robust vs. classical principal component analysis in the presence of outliers.* Applied Economics Letters, **17(6)**, 519–523.

[16] Serneels, S. and Verdonck, T. (2008) *Principal component analysis for data containing outliers and missing elements.* Computat. Statistics & Data Analysis, **52(3)**, 1712–1727.

[17] Shyu, M.-L., Chen, S.-C., Sarinnapakorn, K. and Chang, L. (2003) *A novel anomaly detection scheme based on principal component classifier.* Tech. report, DTIC Document.

[18] Wall, M. E, Rechtsteiner, A. and Rocha, L. M. (2003) *Singular value decomposition and principal component analysis.* A practical approach to microarray data analysis.

[19] Xu, H., Caramanis, C. and Mannor, S. (2013) *Outlier-robust pca: The high-dimensional case.* Local journal of interesting topics research, **59(1)**, 546–572.

[20] Zimek, A., Schubert, E. and Kriegel, H.-P. (2012) *A survey on unsupervised outlier detection in high-dimensional numerical data.* Stat. Anal. and Data Mining, **5(5)**, 363–387.

Supervised-Component based Cox Regression

Xavier Bry, *University of Montpellier*, `xavier.bry@univ-montp2.fr`
Théo Simac, *University of Montpellier*, `theo.simac@umontpellier.fr`
Thomas Verron, *SEITA-ITG*, `thomas.verron@fr.imptob.com`

Abstract. In survival analysis with high-dimensional regressors, the Cox Proportional Hazard Model (CPHM) encounters instability problems owing to regressor-collinearity. Its regularisation is therefore needed. Two family of methods currently perform regularisation of regression models: penalty-based methods such as Ridge and Lasso, and component-based models such as PLS-type regression models. Only the latter enable exploratory analysis of predictive structures by making components lean on the regressors' strong correlation structures. We propose a new way to take into account the structural relevance of components within the estimation procedure of the Cox Model. Our algorithm, called SCCoxR (for Supervised-Component based Cox Regression), is tested on simulated data, and then, applied to life-history data of HIV-positive Thai subjects, in order to model the age at disclosure of their serologic status.

Keywords. Component, Cox Regression, Proportional Hazard Model, PLS Cox Regression, Regularisation, SCGLR, Survival analysis, THEME.

1 Introduction

We consider a survival time Y, depending on a block of covariates X, plus a set of extra-covariates Z. X and Z can be time-dependent. There may be non-informative right-censoring on Y. Variables in Z are few and exhibit low or no correlation. By contrast, variables in X are many and possibly redundant, so that the CPHM demands regularisation with respect to X. Now, there are several approaches to regularisation. One is through penalising the norm of the coefficient vector (in the Ridge and Lasso ways) [9, 11]. Another is using components that maximise, at some point, a trade-off between the model's goodness-of-fit and a measure of the component's structural strength (e.g. its variance, for PLS-related Component-based techniques [4, 2, 8]). Only component-based methods enable a full exploration of the regressor-space in terms of interpretable predictive structures. [1] proposed a practical PLS-extension of Cox's regression. This extension consists in applying the Cox regression on each regressor separately and then building the first component by summing the corresponding linear predictors. Each further component is built likewise after deflating the regressors on the former components. By contrast, we propose to redesign the CPHM estimation algorithm, basing it on a component to be calculated so as to be both structurally strong and predictive. Recently, [5] have proposed a flexible notion of Structural Relevance (SR) of a component, extending its variance. This work proposes a way to combine it with the likelihood of the model in a new CPHM estimation procedure: SCCoxR, yielding a regularised estimation based on a sequence of supervised components.

2 Cox's Proportional Hazard Model

Notations:

- y_i is the survival-time or censoring-time of unit i.
- $x_{i,t}$ and $z_{i,t}$ are the vectors of covariates X and Z respectively for unit i at time t.
- Censoring-indicator δ is defined through: $\forall i, \delta_i = 1$ if the event occurs for i at time y_i , and $\delta_i = 0$ if i is censored at time y_i.
- $R(t)$ denotes the set of all individuals at risk at time t.

The model

Here, Cox's PHM is thus based on the following formulation of the hazard function of unit i at time t: $h(t; x_{i,t}, z_{i,t}) = h_0(t)e^{\beta' x_{i,t}+\gamma' z_{i,t}}$, where $h_0(t)$ is the baseline hazard function. The survival function having hazard function $h_0(t)$ is the baseline survival function (BSF).

The classical estimation of β and γ

The partial likelihood (PL) defined by [6] is, when there are no simultaneous events:

$$l_p(\beta, \gamma; X, Z) = \prod_{i=1}^{n} \left[\frac{e^{\beta' x_{i,y_i}+\gamma' z_{i,y_i}}}{\sum_{j\in R(y_i)} e^{\beta' x_{j,y_i}+\gamma' z_{i,y_i}}} \right]^{\delta_i}$$

The PL can, under some assumptions, be interpreted as a marginal likelihood of events' ranks [6]. It involves only (β, γ), which can be estimated by $(\hat{\beta}, \hat{\gamma})$ maximising $\log l_p(\beta, \gamma; X, Z)$ through a Newton-Raphson algorithm. Then, [7, 3], among others, proposed an estimation of the BSF, based on $(\hat{\beta}, \hat{\gamma})$.

3 Structural relevance

Structural Relevance (SR) has been introduced by [5]. Let weight-matrix $W = n^{-1}I_n$ reflect the a priori uniform importance of units. Let X be a $n \times p$ matrix associated with a variable block, and M an associated $p \times p$ symmetric positive definite (s.p.d.) matrix, the purpose of which is to "weight" X's variables appropriately. M may take various forms according to the types of variables, as well as the type of structures one would like the components to align on. Basically, one may choose M such that PCA of X with metric matrix M and weight-matrix W is relevant. But indeed, the choice of M is flexible and can adapt a wider range of situations (see section "Particular instances of SR measures" for examples). Finally consider component $f = Xu$, where u is constrained by: $\|u\|^2_{M^{-1}} = 1$.

General formula of structural relevance

Given a set of J "reference" s.p.d. matrices $N = \{N_j, j = 1, \ldots, J\}$ encoding types of structures of interest ("target"-spaces, e.g. variables in X), a weight system $\Omega = \{\omega_j, j = 1, \ldots, J\}$, and a scalar $l \geqslant 1$, the associated Structural Relevance (SR) measure is defined as the following function of u:

$$\phi_{N,\Omega,l}(u) := \left(\sum_{j=1}^{J} \omega_j (u' N_j u)^l \right)^{\frac{1}{l}} \tag{1}$$

In (1), the value of l tunes the locality of the bundles of structures coded in N. The larger the value of l, the more local the bundle.

Particular instances of SR measures

- Component Variance:

$$\phi(u) = V(Xu) = \|Xu\|_W^2 = u'(X'WX)u$$

This is the inertia of units along direction $\langle u \rangle$, and is maximised by the first (direct) eigenvector in the PCA of (X, M, W).

In practice, explanatory variables are often a mixture of numeric and nominal variables. Assume that $X = [x^1, \ldots, x^K, X^1, \ldots, X^L]$, where: $x^1, \ldots, x^K$ are column-vectors coding the numeric regressors, and $X^1, \ldots, X^L$ are blocks of centred indicator variables, each block coding a nominal regressor (X^l has $q_l - 1$ columns if the corresponding variable has q_l levels, the removed level being taken as "reference level"). We should then consider the following M, which bridges ordinary PCA of numeric variables with Multiple Correspondence Analysis of nominal variables:

$$M := \text{diag}\left\{ \left(x^{1\prime}Wx^1\right)^{-1}, \ldots, \left((x^{K\prime}Wx^K\right)^{-1}, \left((X^{1\prime}WX^1\right)^{-1}, \ldots, \left((X^{L\prime}WX^L\right)^{-1} \right\}$$

- Variable Powered Inertia (VPI):

We impose $\|f\|_W^2 = 1$ through $M = (X'WX)^{-1}$. For a block X consisting of p standardised numeric variables x^j, the VPI is defined as:

$$\phi(u) = \left(\sum_{j=1}^{p} \omega_j \rho^{2l}(Xu, x^j) \right)^{\frac{1}{l}} = \left(\sum_{j=1}^{p} \omega_j (u'X'Wx^j x^{j\prime} WXu)^l \right)^{\frac{1}{l}}$$

For a block X consisting of p categorical variables X^j, each of which is coded through the set of its centred indicator variables less one, the VPI is:

$$\phi(u) = \left(\sum_{j=1}^{p} \omega_j \cos^{2l}(Xu, \langle X^j \rangle) \right)^{\frac{1}{l}} = \left(\sum_{j=1}^{p} \omega_j \langle Xu | \Pi_{X^j} Xu \rangle_W^l \right)^{\frac{1}{l}}$$

$$\text{where: } \Pi_{X^j} = X^j (X^{j\prime} W X^j)^{-1} X^{j\prime} W$$

4 The Supervised-Component-based Cox Regression

The component-model

The idea is to replace regressor-block X with a block $F = XU$ of R structurally relevant orthogonal components, in the CPHM. The regressors being time-dependent, so will the components be. Let X be the matrix whose columns are the X-regressors and whose N rows are the individuals-at-risk-at-time-points: (i, t). Orthogonality of the f's will be taken with respect to matrix $\Omega = \frac{1}{N} Id$. The hazard function of a unit i at time t will thus be:

$$h(t; x_{i,t}, z_{i,t}) = h_0(t) e^{\delta' f_{i,t} + \gamma' z_{i,t}} = h_0(t) e^{\delta' U' x_{i,t} + \gamma' z_{i,t}}$$

Components will be calculated hierarchically, starting with one, and each extra- component being constrained to be orthogonal to the former ones.

Once the R components have been calculated, a standard Cox-regression is performed on $[F^R, Z]$, where $F^R := [f^1, \ldots, f^R]$, yielding linear predictor:

$$\eta = \mu + F^R\delta + Z\gamma \;=\; \mu + XU^R\delta + Z\gamma \;=\; \mu + X\beta + Z\gamma \text{ , where } \beta = U^R\delta$$

If $\mathbf{X}$ denotes the original *uncentered*-variable matrix, then: $X = \mathbf{X} - 1_N'\Omega\mathbf{X}$, so:

$$\eta = \alpha + \mathbf{X}\beta + Z\gamma, \quad \text{with } \beta = U^R\delta \text{ and } \alpha = \mu - 1_N'\Omega\mathbf{X}\beta \tag{2}$$

Combining the partial-likelihood with the SR

When we look for rank- r-component $f^r = Xu^r$, we consider former components F^{r-1} known, and we impose the following orthogonality constraint:

$$F^{r-1'}\Omega f^r = 0 \;\Leftrightarrow\; D_r'u^r = 0 \text{ with } D_r = X'\Omega'F^{r-1}$$

We combine l_p with the SR $\phi_X(u^r)$ into the following criterion:

$$S(u^r;s) = l_p(u^r,\delta,\gamma)^{1-s}\phi_X^s(u^r)$$

We then solve the program:

$$P: \quad \max_{\substack{u^r,\delta,\gamma \text{ s.t.:} \\ u^{r\prime}M^{-1}u^r=1;\; D_r'u^r=0}} \left[(1-s)\log l_p(u^r,\delta,\gamma) + s\log\phi_X(u^r)\right]$$

by iteratively maximising the criterion with respect to u^r and to (δ,γ), in turn.

- Maximisation with respect to u^r: given (δ,γ), the criterion is maximised on u^r through the PING algorithm given in appendix.
- Maximisation with respect to (δ,γ): given u^r, the criterion is maximised on (δ,γ) exactly as in the classical Cox-Regression on covariates $\{f^r, Z\}$.

The SCCoxR algorithm

A number R of components to be calculated is chosen. Let D_1 = null-matrix and $F^0 = \emptyset$. Then:

Step 1 For $r = 1$ to R:

- Calculate u^r as the solution of P
- Set $F^r = [F^{r-1}; f^r]$ and $D_{r+1} = X'\Omega'F^r$

Step 2 Perform Cox-regression on regressors $[F^R, Z]$.

Step 3 Calculate the coefficients of the original variables through (2)

Model-assessment

There are three tuning-parameters in our method: s (importance of SR-based regularisation), l (bundle-locality), and number R of components. These are not independent, e.g.: $s = 0$ meaning no regularisation should lead to a single component (the predictor of the classical Cox regression). The higher s is, the higher R should be. Therefore, s should never be too high.

For a given (s,l,R), the cross-validation "error" coefficient (CV) is calculated according to the technique proposed for the CPHM in [10]: the sample is split into K parts, and for each part k, we calculate:

$$CV_k(s,l,R) = l(\theta_{-k}(s,l,R)) - l_{-k}(\theta_{-k}(s,l,R))$$

where $\theta = (\alpha,\beta,\gamma)$, l_{-k} is the log-partial likelihood excluding part k of the data, and $\theta_{-k}(s,l,R)$ is the optimal $\theta(s,l,R)$ obtained on the non-left out data. The overall $CV(s,l,r)$ is the average of $\{CV_k(s,l,R); k = 1,\ldots,K\}$.

Parameter l is more of a fine-tuning one with respect to s. So, we propose to first consider the minimum value $l = 1$, and tune s by trying a sequence of values from $s = .5$ downwards, calculating for each s the R^*_{s1} that yields the highest CV. Thus, we get the optimal $\left(s^*, R^*_{s^*1}\right)$ for $l = 1$. Then, keeping $s = s^*$, tune the structure-locality by trying a sequence of values from $l = 1$ upwards, calculating for each l the $R^*_{s^*l}$ that yields the highest CV. Hence the best $\left(l^*_{s^*}, R^*_{s^*l^*_{s^*}}\right)$. Of course, we can consider iterating the procedure starting with $l = l^*_{s^*}$ if the cost is not too high.

5 A short simulation study

Simulation scheme

We consider $n = 100$ independent individuals. Let φ^k, $k \in \{1, 2, 3\}$ be three independent standard normal variables simulated in $\mathbb{R}^{100}$. We first calculate the following three latent variables:

$$f^1 = \varphi^1; \quad f^2 = \varphi^1 \cos\left(\frac{2\pi}{5}\right) + \varphi^2 \sin\left(\frac{2\pi}{5}\right); \quad f^3 = \varphi^3$$

Then, we simulate three regressor-bundles about the f^k's: $B_1 = \left\{x^j, j = 1, \ldots, 10\right\}$, $B_2 = \left\{x^j, j = 11, \ldots, 20\right\}$ and $B_3 = \left\{x^j, j = 21, \ldots, 35\right\}$, as follows. Let $\left\{\varepsilon^j, j = 1, \ldots, 35\right\}$ be i.i.d. variables with distribution $\mathcal{N}(0, \sigma^2)$, $\sigma = 0.2$; we define:

$$\forall x^j \in B_k : x^j = f^k + \varepsilon^j$$

Within each bundle, regressors are thus highly correlated. An exponentially-distributed survival time T is then simulated for each individual:

$$\forall i \in \{1, \ldots, 100\} : T_i \sim \Gamma(1, \lambda_i), \text{ with } \lambda_i = \exp(0.7 f_i^1 - 0.4 f_i^2)$$

Note that B_3 contains more regressors than B_1 and B_2 do, and that its central direction f^3 plays no role in T's model: B_3 is the strongest structure in X, but as far as T is concerned, it is a nuisance. Finally, we simulate an independent exponentially-distributed censoring variable: $\forall i$, $C_i \sim \Gamma(1, 0.2)$, and put: $\forall i$, $Y_i = T_i \wedge C_i$. In the simulated sample, 5 observations happen to be censored.

Model estimation

Classical Cox-regression of Y on f^1, f^2, f^3 yields:

	Estimated coefficient in predictor	P-value
f^1	0.691	1.9e-07
f^2	-0.242	2.5e-02
f^3	0.017	8.8e-01

Now, in practice, only the x's are known, and because of their high number and collinearities, Cox's regression cannot identify the explanatory dimensions. SCCoxR was carried out on X with integer values of l ranging from 1 to 10, and $s = 0.5$. Components 1 and 2 proved, in all cases, to be the only ones having a highly significant effect on the hazard. The correlation-scatterplots of Figure 1 show that, whatever l, the first explanatory plane is that of B_1 and B_2, B_3 being cast behind it. When $l = 1$, the first component lies in-between B_1 and B_2. Increasing l to 5 causes the first component to better align on B_1, which has the greatest explanatory power, and the second component on B_2.

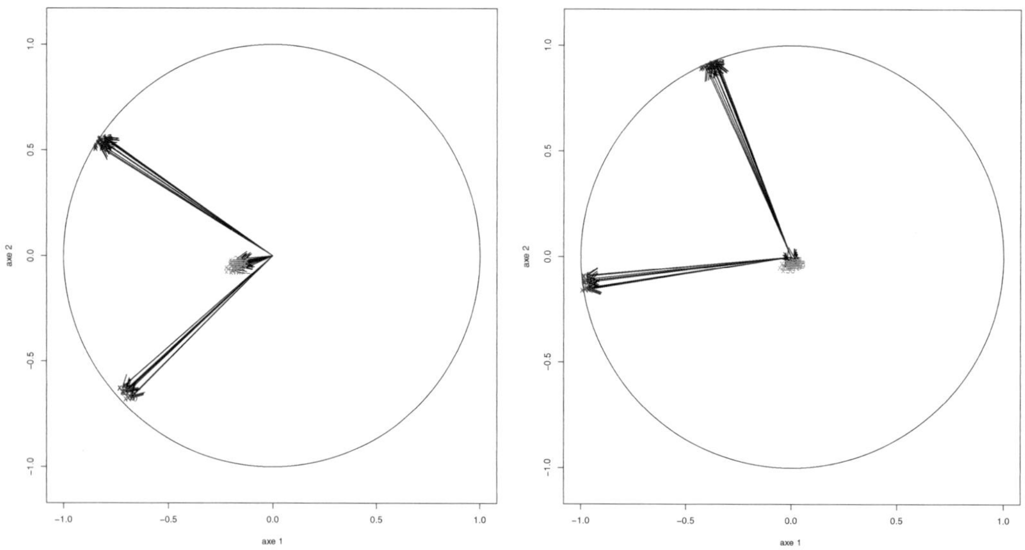

Figure 1. Simulated data ; correlation-scatterplots of the regressors on the first two components, with $l = 1$ (left hand side) and $l = 5$ (right hand side).

6 Application

Our algorithm is now applied to life-histories of 453 HIV-positive Thai subjects investigated in a retrospective survey. The aim is to find out the main explanatory dimensions that can model the age at which their serologic status was disclosed. The datafile contains 75 variables. 269 status-disclosure events have been observed. The date of the survey is a non-informative right-censoring time, since all subjects do not know their status at that time. The high number of variables and their redundancy preclude direct Cox Regression. SCCoxR was carried out with integer values of l ranging from 1 to 10, and $s \in \{0.1, 0.2, \ldots, 0.9\}$. Components 1 and 2 proved, in all cases, to be the only ones having a highly significant effect on the disclosure risk. The best results (components 1 and 2 both close to observed variables and with a very significant effect) were obtained for $s = 0.8$ and $l = 8$. The correlation-scatterplots of Figure 2 show that increasing l helps better focus component 2 on an observed variable bundle.

Acknowledgements

The authors wish to thank Eva Lelièvre (INED, Paris) and Sophie Le Cœur (IRD 174, Chiang Mai, Thailand) for kindly providing the application data, collected in the TEEWA (Teens living With Antiretrovirals) project, funded by Sidaction and Oxfam GB.

Appendix

The Projected Iterated Normed Gradient (PING) algorithm Notation: the current value of any quantity a on iteration t is denoted: $a^{[t]}$. Consider program:

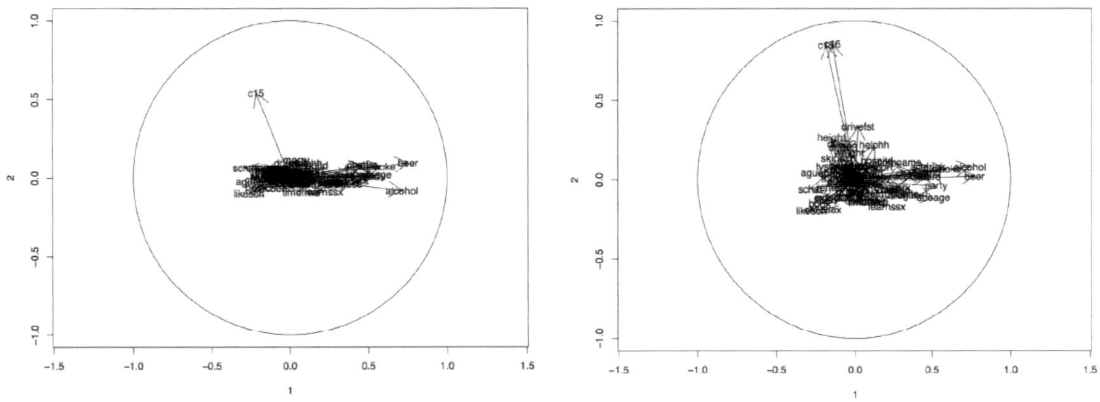

Figure 2. Real data ; correlation-scatterplots of the regressors on the first two components, with $l = 1$ (left hand side) and $l = 8$ (right hand side).

$$\max_{\substack{u \in \mathbb{R}^p, u'M^{-1}u=1 \\ D'u=0}} h(u)$$

Putting $v = M^{-1/2}u$, $g(x) = h(M^{1/2}x)$ and $C = M^{1/2}D$, this is strictly equivalent to:

$$R_C : \max_{\substack{v \in \mathbb{R}^p, v'v=1 \\ C'v=0}} g(v)$$

The corresponding lagrangian is:

$$L(v, \lambda, \mu) = g(v) - \lambda(v'v - 1) - \mu'C'v \tag{3}$$

- $\nabla_{\lambda,\mu} L(v, \lambda, \mu) = 0 \Leftrightarrow$

$$\begin{aligned} v'v &= 1 && (4) \\ C'v &= 0 && (5) \end{aligned}$$

- $\nabla_{v} L(v, \lambda, \mu) = 0$

$$\Leftrightarrow \Gamma(v) - 2\lambda v - C\mu = 0 \tag{6}$$

$$\Leftrightarrow v = \frac{1}{2\lambda}(\Gamma(v) - C\mu) \tag{7}$$

with $\Gamma(v) := \nabla_{v} g(v)$

Note that

$$C'(6) \text{ with } (5) \Rightarrow C'\Gamma(v) = C'C\mu \ \Leftrightarrow \ \mu = (C'C)^{-1}C'\Gamma(v)$$

Put back into (4), this yields:

$$v = \frac{1}{2\lambda}\Pi_{C^\perp}\Gamma(v) \quad \text{where } \Pi_{C^\perp} := I - C(C'C)^{-1}C' \tag{8}$$

Note that in the particular case where $C = 0$, we shall take: $\Pi_{C^\perp} = I$. Finally, (8) and (4) imply:

$$v = \frac{\Pi_{C^\perp}\Gamma(v)}{\|\Pi_{C^\perp}\Gamma(v)\|}$$

This gives the basic iteration of the PING algorithm:

$$v^{[t+1]} = \frac{\Pi_{C^\perp}\Gamma(v^{[t]})}{\|\Pi_{C^\perp}\Gamma(v^{[t]})\|}$$

Let us show that this iteration follows a direction of ascent Since, by construction: $\forall s : v^{[s]} \perp C$, we have:

$$\begin{aligned}\forall s:\ v^{[s]} = \Pi_{C^\perp} v^{[s]} \ \Rightarrow\ \langle v^{[t+1]} - v^{[t]} | \Gamma(v^{[t]})\rangle &= \langle \Pi_{C^\perp}(v^{[t+1]} - v^{[t]}) | \Gamma(v^{[t]})\rangle \\ &= \langle v^{[t+1]} - v^{[t]} | \Pi_{C^\perp}\Gamma(v^{[t]})\rangle\end{aligned}$$

which has the sign of:

$$\langle v^{[t+1]} - v^{[t]} | v^{[t+1]}\rangle = 1 - \langle v^{[t]} | v^{[t+1]}\rangle = 1 - \cos(v^{[t]}, v^{[t+1]}) \geq 0$$

Picking a point on a direction of ascent does not guarantee that g actually increases, since we may "go too far" in this direction. Let $\gamma^{[t]} := \frac{\Pi_{C^\perp}\Gamma(v^{[t]})}{\|\Pi_{C^\perp}\Gamma(v^{[t]})\|}$. Staying "close enough" to the current starting point on the arc $(v^{[t]}, \gamma^{[t]})$ guarantees that g increases. Indeed, let ϖ be the plane tangent to the sphere on $v^{[t]}$ and let w denote the vector tangent to arc $(v^{[t]}, \gamma^{[t]})$ on $v^{[t]}$ (cf. Figure (3)). Then:

$$\exists \tau > 0, w = \tau \Pi_{\varpi}\gamma^{[t]} \ \Rightarrow\ \langle w | \gamma^{[t]}\rangle = \tau\langle \Pi_{\varpi}\gamma^{[t]} | \gamma^{[t]}\rangle = \tau \cos^2(\gamma^{[t]}, \varpi) > 0$$

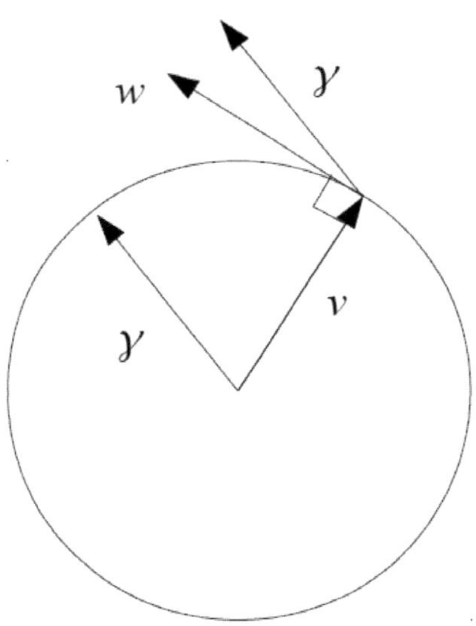

Figure 3. PING vectors.

Yet, staying "too close" to the current starting point on the arc $(v^{[t]}, \gamma^{[t]})$ may make the algorithm too slow to reach the maximum. To avoid that, we propose to use a Gauss-Newton unidimensional maximization to find the maximum of $g(v)$ on the arc $(v^{[t]}, \gamma^{[t]})$, and take it as $v^{[t+1]}$. The fixed point of the resulting algorithm is a critical point of (3), hence a local maximum of g s.t. $C'v = 0$.

Bibliography

[1] Bastien, P. (2007) *Deviance residuals based PLS regression for censored data in high dimensional setting.* Chemometrics and Intelligent Laboratory Systems, 78–86.

[2] Bastien, P., Esposito Vinzi, V. and Tenenhaus, M. (2005) *PLS generalised linear regression.* Computational Statistics & Data Analysis, **48**, 17–46.

[3] Breslow, N.E. and Crowley, J. (1974) *A large-sample study of the life table and product limit estimates under random censorship.* Annals of Statistics, **2**, 437–454.

[4] Bry, X. and Antoine, P. (2004) *Explorer l'explicatif ; application l'analyse biographique.* Population-F, 59, **6**, 909–945.

[5] Bry, X., Verron, T. (2015) *THEME: THEmatic Model Exploration through Multiple Co-Structure maximization.* Journal of Chemometrics, **12**, 637–647.

[6] Cox, D.R. (1975) *Partial Likelihood*, Biometrika, **62**, 269–276.

[7] Kalbfleisch, J.D. and Prentice, R.L. (1973) *Marginal likelihoods based on Cox's regression and life model.* Biometrika **60**, 267–278.

[8] Nygard, S., Borgan, O., Lingjaerde, O.C. and Storvold, H.L. (2008) *Partial least squares Cox regression for genome-wide data.* Lifetime Data Anal, 14, **2**, 179–95.

[9] Simon, N., Friedman, J., Hastie, T. and Tibshirani, R. (2011) *Regularization Paths for Cox's Proportional Hazards Model via Coordinate Descent.* JSS, 39, **5**.

[10] van Houwelingen, H.C., Bruinsma, T., Hart, A.A.M., van't Veer, L.J., Wessels, L.F.A. (2006) *Cross-Validated Cox Regression on Microarray Gene Expression Data.* Statistics in Medicine, **25**, 3201–3216.

[11] Lai, Y., Hayashida, M. and Akutsu, T. (2013) *Survival Analysis by Penalized Regression and Matrix Factorization.* The Scientific World Journal, Article ID 632030.

Test of mean difference in longitudinal data based on block resampling approaches

Hirohito Sakurai, *National Center for University Entrance Examinations*, sakurai@rd.dnc.ac.jp
Masaaki Taguri, *National Center for University Entrance Examinations*, tagurimm@yahoo.co.jp

Abstract. In this paper, we focus on a two-sample problem, and propose two block resampling testing methods with permutation analogy for comparing the difference of two means in longitudinal data when the data of two groups are not paired. In order to detect mean difference of two samples, we consider the following four types of test statistics: (i) sum of absolute values of difference between two mean sequences, (ii) sum of squares of difference between two mean sequences, (iii) estimator of area-difference between two mean curves, and (iv) difference of kernel estimators based on two mean sequences. The block resampling techniques considered in our paper include circular block bootstrap and stationary bootstrap, and are used to approximate the null distributions of the above test statistics. Monte Carlo simulations are conducted to examine the size and power of the testing methods.

Keywords. Circular block bootstrap, Stationary bootstrap, Moving block bootstrap, Two-sample problem, Size and power of test.

1 Introduction

Comparison of two means or regression curves of two samples is one of the important topics in statistics and related fields. Suppose that there are two samples given by $\{Y_i(t)\}_{i=1}^{q_1}$ and $\{X_j(t)\}_{j=1}^{q_2}$ for $t = 1, \ldots, n$, and assume that they are mutually independent, where q_1 and q_2 are numbers of subjects, and n is the number of observed points. We also assume that, for fixed t, $Y_i(t)$ and $X_j(t)$ are independent over q_1 and q_2 subjects, respectively. Then we consider the model

$$\begin{cases} Y_i(t) = p_1(t) + \varepsilon_i(t), & i = 1, \ldots, q_1, \\ X_j(t) = p_2(t) + \eta_j(t), & j = 1, \ldots, q_2, \end{cases} \tag{1}$$

where $p_1(t)$ and $p_2(t)$ are unknown regression functions, and $\varepsilon_i(t)$ and $\eta_j(t)$ are the error terms having means 0 and finite variances, respectively. Then, we are interested in a testing problem

$$H_0 : p_1(t) = p_2(t) \text{ for all } t \quad \text{vs.} \quad H_1 : p_1(t) \neq p_2(t) \text{ for some } t, \tag{2}$$

where H_0 and H_1 denote the null and alternative hypotheses.

An example of (1) is wind velocity data shown in Figure 1. The data are measured by an artificial satellite (left panel) and a radar on the earth (right panel), and are obtained at altitudes from 80 km to 90 km every 1 km (($q_1 = q_2 =$)11 subjects) during 13 days ($t = 1, \ldots, 13$). From Figure 1, we want to know whether the mean behavior of the two devices in measuring wind velocity is equal or not. The problem of our interest is then formulated as (2).

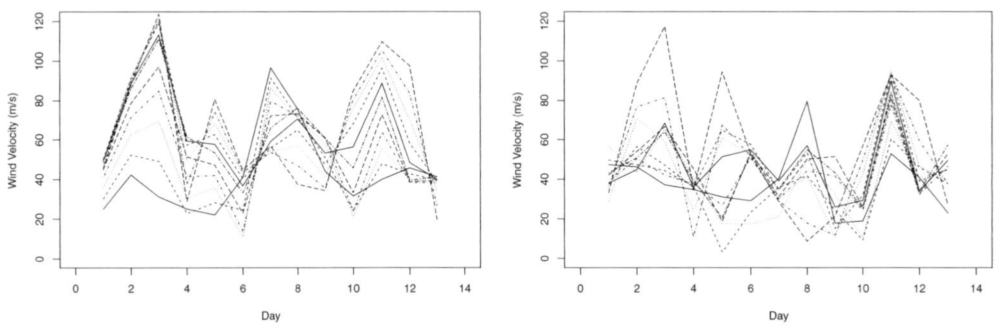

Figure 1. Wind velocity data (left: satellite, right: radar)

There are several ways to address the problem (2). If we cannot assume the normality of the error terms in (1), nonparametric methods are available. For example, graphical method by [2] could be a possible choice. Another approach would be an application of nonparametric bootstrap. In the case of time series analysis, block resampling could be applicable; see, for example, [4], and references therein. Block resampling techniques used for the problem (2) can be found in [8], [10], [11], and [9]. The papers [8], [10] and [11] examine the cases where block resampling is done *with replacement*, and refer to moving block bootstrap, circular block bootstrap and stationary bootstrap, respectively. On the other hand, [9] considers the case where block resampling is done *without replacement*, and refers to moving block bootstrap. However, in the case where resampling is done *without replacement*, the behaviors of testing methods using circular block bootstrap and stationary bootstrap are not examined.

In this paper, we focus on a two-sample problem when longitudinal data from two groups are given, and propose two testing methods for detecting the difference of two means in longitudinal data based on block resampling approach. Our methods generate blocks of observations similar to circular block bootstrap [6] and stationary bootstrap [7], and resampling is done *without replacement*. As seen in the numerical studies given below, our testing methods are superior to [2] in power. Further, we carry out comparison of level and power properties of the above methods including the testing method proposed by [9] in order to clarify the mutual relationships of the level and power properties among the tests using circular block bootstrap, stationary bootstrap and moving block bootstrap.

The rest of the paper is organized as follows. In Section 2, we review the testing method using moving block bootstrap [9], and propose two testing methods using circular block bootstrap and stationary bootstrap, for which p-values (achieved significance levels) are calculated. In order to investigate the properties of size and power of the above three methods, Monte Carlo simulations are conducted in Section 3, and some concluding remarks and results of the above data analysis are summarized in Section 4.

2 Testing methods using block resampling

In order to detect the difference between two means, $p_1(t)$ and $p_2(t)$, in (1), we focus on the behavior of four types of test statistics given below. First, for the testing problem (2), the following statistic proposed by [3] is available:

$$S_n = S_n(D_1, \ldots, D_n) = \left[\sum_{j=0}^{n-1}\left(\sum_{t=j+1}^{j+g} D_t\right)^2\right]\left[n\sum_{t=1}^{n-1}\frac{(D_{t+1}-D_t)^2}{2}\right]^{-1}, \tag{3}$$

where $D_t = Y_t - X_t$ for $t = 1, \ldots, n$ or $D_t = Y_{t-n} - X_{t-n}$ for $t = n+1, \ldots, n+g$, $Y_t = \sum_{i=1}^{q_1} Y_i(t)/q_1$, $X_t = \sum_{j=1}^{q_2} X_j(t)/q_2$, $g = [np]$ is the integer part of np, and p is a tuning constant satisfying $0 < p < 1$ which is determined by the fully data-driven approach; the second approach described in [3, pp.1043–1044]. The statistic (3) is essentially based on kernel estimators of $p_1(t)$ and $p_2(t)$. As another type of test statistics, we can also consider

$$T_{1n} = T_{1n}(D_1, \ldots, D_n) = \sum_{t=1}^{n} |D_t|, \tag{4}$$

$$T_{2n} = T_{2n}(D_1, \ldots, D_n) = \sum_{t=1}^{n} D_t^2, \tag{5}$$

$$T_{3n} = T_{3n}(D_1, \ldots, D_n) = \frac{1}{2}\sum_{t=1}^{n-1}(|D_t| + |D_{t+1}|)I_+ + \frac{1}{2}\sum_{t=1}^{n-1}\frac{|D_t|^2 + |D_{t+1}|^2}{|D_t| + |D_{t+1}|}I_-, \tag{6}$$

where $I_+ = I\{D_tD_{t+1} \geq 0\}$, $I_- = I\{D_tD_{t+1} < 0\}$, and $I\{\cdot\}$ is the indicator function, respectively. The expression of (6) seems to have a complicated form, however it is a naive estimator of area-difference $A = \int |p_1(t) - p_2(t)|\, dt$ constructed by the trapezoidal rule with linear interpolations of adjacent observation values. The values (3), (4), (5) and (6) will be small when H_0 is true, while they are large when H_0 is false. Therefore, these statistics enable us to measure difference between $p_1(t)$ and $p_2(t)$. Since A is 0 under H_0 and positive under H_1, the hypothesis of our interest reduces to testing

$$H_0 : A = 0 \quad \text{vs.} \quad H_1 : A > 0. \tag{7}$$

We now briefly explain the testing method for (2) or (7) using the moving block bootstrap with permutation analogy proposed by [9]. The main ideas of [9] are to generate blocks of observations in each sample similar to moving block bootstrap [5], and to draw resamples *without replacement* from the mixed blocks which are generated by two samples. The latter is motivated from the technique that can reflect the null hypothesis by resampling from a combined sample. Combining observations of two samples and drawing resamples with replacement from the combined sample are previously considered in [1] for test of homogeneity of scale, and in [12] for test of equality of two means.

For simplicity, let T be a generic notation for T_{1n}, T_{2n}, T_{3n} or S_n. For a given significance level α, the testing method, which we call "Mixed Moving Block Resampling test" (for short, Mixed MBR test) in this paper, is summarized in Algorithm 2.

Next we propose a testing algorithm using the idea of circular block bootstrap [6] with permutation analogy. We call it "Mixed Circular Block Resampling test" (Mixed CBR test). In this procedure, blocks of observations in each sample are generated similar to [6], and Steps 2 and 3 of Algorithm 2 are changed as follows.

Further, we propose a testing algorithm using blocks similar to stationary bootstrap [7] with permutation analogy. We call it "Mixed Stationary Resampling test" (Mixed SR test). In this method, blocks of observations in each sample are generated similar to [7], and Steps 2 and 3 of Algorithm 2 are changed as follows.

3 Numerical study

In this section, in order to investigate the finite-sample behavior of the proposed testing methods, we calculate size and power of Mixed CBR and SR tests based on Algorithms 2 and 2 via Monte Carlo

---

Step 1 Calculate $t_{obs} = T(D_1, \ldots, D_n)$.

Step 2 Divide centered samples, $\{C_{y,1}, \ldots, C_{y,n}\}$ and $\{C_{x,1}, \ldots, C_{x,n}\}$, into $k(= n - \ell + 1)$ successive overlapping blocks with each length ℓ, and put the collection of blocks $\xi_y = \{\xi_{y,1}, \ldots, \xi_{y,k}\}$ and $\xi_x = \{\xi_{x,1}, \ldots, \xi_{x,k}\}$, where $C_{y,t} = Y_t - \bar{Y}$, $C_{x,t} = X_t - \bar{X}$, $\bar{Y} = \sum_{t=1}^{n} Y_t/n$, $\bar{X} = \sum_{t=1}^{n} X_t/n$, $\xi_{y,t} = \{C_{y,t}, \ldots, C_{y,t+\ell-1}\}$ and $\xi_{x,t} = \{C_{x,t}, \ldots, C_{x,t+\ell-1}\}$ $(t = 1, \ldots, k)$, respectively. The combined blocks are denoted by

$$\xi_{\text{pooled}} = \{\xi_{y,1}, \ldots, \xi_{y,k},\ \xi_{x,1}, \ldots, \xi_{x,k}\}.$$

Step 3 Draw $2m$ blocks, $\xi_y^{*b} = \{\xi_{y,1}^{*b}, \ldots, \xi_{y,m}^{*b}\}$ and $\xi_x^{*b} = \{\xi_{x,1}^{*b}, \ldots, \xi_{x,m}^{*b}\}$, without replacement from ξ_{pooled} to obtain resamples $Y^{*b} = \{Y_1^{*b}, \ldots, Y_n^{*b}\}$ and $X^{*b} = \{X_1^{*b}, \ldots, X_n^{*b}\}$ $(b = 1, \ldots, B)$, where $m = [n/\ell]$ (if n/ℓ is an integer) or $m = [n/\ell] + 1$ (otherwise), and $[n/\ell]$ is the integer part of a real n/ℓ.

Step 4 Calculate $t^{*b} = T(D_1^{*b}, \ldots, D_n^{*b})$, where $D_t^{*b} = Y_t^{*b} - X_t^{*b}$.

Step 5 Repeating Steps 3 and 4 an appropriate number of times B, calculate $t^{*1}, \ldots, t^{*B}$.

Step 6 From Steps 1 and 5, approximate the achieved significance level by $\widehat{\text{ASL}} = \sum_{b=1}^{B} I\{t^{*b} \geq t_{obs}\}/B$, and reject H_0 when $\widehat{\text{ASL}} \leq \alpha$.

Step 2 Divide centered samples, $\{C_{y,1}, \ldots, C_{y,n}\}$ and $\{C_{x,1}, \ldots, C_{x,n}\}$, into n collections of blocks, $\xi_y = \{\xi_{y,1}, \ldots, \xi_{y,n}\}$ and $\xi_x = \{\xi_{x,1}, \ldots, \xi_{x,n}\}$, where $\xi_{y,t}$ and $\xi_{x,t}$ are blocks of length ℓ, obtained in the manner of circular block bootstrap [6]. The combined blocks are denoted by

$$\xi_{\text{pooled}} = \{\xi_{y,1}, \ldots, \xi_{y,n}\, \xi_{x,1}, \ldots, \xi_{x,n}\}.$$

Step 3 Draw $\xi_y^{*b} = \{\xi_{y,1}^{*b}, \ldots, \xi_{y,m}^{*b}\}$ and $\xi_x^{*b} = \{\xi_{x,1}^{*b}, \ldots, \xi_{x,m}^{*b}\}$ randomly without replacement from ξ_{pooled} to obtain resamples $Y^{*b} = \{Y_1^{*b}, \ldots, Y_n^{*b}\}$ and $X^{*b} = \{X_1^{*b}, \ldots, X_n^{*b}\}$, where m is defined in Algorithm 2.

---

simulations. For comparison, we also examine those of Mixed MBR test based on Algorithm 2. Further, our numerical study includes the comparison with the test [2] for unpaired data (hereafter termed "BY" for short) as a preceding testing method.

All our results are based on independent 2000 pairs of two samples, $\{Y_i(t)\}$ and $\{X_j(t)\}$, where $B = 2000$ replications of resampling in our tests are applied to every two samples, and the nominal level of test is $\alpha = 0.05$. We generate initial two samples according to (1) whose means are specified by $p_1(t) = 0$ and $p_2(t) = c$, where $c = 0, 0.2, 0.4, 0.6, 0.8, 1.0$; $c = 0$ or $c \neq 0$ corresponds to the null hypothesis or the alternative hypothesis being true. As for the error terms, $\varepsilon_i(t)$ and $\eta_j(t)$, we consider the following Gaussian AR(1) errors: $\varepsilon_i(t) = \phi\varepsilon_i(t-1) + z_i(t)$ and $\eta_j(t) = \phi\eta_j(t-1) + z_j(t)$, where $z_i(t) \overset{i.i.d.}{\sim} N(0, \tau_1^2)$, $z_j(t) \overset{i.i.d.}{\sim} N(0, \tau_2^2)$, $\phi = 0, \pm 0.1, \pm 0.2$, $\tau_1^2 = \tau_2^2 = (1 - \phi^2)\text{V}(\varepsilon_i(t))$, and $\text{V}(\varepsilon_i(t)) = 1, 3, 5$. For $n = 10$ points, the cases of $(q_1, q_2) = (10, 10)$, $(10, 20)$, $(10, 30)$, $(20, 20)$, $(20, 30)$, $(30, 30)$ are examined. Due to limitations of space, however, we restrict ourselves to discussing the case of $\text{V}(\varepsilon_i(t)) = 3$ in this paper.

Since it is preferable that the empirical level is nearly equal to the nominal level α, our choice of ℓ in Mixed MBR, CBR and SR tests is made so that the empirical level is close to α. If there are some

Step 2 Divide $\{C_{y,1},\ldots,C_{y,n}\}$ and $\{C_{x,1},\ldots,C_{x,n}\}$ into n collections of blocks as follows:

$$\xi_y=\{\xi_y(1,L_1),\ldots,\xi_y(n,L_n)\},\quad \xi_x=\{\xi_x(1,M_1),\ldots,\xi_x(n,M_n)\},$$

where $\xi_y(t,\ell)$ and $\xi_x(t,\ell)$ are the blocks starting from $C_{y,t}$ and $C_{x,t}$ with length $\ell(\geq 1)$, $L_1,\ldots,L_n,\ M_1,\ldots,M_n$ are independent and identically distributed to a geometric distribution with parameter $1/\ell$. The combined blocks are denoted by

$$\xi_{\text{pooled}}=\{\xi_y(1,L_1),\ldots,\xi_y(n,L_n),\ \xi_x(1,M_1),\ldots,\xi_x(n,M_n)\}.$$

Step 3 Resamples corresponding to two samples are generated as follows.
(a) Draw $K_{y,b}$ and $K_{x,b}$ blocks randomly without replacement from ξ_{pooled}, and put

$$\xi_y^{*b}=\{\xi(I_1^{*b},L_1^{*b}),\ldots,\xi(I_{K_{y,b}}^{*b},L_{K_{y,b}}^{*b})\},\quad \xi_x^{*b}=\{\xi(J_1^{*b},M_1^{*b}),\ldots,\xi(J_{K_{x,b}}^{*b},M_{K_{x,b}}^{*b})\},$$

where $\xi(t,\ell)=\xi_y(t,\ell)$ (if $1\leq t\leq n$), $\xi(t,\ell)=\xi_x(t,\ell)$ (otherwise), $I_1^{*b},\ldots,I_{K_{y,b}}^{*b},\ J_1^{*b},\ldots,J_{K_{x,b}}^{*b}$ are independent and identically distributed to a discrete uniform distribution on $\{1,\ldots,n,\ n+1,\ldots,2n\}$. A pair of random variables, (I_i^{*b},L_i^{*b}) or (J_i^{*b},M_i^{*b}), is one of $\{(1,L_1),\ldots,(n,L_n),$
$(1,M_1),\ldots,(n,M_n)\}$, and $K_{y,b}=\min\{k:\sum_{i=1}^k L_i^{*b}\geq n\}$, $K_{x,b}=\min\{k:\sum_{j=1}^k M_j^{*b}\geq n\}$, respectively.
(b) Construct resamples, $Y^{*b}=\{Y_1^{*b},\ldots,Y_n^{*b}\}$ and $X^{*b}=\{X_1^{*b},\ldots,X_n^{*b}\}$, by putting the first n elements of ξ_y^{*b} and ξ_x^{*b}, respectively.

candidates which have the same level error, we make the conservative choice, viz., we choose ℓ such that the empirical level is less than the nominal level. Further, if there are some candidates whose empirical levels are equal, we select ℓ to maximize the empirical power among them. The resulting choices of such ℓ are given in Table 1. It is worth noting that S_n does not need longer ℓ than T_{rn} $(r=1,2,3)$ to keep the nominal level as is shown in Table 1.

We first summarize the results of the level studies in Table 2. The table shows that the empirical levels of Mixed MBR, CBR and SR tests with T_{rn} $(r=1,2,3)$ and S_n tend to keep the nominal level α when $\phi\leq 0$, however it is not true for most cases of $\phi=0.2$. When $\phi>0$, the level error of T_{1n}, T_{2n} and T_{3n} seems to be slightly larger. From a viewpoint of our level studies, there is no significant difference among Mixed MBR, CBR and SR tests on the whole. On the other hand, BY test shows a tendency to underestimate the nominal level for most cases except for $(q_1,q_2)=(10,10)$.

Next, we present the results of the power studies in Figures 2 and 3 to compare the behaviors of the proposed tests with those of Mixed MBR and BY tests. The vertical and horizontal axes of Figures 2 and 3 are the empirical power and c $(0\leq c\leq 1)$ defined above, respectively. Since we found similar tendencies among the six cases of (q_1,q_2), we show the results for $(q_1,q_2)=(10,30),(20,20)$ with $\phi=\pm 0.2$. Figures 2 and 3 compare the empirical power corresponding to T_{1n},T_{2n},T_{3n},S_n, and BY when the block resampling testing method is fixed. We can observe that the empirical power of T_{3n} in Mixed MBR, MCR and SR tests is the largest of these five statistics, and that the overall relationship of power in each test together with BY test is given by $T_{3n}\geq T_{2n}\geq T_{1n}\geq S_n\geq$ BY. This indicates the numerical superiority of Mixed MBR, CBR and SR tests using T_{rn} $(r=1,2,3)$ and S_n in power. The superiority of T_{3n} in power is especially confirmed from Figures 2 and 3. As the number of subjects increases, the empirical power is improved. For $0\leq c\leq 1$, the empirical power of T_{1n} is nearly equal to that of T_{2n}, however the latter is slightly higher than the former for most cases. Figures 4 and 5 show the comparison of empirical power

			Mixed MBR				Mixed CBR				Mixed SR			
q_1	q_2	ϕ	T_{1n}	T_{2n}	T_{3n}	S_n	T_{1n}	T_{2n}	T_{3n}	S_n	T_{1n}	T_{2n}	T_{3n}	S_n
10	10	−0.2	6	4	4	2	9	9	9	2	4	4	9	4
		−0.1	2	2	2	1	2	9	2	1	4	4	6	2
		0	3	2	2	1	9	9	6	1	9	7	6	1
		0.1	3	2	2	1	7	4	6	1	6	7	6	1
		0.2	2	3	2	1	6	4	6	1	6	5	6	1
10	20	−0.2	6	8	4	2	2	9	9	2	5	4	5	3
		−0.1	8	2	3	2	2	9	4	2	5	5	6	3
		0	3	3	2	1	2	2	3	1	8	7	6	2
		0.1	3	3	2	2	4	3	7	1	6	7	6	1
		0.2	3	3	3	2	7	3	6	2	6	6	9	1
10	30	−0.2	6	6	1	2	3	2	1	2	5	4	6	4
		−0.1	5	7	2	1	3	2	2	1	5	8	9	3
		0	7	4	4	1	2	2	9	1	5	5	9	3
		0.1	4	3	4	1	3	3	9	1	9	9	9	1
		0.2	3	3	3	1	4	9	9	1	9	9	9	1
20	20	−0.2	7	6	6	2	9	9	9	2	3	9	3	3
		−0.1	2	3	2	2	9	9	2	2	4	4	6	2
		0	3	3	2	1	9	5	3	1	7	6	5	2
		0.1	3	4	3	1	3	4	8	1	8	6	6	2
		0.2	2	4	4	1	8	4	8	1	5	5	5	1
20	30	−0.2	6	2	4	2	2	2	9	2	4	4	9	4
		−0.1	7	5	2	1	2	2	6	1	6	7	7	4
		0	3	3	2	1	2	8	7	1	8	5	7	2
		0.1	3	3	2	2	3	6	7	1	5	6	6	1
		0.2	5	3	2	2	7	6	7	2	7	6	6	1
30	30	−0.2	5	8	4	2	2	9	9	2	4	4	3	3
		−0.1	3	4	2	1	9	2	3	2	4	4	5	3
		0	3	2	2	1	9	5	4	1	6	7	5	2
		0.1	3	2	3	1	6	6	9	1	7	6	5	1
		0.2	3	3	4	3	4	6	9	1	7	6	9	1

Table 1. Optimum ℓ in Mixed MBR, CBR and SR tests for $\alpha = 0.05$ and $\mathrm{V}(\varepsilon_i(t)) = 3$

corresponding to Mixed MBR, CBR and SR tests when the test statistic is fixed. From these figures, we can observe that there are some cases where the empirical power of Mixed SR test is superior to that of Mixed MBR and CBR tests, however there is little difference or no significant difference on the whole among the three block resampling tests for most cases.

4 Data analysis and some concluding remarks

In this paper we have proposed two testing methods for detecting the difference of two means in longitudinal data, in which blocks of observations are generated similar to circular block bootstrap and stationary bootstrap with permutation analogy. Further, we have compared these methods with that whose blocks of observations are generated similar to moving block bootstrap. Our simulation results indicate the applicability of Mixed MBR, CBR and SR tests for weakly dependent data even when the sample size is very small.

Applying Mixed MBR, CBR and SR tests with every possible (mean) block length and BY test to the data given in Figure 1, we obtain the results that H_0 is rejected by the following tests when $\alpha = 0.05$: Mixed MBR with S_n for $\ell = 4, \ldots, 9, 11, 12$; Mixed CBR with S_n for $\ell = 4, \ldots, 12$; and Mixed SR with S_n for $\ell = 6, \ldots, 12$. BY test also rejects H_0. Therefore, there is a possibility of the significant

			Mixed MBR				Mixed CBR				Mixed SR				
q_1	q_2	ϕ	T_{1n}	T_{2n}	T_{3n}	S_n	T_{1n}	T_{2n}	T_{3n}	S_n	T_{1n}	T_{2n}	T_{3n}	S_n	BY
10	10	−0.2	0.051	0.050	0.042	0.047	0.030	0.034	0.050	0.045	0.046	0.060	0.049	0.050	0.066
		−0.1	0.048	0.051	0.052	0.054	0.041	0.047	0.051	0.051	0.049	0.049	0.054	0.054	0.067
		0	0.049	0.064	0.064	0.049	0.051	0.049	0.058	0.046	0.050	0.051	0.074	0.056	0.060
		0.1	0.066	0.079	0.087	0.057	0.048	0.049	0.080	0.058	0.053	0.062	0.102	0.040	0.063
		0.2	0.086	0.106	0.118	0.065	0.059	0.071	0.107	0.066	0.075	0.086	0.129	0.044	0.063
10	20	−0.2	0.043	0.046	0.052	0.054	0.027	0.026	0.051	0.049	0.064	0.038	0.051	0.047	0.040
		−0.1	0.049	0.050	0.051	0.060	0.034	0.034	0.050	0.057	0.040	0.046	0.058	0.045	0.040
		0	0.053	0.058	0.074	0.051	0.043	0.046	0.059	0.049	0.051	0.049	0.076	0.060	0.041
		0.1	0.068	0.075	0.102	0.064	0.050	0.050	0.076	0.060	0.060	0.062	0.098	0.041	0.041
		0.2	0.088	0.101	0.133	0.066	0.051	0.064	0.098	0.067	0.078	0.086	0.127	0.050	0.041
10	30	−0.2	0.049	0.048	0.049	0.052	0.026	0.023	0.042	0.047	0.036	0.059	0.050	0.052	0.045
		−0.1	0.056	0.046	0.051	0.043	0.035	0.030	0.049	0.044	0.047	0.044	0.057	0.042	0.046
		0	0.050	0.051	0.070	0.059	0.045	0.044	0.050	0.059	0.056	0.048	0.071	0.050	0.045
		0.1	0.069	0.071	0.091	0.065	0.045	0.049	0.066	0.065	0.054	0.053	0.092	0.045	0.045
		0.2	0.089	0.086	0.119	0.070	0.051	0.053	0.088	0.070	0.072	0.076	0.121	0.057	0.046
20	20	−0.2	0.046	0.050	0.050	0.057	0.037	0.033	0.051	0.057	0.053	0.049	0.049	0.045	0.024
		−0.1	0.050	0.048	0.051	0.059	0.043	0.043	0.052	0.056	0.054	0.046	0.052	0.050	0.027
		0	0.050	0.059	0.068	0.050	0.051	0.051	0.058	0.049	0.048	0.050	0.074	0.058	0.027
		0.1	0.067	0.075	0.094	0.061	0.049	0.053	0.083	0.061	0.056	0.067	0.101	0.040	0.026
		0.2	0.086	0.102	0.118	0.076	0.064	0.079	0.108	0.077	0.077	0.090	0.127	0.052	0.024
20	30	−0.2	0.050	0.052	0.053	0.059	0.025	0.027	0.042	0.051	0.049	0.050	0.049	0.049	0.027
		−0.1	0.051	0.051	0.053	0.048	0.032	0.035	0.050	0.046	0.058	0.044	0.047	0.048	0.026
		0	0.047	0.051	0.070	0.053	0.043	0.049	0.060	0.051	0.051	0.055	0.067	0.037	0.025
		0.1	0.072	0.074	0.098	0.062	0.047	0.048	0.082	0.061	0.057	0.061	0.091	0.044	0.026
		0.2	0.088	0.100	0.137	0.080	0.059	0.070	0.101	0.080	0.073	0.082	0.125	0.051	0.026
30	30	−0.2	0.043	0.052	0.051	0.050	0.035	0.035	0.050	0.047	0.044	0.040	0.050	0.048	0.025
		−0.1	0.051	0.055	0.056	0.056	0.043	0.049	0.048	0.053	0.050	0.051	0.052	0.059	0.024
		0	0.058	0.070	0.078	0.052	0.051	0.050	0.062	0.053	0.050	0.051	0.066	0.054	0.023
		0.1	0.071	0.087	0.101	0.070	0.051	0.057	0.080	0.072	0.059	0.067	0.090	0.048	0.022
		0.2	0.091	0.112	0.127	0.077	0.065	0.074	0.097	0.077	0.073	0.084	0.113	0.053	0.021

Table 2. Empirical levels for $\alpha = 0.05$ and $\mathrm{V}(\varepsilon_i(t)) = 3$

difference between the satellite and radar in measuring wind velocity. However, the problem on block length selection in the block resampling is very important, and the development of a fully data-driven approach to selecting (mean) block length in the above tests will be needed for practical data analyses. Further, extending the testing methods to several curves and/or grouped data problems would be required in the future.

Acknowledgement

The authors would like to thank two anonymous referees for their constructive comments that led to a clearer presentation of this paper. The research of the first author was supported in part by Grant-in-Aid for Scientific Research (15K00065) from Japan Society for the Promotion of Science.

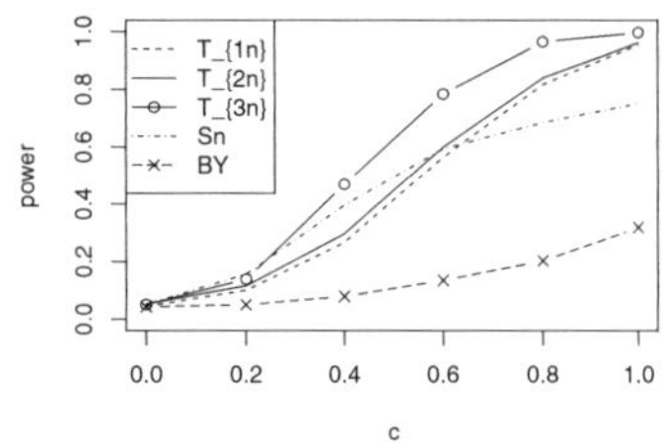

(a) Mixed MBR ($\phi = -0.2$)

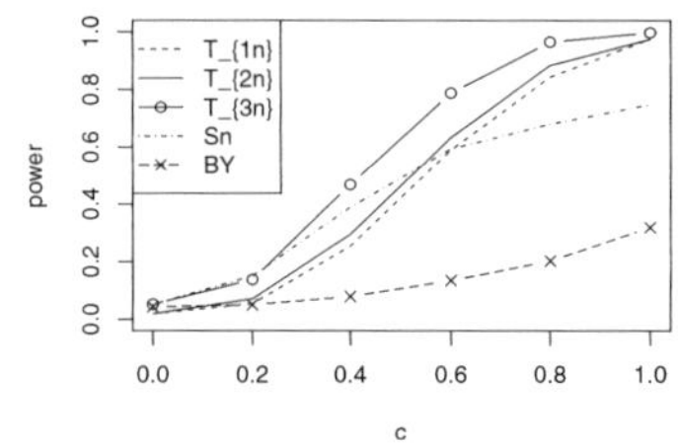

(b) Mixed CBR ($\phi = -0.2$)

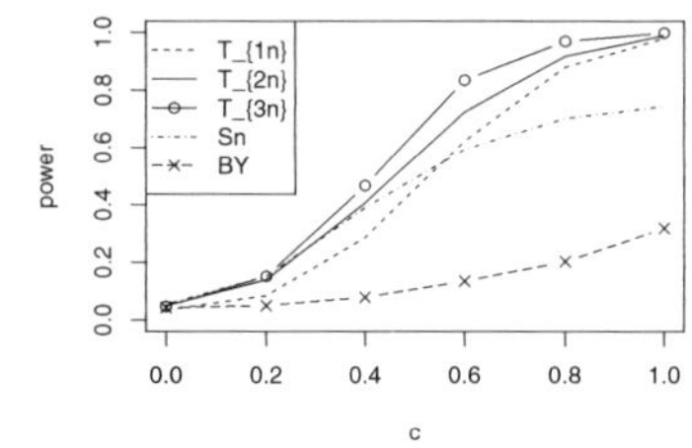

(c) Mixed SR ($\phi = -0.2$)

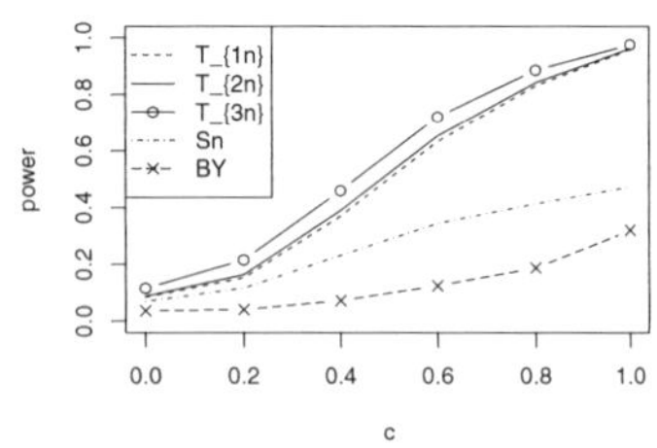

(d) Mixed MBR ($\phi = 0.2$)

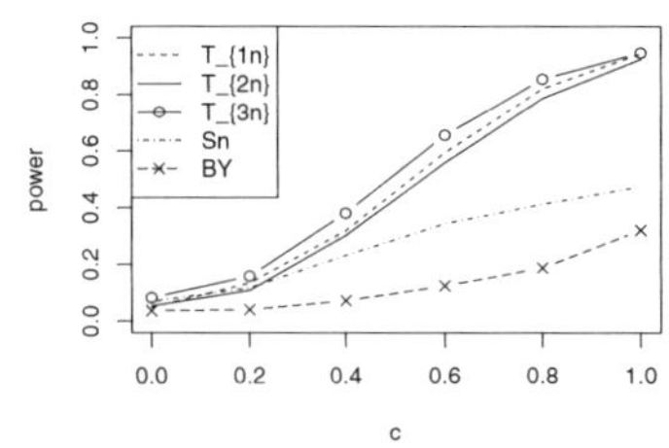

(e) Mixed CBR ($\phi = 0.2$)

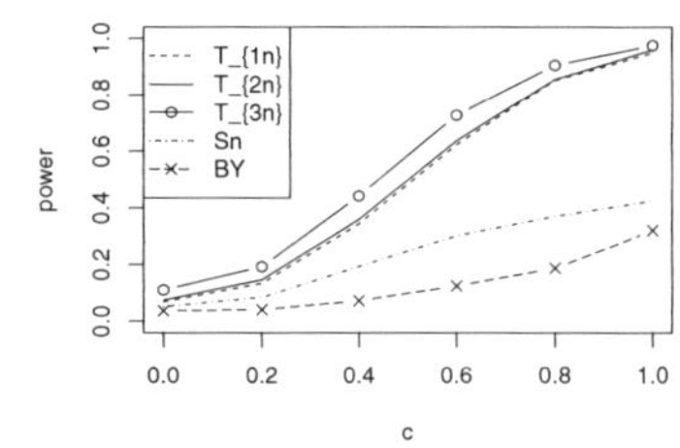

(f) Mixed SR ($\phi = 0.2$)

Figure 2. Comparison of T_{1n}, T_{2n}, T_{3n}, S_n and BY for $(q_1, q_2) = (10, 30)$ and $\mathrm{V}(\varepsilon_i(t)) = 3$

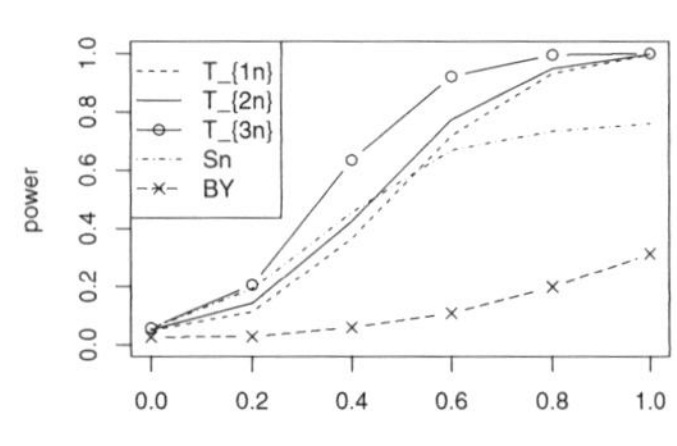

(a) Mixed MBR ($\phi = -0.2$)

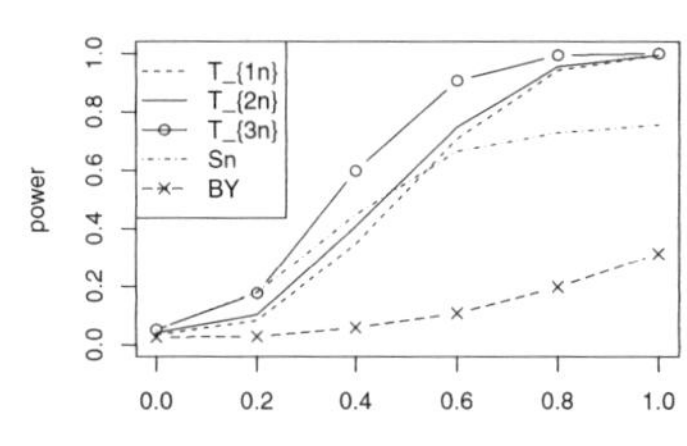

(b) Mixed CBR ($\phi = -0.2$)

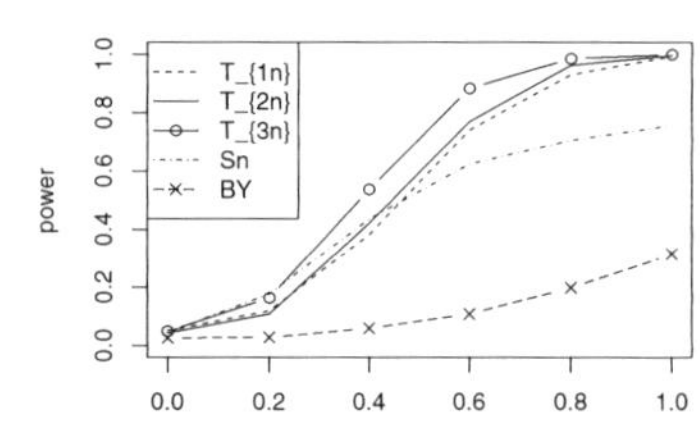

(c) Mixed SR ($\phi = -0.2$)

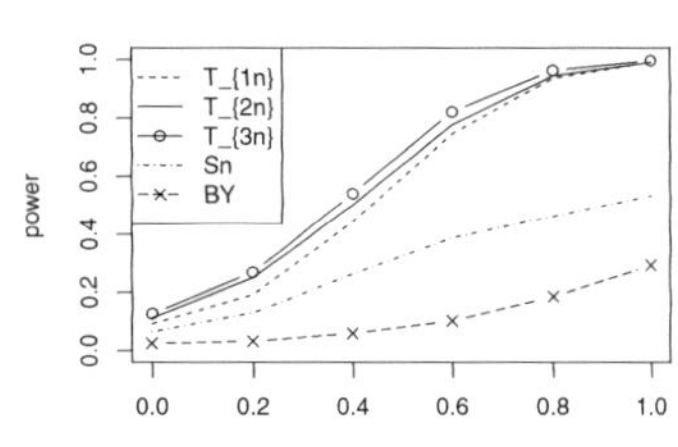

(d) Mixed MBR ($\phi = 0.2$)

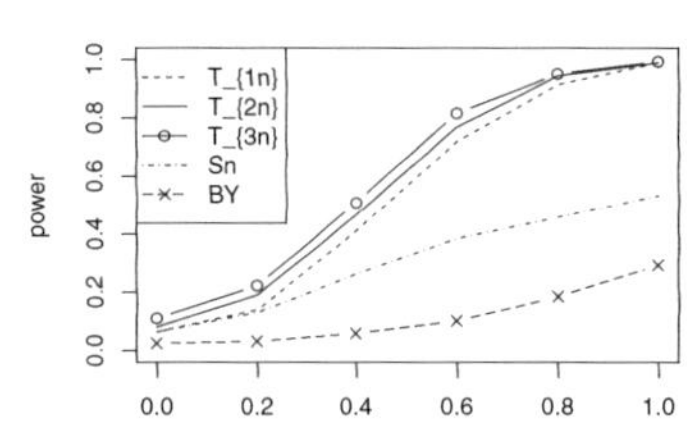

(e) Mixed CBR ($\phi = 0.2$)

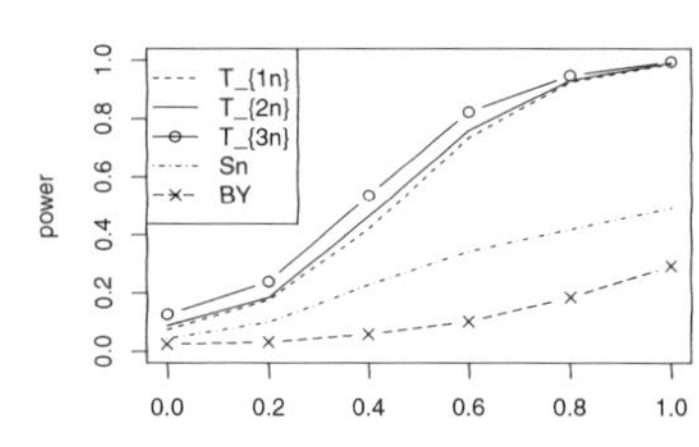

(f) Mixed SR ($\phi = 0.2$)

Figure 3. Comparison of T_{1n}, T_{2n}, T_{3n}, S_n and BY for $(q_1, q_2) = (20, 20)$ and $\mathrm{V}(\varepsilon_i(t)) = 3$

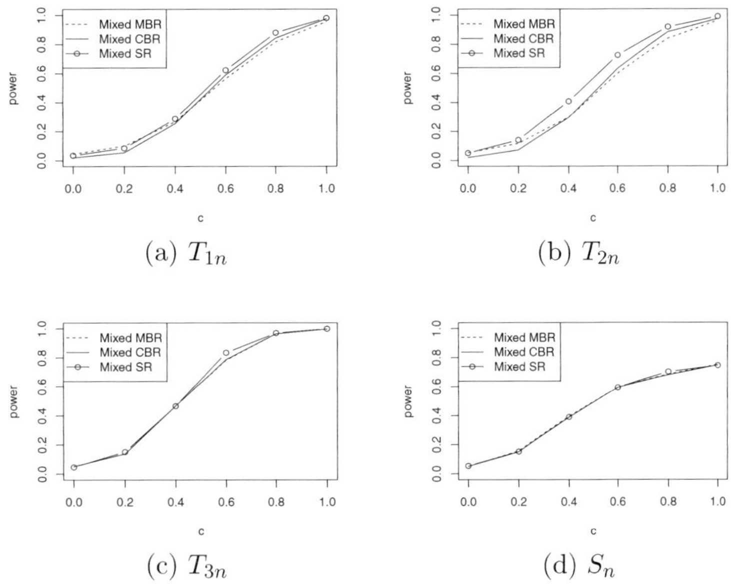

(a) T_{1n} (b) T_{2n}

(c) T_{3n} (d) S_n

Figure 4. Comparison of Mixed MBR, CBR and SR tests for $(q_1, q_2) = (10, 30)$, $\phi = -0.2$ and $\mathrm{V}(\varepsilon_i(t)) = 3$

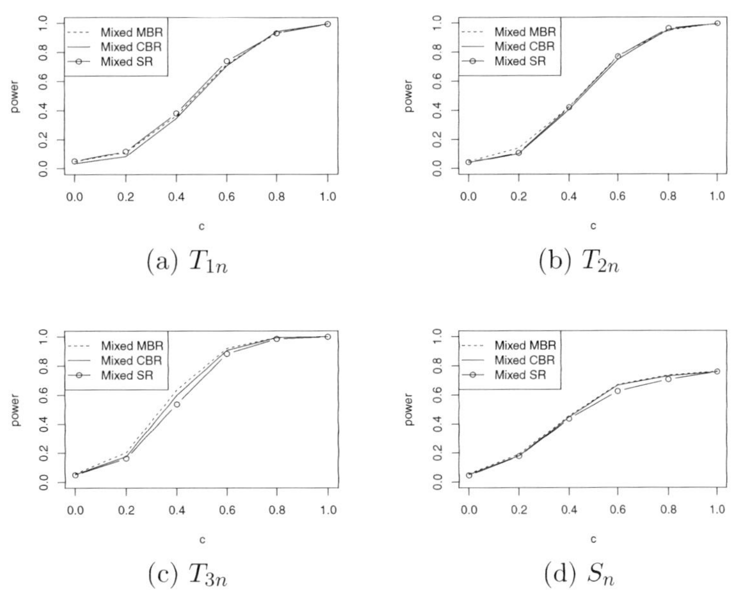

(a) T_{1n} (b) T_{2n}

(c) T_{3n} (d) S_n

Figure 5. Comparison of Mixed MBR, CBR and SR tests for $(q_1, q_2) = (20, 20)$, $\phi = -0.2$ and $\mathrm{V}(\varepsilon_i(t)) = 3$

Bibliography

[1] Boos, D., Janssen, P. and Veraverbeke, N. (1989). *Resampling from centered data in the two-sample problem.* Journal of Statistical Planning and Inference, **21**, 327–345.

[2] Bowman, A. and Young, S. (1996). *Graphical comparison of nonparametric curves.* Applied Statistics, **45**, 83–98.

[3] Hall, P. and Hart, J. D. (1990). *Bootstrap test for difference between means in nonparametric regression.* Journal of the American Statistical Association, **85**, 1039–1049.

[4] Kreiss, J. P. and Lahiri, S. N. (2012). *Bootstrap methods for time series.* Handbook of Statistics, Volume 30, Time Series Analysis: Methods and Applications (eds. Rao, T. S., Rao, S. S. and Rao, C. R.), North-Holland, Chapter 1, 3–26.

[5] Künsch, H. R. (1989). *The jackknife and the bootstrap for general stationary observations.* Annals of Statistics, **17**, 1217–1241.

[6] Politis, D. N. and Romano, J. P. (1992). *A circular block-resampling procedure for stationary data.* Exploring the Limit of Bootstrap (eds. LePage, R. and Billard, L.), Wiley, 263–270.

[7] Politis, D. N. and Romano, J. P. (1994). *The stationary bootstrap.* Journal of the American Statistical Association, **89**, 1303–1313.

[8] Sakurai, H. and Taguri, M. (2005). *Test for difference of two means in longitudinal data using moving block bootstrap.* Proceedings of the 5th IASC Asian Conference on Statistical Computing, 139–142.

[9] Sakurai, H. and Taguri, M. (2006). *Test of mean difference in longitudinal data based on block resampling.* COMPSTAT2006 Proceedings in Computational Statistics (eds. Rizzi, A. and Vichi, M.), Physica-Verlag, 1087–1094.

[10] Sakurai, H. and Taguri, M. (2010). *Test of mean difference for longitudinal data using circular block bootstrap.* COMPSTAT2010 Proceedings in Computational Statistics (eds. Lechevallier, Y. and Saporta, G.), Physica-Verlag, 1581–1588.

[11] Sakurai, H. and Taguri, M. (2013). *Testing methods of mean difference for longitudinal data based on stationary bootstrap.* Proceedings of the 59th World Statistics Congress of the International Statistical Institute, 2013, 5303–5308.

[12] Wang, J. and Taguri, M. (1996). *Bootstrap method — an introduction from a two sample problem (in Japanese).* Proceedings of the Institute of Statistical Mathematics, **44**, 3–18.

Expanded alternating optimization for matrix factorization and penalized regression

W. James Murdoch, *University of California at Berkeley*, jmurdoch@berkeley.edu
Mu Zhu, *University of Waterloo*, mu.zhu@uwaterloo.ca

Abstract. We propose a general technique for improving alternating optimization (AO) of nonconvex functions. Starting from the solution given by AO, we conduct another sequence of searches over subspaces that are both meaningful to the optimization problem at hand and different from those used by AO. To demonstrate the utility of our approach, we apply it to the matrix factorization (MF) algorithm for recommender systems and the coordinate descent algorithm for penalized regression (PR), and show meaningful improvements using both real-world (for MF) and simulated (for PR) data sets. Moreover, we demonstrate for MF that, by constructing search spaces customized to the given data set, we can significantly increase the convergence rate of our technique.

Keywords. Coordinate descent, MC+ penalty, Recommender system, Saddle points, SparseNet

1 Introduction

Alternating optimization (AO) is a commonly used technique for finding the extremum of a multivariate function, $f(\boldsymbol{z})$, where $\boldsymbol{z} \in \mathbb{R}^d$. In this approach, one breaks up the (multi-dimensional) input variable $\boldsymbol{z}$ into a few blocks, say, $\boldsymbol{z}_1, \boldsymbol{z}_2, ..., \boldsymbol{z}_B$, and successively optimizes the objective function over each block of variables while holding all other blocks fixed. That is, one solves

$$\min_{\boldsymbol{z}_b} \quad f(\boldsymbol{z}_1, \boldsymbol{z}_2, ..., \boldsymbol{z}_B) \tag{1}$$

successively over $b = 1, 2, ..., B, 1, 2, ..., B, ...$ until convergence is achieved. This is an especially natural approach when each individual optimization problem (1) over $\boldsymbol{z}_b$ is relatively easy to solve. Two well-known examples in statistics are: matrix factorization and penalized regression, but there are many others.

The Netflix contest drew much attention to the matrix factorization problem [6, 14]. Given a user-item rating matrix $\boldsymbol{R}$, whose element r_{ui} is the rating of item i by user u, the goal is to find low-rank matrices $\boldsymbol{P}$ and $\boldsymbol{Q}$, such that

$$\boldsymbol{R} \approx \boldsymbol{P}\boldsymbol{Q}^{\mathrm{T}} = \underbrace{\begin{bmatrix} \boldsymbol{p}_1^{\mathrm{T}} \\ \boldsymbol{p}_2^{\mathrm{T}} \\ \vdots \\ \boldsymbol{p}_n^{\mathrm{T}} \end{bmatrix}}_{n \times K} \underbrace{\begin{bmatrix} \boldsymbol{q}_1 & \boldsymbol{q}_2 & \cdots & \boldsymbol{q}_m \end{bmatrix}}_{K \times m}.$$

The vector $\boldsymbol{p}_u$ can be viewed as the coordinate of user u in a K-dimensional map and the vector $\boldsymbol{q}_i$, the coordinate of item i. With these coordinates, it is then possible to recommend item i to user u if $\boldsymbol{q}_i$ and $\boldsymbol{p}_u$ are closely aligned. Since we don't know every user's preferences on every item, many entries of $\boldsymbol{R}$ are missing. Let

$$T = \{(u, i) : \quad r_{ui} \text{ is known}\}$$

be the set of observed ratings. In order to estimate these user- and item-coordinates, we can solve the following optimization problem:

$$\min_{\boldsymbol{P},\boldsymbol{Q}} \quad L(\boldsymbol{P}, \boldsymbol{Q}) \equiv \sum_{(u,i)\in T} \left(r_{ui} - \boldsymbol{p}_u^{\mathrm{T}}\boldsymbol{q}_i\right)^2 + \lambda \left[\sum_{u=1}^{n} \|\boldsymbol{p}_u\|^2 + \eta \sum_{i=1}^{m} \|\boldsymbol{q}_i\|^2\right] \tag{2}$$

where the bracketed terms being multiplied by $\lambda > 0$ are penalties on the parameters being estimated, introduced to avoid over-fitting, because n and m are typically quite large relative to the number of known ratings (or the size of the set T). Here, we follow the work of Nguyen and Zhu [10] and use an extra factor $\eta = n/m$ to balance the penalties imposed on the two matrices, $\boldsymbol{P}$ and $\boldsymbol{Q}$. It is natural to use AO for solving (2). With both $\boldsymbol{p}_u$ and $\boldsymbol{q}_i$ being unknown, (2) is not a convex problem, but once we fix all $\boldsymbol{p}_v$ ($v \neq u$) and $\boldsymbol{q}_i$, the individual problem of minimizing $L(\boldsymbol{P}, \boldsymbol{Q})$ over $\boldsymbol{p}_u$ is convex and hence easy to solve.

During the last decade, penalized regression techniques have attracted much attention in the statistics literature [12, 2, 13]. Suppose that $\boldsymbol{y}, \boldsymbol{x}_1, ..., \boldsymbol{x}_d \in \mathbb{R}^n$ are all properly standardized to have mean zero ($\mathbf{1}^{\mathrm{T}}\boldsymbol{y} = 0$, $\mathbf{1}^{\mathrm{T}}\boldsymbol{x}_j = 0$) and variance one ($\|\boldsymbol{y}\| = 1$, $\|\boldsymbol{x}_j\| = 1$). The prototypical problem can be expressed as follows:

$$\min_{\boldsymbol{\beta}} \quad L(\boldsymbol{\beta}) \equiv \|\boldsymbol{y} - (\beta_1\boldsymbol{x}_1 + \beta_2\boldsymbol{x}_2 + \cdots + \beta_d\boldsymbol{x}_d)\|^2 + \sum_{j=1}^{d} J(\beta_j), \tag{3}$$

where $J(\cdot)$ is a penalty function. Many different penalty functions have been proposed. A widely used class of penalty functions is $J(\beta_j) = \lambda|\beta_j|^\alpha$. The case of $\alpha = 2$ is known as the ridge penalty [5], and that of $\alpha = 1$ is known as the LASSO [12]. In both of these cases, the function $J(\cdot)$ is convex. In recent years, *nonconvex* penalty functions have started to garner the attention of the research community, e.g., the SCAD [2], and the MC+ [13]:

$$J(\beta_j) = \lambda \int_0^{|\beta_j|} \left(1 - \frac{x}{\gamma\lambda}\right)_+ dx = \begin{cases} \lambda|\beta_j| - \frac{\beta_j^2}{2\gamma}, & |\beta_j| \leq \gamma\lambda; \\ \frac{1}{2}\gamma\lambda^2, & |\beta_j| > \gamma\lambda. \end{cases} \tag{4}$$

We will focus on the MC+ in this paper; hence, details of the SCAD are omitted, and we refer the readers to the original papers [2, 13] for explanations of for why these particular types of penalty functions are interesting. Currently, the preferred algorithm for fitting these penalized regression models is the coordinate descent algorithm [3]. One can view the coordinate descent algorithm as the "ultimate" AO strategy. For given $\lambda > 0$, the coordinate descent algorithm minimizes $L(\boldsymbol{\beta})$ over β_j while fixing all β_k ($k \neq j$), successively over $j = 1, 2, ..., d, 1, 2, ...d, ...$ until convergence. In fact, sometimes the general AO algorithm, with which we started this article, is dubbed the "blockwise coordinate descent" algorithm.

For the ridge penalty and the LASSO, the penalty function $J(\cdot)$ is convex, so the coordinate descent algorithm behaves "well". But for nonconvex penalty functions such as the SCAD and the MC+, there is no guarantee that the coordinate descent algorithm can reach the global solution of (3). Exactly the same point can be made about the AO algorithm for solving (2). In fact, for these nonconvex problems, not only can the AO strategy get stuck at inferior local solutions, but it also can be trapped at saddle points. Dauphin et al. [1] found that getting stuck at a saddle point can be a far more serious problem than getting trapped at a local minimum. Tayal et al. [11] proposed an intriguing method to improve AO by facilitating AO algorithms to escape saddle points. They introduced the concept of a shared "perspective variable" (more details in Section 2), but lacked intuition for why this was a good idea.

In this article, we begin by interpreting the approach of [11] geometrically — in particular, sharing a "perspective variable" results in an expanded search space at each step. However, searching over a slightly larger space at each step comes with a higher computational cost, which should be avoided if possible. Thus, our proposal is as follows: first, run the faster AO iterations until convergence; then, try to escape being trapped at an undesirable location by searching over a different space. Of critical importance is the choice of the search space. To this end, we introduce the important idea of defining search spaces that depend upon the particular data observed, as opposed to traditional techniques, including AO and those in [11], that fix the search spaces a priori. We apply these ideas to improve the AO algorithm for solving (2) as well as the coordinate descent algorithm for solving (3), with a focus on the MC+ penalty. However, we stress that this is a general algorithm that can be applied to other AO problems as well.

2 Main idea

It is convenient for us to motivate our key ideas with a very simple example. Consider the function,

$$f(x, y) = (x - y)^2 - x^2y^2.$$

Suppose that we attempt to minimize $f(x, y)$ with AO, and that, at iteration t, we have reached the point $(x_t, y_t) = (0, 0)$. While fixing $x_t = 0$, $f(0, y) = y^2$ is minimized at $y = 0$. Similarly, $x = 0$ is the optimal point when y_t is fixed at 0. Thus, the AO algorithm is stuck at $f(0, 0) = 0$. However, it is easy to see that $f(x, y)$ is actually unbounded below, and that $(0, 0)$ is a saddle point. At $(x_t, y_t) = (0, 0)$, the search space defined by AO is

$$\mathcal{S}_{AO}^{(t)} = \{(x, y) : x = 0\} \cup \{(x, y) : y = 0\}.$$

In this case, we can see that being restricted to this particular search space is the very reason why we are trapped at the saddle point. Therefore, if we could use a slightly different search space, we might be able to escape this saddle point. For example, Figure 1 shows that it would suffice to use the search space,

$$\mathcal{S}_{escape}^{(t)} = \{(x, y) : x = y\}.$$

The main lesson from the simple example above is that the restricted search space defined by the AO strategy, $\mathcal{S}_{AO}$, may cause the search to be trapped at undesirable locations such as saddle points, and that we may escape such traps by conducting the search in a slightly different space, $\mathcal{S}_{escape}$. This observation naturally leads us to propose the following strategy.

First, we run standard AO, i.e., search in $\mathcal{S}_{AO}^{(1)}$, $\mathcal{S}_{AO}^{(2)}$, ... until convergence, say, at $\mathcal{S}_{AO}^{(\tau)}$. Then, we continue searching in a different sequence of spaces, i.e., $\mathcal{S}_{escape}^{(\tau+1)}$, $\mathcal{S}_{escape}^{(\tau+2)}$, ... until convergence. If we see a sufficient amount of improvement in the objective function — an indication that the escaping strategy "worked", then we repeat the entire process using the improved result as the new starting point; otherwise, the algorithm terminates.

Needless to say, the key to the strategy we outlined above lies in the definition of the escaping sequence, $\mathcal{S}_{escape}^{(t)}$. We now describe in more detail how these different search spaces can be specified. We will use the notation $\boldsymbol{z}_{-b}$ to denote all other components except those in block b.

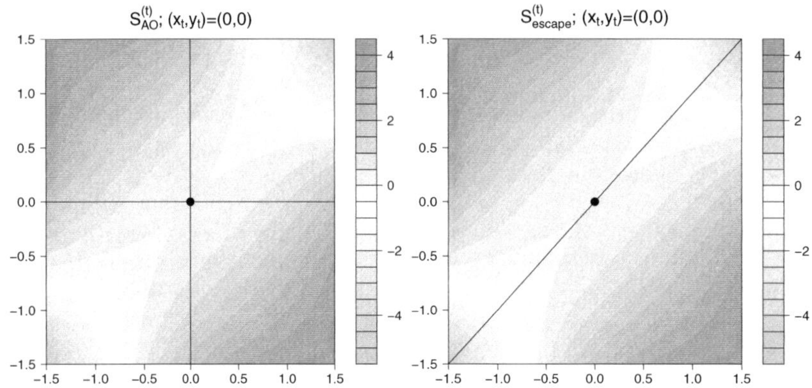

Figure 1. Different search spaces, $\mathcal{S}^{(t)}_{AO}$ and $\mathcal{S}^{(t)}_{escape}$, at $(x_t, y_t) = (0, 0)$, superimposed on a contour plot of $f(x, y) = (x - y)^2 - x^2y^2$.

Scaling

The proposal by [11] of sharing a "perspective variable", which we alluded to earlier in Section 1, essentially amounts to the following. At each alternating step, rather than solving (1), they proposed that we solve

$$\min_{\boldsymbol{z}_b, v_b} \quad f\left(v_b\boldsymbol{z}_1, ..., v_b\boldsymbol{z}_{b-1},\ \boldsymbol{z}_b,\ v_b\boldsymbol{z}_{b+1}, ..., v_b\boldsymbol{z}_B\right) \tag{5}$$

instead, where $v_b \in \mathbb{R}$ is the so-called "perspective variable". That is, we no longer just search for the optimal $\boldsymbol{z}_b$ while keeping $\boldsymbol{z}_{-b}$ fixed. When searching for the optimal component $\boldsymbol{z}_b$, we are free to *scale* all other components as well. Suppose the optimal scaling variable coming out of solving (5) is v_b^*. The component $\boldsymbol{z}_{-b}$ is then adjusted accordingly, i.e.,

$$\boldsymbol{z}_{-b} \quad \longleftarrow \quad v_b^*\boldsymbol{z}_{-b},$$

before the next alternating step (for optimizing over $\boldsymbol{z}_{b+1}$ and scaling $\boldsymbol{z}_{-(b+1)}$) begins. When so described, it is somewhat mysterious why it helps to scale $\boldsymbol{z}_{-b}$ when solving for $\boldsymbol{z}_b$. However, when viewed in terms of their respective search spaces, we can interpret this proposal as one particular way to define the search space $\mathcal{S}_{escape}$. As illustrated in Figure 2(a), at iteration t, the search space defined by AO is

$$\mathcal{S}^{(t)}_{AO} = \{(\boldsymbol{z}_b, \boldsymbol{z}_{-b}) : \boldsymbol{z}_{-b} = \boldsymbol{z}^{(t-1)}_{-b}\} = \boldsymbol{z}^{(t-1)} + \text{span}\{\boldsymbol{z}_b\};$$

whereas, if we are free to scale $\boldsymbol{z}_{-b}$ at the same time, the search space becomes

$$\begin{aligned}\mathcal{S}^{(t)}_{escape} &= \{(\boldsymbol{z}_b, \boldsymbol{z}_{-b}) : \exists\ v \in \mathbb{R} \text{ such that } \boldsymbol{z}_{-b} = v\boldsymbol{z}^{(t-1)}_{-b}\} \\ &= \{\lambda\boldsymbol{z}^{(t-1)}_{-b} + \boldsymbol{x} | \lambda \in \mathbb{R}, \boldsymbol{x} \in \text{span}\{\boldsymbol{z}_b\}\}.\end{aligned}$$

Clearly, $\mathcal{S}^{(t)}_{escape}$ is larger than $\mathcal{S}^{(t)}_{AO}$ (but still much smaller than the entire space $\mathbb{R}^d$). Thus, one way to understand the idea of freely scaling $\boldsymbol{z}_{-b}$ while optimizing over $\boldsymbol{z}_b$ is that it allows us to search in a slightly larger subspace, thereby improving the chance of finding a better solution.

Restricted joint search

This particular point of view immediately suggests that there are many other ways to expand, or simply alter, the search space. For example, once the AO steps have converged, we can try to escape by solving

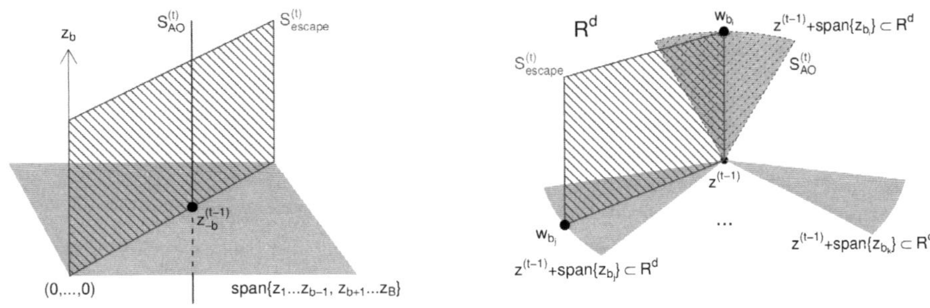

(a) Scaling (b) Restricted Joint Search

Figure 2. (a) Illustration of $\mathcal{S}_{AO}^{(t)}$ vs. $\mathcal{S}_{escape}^{(t)}$ as implied by the idea of sharing a "perspective variable" — equation (5). (b) Illustration of $\mathcal{S}_{AO}^{(t)}$ vs. $\mathcal{S}_{escape}^{(t)}$ as implied by the restricted joint search problem (6). The three shaded areas denote three different subspaces of $\mathbb{R}^d$. The ellipsis $(\cdots)$ denotes the fact that many other subspaces are not shown. One of these subspaces would correspond to $\mathcal{S}_{AO}^{(t)}$, e.g., $\boldsymbol{z}^{(t-1)} + \text{span}\{\boldsymbol{z}_{b_i}\}$. In this illustration, $I_{b_i} = I_{b_j} = 1$ and $I_b = 0$ for all $b \neq b_i, b_j$, including b_k.

a *restricted* joint optimization problem such as

$$\min_{\alpha_1,\ldots,\alpha_B} \quad f\left(\boldsymbol{z}_1 + \alpha_1 \boldsymbol{w}_1 I_1,\ \boldsymbol{z}_2 + \alpha_2 \boldsymbol{w}_2 I_2,\ \ldots,\ \boldsymbol{z}_B + \alpha_B \boldsymbol{w}_B I_B\right), \tag{6}$$

where $\boldsymbol{w}_b \in \text{span}\{\boldsymbol{z}_b\}$ $(b = 1, \ldots, B)$ are some pre-chosen directions (more about these later), and

$$I_b = \begin{cases} 1, & \text{if component } b \text{ is chosen to participate in this restricted joint optimization step;} \\ 0, & \text{otherwise.} \end{cases}$$

The kind of search spaces generated by (6) can be described as

$$\mathcal{S}_{escape}^{(t)} = \boldsymbol{z}^{(t-1)} + \text{span}\{\boldsymbol{w}_b : I_b \neq 0\};$$

see Figure 2(b) for an illustration. The restricted joint search problem (6) can be viewed as a compromise between using a different search space — i.e., $\mathcal{S}_{escape}^{(t)}$ rather than $\mathcal{S}_{AO}^{(t)}$ — and avoiding a full-scale, simultaneous search over the entire space $\mathbb{R}^d$.

3 Improved AO for matrix factorization

In this section, we apply our escaping strategies to the matrix factorization problem.

Scaling

As we described in Section 2, introducing a shared variable allows us to search over a slightly expanded space. For fixed $\boldsymbol{Q}$, this strategy minimizes

$$L(\boldsymbol{P}, v) = \sum_{u,i \in T} \left[r_{ui} - \boldsymbol{p}_u^{\mathrm{T}}(v\boldsymbol{q}_i)\right]^2 + \lambda \left[\sum_{u=1}^{n} \|\boldsymbol{p}_u\|^2 + \eta \sum_{i=1}^{m} \|v\boldsymbol{q}_i\|^2\right] \tag{7}$$

over $\boldsymbol{P}$ and v simultaneously, which we solve using a quasi-Newton algorithm with BFGS updates, e.g., [4]. Analogously, for fixed $\boldsymbol{P}$, we also numerically optimize over $\boldsymbol{Q}$ and a scaling variable u for $\boldsymbol{P}$.

Restricted joint search

As the loss function for matrix factorization is generally quite high-dimensional, we expect that only increasing the dimensionality of the search space by one can have only a limited effect. In addition, expanding the search space by introducing a scaling variable also limits the types of subspaces we can search over. These are the reasons why we proposed the restricted joint search problem (6) in Section 2. For the matrix factorization problem, this proposal amounts to constructing a family of search vectors $\{\boldsymbol{w}_p^i\}_{i=1}^n \cup \{\boldsymbol{w}_q^i\}_{i=1}^m$, each corresponding to a user- or item-vector, and searching over all of these directions simultaneously. Mathematically, this corresponds to solving

$$\min_{\boldsymbol{\alpha},\boldsymbol{\beta}} \quad L(\boldsymbol{\alpha},\boldsymbol{\beta}) = \sum_{u,i\in T} \left[r_{ui} - (\boldsymbol{p}_u + \alpha_u \boldsymbol{w}_p^u)^{\mathrm{T}} (\boldsymbol{q}_i + \beta_i \boldsymbol{w}_q^i)\right]^2 + \lambda \left[\sum_{u=1}^{n} \|\boldsymbol{p}_u + \alpha_u \boldsymbol{w}_p^u\|^2 + \eta \sum_{i=1}^{m} \|\boldsymbol{q}_i + \beta_i \boldsymbol{w}_q^i\|^2\right], \quad (8)$$

over $\boldsymbol{\alpha} \in \mathbb{R}^n$ and $\boldsymbol{\beta} \in \mathbb{R}^m$. To make this approach computationally feasible for larger problems, we generally require that $\boldsymbol{w}_q^i, \boldsymbol{w}_p^u = \boldsymbol{0}$ for all but a relatively small number of indices u, i. To do so, we sample each u with probability s/n and each i with probability s/m, for some $s \ll n, m$. That is, on average, we randomly choose s user-vectors and item-vectors to participate in the restricted joint optimization (8). The important remaining question is: how to choose our search vectors $\boldsymbol{w}_q^i, \boldsymbol{w}_p^u$? This requires a notion of what an informative subspace is to search over. Below, we describe two different approaches.

Random choices of $\boldsymbol{w}_p^u$ and $\boldsymbol{w}_q^i$ Trying to determine what set of vectors $\boldsymbol{w}_p^u, \boldsymbol{w}_q^i$ will produce the largest decrease in the loss function (8) is a challenging task. As such, it is a good idea to establish a simple baseline against which to measure more sophisticated spaces. Having such a baseline space will also serve to illustrate the power of our method in its simplest form. For our baseline, we simply choose our search vectors at random. Specifically, for each chosen index u, we sample $\boldsymbol{w}_p^u \sim N(\boldsymbol{0}, \boldsymbol{I})$ from the K-variate standard normal distribution, and likewise for each chosen index i. This procedure incorporates no information about the optimization problem at hand, nor the data. But, as we shall report below (Section 5), even these random choices of directions can lead to better solutions. Next, we provide a more sophisticated subspace that uses such information to produce better results.

Greedy choices of $\boldsymbol{w}_p^u$ and $\boldsymbol{w}_q^i$ In general, the optimal search space for a given loss function would depend upon specific properties of the function itself. However, none of our previous approaches explicitly took such information into account. We now describe an approach that does. Suppose that, for $\boldsymbol{p}_u$, we are searching for the optimal step size α in a given direction $\boldsymbol{w}$, while keeping everything else fixed. The objective function for this particular search is

$$L(\alpha) = \sum_{i\in T_u} \left[r_{ui} - (\boldsymbol{p}_u + \alpha \boldsymbol{w})^{\mathrm{T}} \boldsymbol{q}_i\right]^2 + \lambda \|\boldsymbol{p}_u + \alpha \boldsymbol{w}\|^2 + (\text{terms not depending on } \alpha), \quad (9)$$

where $T_u = \{i : r_{ui} \text{ is known}\}$. Differentiating (9) with respect to α and setting it equal to zero, we can solve for the optimal α as a function of $\boldsymbol{w}$:

$$\widehat{\alpha}(\boldsymbol{w}) = \left[\sum_{i\in T_u} \boldsymbol{w}^{\mathrm{T}} \boldsymbol{q}_i (r_{ui} - \boldsymbol{p}_u^{\mathrm{T}} \boldsymbol{q}_i) - \lambda \boldsymbol{w}^{\mathrm{T}} \boldsymbol{p}_u\right] \Big/ \left[\sum_{i\in T_u} (\boldsymbol{w}^{\mathrm{T}} \boldsymbol{q}_i)^2 + \lambda \|\boldsymbol{w}\|^2\right].$$

By plugging in the optimal step size $\widehat{\alpha}(\boldsymbol{w})$ into (9), the function $L(\alpha)$ becomes a function of $\boldsymbol{w}$, $L(\widehat{\alpha}(\boldsymbol{w}))$. We can now solve for the optimal search direction $\boldsymbol{w}$, using standard numerical optimization techniques — again, we use quasi-Newton with BFGS updates. In doing so, we have solved for a search direction $\boldsymbol{w}$ such that letting $\boldsymbol{p}_u$ take an optimal step in its direction will produce the maximal decrease in the overall loss function. We construct our set of search vectors $\{\boldsymbol{w}_p^u\}$ by repeating this process for each chosen $\boldsymbol{p}_u$, and the set of search vectors $\{\boldsymbol{w}_q^i\}$ is obtained in the same fashion.

4 Improved AO for MC+ regression

In this section, we apply our escaping strategies to the penalized regression problem. We focus on the MC+ penalty function, but our strategies can be applied to other nonconvex penalty functions as well. We also investigate the application of our method to fitting an entire regularization surface, as introduced by [7]. Although the singularity of the MC+ penalty function $J(\cdot)$ at 0 places certain limits on the types of subspaces that can be feasibly optimized over, applying our ideas in their simple forms still produces notable improvements.

Scaling

Again, the idea of using a shared variable applies. In this setting, this amounts to taking a number of expanded coordinate descent steps (after the standard coordinate descent steps have converged), so that we simultaneously search over a single coefficient β_j, as well as a scaling variable for the rest of the coefficient vector, $\boldsymbol{\beta}_{-j}$. Mathematically, these expanded coordinate descent steps solve

$$\min_{\beta_j, v} \quad L(\beta_j, v) = \left\| \boldsymbol{y} - \beta_j \boldsymbol{x}_j - \sum_{k \neq j} (v\beta_k)\boldsymbol{x}_k \right\|^2 + J(\beta_j) + \sum_{k \neq j} J\left(v\beta_k\right), \tag{10}$$

for $j = 1, 2, ..., d$. We can actually solve for the optimal β_j and v explicitly in this case. This is attractive because it allows us to avoid using numerical optimization for these steps, which would have been more difficult due to the non-differentiability of the penalty function at zero. The technical details for these steps are available in the pre-print version of this article [9].

Selective scaling

Using our general notation (Sections 1–2), coordinate descent corresponds to each "block" $\boldsymbol{z}_b$ being one-dimensional. This puts a certain limit on the kind of restricted joint search operations we can implement. In particular, for any given $\boldsymbol{z}_b$, the only available choice of $\boldsymbol{w}_b$ is $\boldsymbol{z}_b$ itself. However, we are still free to determine which $I_b = 1$.

As we pointed out in Section 3, tailoring the search space to the observed data can yield improved results. In the context of regression, it is useful to consider how changing the coefficient in front of $\boldsymbol{x}_j$ could affect the coefficient in front of another variable, say $\boldsymbol{x}_k$. If these two variables are independent, then a change in β_j would not result in a change to the optimal β_k. However, if these two variables are highly correlated, then one would expect that a decrease in β_j could lead to an increase or decrease in β_k depending on whether their correlation is positive or negative, as some of the variance in $\boldsymbol{y}$ previously accounted for by $\boldsymbol{x}_j$ can be "taken over" by $\boldsymbol{x}_k$.

Thus, we implement a selective scaling strategy. While searching over β_j, we only allow the scaling of β_k if the correlation between $\boldsymbol{x}_j$ and $\boldsymbol{x}_k$ is above some threshold, instead of scaling all $\boldsymbol{\beta}_{-j}$. Let E_j denote the set of variables that are sufficiently correlated with $\boldsymbol{x}_j$, i.e.,

$$E_j \quad = \quad \{k \neq j : \quad |\text{corr}(\boldsymbol{x}_k, \boldsymbol{x}_j)| > \rho_{min}\}, \tag{11}$$

for some pre-chosen $\rho_{min} > 0$. The selective scaling steps solve

$$\min_{\beta_j, v} \quad L(\beta_j, v) = \left\| \boldsymbol{y} - \beta_j \boldsymbol{x}_j - \sum_{\ell \in E_j^C \setminus \{j\}} \beta_\ell \boldsymbol{x}_\ell - \sum_{k \in E_j} (v\beta_k) \boldsymbol{x}_k \right\|^2 + \\ J(\beta_j) + \sum_{\ell \in E_j^C \setminus \{j\}} J(\beta_\ell) + \sum_{k \in E_j} J(v\beta_k), \quad (12)$$

for $j = 1, 2, ..., d$. Again, here we can compute the optimal β_j and v explicitly [9].

Fitting entire regularization surfaces with multiple warm starts

Using the MC+ penalty (4), the optimal solution $\widehat{\boldsymbol{\beta}}$ to (3) depends on two regularization parameters, λ and γ. For different values of (λ, γ), one can think of $\widehat{\boldsymbol{\beta}}(\lambda, \gamma)$ as tracing out an entire regularization surface. Mazumder et al. [7] provided a nice algorithm, called SparseNet, for fitting the entire regularization surface. Fitting an entire surface of solutions, rather than just a single solution, introduces an interesting set of challenges for our work. When fitting a single solution, we are only concerned with how to best find a good solution for a given pair of (λ, γ). However, SparseNet fits the entire surface of solutions sequentially, using each point on the surface, $\widehat{\boldsymbol{\beta}}(\lambda, \gamma)$, as a warm start for fitting the "next" point. Thus, improving the solution at (λ, γ) *may* not be desirable if the improved solution provides an inferior warm start for the next point, resulting in worse solutions further down the surface. Empirically, we have found this to be a common occurrence. To remedy this problem, our strategy is to keep track of a few different solution surfaces:

- $\widehat{\boldsymbol{\beta}}_A(\lambda, \gamma)$ — this is the "usual" surface obtained by SparseNet, i.e., each point on this surface is obtained by running the coordinate descent algorithm (an ultimate AO strategy), using the "previous" point on this surface (A) as a warm start;
- $\widehat{\boldsymbol{\beta}}_B(\lambda, \gamma)$ — each point on this surface is obtained by first running the coordinate descent algorithm and then switching over to search in a different space, but each point also uses the "previous" point on this surface (B) as a warm start;
- $\widehat{\boldsymbol{\beta}}_C(\lambda, \gamma)$ — like surface B above, each point on this surface also is obtained by first running the coordinate descent algorithm and then switching over to search in a different space, *except* that each point uses the "previous" point from the surface $\widehat{\boldsymbol{\beta}}_A(\lambda, \gamma)$ as a warm start.

At each point (λ, γ), we keep the better of $\widehat{\boldsymbol{\beta}}_B(\lambda, \gamma)$ or $\widehat{\boldsymbol{\beta}}_C(\lambda, \gamma)$ as our solution.[15]

5 Experimental results

We now present some experimental results to demonstrate the effectiveness of our method.

Matrix factorization

To demonstrate our method for matrix factorization, we used a data set compiled by [8], which consists of approximately 35.3 million reviews from `www.amazon.com` between 1995 and 2013. We took a dense subset of their data consisting of approximately 5.5 million reviews, such that all users in our subset have rated ≥ 55 items, and all items have been rated ≥ 24 times.

[15] In the actual implementation, it is clear that we need not start from scratch in order to obtain the surface $\widehat{\boldsymbol{\beta}}_C(\lambda, \gamma)$; we can simply start with the surface $\widehat{\boldsymbol{\beta}}_A(\lambda, \gamma)$, and apply our escaping strategies directly at each point. Conceptually, we think it is easier for the reader to grasp what we are doing if we describe three separate surfaces rather than two, but this does *not* mean we have to triple the amount of computation.

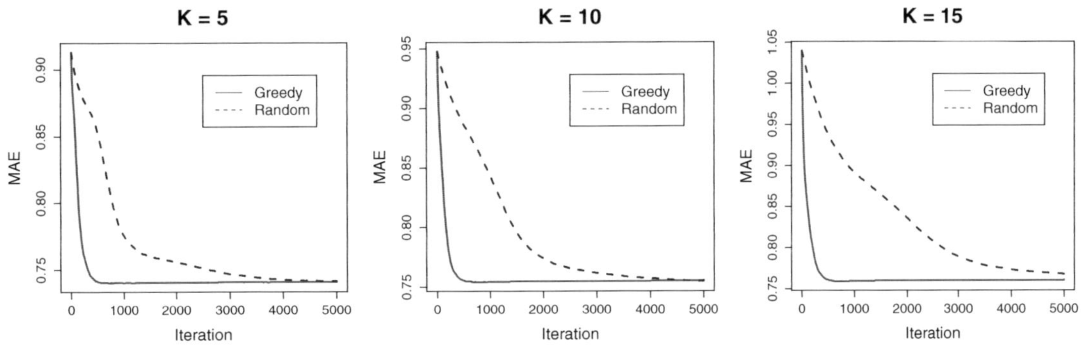

Figure 3. Matrix factorization example. The MAE on test data vs. number of iterations.

For our restricted joint optimization approach, we allowed only a small number ($s \ll n, m$) of user- and item-vectors in each round to participate in the joint optimization (see Section 3). Empirically, we obtained reasonably good and comparable performance results with a wide range of $s \in [20, 200]$, but results reported here are for $s = 50$.

We tested our method by randomly splitting the ratings into two halves, using one half as the training set T, and the other half as the test set V. All statistics were averaged over ten runs. As our metric, we used the mean absolute error (MAE),

$$\text{MAE} = \frac{1}{|V|} \sum_{(u,i) \in V} |\widehat{r}_{ui} - r_{ui}|. \tag{13}$$

McAuley and Leskovec [8] reported a mean squared error (MSE) of about 1.42 on their full Amazon data set using the baseline matrix factorization method with $K = 5$. This would translate to about 1.19 on the root mean squared error (RMSE) scale, which is more comparable with the MAE. Our baseline AO produced slightly better results (Table 1) because we used a dense subset, so there is presumably more information about each user and item in our subset.

Table 1. Matrix factorization example. MAEs (and their standard errors) on the test set.

	Baseline AO	Scaling Only	Random Subspace	Greedy Subspace
K = 5	0.856 (0.0002)	0.763 (0.0042)	0.747 (0.0009)	0.740 (0.0004)
K = 10	0.859 (0.0002)	0.756 (0.0030)	0.760 (0.0006)	0.754 (0.0003)
K = 15	0.861 (0.0005)	0.760 (0.0019)	0.769 (0.0006)	0.760 (0.0002)

In order to produce fair comparisons between different methods, for given $K = 5, 10$ and 15 we used cross-validation to choose an optimal value of λ for each method; the selected values ranged from 1 to 15. As can be seen from Table 1, our approach produces meaningfully better models. Figure 3 shows that, while there appeared to be little difference (Table 1) between using a random choice and using a greedy choice of $\{\boldsymbol{w}_p^u, \boldsymbol{w}_q^i\}$ to conduct the restricted joint search, the greedy strategy was much faster and more efficient at improving the results.

MC+ regression

To demonstrate our method for MC+ regression, we used a simulated data set from [7] — more specifically, their model M_1. The sample size is $n = 100$, with $d = 200$ predictors generated from the Gaussian

Table 2. MC+ regression example. Percent changes in the terminal values of the loss function [$\%\Delta_L = (L_{new} - L_{old})/L_{old}$, top two rows] and in the variable-selection error [$\%\Delta_e = (\text{error}_{new} - \text{error}_{old})/\text{error}_{old}$, bottom two rows], over the entire regularization surface.

	Small γ's	Large γ's	All γ's
fraction($-0.005 \leq \%\Delta_L < 0$)	0.715	0.730	0.723
average($\%\Delta_L \| \%\Delta_L < -0.005$)	−0.063	−0.036	−0.050
fraction($\%\Delta_e = 0$)	0.575	0.205	0.390
average($\%\Delta_e \| \%\Delta_e \neq 0$)	−0.011	−0.027	−0.021

distribution with mean zero and covariance matrix $\mathbf{\Sigma}$, whose (j,k)-th entry is equal to $0.7^{|j-k|}$. The response is generated as a linear function of only 10 of the 200 predictors plus a random noise; in particular, $\boldsymbol{y} = \boldsymbol{x}_1 + \boldsymbol{x}_{21} + \boldsymbol{x}_{41} + ... + \boldsymbol{x}_{161} + \boldsymbol{x}_{181} + \boldsymbol{\varepsilon}$. That is, $\beta_{20j+1} = 1$ for $j = 0, 1, 2, ..., 9$ and $\beta_j = 0$ otherwise. Mazumder et al. [7] took $\varepsilon_i \sim N(0, \sigma^2)$ with $\sigma = \sqrt{\boldsymbol{\beta}^{\mathrm{T}}\mathbf{\Sigma}\boldsymbol{\beta}/3}$ so that the signal-to-noise ratio is 3.

Using $\rho_{min} = 0.3$ in equation (11), we estimated $\widehat{\boldsymbol{\beta}}(\lambda, \gamma)$ on a grid consisting of 8 different γ's and 50 different λ's. The γ's were equally spaced on the logarithmic scale between $\gamma = 1.000001$ and $\gamma = 150$. The λ's were equally spaced on the logarithmic scale between λ_{max} — the smallest λ such that $\widehat{\beta}_j = 0$ for all j — and $0.01\lambda_{max}$.

For each point on the grid, we computed the percent decrease in the value of the objective function, i.e., $\%\Delta_L = (L_{new} - L_{old})/L_{old}$, where L_{old}, L_{new} are the values of the objective function when the coordinate descent algorithm converged, and after our restricted joint search, respectively. The top two rows of Table 2 show that, for about 72% of points on the grid, our strategy made little difference ($\%\Delta_L$ no more than 0.5 percentage points), indicating that the original coordinate descent algorithm already found relatively good solutions at those points. For the remaining 28% of the points, however, our strategy found a better solution — searching in a slightly expanded space further reduced the value of the objective function by an average of 5%. For the smaller half of γ's (more nonconvex objective functions), the average percent decrease was a little over 6%; for the larger half (less nonconvex objective functions), the average percent decrease was close to 4%.

For each point on the grid, we also computed the percent decrease in the variable-selection error, i.e., $\%\Delta_e = (\text{error}_{new} - \text{error}_{old})/\text{error}_{old}$, where error_{old}, error_{new} are the errors of the coordinate descent solution and of our solution, respectively. The variable-selection error was measured in terms of

$$\text{error}(\widehat{\boldsymbol{\beta}}) = \frac{1}{d}\sum_{j=1}^{d} I(\beta_j = 0 \text{ and } \widehat{\beta}_j \neq 0) + I(\beta_j \neq 0 \text{ and } \widehat{\beta}_j = 0). \tag{14}$$

The bottom two rows of Table 2 show that, overall, our strategy led to improved variable-selection results as well — a 2% reduction in error on average.

6 Summary

We have proposed a general framework for improving alternating optimization of nonconvex functions. The main idea is that, once standard AO has converged, we switch to conduct our search in a different subspace. We have provided general guidelines for how these different subspaces can be defined, as well as illustrated with two concrete statistical problems — namely, matrix factorization and regression with the MC+ penalty — how problem-specific information can (and should) be used to help us identify good and meaningful search subspaces. By carefully selecting a relevant space to search over, we can escape undesirable locations such as saddle points and produce notable improvements. In addition to serving as examples of our general idea, we think that these improved AO algorithms, for matrix factorization and for regression with the MC+ penalty, are meaningful contributions on their own.

Acknowledgement

This research is partially supported by the Natural Sciences and Engineering Research Council (NSERC) of Canada and by the University of Waterloo. The first author did most of the work as an undergraduate research assistant at Waterloo, prior to becoming a PhD student at Berkeley.

Bibliography

[1] Dauphin, Y. N., Pascanu, R., Gulcehre, C., Cho, K., Ganguli, S., and Bengio, Y. (2014). Identifying and attacking the saddle point problem in high-dimensional non-convex optimization. Preprint, arXiv:1406.2572.

[2] Fan, J. and Li, R. (2001). Variable selection via nonconcave penalized likelihood and its oracle properties. *Journal of the American Statistical Association*, 96, 1348–1360.

[3] Friedman, J. H., Hastie, T. J., and Tibshirani, R. J. (2010). Regularization paths for generalized linear models via coordinate descent. *Journal of Statistial Software*, 33, 1–22.

[4] Gill, P. E., Murray, W., and Wright, M. H. (1981). *Practical Optimization*. Academic Press.

[5] Hoerl, A. E. and Kennard, R. W. (1970). Ridge regression: Biased estimation for nonorthogonal problems. *Technometrics*, 12, 55–67.

[6] Koren, Y., Bell, R., and Volinsky, C. (2009). Matrix factorization techniques for recommender systems. *Computer*, 42, 30–37.

[7] Mazumder, R., Friedman, J. H., and Hastie, T. J. (2011). SparseNet: Coordinate descent with nonconvex penalties. *Journal of the American Statistical Association*, 106, 1125–1138.

[8] McAuley, J. and Leskovec, J. (2013). Hidden factors and hidden topics: Understanding rating dimensions with review text. In *Proceedings of the 7th ACM Conference on Recommender Systems* (RecSys '13), pages 165–172, New York, NY, USA.

[9] Murdoch, W. J. and Zhu, M. (2014). Expanded alternating optimization of nonconvex functions with applications to matrix factorization and penalized regression. Preprint, arXiv:1412.4128.

[10] Nguyen, J. and Zhu, M. (2013). Content-boosted matrix factorization techniques for recommender systems. *Statistical Analysis and Data Mining*, 6, 286–301.

[11] Tayal, A., Coleman, T. F., and Li, Y. (2014). Primal explicit max margin feature selection for nonlinear support vector machines. *Pattern Recognition*, 47, 2153–2164.

[12] Tibshirani, R. (1996). Regression shrinkage and selection via the Lasso. *Journal of the Royal Statistical Society Series B*, 58, 267–288.

[13] Zhang, C.-H. (2010). Nearly unbiased variable selection under minimax concave penalty. *The Annals of Statistics*, 38, 894–942.

[14] Zhu, M. (2014). Making personalized recommendations in e-commerce. In J. F. Lawless, editor, *Statistics in Action: A Canadian Outlook*, pages 259–268. Chapman & Hall.

NPC to assess effects of maternal iodine nutrition and thyroid status on children cognitive development

Angela Alibrandi, *University of Messina*, aalibrandi@unime.it
Agata Zirilli, *University of Messina*, azirilli@unime.it
Massimiliano Giacalone, *University of Naples Federico II*, massimiliano.giacalone@unina.it
Mariacarla Moleti, *University of Messina*, mmoleti@unime.it

Abstract. Maternal iodine nutrition and thyroid status may influence neurocognitive development in children. This study investigated the effects on the intelligence quotient (IQ) of children born to mothers with different levels of iodine supplementation, with or without the administration of levothyroxine (LT4), prior to and during pregnancy. From a methodological point of view, we used the Non Parametric Combination test or NPC test, based on permutation solution. It was chosen for the several optimal properties of which it is characterized, that make it very flexible and widely applicable in many fields; in particular, it allows stratified analyses and represents an effective solution for problems concerning the testing of multidimensional hypotheses, that are difficult to face in a parametric context.

Keywords. NPC test, Neurocognitive Development, Iodine Nutrition, Thyroid Status.

1 Introduction

Thyroid hormone (TH) is required for normal brain development. Prior to the onset of fetal thyroid function, the mother is the only source of TH for the developing brain. By 16-20 weeks post-conception the fetal thyroid is mature enough, and from this point in time onwards the fetus cooperate to make up his own TH pool [34]. However, the relative contribution of the mother and the fetus to the regulation of TH-dependent processes within the brain is not yet fully established.

Since iodine is essential for TH synthesis [24], a gestational intake of this micronutrient that fails to meet the needs of pregnancy may simultaneously impair both maternal and fetal TH production [15]. Accordingly, the most serious consequence of gestational iodine deficiency (ID) is endemic cretinism, due to severely impaired TH synthesis in the mother and fetus from early pregnancy onwards [15]. In conditions of mild to moderate ID, less severe degrees of maternal thyroid insufficiency may occur over gestation [19], [22], and several studies have shown these conditions to be associated to minor neuropsychiatric and intellectual deficits in progeny [27], [15], [33], [20] and [23]. On the other hand, other studies carried out in iodine sufficient regions failed to confirm these associations, with some reports anecdotally reporting normal neurodevelopment in children born to mothers who were severely hypothyroid for causes other than ID [18] and [26]. Finally, a growing body of evidence has recently been provided, overall indicating that gestational iodine supplementation, while not having a clear impact on maternal thyroid function [35], [21] is actually effective in improving infant cognitive development [6]. By contrast, iodine supplementation given in later stages of life, i.e. during adulthood, while improving iodine status, proved to have no impact on cognitive function [31]. Since potential mechanisms by which iodine affects cognition include white matter maturation, the observed lack of effect of iodine supplementation on the cognitive scores of mildly iodine-deficient young adults has been attributed to the fact that the process of myelination is more complete in adulthood compared to fetal life and infancy, and therefore less malleable [31].

Taken as a whole, these data might be consistent with the hypothesis that maternal iodine status primarily influences fetal thyroid function, which may play a more critical role than maternal thyroid function in determining neuro-intellectual outcomes in progeny. In order to assess this assumption, a prospective study was carried out on schoolchildren living in an ID area and born to mothers exposed to different iodine supplementation regimens, half of whom had been receiving levo-thyroxine (LT4) prior to and during pregnancy in order to guarantee maternal euthyroidism throughout gestation [25]. The main results of this study were that children born to mothers with similar iodine intake during pregnancy had comparable intellectual abilities, regardless of their mothers' thyroid function. Conversely, children born to mothers with comparable thyroid function during pregnancy, namely those born to mothers on LT4 therapy, performed differently on intelligence quotient (IQ) tests, with those born to mothers who had been receiving iodine supplementation during pregnancy showing significantly higher IQ scores than those born to unsupplemented mothers. Indeed, logistic regression models designed to assess the dependence of suboptimum cognitive outcomes (IQ<85 points) on various explanatory variables failed to show a significant association with maternal thyroid parameters at any stage in pregnancy, whereas maternal iodine status proved to be positively associated with cognitive outcomes.

In this study we aimed at evaluating the intelligence quotient (IQ) of children born to mothers with different levels of iodine supplementation, with or without the administration of levothyroxine (LT4), prior to and during pregnancy. In particular we focused our attention on some mother-child pairs and we compared them according to iodized salt consumption and LT4 treatment.

2 The Data

The examined sample included four groups, each comprising 25 mother-child pairs, identified on the basis of maternal histories of iodized salt consumption and LT4 treatment prior to and during pregnancy. The groups were labeled as follows: iodine (I), no iodine (no-I), iodine + LT4 (I+T4) and no iodine + LT4 (no-I+T4). Inclusion criteria for the mothers were:

a) age>18 years;
b) singleton and uncomplicated pregnancy;
c) term delivery;
d) no severe or chronic diseases (including thyroid autoimmune diseases);
e) no major post-partum complications (including post-partum depression);
f) thyroid function evaluation throughout gestation;
g) full diet and lifestyle information during pregnancy and afterwards;
h) informed consent.

Inclusion criteria for the children were:

a) age between 6 and 14 years;
b) no major neonatal complications (including birth trauma);
c) no congenital hypothyroidism;
d) no severe or chronic diseases;
e) no ascertained major cognitive deficits;
f) regular education;
g) approval to cognitive test administration.

Child Intelligence Quotients (IQ) was assessed with the use of Wechsler Intelligence Scale for Children - Third Edition (WISC-III), which was administered by trained psychologists who were blinded as to which group subjects were allocated. The Full-Scale IQ (FSIQ), the Verbal IQ (VIQ) and the Performance IQ (PIQ) were calculated for each child and used into analysis.

For each mother the following information was collected: Triiodothyronine (T3), Thyroxine (T4), Thyroid-stimulating hormone (TSH), Free Triiodothyronine (FT3) and Free Thyroxine (FT4) for assessing maternal thyroid status at each point in time during gestation (recorded in the following weeks range: 4-12, 13-18, 19-24, 25-30, 31-36), Urinary Iodine Concentrations (UIC), family Socio-Economic Status (S.E.S.) evaluated by means of Hollingshead Index, iodized salt consumption (yes or no) and L-T4 treatment (yes or no).

3 The Non Parametric Combination Test (NPC)

The non-normality in the distribution of the considered phenomena (as verified by Kolmogorov Smirnov test) does not guarantee valid asymptotic results; consequently the non-parametric approach has been used. In particular we used the Non Parametric Combination (NPC) test, based on permutation test [30], [10], chosen for the several optimal properties of which it is characterized. Permutation tests [29], [16] represent an effective solution for problems concerning the verifying of multidimensional hypotheses, because they are difficult to face in parametric context. This multivariate and multistrata procedure allows to reach effective solutions concerning problems of multidimensional hypotheses verifying within the non parametric permutation inference [28]; it is used in different application fields that concern verifying of multidimensional hypotheses with a complexity that can not be managed in parametric context. In comparison to the classical approach, NPC test is characterized by several advantages:

- it doesn't request normality and homoscedasticity assumption;
- it draws any type of variable;
- it also assumes a good behavior in presence of missing data;
- it is also powerful in low sampling size;
- it resolves multidimensional problems, without the necessity to specify the structure of dependence among variables;
- it allows to test multivariate restricted alternative hypothesis (allowing the verifying of the directionality for a specific alternative hypothesis);
- it allows stratified analysis;
- it can be applied also when the sampling number is smaller than the number of variables.

All these properties make NPC test very flexible and widely applicable in several fields; in particular we cite applications in sociological context [5], [3], [7], in medical context [37],[4], [1], [8], [36], [2], [32], [9] and in genetics [13], [12].

We supposed to notice K variables on N observations (dataset N×K) and that an appropriate K-dimensional distribution P exists. The null hypothesis postulates the equality in distribution of k-dimensional distribution among all C groups

$$H_0 = [P_1 = ... = P_C] = [X_1 \stackrel{d}{=} ... \stackrel{d}{=} X_C]$$
$$\text{i.e. } H_0 = \cap_{i=1}^{k} X_{1i} \stackrel{d}{=} ... \stackrel{d}{=} X_{Ci}] = [\cap_{i=1}^{k} H_{0i}]$$

against the alternative hypothesis

$$H_1 = \cup_{i=1}^{k} H_{1i}.$$

Let's assume that, without loss of generality, the partial tests assume real values and they are marginally correct, consistent and significant for great values; the NPC procedure (based on Conditional Monte Carlo resampling) develops into the following phases, such as illustrated in Figure 1.

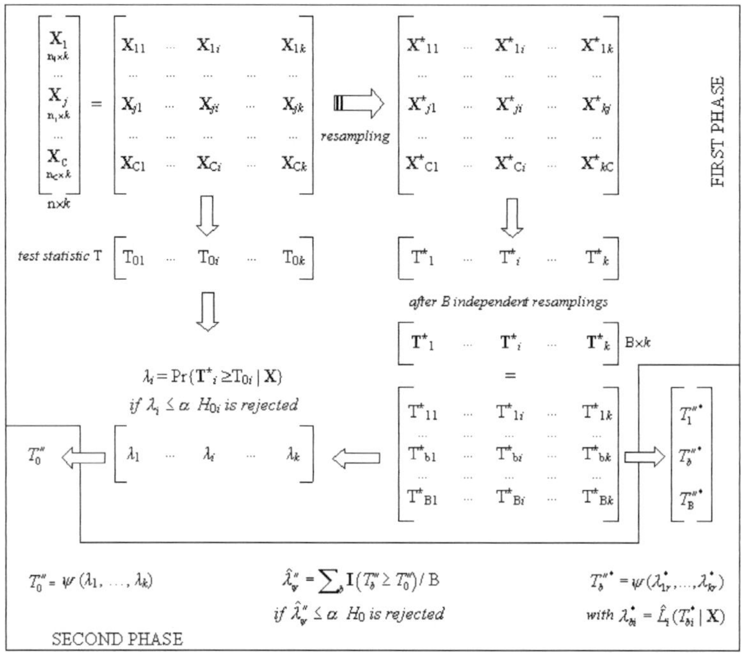

Figure 1. Two-phases NPC algorithm

The null hypothesis, that postulates the indifference among the distributions, and the alternative one are expressed as follows:

$$H_0 : \left\{ X_{11} \overset{d}{=} X_{12} \right\} \cap \cdots \cap \left\{ X_{n1} \overset{d}{=} X_{n2} \right\} \tag{1}$$

$$H_1 : \left\{ X_{11} \overset{d}{\neq} X_{12} \right\} \cup \cdots \cup \left\{ X_{n1} \overset{d}{\neq} X_{n2} \right\} \tag{2}$$

In presence of a stratification variable, the hypotheses system is:

$$H_{0i} : \left\{ X_{11i} \overset{d}{=} X_{12i} \right\} \cap \cdots \cap \left\{ X_{n1i} \overset{d}{=} X_{n2i} \right\} \tag{3}$$

$$H_{1i} : \left\{ X_{11i} \overset{d}{\neq} X_{12i} \right\} \cup \cdots \cup \left\{ X_{n1i} \overset{d}{\neq} X_{n2i} \right\} \tag{4}$$

The hypotheses systems are verified by the determination of partial tests (first order) that allow to evaluate the existence of statistically significant differences. By means of this methodology we can preliminarily define a set of k (k>1) unidimensional permutation tests (partial tests); they allow to examine every marginal contribution of answer variable, in the comparison among the examined groups. The partial tests are combined, in a non parametric way, in a second order test that globally verifies the existence of differences among the multivariate distributions. A procedure of conditioned resampling CMC (Conditional Monte Carlo) allows to estimate the p-values, associated both to partial tests and to second order tests.

Under the exchangeability data among groups condition, according to null hypothesis, NPC test is characterized by two properties:

- similarity: whatever the underlying distribution data, the probability to refute the null hypothesis is invariant to the actually observed dataset, whatever the type of data collection;
- for each α, for each distribution and for each set of observed data, if under the alternative hypothesis, the distribution dominates the null hypothesis, then an unbiased conditional test exists and, therefore, the probability of refuting the null hypothesis is always no less than the α significance level.

4 The results

The analysis was performed comparing four groups, defined on the basis of maternal histories of iodized salt consumption and LT4 treatment prior to and during pregnancy:

- Group A= iodine (I);
- Group B= iodine + LT4 (I + T4);
- Group C= no iodine and no LT4 (no-I);
- Group D= no iodine + LT4 (no-I + T4).

In Table 1 we report the results of the NPC test; all combined p-values are obtained using Fisher's combining function.

Table 1. Mean values and p-values for groups comparison

VARIABLES	Group A	Group B	Group C	Group D	p-value
Week birth	38.40	38.36	38.73	38.33	0.780
Weight birth	3241.1	3164.2	3171.4	3292.2	0.646
FSIQ	93.13	96.07	81.73	81.27	**0.012**
VIQ	90.07	97.29	80.33	79.60	**0.006**
PIQ	98.20	96.43	87.33	87.53	0.111
Child age	9.247	9.825	10.52	10.20	0.390
Mother age	29.53	29.07	27.67	28.87	0.622
UIC	90.07	124.7	51.27	73.87	**0.001**
S.E.S.	18.53	19.82	16.83	21.67	0.374
T3(4-12)	171.7	156.6	168.9	147.5	0.332
T4 (4-12)	11.92	12.60	12.19	11.95	0.913
FT3 (4-12)	3.791	4.180	3.856	3.585	0.072
FT4 (4-12)	17.96	19.15	14.74	17.69	**0.002**
TSH (4-12)	0.613	0.526	0.828	0.722	0.254
T3(13-18)	185.3	178.9	186.7	168.7	0.690
T4 (13-18)	12.39	12.66	11.70	12.45	0.726
FT3 (13-18)	3.932	4.426	3.866	3.783	0.093
FT4 (13-18)	16.42	17.40	12.53	15.80	**0.000**
TSH (13-18)	0.776	0.556	1.193	0.829	**0.009**
T3(19-24)	193.7	188.5	181.1	168.9	0.447
T4 (19-24)	11.69	13.91	11.81	13.49	0.062
FT3 (19-24)	4.063	3.913	3.659	3.860	0.334
FT4 (19-24)	14.72	16.40	12.64	16.63	**0.000**
TSH (19-24)	0.982	0.495	1.225	0.583	**0.000**
T3(25-30)	188.1	191.8	184.7	181.4	0.931
T4 (25-30)	11.67	12.95	12.04	13.81	0.094
FT3 (25-30)	3.824	3.986	3.620	3.645	0.362
FT4 (25-30)	14.10	15.80	12.28	16.47	**0.000**
TSH (25-30)	0.995	0.449	1.302	0.376	**0.000**
T3(31-36)	178.3	188.1	174.9	172.6	0.651
T4 (31-36)	11.88	14.05	11.94	13.06	0.059
FT3 (31-36)	3.620	3.819	3.579	3.505	0.594
FT4 (31-36)	13.77	16.62	12.22	16.41	**0.000**
TSH (31-36)	0.959	0.355	1.454	0.330	**0.000**
					↓
Combined					**0.000**

As we can see, significant differences exists among the four analyzed groups, with reference to FSIQ, VIQ, maternal UIC and the serum of TSH and TF4 in different times of observation. So, for only these variables, we performed the two-by-two comparison between groups. For these multiple comparisons, we had to apply Bonferroni's correction; the number of possible comparisons that can performed with four groups are 6, so the adjusted significance level for this analysis was equal to $\frac{0.050}{6} = 0.008$.

Table 2. Partial and Combined p-value for two-by-two comparisons

VARIABLES	**A vs B**	**A vs C**	**A vs D**	**B vs C**	**B vs D**	**C vs D**
FSIQ	0.603	0.022	0.014	0.026	0.017	0.942
VIQ	0.250	0.058	0.048	**0.007**	**0.006**	0.895
UIC	**0.001**	**0.000**	0.074	**0.000**	**0.000**	0.026
FT4 (4-12)	0.515	**0.000**	0.763	**0.001**	0.419	**0.002**
FT4 (13-18)	0.401	**0.000**	0.566	**0.001**	0.267	**0.003**
TSH (13-18)	0.182	0.012	0.790	**0.002**	0.199	0.078
FT4 (19-24)	**0.006**	**0.004**	0.018	**0.000**	0.812	**0.000**
TSH (19-24)	0.010	0.314	0.016	**0.002**	0.538	**0.001**
FT4 (25-30)	**0.007**	0.013	**0.005**	**0.000**	0.452	**0.000**
TSH (25-30)	**0.005**	0.211	**0.001**	**0.000**	0.558	**0.000**
FT4 (31-36)	**0.000**	0.045	**0.006**	**0.000**	0.832	**0.001**
TSH (31-36)	**0.001**	0.020	**0.000**	**0.001**	0.821	**0.000**
	↓	↓	↓	↓	↓	↓
Combined	**0.000**	**0.000**	**0.000**	**0.000**	**0.002**	**0.001**

The result highlight that offspring of mothers belonging to groups A (Iodine) and B (iodine + LT4) had similar Verbal, Performance and Full-Scale Intelligence Quotients; all these IQ were higher than children born to no-I and no-I + T4 mothers (groups C and D, respectively). Moreover, a further analysis was also performed aggregating Groups A and B (iodine) and Groups C and D (No-iodine) in order to evaluate effects of maternal iodine nutrition.

Table 3. Mean values and p-values for comparison between Iodio vs No-Iodio groups

VARIABLES	**Iodine**	**No-Iodine**	**p-value**
FSIQ	94.55	81.50	**0.001**
VIQ	93.55	79.97	**0.001**
UIC	106.8	62.57	**0.000**
FT4 (4-12)	18.53	16.22	**0.005**
FT4 (13-18)	16.89	14.16	**0.001**
TSH (13-18)	0.670	1.011	**0.013**
FT4 (19-24)	15.53	14.64	0.209
TSH (19-24)	0.747	0.904	0.315
FT4 (25-30)	14.92	14.38	0.438
TSH (25-30)	0.731	0.839	0.516
FT4 (31-36)	15.15	14.31	0.284
TSH (31-36)	0.667	0.892	0.188
			↓
Combined			**0.001**

Defective cognitive function (in term of FSIQ and VIQ) was significantly higher in the children of mothers not using iodized salt than of those mothers using it. Also UIC results to be significantly higher in mothers who use iodized salt. The use of iodized salt also implies a significant increase in FT4 values in the period between 4 and 18 weeks of pregnancy.

We underline also that the TSH values are significantly reduced in the weeks 13-18 for the only women who consume iodized salt. The other examined variables show no significant difference between the two compared groups.

5 Final remarks

Thyroid hormone is essential for normal pregnancy progression and brain development. Impairments in maternal thyroid function are associated to several obstetrical complications [17],[11]. They require prompt intervention, bearing in mind that both maternal thyroid disease per se and related treatments may adversely affect the newborn's health [17] and [14].

From a methodological point of view, in this paper we aim to show as the permutation tests are very helpful in medical contexts, in particular in the endocrinological research. We applied permutation tests to perform comparison between four groups of children, defined on the basis of maternal histories of iodized salt consumption and LT4 treatment. Examining the results achieved by applying NPC tests, we have to notice an interesting result: defective cognitive function (in term of FSIQ and VIQ) was significantly higher in the children of mothers not using iodized salt, than of those mothers using it. Also UIC results to be significantly higher in mothers who use iodized salt.

So, our research emphasizes the importance of taking iodine in pregnancy to the child's future welfare. It also shows how much a lack of iodine may hinder children into reaching their full intellectual potential. An inadequate maternal iodine intake during pregnancy may result in cognitive impairment in later life, likely because of an insufficient fetal TH output due to reduced iodine storage in the fetal gland. Therefore, based on the obtained results, we can affirm that children, whose mothers took the right amount of iodine during pregnancy, exhibit more learning and cognitive abilities and are more intelligent than children whose fetal life was marked by a low iodine intake.

Bibliography

[1] Alibrandi A. and Zirilli A. (2007). A statistical evaluation on high seric levels of D-Dimer: a case control study to test the influences of ascites. In *Atti S.Co.2007 Conference*, CLEUP, Padova, 9–14.

[2] Arboretti Giancristofaro R., Marozzi M. and Salmaso L. (2005). Repeated measures designs: a permutation approach for testing for active effects. *Far East Journal of Theoretical Statistics, Special Volume on Biostatistics* **16**, 2,303–325.

[3] Arboretti Giancristofaro R., Pesarin F., Salmaso L. and Solari A. (2007). Nonparametric procedure for testing for dropout rates on university courses with application to an Italian case study. In: S. Sawilowsky (ed.), *Real Data Analysis*. Charlotte, NC: Information Age Publishing, 355–385.

[4] Arboretti Giancristofaro R., Brombin C., Pellizzari S., Salmaso L. and Mozzanega B. (2008). Nonparametric methods applied to nuchal translucency and fetal macrosomia. *Journal of Biostatistics* **2**, 19–36.

[5] Arboretti Giancristofaro R., Bonnini S. and Salmaso L. (2009). Employment status and education/employment relationship of PhD graduates from the University of Ferrara. *Journal of Applied Statistics*. http://dx.doi.org/10.1080/02664760802638108

[6] Bath S.C. and Rayman M.P. (2015). A review of the iodine status of UK pregnant women and its implications for the offspring.*Environmental Geochemistry and Health*, **37**, 4, 619-629.

[7] Bonnini S. ,Salmaso L. and Solari A. (2005). Multivariate permutation tests for evaluating effectiveness of universities through the analysis of students dropouts. *Statistica & Applicazioni* **3**, 37–44.

[8] Bonnini S., Corain L., Munaò F. and Salmaso L. (2006). Neurocognitive Effects in Welders Exposed to Aluminium: An Application of the NPC Test and NPC Ranking Methods. *Statistical Methods and Applications, Journal of the Statistical Society* **15**, 2, 191–208.

[9] Callegaro A., Pesarin F. and Salmaso L. (2003). Test di permutazione per il confronto di curve di sopravvivenza. *Statistica Applicata* **15**, 2, 241–261.

[10] Corain L. and Salmaso L. (2004). Multivariate and multistrata nonparametric tests: the nonparametric combination method. *Journal of Modern Applied Statistical Methods* **3**, 443–461.

[11] De Vivo A., Mancuso A., Giacobbe A., Moleti M., Maggio Savasta L., De Dominici R., Priolo A.M. and Vermiglio F. (2010). Thyroid function in women found to have early pregnancy loss. *Thyroid*, **20**, 6, 633-637.

[12] Di Castelnuovo A., Mazzaro D., Pesarin F. and Salmaso L.(2000). Test di permutazione multidimensionali in problemi d'inferenza isotonica: un'applicazione alla genetica. *Statistica* **60**, 4, 691–700.

[13] Finos L., Pesarin F., Salmaso L. and Solari A.(2004). *Nonparametric iterated procedure for testing genetic differentiation*, Atti XLIII Riunione Scientifica SIS, CLEUP, Padova.

[14] Gianetti E., Russo L., Orlandi F., Chiovato L., Giusti M., Benvenga S., Moleti M., Vermiglio F., Macchia P.E., Vitale M., Regalbuto C., Centanni M., Martino E., Vitti P. and Tonacchera M. (2015). Pregnancy outcome in women treated with methimazole or propylthiouracil during pregnancy. *Journal of Endocrinological Investigation*,**38**,9, 977-985.

[15] Glinoer D. and Delange F. (2000). The potential repercussions of maternal, fetal, and neonatal hypothyroxinemia on the progeny. *Thyroid*, **10**, 10, 871-887.

[16] Good P. (2000). *Permutation test.* 2nd Edition, Springer-Verlag, New York.

[17] Krassas G.E., Poppe K. and Glinoer D. (2010). Thyroid function and human reproductive health.*Endocrine Reviews*,**31**, 5,702-755

[18] Liu H., Momotani N., Noh J.Y., Ishikawa N., Takebe K. and Ito K. (1994). Maternal hypothyroidism during early pregnancy and intellectual development of the progeny. *Archives of Internal Medicine*, 154, 7, 785-787

[19] Moleti M., Lo Presti V.P., Campolo M.C., Mattina F., Galletti M., Mandolfino M., Violi M.A., Giorgianni G., De Domenico D., Trimarchi F. and Vermiglio F. (2008). Iodine prophylaxis using iodized salt and risk of maternal thyroid failure in conditions of mild iodine deficiency. *Journal of clinical endocrinology and metabolism*,**93**, 7, 2616-2621.

[20] Moleti M., Vermiglio F. and Trimarchi F. (2009a). Maternal isolated hypothyroxinemia: To treat or not to treat? *Journal of Endocrinological Investigation*, **32**, 9:780-782.

[21] Moleti M., Lo Presti V.P., Mattina F., Mancuso A., De Vivo A., Giorgianni G., Di Bella B., Trimarchi F. and Vermiglio F. (2009b). Gestational thyroid function abnormalities in conditions of mild iodine deficiency: early screening versus continuous monitoring of maternal thyroid status. *European Journal of Endocrinology*, **160**, 4, 611-617.

[22] Moleti M., Di Bella B., Giorgianni G., Mancuso A., De Vivo A., Alibrandi A., Trimarchi F. and Vermiglio F. (2011a). Maternal thyroid function in different conditions of iodine nutrition in pregnant women exposed to mild-moderate iodine deficiency: an observational study. *Clinical Endocrinology (Oxford)***74**, 6, 762-768.

[23] Moleti M., Trimarchi F. and Vermiglio F. (2011b). Doubts and Concerns about Isolated Maternal Hypothyroxinemia.*Journal of Thyroid Research*, 2011:463029

[24] Moleti M., Trimarchi F. and Vermiglio F. (2014). Thyroid physiology in pregnancy.*Endocrine Practice*, **20**, 6, 589-596.

[25] Moleti M., Trimarchi F., Tortorella G., Candia Longo A., Giorgianni G., Sturniolo G., Alibrandi A. and Vermiglio F. (2016). Effects of Maternal Iodine Nutrition and Thyroid Status on Cognitive Development in Offspring: A Pilot Study. *Thyroid*, 26, 2, 296-305.

[26] Momotani N., Iwama S. and Momotani K. (2012). Neurodevelopment in children born to hypothyroid mothers restored to normal thyroxine (T4) concentration by late pregnancy in Japan: no apparent influence of maternal T4 deficiency. *Journal of clinical endocrinology and metabolism*, **97**, 4, 1104-1108.

[27] Morreale De Escobar G., Obregn M.J. and Escobar Del Rey F. (2000). Is neuropsychological development related to maternal hypothyroidism or to maternal hypothyroxinemia? *Journal of clinical endocrinology and metabolism*,**85**, 11, 3975-3987.

[28] Pesarin F.(1997). *Permutation testing of multidimensional Hypotheses*, CLEUP, Padova.

[29] Pesarin F. (2001). *Multivariate permutation tests with applications in biostatistics.* Wiley, Chichester.

[30] Pesarin F.and Salmaso L. (2010). *Permutation Tests for Complex Data: Theory, Applications and Software.* Wiley Series in Probability and Statistics, Chichester.

[31] Redman K.F. (2011). *Iodine and Cognition in Young Adults: A Randomised, Placebo-Controlled Trial.* Thesis submitted for the degree of Master of Science at the University of Otago, Dunedin, New Zealand. December 2011.

[32] Salmaso L. (2005). Permutation tests in screening two-level factorial experiments. *Advances and Applications in Statistics*, **5**, 1, 91–110

[33] Vermiglio F., Lo Presti V.P., Moleti M., Sidoti M., Tortorella G., Scaffidi G., Castagna M.G., Mattina F., Violi M.A., Cris A., Artemisia A. and Trimarchi F. (2004). Attention deficit and hyperactivity disorders in the offspring of mothers exposed to mild-moderate iodine deficiency: a possible novel iodine deficiency disorder in developed countries. *Journal of clinical endocrinology and metabolism*, **89**, 12, 6054-6060.

[34] Williams G.R. (2008). Neurodevelopmental and neurophysiological actions of thyroid hormone. *Journal of Neuroendocrinology*, **20**, 6, 784-794.

[35] Zimmermann M.B. (2009). Iodine deficiency.*Endocrine Reviews*, **30**, 4, 376-408.

[36] Zirilli A., Alibrandi A., Spadaro A. and Freni M.A.(2005). *Prognostic factors of survival in the cirrhosis of the liver: A statistical evaluation in a multivariate approach.* In Atti S.Co.2005 Conference, CLEUP, Padova, 173–178

[37] Zirilli A. and Alibrandi A. (2009). *A permutation approach to evaluate hyperhomocysteinemia in epileptic patients.* In: Supplemento ai rendiconti del circolo matematico di palermo. VII International Conference in "Stochastic Geometry, Convex Bodies, Empirical Measures and application to mechanics and Engineering train-transport", Messina, 22-24 April 2009, 369–378.

Using intraclass correlation coefficients to quantify spatial variability of catastrophe model errors

Baldvin Einarsson, *Financial Modeling, AIR Worldwide*, beinarsson@air-worldwide.com
Rafał Wójcik, *Financial Modeling, AIR Worldwide*, rwojcik@air-worldwide.com
Jayanta Guin, *Financial Modeling, AIR Worldwide*, jguin@air-worldwide.com

Abstract. Systematic spatial errors of natural catastrophe (CAT) models are quantified using hierarchical linear models. Insurance claims are grouped into spatial bins on a regular grid, which avoids computationally expensive distance calculations when estimating spatial covariances. For insurance claims and CAT model estimates, damage ratios are used to determine the model errors. The spatial structure of claims distributions around a model estimate is determined via intraclass correlation coefficient (ICC). A methodology is introduced to incorporate all claims, which greatly enhances the usability and robustness of the statistical models. These statistical models can have a hierarchy of spatial bins nested within larger bins, and both the number of such hierarchies, as well as the sizes of the rectangular bins at each layer, are investigated. Furthermore, several validation procedures are presented using the claims data from a major earthquake. The results are obtained with the R-package lme4.

Keywords. Hierarchical linear models, Intraclass correlation coefficients, Catastrophe models, Model errors, Spatial correlation

1 Introduction

Catastrophe Models and Spatial Correlation of Model Errors Insurance for natural catastrophes is quite different from car insurance. In the latter, an insurance company has vast amounts of data to infer with considerable accuracy how risky it is to sell car insurance to an adolescent male. However, when it comes to home insurance, natural catastrophes pose several difficulties. First, absence of data does not necessarily mean that an event is unlikely to occur, e.g. an earthquake can occur on a previously unknown fault line. But, even if data is plentiful, it might still not give an accurate estimate of the frequency of events. Furthermore, relying on historical events might give a skewed view of the potential event intensities. This was evident with hurricane Andrew in 1992, which caused industry wide losses around four times that of the previous significant event, which was hurricane Hugo. In 2005, hurricane Katrina added another factor of four to hurricane Andrew's losses. This level of uncertainty has to be dealt with in some way. Several decades ago, insurers relied on rules of thumb and conventions. Today, insurance companies use catastrophe (CAT) models to quantify their risk. The purpose of CAT modeling is to anticipate the likelihood and severity of losses from earthquakes and hurricanes to terrorism and crop failure, so that companies (and governments) can appropriately prepare for their financial impact. This is achieved by simulating a large ensemble of hypothetical events with varying intensity, e.g. maximum wind speed for hurricane, attenuation and shake magnitude for earthquake, water level height for flood etc., followed by loss estimation for each simulated event at every location.

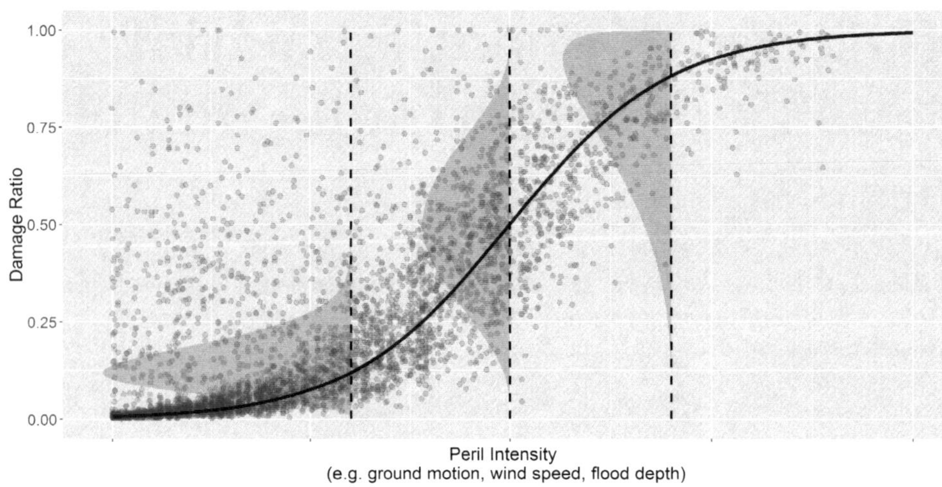

Figure 1. Damage function converting event intensity to a modeled mean damage ratio (blue solid curve). Also shown are hypothetical claims (red points) for these intensities, which yield a distribution of claims damage ratios around the model mean damage ratio. Here, three such distributions are shown (grey areas).

When aggregating losses at different locations to a portfolio level it is important to account for spatial correlation between these losses, i.e. how the losses at nearby locations are related to each other. There are several different components contributing to spatial correlation. For example in a hurricane model, some aspects of spatial correlation are implicitly accounted for by the predicted wind field where nearby locations experience similar wind speeds. In what follows, we address another type of correlation: correlation conditional on the CAT model estimate. Intuitively, if the predicted wind speed for an area is too high, then the losses in that area are likely to be lower than expected. This means the losses will be correlated with each other, relative to the model estimate. Such conditional dependency is designated as the correlation of model errors. Losses may also be correlated at a local level due to common building practices. If an entire neighborhood is built by the same construction company, using the same substandard materials, the losses in that neighborhood will likely be higher than modeled and correlated within the neighborhood. In this paper we assess these non-modeled sources of spatial correlation.

Claims and Model Estimates An insurance *claim* amount is dependent on several aspects which include the replacement value (i.e. the property's worth), *exposure* information e.g. construction and occupancy type with other building characteristics (e.g. year built and height), location (geographical and/or address), and financial terms (e.g. deductibles and limits). For each available claim, we obtain the modeled losses for the given peril (and event) at the specified location, using all available building characteristics. In both cases, we work with damage ratios (DR), which are defined as loss divided by replacement value. Note that in this paper we are only interested in *ground up* losses, i.e. losses before any financial terms have been applied.

The modeled losses are obtained from the modeled peril intensity of the particular event, e.g. ground motion, flood depth, or wind speed. Thus, damage functions are created to convert intensities to model mean damage ratios (MMDR). In an ideal world, two identical structures experiencing the same peril of the same intensity should experience the exact same damage. However, for exposures with the same predicted MMDR, usually at different geographical locations, we have a distribution of claims damage ratios, as seen in Figure 1. It is possible for identical buildings to experience different levels of damage when impacted by the same intensity of a particular peril. This is due to differences in construction and local site-specific effects. To capture this variability we group claims DRs using MMDR and derive the whole distribution of possible values. It is natural to require the MMDRs to be close to the means of the distributions of the corresponding claims damage ratios.

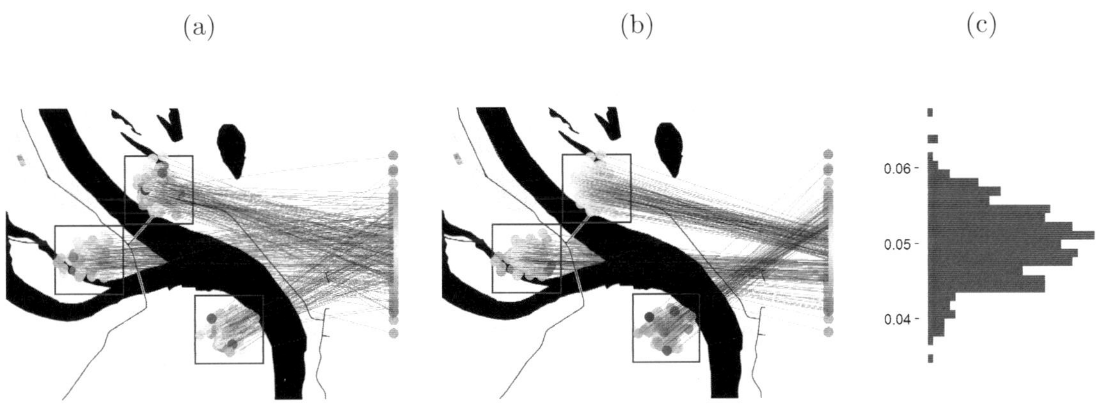

Figure 2. In (a) and (b) are two hypothetical scenarios from 3 geographical locations, shown as rectangular bins, where the black regions are water bodies. The points denote exposures, the colors of which indicate the (hypothetical) claims DRs. The blue lines connect a location to its corresponding DR value in the color bars. See text for further discussion on the differences between the two scenarios. The overall distribution of claims (c) is the same for both scenarios, and the MMDR is assumed to be 0.05 for all claims.

Spatial Structure of Claims Distributions Figure 2 (a)-(b) shows two hypothetical scenarios which highlight what we are trying to capture. Shown are three areas (green rectangles), each of which has several claims. All claims are assumed to have MMDR of 0.05. The overall distribution of claims, Figure 2 (c), is the same in both cases.

In Figure 2 (a) we see claims distributions which do not exhibit any obvious underlying spatial structure determined by the three different locations. This means that for any of the three spatial bins, the claims therein are dissimilar. Thus, inspecting the value of one claim does not provide any information on the values of the remaining claims in that spatial bin. This implies low spatial correlation of model errors.

However, in Figure 2 (b) we see that the distribution of claims has a certain underlying spatial structure, with the claims in each spatial bin having lower variability i.e. two randomly selected claims will have similar values and therefore high spatial correlation.

It is important to note that we are *not* attempting to capture spatial correlation of a specific peril in the classical geostatistical sense [6], i.e. we are not attempting to develop a simulation or interpolation procedure for claims using classical geostatistical tools. We are quantifying in an efficient way the spatial variability of model errors, later to be used for portfolio loss aggregation algorithms; the details of which are not our main focus of attention.

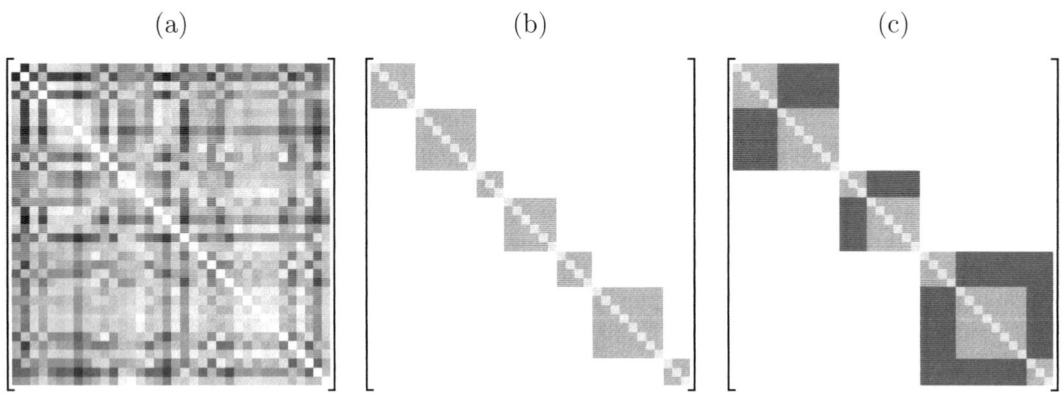

Figure 3. Illustration of hypothetical covariance matrices (of CAT model errors) for an exposures portfolio. (a) Classical distance based, full covariance matrix. (b) A block diagonal covariance matrix from hierarchical linear models, with one spatial scale. The block structure is determined by the spatial bins, and the correlation is assumed to be zero for exposures in different spatial bins. (c) Similar matrix as in (b), but with two spatial scales.

We prefer avoiding computationally expensive distance calculations, which yield full covariance matrices as illustrated in Figure 3 (a). We have adopted hierarchical linear models, and assume that correlations between locations are fully determined by the spatial bins they reside in. Figures 3 (b) and (c) show examples of such computationally efficient block diagonal covariance matrices. Diagonal blocks at one spatial scale are comprised of compound covariance matrices. This implies that correlation between any two locations within the same spatial block is the same. Likewise, diagonal blocks at the second spatial scale form another set of compound covariance matrices which represent the situation when correlation between any two locations within the same spatial block at the second scale but different blocks at the first scale, is the same. Finally, we note that it is possible for two locations to be very close in space, but fall into distinct spatial bins and thus be considered independent. This is a simplifying assumption for computational efficiency.

2 Hierarchical Linear Models

Hierarchical linear models [20, 5, 17, 16] are popular in the study of group effects, especially in the social sciences where they are e.g. used to assess differences between students across different classes. Here, we employ these models to the claims DR distributions around each MMDR ratios (Figure 1), with the grouping effects determined by the spatial bins (Figure 2). We are investigating how much of the variability in the claims distribution can be attributed to spatial variability. By grouping the claims DRs by their corresponding model MMDRs, we obtain multiple subsets of the data, where each group corresponds to an MMDR interval of fixed width. We now describe our methodology which allows us to work with *all* available claims, regardless of the model estimates.

Methodology We let $\mathcal{D} = \{(c_i, m_i)\}_{i=1}^{M}$ denote the collection of claims DRs, c_i, and the corresponding MMDRs, m_i, where M is the total number of available claims. As outlined above, we want to model the variability of claims around a mean as determined by the CAT model MMDR. With one spatial scale, we model claims as follows:

$$c_{ij} = a(m_{ij}, \mathcal{D}) + \alpha_j + \epsilon_{ij}. \tag{1}$$

Here, c_{ij} denotes *claim i in bin j*, with $i = 1, \ldots, n_j$, which clearly shows the hierarchy between claims within bins. In multilevel approach, the model in Equation (1) is called the *empty model* because there are no explanatory variables. The value m_{ij} is the corresponding MMDR, and $a(m_{ij}, \mathcal{D})$ (called the *fixed part*) is calculated as the mean of claims which have MMDR values similar to m_{ij}. We expect this mean to be the same (or similar) to the MMDR value, i.e.

$$a(m, \mathcal{D}) \simeq m, \forall m. \tag{2}$$

If this is not satisfied, then care has to be taken to modify the MMDRs, either via CAT model calibrations or transformations. These considerations are beyond the scope of this paper, and we assume that Equation (2) holds. The terms α_j are the grouping effects, and the ϵ_{ij} are the residuals. The sum of those two terms is called the *random part*, which highlights that we express claims as random variability around CAT model estimates. We allow for different numbers of claims in the spatial bins (i.e. the n_j's can vary), thus making the design *unbalanced*. The binning effects, α_j, are assumed i.i.d. for all bins, $\alpha_j \sim N(0, \sigma_\alpha^2)$. The residuals, $\epsilon_{ij} \sim N(0, \sigma_\epsilon^2)$ are also i.i.d. with σ_ϵ^2 independent of the bin choice, an assumption called *homoscedasticity*. Furthermore, we assume that the α_j's and ϵ_{ij}'s are independent for all i and j. These statements yield a block diagonal covariance matrix as illustrated in Figure 3 (b).

The model in Equation (1) can be re-written to include two spatial scales:

$$c_{ijk} = a(m_{ijk}, \mathcal{D}) + \beta_k + \alpha_{jk} + \epsilon_{ijk}, \tag{3}$$

where c_{ijk} is *claim i in bin j nested in bin k*. Similarly, m_{ijk} is the corresponding MMDR value of claim ijk, and $a(m_{ijk}, \mathcal{D})$ is determined the same way as before. The effects $\beta_k \sim N(0, \sigma_\beta^2)$ are now the grouping effects at the larger scale, with other quantities having similar interpretations and assumptions as before. The covariance matrix would in this case be as shown in Figure 3 (c).

The formulation above allows us to work with entire datasets, and obtain a single result for each spatial scale, as opposed to one for each MMDR group (which then has to be aggregated in some way). In fact, we are seeking a single estimate of spatial variability to distinguish between the two scenarios in Figure 2 (a)-(b). Thus, we are not interested in differences between MMDR groups. This approach works well with small datasets in terms of obtaining more robust results.

Intraclass Correlation Coefficients

One spatial scale Under the framework above, we define the *intraclass correlation coefficient* (ICC), also known as *repeatability*, for one spatial scale as:

$$ICC_\alpha^{(1)} := \frac{\sigma_\alpha^2}{\sigma_\alpha^2 + \sigma_\epsilon^2}. \tag{4}$$

The superscript $^{(1)}$ denotes that we are using one spatial scale. From Equation (1), we see that σ_α^2 is the variance from the grouping effects, and $\sigma_\alpha^2 + \sigma_\epsilon^2$ is the total variance. Therefore, as a ratio of variances, the intraclass correlation coefficient signifies how much of the claims variance can be explained by the binning effect. The ICC values range from 0 to 1.

Another interpretation of the *ICC* is that it denotes the usual Pearson correlation (see e.g. [9]) between any two randomly chosen claims y_{ij} and y_{lj} from the same (arbitrary) bin j, that is

$$ICC_\alpha^{(1)} = \text{Corr}(y_{ij}, y_{lj}). \tag{5}$$

For further discussions on the intraclass correlations coefficient and its interpretations, see e.g. [8, 12, 5, 20, 21].

The ICC value can be used to discriminate between the two hypothetical scenarios in Figure 2. The ICC value for scenario (a) is close to zero, but the ICC value for scenario (b) is close to 1. Thus, for the remainder of this paper, the ICC values are our main focus of interest.

Two spatial scales The intraclass correlation coefficients extend naturally for two levels of hierarchy in the following way:

$$ICC_{\alpha}^{(2)} := \frac{\sigma_{\beta}^2 + \sigma_{\alpha}^2}{\sigma_{\beta}^2 + \sigma_{\alpha}^2 + \sigma_{\epsilon}^2} \tag{6}$$

and

$$ICC_{\beta}^{(2)} := \frac{\sigma_{\beta}^2}{\sigma_{\beta}^2 + \sigma_{\alpha}^2 + \sigma_{\epsilon}^2}. \tag{7}$$

The superscript $^{(2)}$ now signifies that we have two spatial scales, see Equation (3). The interpretations of these coefficients are similar to the case with one spatial scale. Here, $ICC_{\alpha}^{(2)}$ measures similarity of claims withing the smaller spatial scale bins, while $ICC_{\beta}^{(2)}$ measures the similarity within the larger spatial bins. Note that if including this larger spatial scale does not capture any additional variance, i.e. if $\sigma_{\beta}^2 = 0$, then we have $ICC_{\beta}^{(2)} = 0$ and $ICC_{\alpha}^{(2)} = ICC_{\alpha}^{(1)}$. The last equality states that if no additional variance is found, then the ICC estimates at the smaller scale remains unchanged. In principle, Equation (1) can be extended to an arbitrary number of spatial scales, and the ICC calculations and interpretations follow naturally. However, apart from the nature of each peril, practical issues like data availability or compute speed of loss aggregation procedure are a limiting factor.

Stationarity and isotropy

An important assumption of the hierarchical linear models for claims is that of *stationarity*. This means that estimates of the parameters of the model are invariant to shifts of the underlying CAT model grid which determines the spatial grouping of claims. If further the estimates are invariant to all rigid motions the model is *stationary and isotropic*. To verify these assumptions, we implemented the following procedure:

1. For the regular non-shifted grid, obtain the ICC results.
2. Shift the grid some fixed distance in any of several directions; we chose the 16 directions $m \cdot 360°/16$, with $m = 0, \ldots, 15$. For each of these shifted scenarios:
   a) Regroup claims into bins based on the shifted grid.
   b) Calculate ICC estimates as before.

In steps 1 and 2 b) above, we note that we obtain ICC estimates for each of the spatial scales involved. If using one spatial scale, we find $ICC_{\alpha}^{(1)}$; if using two spatial scales we obtain $ICC_{\alpha}^{(2)}$ and $ICC_{\beta}^{(2)}$, and so on.

We plot the results in polar coordinates, where the radius denotes the ICC estimate and the angle denotes the direction of the shift. To compare the results with those from the regular grid, we draw circles, one for each spatial scale involved, radii of which are the ICC estimates from step 1, see Figure 5. If stationarity and isotropy assumptions hold, then the points should fall on those circles. Our results indicate that the assumptions hold; if they did not hold, then additional algorithms, that are beyond the scope of this paper, would be required.

Bootstrapping

Determining confidence intervals for the output of hierarchical linear models, and the intraclass correlation coefficients, is not straightforward. For a review of methods under different model frameworks, see [8] and [9], respectively. Those theoretical results are for one spatial scale only, and it is not obvious how to extend these to multiple spatial scales.

One complicating factor is that the claims data are almost always *unbalanced*, meaning that the spatial bins can have varying numbers of claims. However, the main issue is that the computational methods of the **lme4** package (penalized least squares and REML) produce estimates of variances which are not entirely based on mean squared errors. This makes their theoretical distributions unclear, and so our task is not to simply employ traditional methods of finding the appropriate degrees of freedom for an F-distribution, such as those of Satterthwaite [19] or Kenward-Roger [10]. See [2] for an entertaining elaboration on this issue.

Following the advice of the **lme4** package authors [3, pp. 35-36], we use bootstrapping techniques (see e.g. [7]) to obtain confidence intervals for the ICCs. However, fully non-parametric, spatial bootstrapping is non-trivial, and beyond the scope of this paper. Furthermore, semi-parametric bootstrapping (by sampling from the predicted binning effects and residuals) in [14] is biased. We therefore use parametric bootstrapping as follows:

1. Obtain the variance estimates of the original dataset. This gives us the distributions of the binning effects and residuals to draw from in step 2.
2. Using the estimated distributions, generate a new dataset $\mathcal{D}' = \{(c'_i, m_i)\}_{i=1}^{M}$ according to the model assumptions (e.g. Equations (1) and (3)).
3. Refit the new data according to the hierarchical linear models as before and extract the parameters of interest needed to calculate ICC values.
4. Repeat steps 2 and 3 *ad libitum*.

The procedure is easy to implement, and the **lme4** package in R includes a function called `bootMer()`, which carries out the above steps. However, these computations are intense, and obtaining the results of 10,000 iterations took over one week to run.

3 Results

In all instances, the R-package **lme4** was used to obtain the ICC results. A sample function call, with two spatial scales is the following:

```
R> lme4::lmer(GUCMDR_A ~ 0 + offset(MMDR.mu) +
                         (1|kmbinLarge) +
                         (1|kmbinLarge:kmbinSmall),
              data=allData, REML=TRUE)
```

The data (in data frame `allData`) comes from a large earthquake off the coast of Japan, and has nearly 2 million entries. The column `GUCMDR_A` denotes the claims DR, and `MMDR.mu` contains the corresponding $a(\cdot, \mathcal{D})$ values. Furthermore, `kmbinSmall` and `kmbinLarge` represent the spatial bins at the smaller and larger spatial scales, respectively. The relationship between the formula object and Equation (3) should be evident, with the last two terms signifying contributions from the smaller bins nested inside the larger bins. To give an idea of the spatial extent of the earthquake, there are approximately sixteen thousand 1 km bins (which have a minimum of 20 claims each), and nearly two hundred 25 km bins (each with at least three 1 km bins).

Analyzing ICC values for a specific peril generally involves determining both how many spatial scales to use and dimensions of each such spatial scale. However, for practical reasons, we don't exceed two spatial scales, although both the theory and tools are available.

First, we analyze the data using only one spatial scale of varying sizes and obtain ICC results at each step. This process can indicate the extent of spatial correlation in the data. Results from this procedure can be seen in Figure 4 (a), where the width of the spatial bins ranged from 1 km to 64 km. We see that the $ICC_\alpha^{(1)}$ decreases as the bin sizes increase, as expected. These results appear to be fairly smooth up to at least 25 km; after which, the results become more noisy.

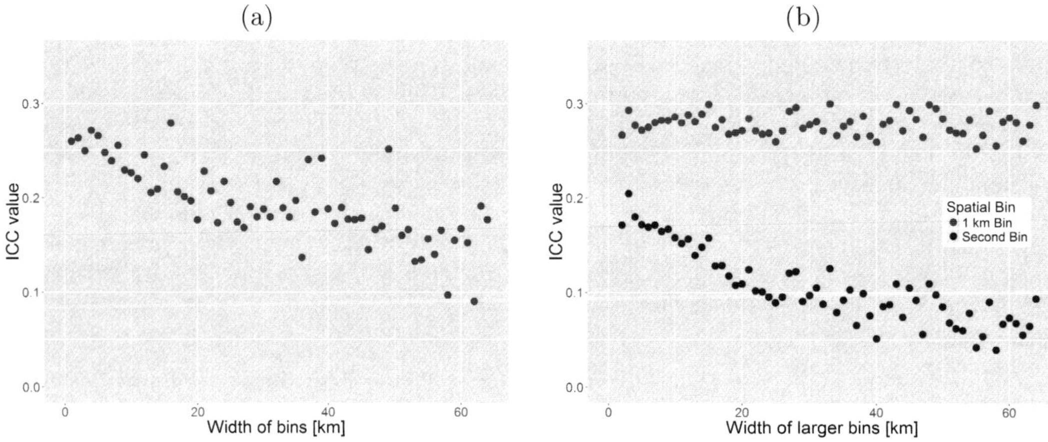

Figure 4. ICC results from the earthquake peril. (a) $ICC_\alpha^{(1)}$ with bin widths ranging from 1 km to 64 km. (b) $ICC_\alpha^{(2)}$ and $ICC_\beta^{(2)}$ for the smaller bins fixed at 1 km (red) and the larger bins ranging from 2 km to 64 km (blue).

Then, we add a second spatial scale by fixing the smaller scale. The size of the larger scale is then varied. In Figure 4 (b) we have fixed the smaller spatial scale at 1 km, and varied the larger bins from 2 km to 64 km. The main reason for this is that at 1 km we capture most of the small scale correlations. Another reason is that a grid cell of AIR's CAT modeling software has length and width of $\frac{1}{120}$ decimal degrees which is $\sim 1\ km^2$ at the equator, making 1 km resolution a natural first choice. The results in Figure 4 (b) seem to be fairly reliable up to 25 km width of the larger bins. These results are intuitive: the intraclass correlation at the larger scale, $ICC_\beta^{(2)}$, decreases as the larger bins increase in size; the intraclass correlation at the smaller scale, $ICC_\alpha^{(2)}$, remains fairly constant as the larger bins increase in size, as well as having a similar value to that of $ICC_{alpha}^{(1)}$ for 1 km bins.

Next, we verified if the stationarity and isotropy assumptions hold. For the earthquake peril, we show in Figure 5 results using two spatial scales of dimensions 1 km and 25 km, grid shifts of 350 m and 1.5 km, and rotations in 22.5° increments. In both cases, the assumptions are fulfilled. Note that we have obtained similarly good results for different shifts and different scales.

For those two spatial scales, 1 km and 25 km, we computed confidence intervals with parametric bootstrapping, as described above. The ICC estimates were $ICC_\alpha^{(2)} = 0.260$ at 1 km, and $ICC_\beta^{(2)} = 0.089$ at 25 km, and after obtaining 10,000 bootstrapped ICC estimates, the 95% confidence intervals are $[0.247, 0.272]$ and $[0.074, 0.105]$, respectively. This procedure is computationally expensive; the simulations above ran for 8 days.

4 Discussion

We have presented a methodology to use hierarchical linear models on a regular spatial grid to quantify the spatial variability of claims around CAT model estimates. The results presented here used insur-

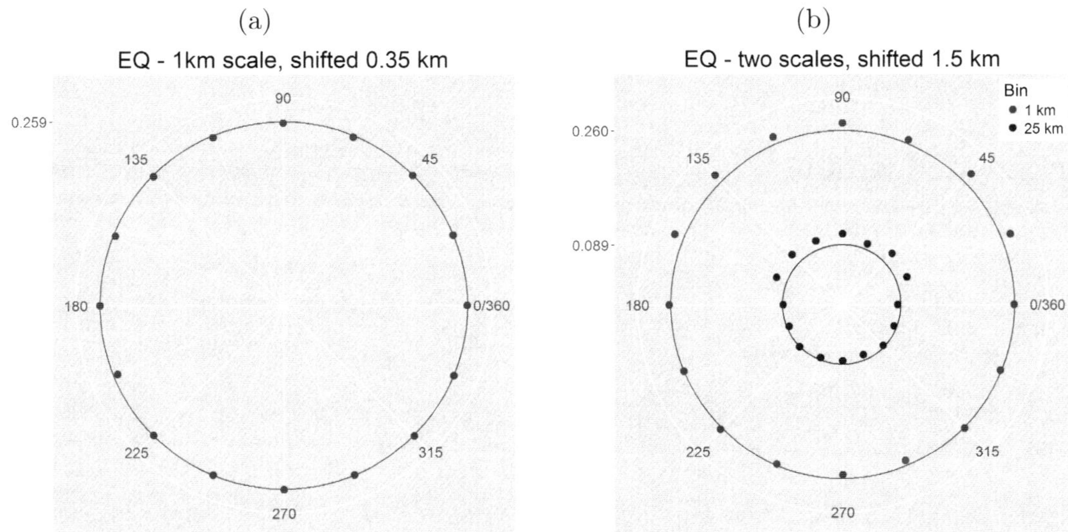

Figure 5. Stationarity and isotropy results for the earthquake peril. See page 248 for details on the procedure. (a) One spatial scale with 1 km bins and a grid shift of 350 m. (b) Two spatial scales with 1 km (red outer circle) and 25 km (blue inner circle) bins, and a grid shift of 1.5 km.

ance claims data from an undisclosed insurance company for a particular earthquake of significant size. However, this methodology can be employed to various other perils.

These statistical models have several important and easily understood interpretations, however, one disadvantage is that they make many assumptions on the structure of the underlying data. One such assumption is that the terms in the random part be normally distributed, which we know is not necessarily the case. However, one of the conclusions of [13] is that if the number of groups exceeds 100, the variance estimates are not very sensitive to the distributions of the data. This is especially the case with our approach of using all claims at once. The same paper [13] also recommends parametric bootstrapping for error estimates, which is the approach we adopted.

We did look into using generalized hierarchical linear models [15, 11], in order to account for non-normal distributions of the random parts. There is an R-package called **hglm** [18, 1], which uses similar syntax as the **lme4** package. One advantage the linear models used in this paper have is their interpretation as a variation around the CAT model estimates. If we allow the random parts to have certain distributions which do not have zero means, then this interpretation is less meaningful. For example, it has been suggested that the residuals follow a gamma distribution, but, its mean is always non-negative. In addition, calculating the ICC values (which is the main objective of this paper) becomes difficult, or even meaningless.

Note that we are investigating the feasibility of using other classical models [6, 4] to quantify the spatial variability of model errors. However, most other approaches require distance calculations, which are computationally impractical because insurance portfolios can consist of millions of exposures. We have inspected e.g. Moran's I but the question remains how to efficiently incorporate it into loss aggregation of an insurance portfolio. The benefit of our approach is obtaining a single value at each scale to determine the spatial correlation of CAT model errors. Furthermore, AIR's CAT modeling software already has in place efficient algorithms which use the ICC values and spatial binning procedures. The computational concerns are not regarding analyzing claims data with sophisticated statistical models; such an exercise only needs to be done once. However, an underwriter might not have the time to wait for hours (or even minutes!) each time a new portfolio needs to be analyzed. Additionally, any algorithm which incorporates spatial correlation in loss aggregation will increase the analysis time and thus needs to be efficient. Such an algorithm using distance calculations has yet to be devised.

Acknowledgement

We would like to thank an undisclosed insurance company for use of their claims data. We also thank two anonymous reviewers for helpful comments.

Bibliography

[1] Alam, M., Rönnegård, L. and Shen, X. (2015) *Fitting conditional and simultaneous autoregressive spatial models in hglm.* The R Journal, **7**(2), 5–18.

[2] Bates, D. (2006) *lmer, p-values and all that.* R-help mailing list, May 2006. Available at stat.ethz.ch/pipermail/r-help/2006-May/094765.html. Last accessed 2/11/2016.

[3] Bates, D., Mächler, M., Bolker, B. and Walker, S. (2015) *Fitting linear mixed-effects models using lme4.* Journal of Statistical Software, **67**(1).

[4] Bivand, R., Pebesma, E. and Gómez-Rubio, V. (2013) *Applied Spatial Data Analysis with R.* UseR! Springer.

[5] Bliese, P. (2000). *Within-Group Agreement, Non-Independence, and Reliability.* In Klein, K. and Kozlowski, S., editors, *Multilevel Theory, Research, and Methods in Organizations*, chapter 8, pages 349–381. Jossey-Bass.

[6] Cressie, N. (1993) *Statistics for Spatial Data.* Wiley series in probability and mathematical statistics: Applied probability and statistics.

[7] Davison, A. and Hinkley, D. (1997) *Bootstrap Methods and their Application.* Cambridge University Press.

[8] Donner, A. (1986) *A review of inference procedures for the intraclass correlation coefficient in the one-way random effects model.* International Statistical Review, **54**, 67–82.

[9] Donner, A. and Wells, G. (1986) *A comparison of confidence interval methods for the intraclass correlation coefficient.* Biometrics, **42**(4), 401–412.

[10] Kenward, M. and Roger, J. (1997) *Small sample inference for fixed effects from restricted maximum likelihood.* Biometrics, **53**(3), 983–997.

[11] Lee, Y. and Nelder, J. (1996) *Hierarchical generalized linear models.* Journal of the Royal Statistical Society B, **58**(4), 619–678.

[12] Lessells, C. and Boag, P. (1987) *Unrepeatable repeatabilities: A common mistake.* The Auk, **104**(1), 116–121.

[13] Maas, C. and Hox, J. (2004) *The influence of violations of assumptions on multilevel parameter estimates and their standard errors.* Computational Statistics and Data Anlaysis, **46**, 427–440.

[14] Morris, J. (2002) *The BLUPs are not "best" when it comes to bootstrapping.* Statistics and Probability Letters, **56**(4), 425–430.

[15] Nelder, J. and Wedderburn, R. (1972) *Generalized linear models.* Journal of the Royal Statistical Society A, **135**(3), 370–384.

[16] Pinheiro, J. and Bates, D. (2000) *Mixed-Effects Models in S and S-PLUS.* Springer.

[17] Raudenbush, S. (1993) *Hierarchical Linear Models and Experimental Design.* In Edwards, L., editor, *Applied Analysis of Variance in Behavioral Science*, chapter 13, pages 459–496. Marcel Dekker.

[18] Rönnegård, L., Shen, X. and Alam, M. (2010) *hglm: A package for fitting hierarchical generalized linear models.* The R Journal, **2**(2), 20–28.

[19] Satterthwaite, F. (1946) *An approximate distribution of estimates of variance compononents.* Biometrics Bulletin, **3**(6), 110–114.

[20] Snijders, T. and Bosker, R. (2011) *Multilevel Analysis: An Introduction to Basic and Advanced Multilevel Modeling.* SAGE Publications.

[21] Taylor, P. (2009) *An Introduction to Intraclass Correlation that Resolves Some Common Confusion.* Available at www.faculty.umb.edu/pjt/TL-TOC.html. Last accessed 2/24/2016.

Detection of exceptional genomic words: a comparison between species

Ana Tavares, *University of Aveiro*, ahtavares@ua.pt
João Rodrigues, *University of Aveiro*, jmr@ua.pt
Carlos Bastos, *University of Aveiro*, cbastos@ua.pt
Armando Pinho, *University of Aveiro*, ap@ua.pt
Paulo Ferreira, *University of Aveiro*, pjf@ua.pt
Paula Brito, *University of Porto*, mpbrito@fep.up.pt
Vera Afreixo, *University of Aveiro*, vera@ua.pt

Abstract. In this study we explore the potentialities of the inter-word distances to detect exceptional genomic words (oligonucleotides) in several species, using whole-genome analysis. We confront the empirical results obtained from the complete genomes with the corresponding results obtained from the random background. We develop a procedure, based on some statistical properties of the global distance distributions in DNA sequences, to discriminate words with exceptional inter-word distance distribution and to identify distances with exceptional frequency of occurrence. We identify the statistically exceptional words in whole-genomes, i.e., words with unexpected inter-word distance distributions, and we suggest species signatures based on exceptional word profiles.

Keywords. inter-oligonucleotide distances, DNA sequence, exceptional genomic word, stochastic model, goodness of fit.

1 Introduction

Several authors tried to identify exceptional words using different statistical criteria. A standard approach to detect exceptional words relies on their frequency. For example, based on genomic word frequencies and on comparisons between those frequencies and the random background (e.g. [10, 15, 16]).

The distance between two successive occurrences of a pattern in strings has been thoroughly studied and theoretical results have been deduced, in particular the generating functions of the waiting times to return to a specific pattern (e.g., [14, 18]). The probability mass function of the waiting times to return for the first time to a specific genomic word, or inter-word distance distribution, can be obtained by the Markov chain embedding technique, first developed by Fu (see, for example, [6]).

There are some interesting and counter-intuitive relations between frequency and distance distributions. Thus, the two perspectives are worth of separate investigation.

The inter-nucleotide distance (i.e., the distance between successive occurrences of the same nucleotide) has been previously explored to compare the complete genomes of several organisms; this comparison was based on genome distance distributions explored by [2]. The inter-nucleotide distance was also explored in the context of genome annotation by [11]. In [3], the inter-dinucleotide distance distribution was proposed and a comparison between all dinucleotide distributions in the human genome was performed. Note that in [3] overlapping dinucleotides were excluded from analysis, so that the expected distance distribution under an independent nucleotide model is a geometric distribution. Based on an inter-CpG distance, a CpG-island detection algorithm was proposed by [8], where a geometric distribution was used as a reference for comparison.

In this paper, we describe a procedure to highlight exceptional words that is based on inter-word distance distributions, rather than word frequencies. The subtraction of the random background from the counting result (under an independent nucleotide placement assumption) has been suggested as a way of emphasizing the contribution of selective evolution ([12, 5]). Based on this biologic perspective, we take a nucleotide independent model as the departing point and evaluate the discrepancy between real sequences and random background.

2 Materials and methods

Materials

In this study, we used the complete DNA sequences of 30 species, listed in Table 1, downloaded from the website of the National Center for Biotechnology Information (`http://www.ncbi.nlm.nih.gov/genomes`). For each species, we processed the available assembled chromosomes as separate sequences. In each sequence, we studied every word formed by k consecutive unambiguous nucleotides, with $1 < k \leq 5$. The analysis included words partially overlapping preceding or succeeding words. All ambiguous or unsequenced nucleotides, i.e., all non-ACGT symbols, are considered word delimiters.

Methods

Inter-word distance

Consider the alphabet formed by the four nucleotides $\mathcal{A} = \{A, C, G, T\}$, and let s be a symbolic sequence of length N defined in $\mathcal{A}$. For each nucleotide $x \in \mathcal{A}$, consider a numerical sequence, d^x (or simply d), that represents the inter-nucleotide distances between each occurrence of symbol x and the previous occurrence of the same symbol, i.e., the differences between the positions occupied by successive occurrences of symbol x. As an example, we show the four inter-nucleotide distance sequences for $s = AAACGTCGATCCGTG$:

$$d^A = (1, 1, 6),\ d^C = (3, 4, 1),\ d^G = (3, 5, 2),\ d^T = (4, 4).$$

A genomic word, or oligonucleotide (w), is a sequence of length k defined in $\mathcal{A}$. We can extend the notion of inter-nucleotide distance to the case of oligonucleotides. Assuming that the sequence is read through a sliding window of length k, we can define the inter-oligonucleotide (inter-w) distance sequence d^w as the differences between the positions of the first symbol of consecutive occurrences of that oligonucleotide. For example, the inter-CG distance sequence for the short DNA segment above is $d^{CG} = (3, 5)$.

Reference distribution under a nucleotide independence model

Let $w = x_1x_2x_3 \ldots x_k \in \mathcal{A}^k$ be a generic oligonucleotide and D be the random variable that represents the inter-oligonucleotide distance, from a sequence whose nucleotides are independently generated.

The reference distribution of inter-w distances can be deduced using a state diagram, which represents the progress made towards identifying w as each symbol is read from the sequence. The state diagram has $k + 1$ states. The first k states, $S_0, S_1, \ldots, S_{k-1}$, represent intermediate points in the process and state S_k is the final, absorbing state. In the diagram, being in state S_i means that the last i symbols

Species	Biological taxonomy	Abbr.
Homo sapiens (human)	animalia	H.sapiens
Macaca mulatta (Rhesus macaque)	animalia	M.mulatta
Pan troglodytes (chimpanzee)	animalia	P.troglodytes
Mus musculus (mouse)	animalia	M.musculus
Rattus norvegicus (brown rat)	animalia	R.norvegicus
Eqqus caballus (horse)	animalia	E.caballus
Cannis lupus familiaris (dog)	animalia	C.lupus
Bos taurus (cow)	animalia	B.taurus
Monodelphis domesticus (opossum)	animalia	M.domesticus
Ornithorhynchus anatinus (platypus)	animalia	O.anatinus
Danio rerio (zebrafish)	animalia	D.rerio
Apis mellifera (honey bee)	animalia	A.mellifera
Arabidopsis thaliana (thale cress)	plantae	A.thaliana
Vitis vinifera (grape vine)	plantae	V.vinifera
Saccharomyces cerevisiae str	fungi	S.cerevisiae
Schizosaccharomyces pombe	fungi	C.pombe
Escherichia coli	bacteria	E.coli
Helicobacter pylori	bacteria	H.pylori
Streptococcus pneumoniae	bacteria	S.pneumoniae
Streptococcus mutans LJ23	bacteria	S.mutansLJ
Streptococcus mutans GS	bacteria	S.mutansGS
Aeropyrum pernix str.K1	archaea	A.pernix
Nanoarchaeum equitans	archaea	N.equitans
Candidatus korarchaeum	archaea	C.korarchaeum
Caldisphaera lagunensis	archaea	C.lagunensis
Aeropyrum camini	archaea	A.camini
NC001341 virus	virus	vir.001341 virus
NC001447 virus	virus	vir.001447 virus
NC004290 virus	virus	vir.004290 virus
NC011646 virus	virus	vir.011646 virus

Table 1. List of DNA builds used for each species

read from the sequence match a prefix of w. As each new symbol is read, a transition occurs from S_i to a new state S_j, until the final, or absorbing, state S_k is reached, meaning that a new occurrence of w has just been identified in the sequence.

We define the distance to the next occurrence of w, starting from an initial state S_I $(I < k)$, as the number of steps (transitions) it takes to walk through the diagram from S_I until the final state S_k is reached. The initial state is given by the longest word overlap of w, different from w.

To illustrate this procedure, we present the state diagram for inter-ACG distances in Figure 1. In this specific case, the probability of transition between two non-absorbing states, S_i to S_j, is given by element m_{ij} $(0 \leq i, j \leq 2)$ of the the transition matrix

$$M_{ACG} = \begin{bmatrix} 1 - p_A & p_A & 0 \\ 1 - p_A - p_C & p_A & p_C \\ 1 - p_A - p_G & p_A & 0 \end{bmatrix}.$$

where p_x denotes the nucleotide probability $(x \in \mathcal{A})$. Distance one between two occurrences of ACG is only possible from state S_2. Thus, the probabilities of distance one, from each non-absorbing state are

$$P(D = 1) = \begin{bmatrix} P(D = 1|S_0) \\ P(D = 1|S_1) \\ P(D = 1|S_2) \end{bmatrix} = \begin{bmatrix} 0 \\ 0 \\ p_G \end{bmatrix}.$$

For higher distances, $d > 1$, the probabilities can be found by combining the transition probabilities for the first step with the probabilities for distance $d - 1$, which leads to the recurrence relation

$$\begin{bmatrix} P(D=d|S_0) \\ P(D=d|S_1) \\ P(D=d|S_2) \end{bmatrix} = M_{ACG} \times \begin{bmatrix} P(D=d-1|S_0) \\ P(D=d-1|S_1) \\ P(D=d-1|S_2) \end{bmatrix}$$

where M_{ACG} is the transition matrix of non-absorbing states. Since ACG has only null word overlap besides itself, we must consider S_0 as the initial state. Therefore, under an independent symbol model, the reference probability distribution of inter-ACG distances is given by

$$f(d) = P(D = d|S_0).$$

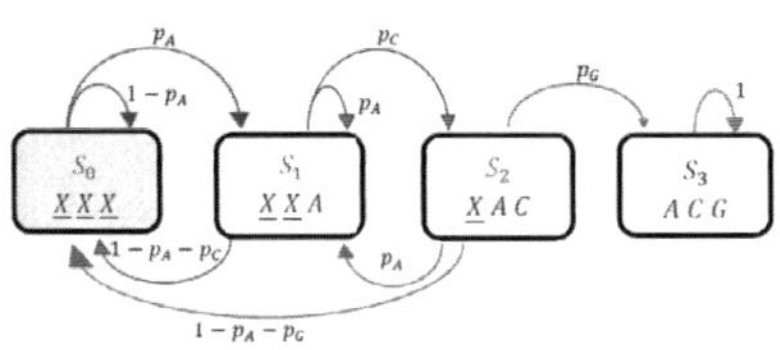

Figure 1. State diagram associated to inter-ACG distances (initial state S_0).

For the generic word w, the reference distance distribution under the independent nucleotide model is given by $f(d) = P(D = d|S_I)$, with

$$\begin{bmatrix} P(D=d|S_0) \\ \vdots \\ P(D=d|S_{k-1}) \end{bmatrix} = M^{d-1} \times \begin{bmatrix} P(D=1|S_0) \\ \vdots \\ P(D=1|S_{k-1}) \end{bmatrix},$$

and

$$P(D = 1) = \begin{bmatrix} 0 & \dots & 0 & p_{x_k} \end{bmatrix}^T.$$

where p_{x_k} is the occurrence probability of nucleotide x_k and M is the transition matrix of non-absorbing states.

Our approach to obtain the exact distribution of inter-word distances is a special case of Fu's procedure based on finite Markov chain embedding [6, 7]. To find the transition matrix for a given word requires "a deep understanding of the structure of the specified pattern" [6]. Next, we propose a general expression to compute the transition matrix of non-absorbing states $M = [m_{ij}]$, with $i, j = 0, \dots, k-1$, based on the concept of word overlap.

Let us denote by $\mathcal{L}(w_1, w_2)$ the length of the longest overlap (a suffix of w_1 that matches with a prefix of w_2) between words w_1 and w_2. Being in S_i means we have just read symbols that match w^i. The next symbol x, appended to w^i, determines the next state. A transition from S_i to S_j with $j > 0$ is only possible if $\mathcal{L}(w^i x_j, w) = j$, so its probability is

$$\text{for } j > 0, \quad m_{ij} = \begin{cases} p_{x_j} & , \quad \mathcal{L}(w^i x_j, w) = j \\ 0 & , \quad \text{otherwise} \end{cases}.$$

And the probability of a transition from S_i to S_0 ($j = 0$) is given by the complementary probability

$$m_{i0} = \begin{cases} 1 - p_{x_{i+1}} - \sum_{s=1}^{i} m_{is} & , \quad i \geq 1 \\ 1 - p_{x_{i+1}} & , \quad i = 0 \end{cases}.$$

The reference distribution under independent nucleotide structure, that we just described, can easily be computed for any whole-genome and for any genomic word, using only four input parameters: the nucleotide frequencies in the sequence.

Measures

To evaluate the goodness of fit between the inter-oligonucleotide distance distribution and the corresponding reference distribution we used the chi-square statistic and the phi coefficient. We also used an effect size measure, Cohen's d, to identify the existence of exceptional distances inside the distribution of a single word.

Due to the sensitivity of these measures to low frequencies that occur for longer distances, we made a cutoff at the 99th percentile of the empirical distribution, $d_{0.99}$. Then, we grouped all distances larger than $d_{0.99}$ in one residual class, $\widetilde{d} = d_{0.99} + 1$.

The empirical distance distribution is given by

$$q_i = \frac{n_i}{N'}, \text{ for } i = 1, \ldots, d_{0.99}$$

and the remaining frequency, $q_{\widetilde{d}}$, where n_i is the number of occurrences of distance i and N' is the total number of inter-w distances. In order to match the size of the reference distribution to the empirical distribution we also made a cutoff in the reference distribution, at $d_{0.99}$.

To extract the exceptional words of each species, we compare the empirical distribution to the corresponding reference distribution under the nucleotide independence (model I). A word is considered exceptional if the empirical inter-word distance and the reference distribution are distinct in a statistically precise way. There are two cases to consider: either the two distributions show a global misfit or there is at least one distance value that deviates significantly from the reference distribution. In the first case, the empirical distribution shows a global misfit to the random background; in the second case, the misfit is more noticeable for specific distances.

To test the goodness of the fit between the empirical and the reference distributions, for each oligonucleotide w, we can use a chi-square statistic, denoted by X_w^2,

$$X_w^2 = \sum_{i=1}^{d} \frac{(n_i - f_i \cdot N')^2}{f_i \cdot N'}.$$

To obtain an effect size measure to evaluate the lack of goodness of fit, we use the phi coefficient, denoted by φ_w,

$$\varphi_w = \sqrt{\frac{X_w^2}{N'}}.$$

A perfect fit between the distributions corresponds to $\varphi_w = 0$. We consider a value above 0.10 as a descriptor for small effect size, above 0.30 for medium effect size, above 0.50 for large effect size ([4]), above 0.60 for strong effect size and above 0.80 for a very strong effect size ([13])

For each inter-w distance distribution we are interested in identifying and evaluating the existence of exceptional distances, i.e., distances that occur with a frequency much higher than the expected value. In order to obtain a standard score able to compare how exceptional a distance is over all oligonucleotides of the same length, we use Cohen's d given by

$$CD_i = \frac{q_i - f_i}{\sqrt{f_i(1 - f_i)}}.$$

For reporting and interpreting Cohen's d, we considered a value above 0.20 as a descriptor for small effect size, above 0.50 for medium effect size and above 0.80 for large effect size ([4]). We established those acceptance thresholds as the levels above which the distance is considered exceptional or very exceptional, respectively.

To identify the most exceptional distance inside a distribution, if there is one, we use Cohen's d effect size. After computing Cohen's d for all distances up to the 99th percentile, we identify the distance d for which the maximum Cohen's d is attained and consider it the candidate to the most exceptional distance of the distribution, i.e., $C_d = \max\{CD_i : i = 1, \ldots, d_{0.99}\}$.

The expected values for distances less than or equal to k (the word length) can be null for certain words. For example, the distances between the word AAA in the text $AAAAAAA\cdots$ can never be 2 or 3. Such zero distances were not considered in the computation of the mentioned measures.

3 Results and discussion

Exceptional distance distributions in human genome

We are interested in exceptional distributions, i.e., empirical distributions that either show a significant global misfit to the reference distribution or that exhibit frequencies much higher than expected for specific distances. For all words, we observe the existence of statistical significant differences between empirical and reference distributions (p-$value < 0.001$).

In order to evaluate the lack of fit phenomenon over all words of the same length, we computed the phi coefficient, φ_w, and sorted the word distance distributions according to the value of φ_w. We observe that CG-rich words (i.e., words comprising one or more CG) and words with long word overlap lead to the poorest goodness of fit, in relation to the reference model (see Table 2). This means that these word distributions have a global misfit or a few distances with exceptional misfit to the reference distribution, in a whole-genome analysis. Let us note that the top-two dinucleotides correspond to well known local motifs (recurrent CG pairs in CpG islands and the TATA binding boxes on transcription start sites). Other high-scoring words may be related to biological motifs.

Conversely, we observe that words with no overlap and without CGs attained the lowest divergences.

k	1	2	3	4	5
max(φ_w)	0.191	1.72e+05	3.11e+05	3.84e+12	8.84e+19
min(φ_w)	0.136	0.209	0.116	0.101	0.127
highest φ_w	C	CG	CGA	CGCG	ACGCG
2nd highest	G	TA	TCG	CGAC	CGCGT
3rd highest	-	CC	CGC	GTCG	CGTCG
4th highest	-	GG	GCG	ATCG	CGACG
5th highest	-	GC	ACG	TACG	CGCGA
6th highest	-	AT	CGT	CGTA	TCGCG
7th highest	-	AC	CCG	TCGA	CGGCG
8th highest	-	GT	CGG	TTCG	CGCCG
9th highest	-	-	ATA	CGAA	CGATA
10th highest	-	-	TAT	TCGT	TATCG
⋮					
10th lowest	-	-	TGT	ACTT	CTCTA
9th lowest	-	-	ACA	AAGT	TAGAG
8th lowest	-	AA	CAA	GACA	TCAGT
7th lowest	-	TT	TTG	TGTC	TGACT
6th lowest	-	AG	ACT	ATCT	AGTCA
5th lowest	-	CT	AGT	AGAT	ACTGA
4th lowest	-	TC	TCA	ATGC	AAGCT
3rd lowest	-	GA	TGA	GCAT	AGCTT
2nd lowest	T	CA	ATG	GCTT	AGAGT
lowest φ_w	A	TG	CAT	AAGC	ACTCT

Table 2. Phi coefficient between empirical and reference distributions, in the *Homo Sapiens* genome. The maximum and minimum φ_w, the words distributions which present the ten largest and the ten smallest values of φ_w, organized by word length (k).

It is known that the human genome has low CG content ([9]). For inter-oligonucleotide distances, the information about CG content ($k = 2$) or CG-rich word ($k > 2$) contents in the sequence is not included in model I. Under this assumption, CG-rich words reach higher phi coefficients and, as a consequence, these words will be identified as exceptional words.

Using Cohen's d, we explored the existence of exceptional distances inside a single distribution, i.e., specific distances with an occurrence probability much higher than expected. Consider, for example, the unexpected spike at distance 24 in the inter-TGCA distance distribution, $C_{24} = 0.616$ (Figure 2).

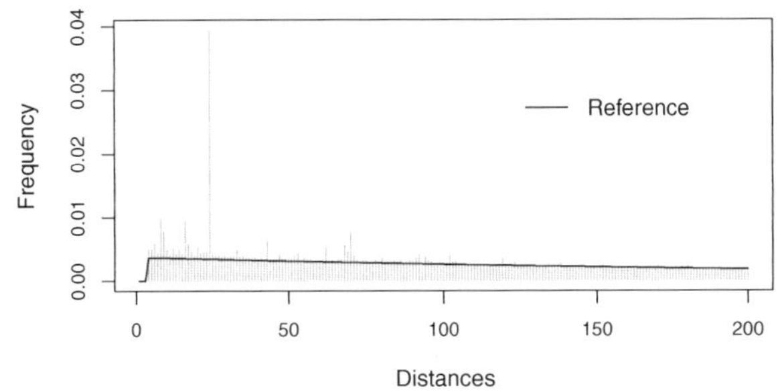

Figure 2. Empirical distance distribution *vs* reference distribution: w =TGCA, $\varphi_w = 0.694$, $C_{24} = 0.616$.

Note that a high Cohen's d could result from a generalized misfit between the empirical and the reference distribution, rather than from a genuine exceptionality of that distance. Thus, we suggest a practical decision based on the goodness of fit between empirical and reference distance distributions: for one empirical distance distribution that presents moderate to strong discrepancy ($0.2 < \varphi_w < 0.8$) we use 0.5 as the cut point on Cohen's d to identify exceptional distances. For the human genome, only eleven inter-word distributions have been identified as comprising exceptional distances. We do not observe the presence of exceptional distances in distance distributions for word lengths less than 4 (see Table 3). Figure 3 shows two inter-word distance distributions that comprise an exceptional distance, by our criteria. This procedure detects exceptional words based on their atypical distance distribution along the sequence and not on their frequency of occurrence.

Strength of Cohen's d maximum	word length			
	2	3	4	5
medium effect size ($0.5 \leq C_d < 0.8$)	0	0	1	10
large effect size ($C_d \geq 0.8$)	0	0	0	0

Table 3. Number of distance distributions with moderate or strong lack of fit ($0.2 < \varphi_w < 0.8$) that present an exceptional distance, organized by strength of effect size and word length9.

This procedure may lead to the identification of new motifs. For example, a word with a perfectly ordinary overall frequency of occurrence may exhibit an abnormal "preference" for occurring at a distance d from the previous occurrence and a slightly decreased preference for occurring at other distances.

Analysis of multiple organisms

Taking into account the empirical distance behaviour and the random background (model I), we introduce exceptionality word criteria and define dichotomic vectors, that may be used as a genomic signature of species.

Consider the following exceptionality word criteria:

- Misfit criterion: the word shows a very strong dissimilarity effect between distributions, $\varphi_w > 0.8$, highlighting the contribution of selective evolution [12];

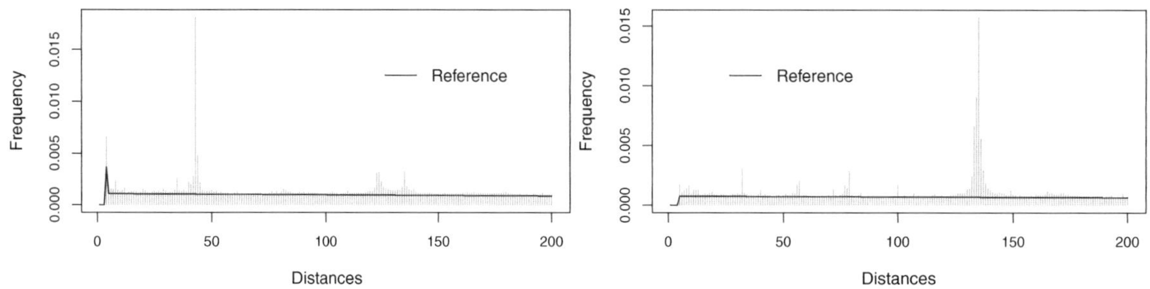

Figure 3. Empirical distance distribution *vs* reference distribution: w =TCACT, $\varphi_w = 0.633$, $C_{43} = 0.533$ (left); w =ATCCC, $\varphi_w = 0.791$, $C_{135} = 0.577$ (right).

- Peak criterion: the word has a small or medium dissimilarity effect between distributions and presents a peak with medium or large effect size, $0.2 < \varphi_w < 0.8 \wedge C_d > 0.5$.

Consider, for each specie, a dichotomic vector that marks as nonzero the words identified as exceptional accordingly to one of the criteria. These vectors allows to build dendrograms, which could then be interpreted as phylogenetic trees.

We performed a hierarchical analysis of the 30 species listed in Table 1, considering each one of the exceptionality criteria. The dendrograms were build using the average linkage method. The similarity matrix was computed using the Euclidean distance. In the case of the *misfit criterion*, the dendrogram displays a first branching between eukaryotes and non-eukaryotes (Figure 4*a*). Inside the eukaryote cluster, we observe that some related species are grouped in the same branch. For instance, primates (*H.sapiens*, *P.troglodytes* and *M.mulatta*), the rodentia (*M.musculus* and *R.norvegicus*) and the fungi (*S.cerevisiae* and *C.pombe*). In the second branch it is observed that, in general, bacteria and archaeotas are closer to each other and separated from the virus. We also notice that the bacteria *S.mutansLJ*, *S.mutansSG* and *S.pneumoniae* are in the same cluster. We emphasize that only the animal organisms reveal distance distributions that verify the *peak criterion*. Restricting the analysis to animal organisms, we obtain a dendrogram which reveals the group of primates and the group of rodentia (Figure 4*b*).

Thus, the binary vector of exceptional words defined by the *misfit criterion* may be used as a genomic signature in all the studied species, while the *peak criterion* can only be used as genomic signature in animal species.

We also constructed dendrograms for the 10 mammal species, using both criteria separately. The obtained dendrograms present some similarities (the split distance between dendrograms is 0.43). We observe that primates are clustered together, as well as the rodentia (Figure 5). These dendrograms support several evolutionary relationships between species. For example, the split distance between our dendrograms and those presented in [17], based in alignment and non-alignment algorithms, is around 50%, which is lower than in random scenarios (see [1]).

4 Conclusions and future research

In this work we studied the inter-word distances in the complete genomes of up to 30 species, for word length k varying between 1 and 5.

We intended to detect exceptional words by comparing the empirical distribution of the inter-word distances with the theoretical one under independent nucleotide model, taking the word overlap structure into account. We evaluated the discrepancy between real sequences and the random background, as a way of emphasizing the contribution of selective evolution. The comparison of the empirical distance frequencies with those that would be observed if the random background model were valid, allowed us to highlight distinct distance distributions for classes of genomic words.

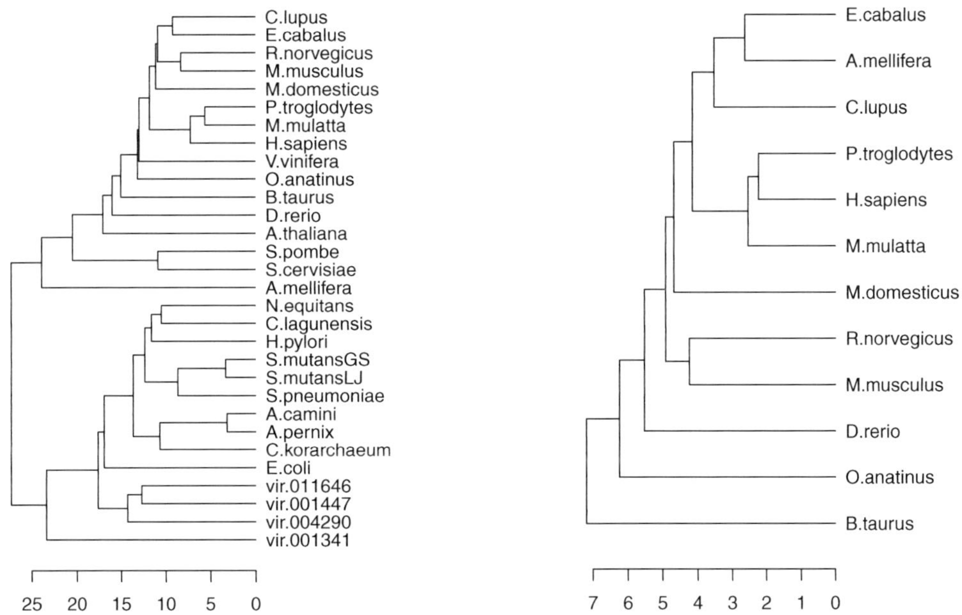

(a) misfit criterion *(b) peak criterion*

Figure 4. Dendrogram of the 30 organisms, with binary vector of exceptional words defined by all words of length 2 to 5 by *misfit criterion* (left); and of the animals by *peak criterion* (rigth).

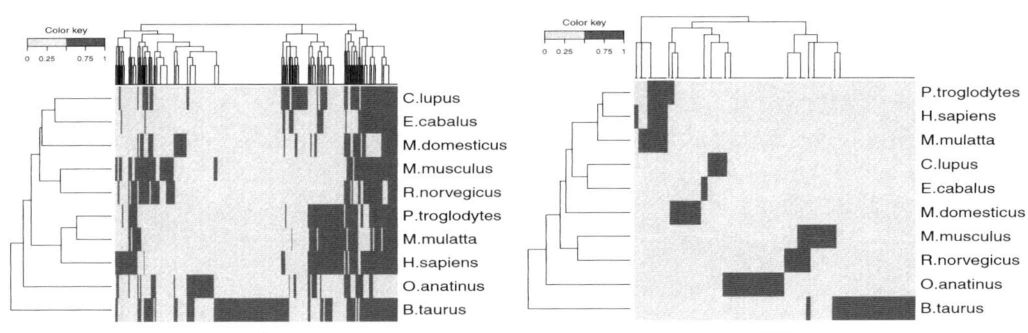

Figure 5. Heat map of mammal species *vs* exceptional words. Binary vectors of exceptional words defined by *misfit criterion* (left) and by *peak criterion* (rigth), considering only vectors with variation.

We introduced a statistical procedure to automatically identify genomic words whose distance distributions show a significant discrepancy from the random background. Our procedure allows to detect some words with a very high lack of fit. These were, in general, words with CG-rich content (as expected). Moreover, we found words with a moderate to strong lack of fit and an unexpected strong spike. Only less than 1 percent of the words of length 4 and 5 show this kind of exceptional distance distribution.

We believe that this procedure, which detects statistically exceptional distributions, may lead to the identification of new motifs. For example, a word with a perfectly ordinary overall frequency of occurrence may exhibit an abnormal "preference" for occurring at a distance d from the previous occurrence and a slightly decreased preference for occurring at other distances.

We also found that the differences mimic, to a certain extent, the evolutionary relationships between the species, which were used to construct dendrograms and perform evolutionary comparisons. In the mammalian organisms, we found matching word dissimilarity values.

In future we intend to extend our procedure to longer words, and evaluate if the method allow to point out known patterns with biological significance. Furthermore, since whole genome are highly heterogeneous, we also expect to perform analysis for detection of regions with exceptional inter-nucleotide distances.

5 Acknowledgements

This work was supported by Portuguese funds through the iBiMED - Institute for Biomedicine, IEETA - Institute of Electronics and Informatics Engineering of Aveiro and the Portuguese Foundation for Science and Technology ("FCT–Fundação para a Ciência e a Tecnologia"), within projects UID/BIM/04501/2013 and UID/CEC/00127/2013.

Bibliography

[1] V. Afreixo, C. A. Bastos, A. J. Pinho, S. P. Garcia, and P. J. Ferreira. Genome analysis with distance to the nearest dissimilar nucleotide. *Journal of theoretical biology*, 275(1):52–58, 2011.

[2] V. Afreixo, C. A. C. Bastos, A. J. Pinho, S. P. Garcia, and P. J. S. G. Ferreira. Genome analysis with inter-nucleotide distances. *Bioinformatics*, 25(23):3064–3070, Dec. 2009.

[3] C. A. C. Bastos, V. Afreixo, A. J. Pinho, S. P. Garcia, J. a. M. O. S. Rodrigues, and P. J. S. G. Ferreira. Inter-dinucleotide distances in the human genome: an analysis of the whole-genome and protein-coding distributions. *Journal of Integrative Bioinformatics*, 8(3):172, 2011.

[4] J. Cohen. *Statistical Power Analysis for the Behavioral Sciences.* Lawrence Erlbaum, 1988.

[5] S. Ding, Q. Dai, H. Liu, and T. Wang. A simple feature representation vector for phylogenetic analysis of dna sequences. *Journal of Theoretical Biology*, 265(4):618–623, Aug. 2010.

[6] J. C. Fu. Distribution theory of runs and patterns associated with a sequence of multi-state trials. *Statistica Sinica*, 6(4):957–974, 1996.

[7] J. C. Fu and W. W. Lou. *Distribution theory of runs and patterns and its applications: a finite Markov chain imbedding approach.* World Scientific, 2003.

[8] M. Hackenberg, C. Previti, P. L. Luque-Escamilla, P. Carpena, J. Martínez-Aroza, and J. L. Oliver. CpGcluster: a distance-based algorithm for CpG-island detection. *BMC Bioinformatics*, 7(1):446, 2006.

[9] J. Karro, M. Peifer, R. Hardison, M. Kollmann, and H. von Grünberg. Exponential decay of GC content detected by strand-symmetric substitution rates influences the evolution of isochore structure. *Molecular biology and evolution*, 25(2):362–374, 2008.

[10] M. Lothaire. *Applied combinatorics on words*, volume 105. Cambridge University Press, 2005.

[11] A. S. S. Nair and T. Mahalakshmi. Visualization of genomic data using inter-nucleotide distance signals. In *Proceedings of IEEE Genomic Signal Processing*, 2005.

[12] J. Qi, B. Wang, and B.-I. Hao. Whole proteome prokaryote phylogeny without sequence alignment: A K-string composition approach. *Journal of molecular evolution*, 58:1–11, 2004.

[13] L. Rea and R. Parker. *Designing and conducting survey research: a comprehensive guide.* Public Administration Series. Jossey-Bass Publishers, 1992.

[14] S. Robin. A compound Poisson model for word occurrences in DNA sequences. *Applied Statistics*, 51, Part 4:437–451, Aug. 2002.

[15] S. Robin, F. Rodolphe, and S. Schbath. *DNA, Words and Models: Statistics of Exceptional Words.* Cambridge University Press, 2005.

[16] S. Robin, S. Schbath, and V. Vandewalle. Statistical tests to compare motif count exceptionalities. *BMC Bioinformatics*, 8(1):84, 2007.

[17] G. E. Sims, S.-R. Jun, G. A. Wu, and S.-H. Kim. Whole-genome phylogeny of mammals: evolutionary information in genic and nongenic regions. *Proceedings of the National Academy of Sciences*, 106(40):17077–17082, 2009.

[18] T. V. Stefanov. The intersite distances between pattern occurrences in strings generated by general discrete- and continuous-time models: an algorithmic approach. *Journal of Applied Probability*, 40:881–892, 2003.

Cluster detection of disease mapping data based on latent Gaussian Markov random field models

Wataru Sakamoto, *Okayama University*, w-sakamoto@okayama-u.ac.jp

Abstract. Detecting clusters of higher prevalence in spatial data is of primary interest. Most of the existing methods use spatial scan statistics based on the likelihood-ratio test. The echelon scan based on the echelon analysis is useful in detecting significant clusters of non-circular shape effectively. Bayesian analysis methods for spatial data have been also studied. The latent Gaussian models, in which a Gaussian Markov random field prior is assumed on the spatial effect, provide very flexible tools. A method of detecting clusters was proposed using Poisson models with the latent Gaussian Markov random field. The clusters are scanned on the echelons constructed from the posterior means of the spatial effect, and the clusters giving maximum marginal likelihood were detected. It can be easily extended to adjustment for covariates and the random effect. An example of applying to disease mapping data illustrated that the proposed method constructed more aggregated echelons and clusters than the echelon scan based on empirical Bayes estimates of relative risk, and that detected clusters provided smallest deviance information criterion values.

Keywords. Bayesian statistical models, Echelon analysis, Integrated nested Laplace approximation, Spatial data

1 Introduction

In the analysis of spatial data such as in disease mapping, our interest is to extract primary information from their spatial structure, and then to detect clusters (so called *hotspots*) with higher prevalence of some disease. The choropleth map is a method of visualizing variation of spatial data. However, it is hard to detect such clusters objectively from the choropleth map for observed counts, which may involve large variability and complicated spatial correlation.

To detect such clusters, Kulldorff [6] proposed scanning regions within a circular window using a spatial scan statistic based on the likelihood-ratio test. Some extensions of the spatial scan statistic have been considered for covariate adjustment [5] and over-dispersion [17]. Also, some variants of scanning without using circular windows have been proposed, such as a flexibly shaped scan [14]. The echelon scan [3, 4, 7], which is based on the echelon analysis [8, 10], is useful in detecting such significant clusters of non-circular shape effectively.

Bayesian analysis methods for spatial data have been also studied. The latent Gaussian models [11, 12] provide very flexible tools to extract useful information behind spatial structure. A Gaussian

Markov random field (GMRF) prior is assumed on the spatial effect using a sparse adjacency matrix based on neighborhood relationship, and its posterior means are visualized on a choropleth map. They can also take account of over-dispersion and covariance adjustment in natural ways. Recently, a Bayesian approach for cluster detection using the Markov chain Monte Carlo was also proposed [16].

In this paper we propose a method of detecting significant clusters by the echelon scan based on the estimated spatial effect in the Poisson models with the latent GMRF. Our method can be easily extended to adjustment for covariates and the random effect. The idea is intended to present a natural procedure to detect more reasonable clusters from estimated relative risk. An example of our approach to disease mapping data is illustrated.

2 Model-based analysis of disease mapping data

Poisson models for disease mapping data

Suppose that a study area is separated into n regions. Let y_i be the number of cases observed in region i $(i = 1, \ldots, n)$, and let e_i be the at-risk population size in region i. Consider the Poisson model

$$y_i \sim \mathrm{Po}(\lambda_i e_i), \quad i = 1, \ldots, n, \tag{1}$$

where λ_i is the unknown relative risk in region i. In standard detection of disease clusters, such as spatial scan statistic [6], the following two-phase model for the relative risk is often assumed:

$$\lambda_i = \begin{cases} \lambda_C, & i \in C, \\ \lambda_N, & i \notin C, \end{cases} \tag{2}$$

where λ_C and λ_N are unknown constants, and C is a given cluster.

Latent Gaussian Markov random field models

As a general model for the relative risk λ_i in (1), consider the following model:

$$\log \lambda_i = \mu + f_i + u_i, \quad i = 1, \ldots, n, \tag{3}$$

where μ is an unknown constant, f_i is the spatial effect, and u_i is the random effect that represents variability between regions, which is assumed to follow independently as $u_i \sim \mathrm{N}(0, \theta_u^{-1})$.

For the spatial effect f_i, we assume the first-order intrinsic GMRF [1, 11]:

$$f_i | \boldsymbol{s}f_{-i} \sim \mathrm{N}(\bar{f}_i, (n_i \theta_f)^{-1}), \tag{4}$$

where $\boldsymbol{s}f_{-i}$ is the vector of $\{f_j, j \neq i\}$,

$$\bar{f}_i = \frac{1}{n_i} \sum_{i \sim j} f_j$$

is the average of f_j's at regions adjacent to the region i, n_i is the number of regions adjacent to the region i, and θ_f is an unknown precision parameter to control complexity of the spatial effect. The vector $\boldsymbol{s}f = (f_1, \ldots, f_n)^T$ follows the multivariate normal distribution $\mathrm{N}(\boldsymbol{s}0, \theta_f^{-1} \boldsymbol{s}Q_f^{-1})$, where $\boldsymbol{s}Q_f$ is a given sparse adjacency matrix.

The Poisson model (1) with (3) belongs to the class of latent Gaussian Markov random field (GMRF) models [12], which is characterized as follows:

- $\boldsymbol{s}y = (y_1, \ldots, y_n)^T$ is a vector of observations, each of which follows independently the distribution $p(y_i | z_i, \boldsymbol{s}\theta_y)$ for given latent variables $\boldsymbol{s}z = (z_1, \ldots, z_n)^T$ as defined next. Let $p(\boldsymbol{s}y | \boldsymbol{s}z, \boldsymbol{s}\theta_y) = \prod_{i=1}^n p(y_i | z_i, \boldsymbol{s}\theta_y)$.

- $\boldsymbol{sz}$ is a high-dimensional vector of latent variables, of which the prior forms a GMRF that follows the multivariate normal distribution with mean vector $\boldsymbol{s0}$ and sparse precision matrix $\boldsymbol{sQ}(\boldsymbol{s\theta}_z)$. Let $p(\boldsymbol{sz}|\boldsymbol{s\theta}_z)$ be its density.
- $\boldsymbol{s\theta} = (\boldsymbol{s\theta}_y, \boldsymbol{s\theta}_z)$ is a hyper-parameter vector of relatively lower dimension. Let $\pi(\boldsymbol{s\theta})$ be a hyper-prior on $\boldsymbol{s\theta}$.

Integrated nested Laplace approximation

The latent GMRF model described above is regarded as a hierarchical Bayes model, and we are mainly interested in predicting each latent variable z_i, that is, obtaining its marginal posterior distribution:

$$p(z_i|\boldsymbol{sy}) = \iint \pi(\boldsymbol{sz}, \boldsymbol{s\theta}|\boldsymbol{sy}) d\boldsymbol{sz}_{-i}\, d\boldsymbol{s\theta}, \tag{5}$$

where $\boldsymbol{sz}_{-i}$ is the vector of $\{z_j, j \neq i\}$, and

$$\pi(\boldsymbol{sz}, \boldsymbol{s\theta}|\boldsymbol{sy}) \propto p(\boldsymbol{sy}|\boldsymbol{sz}, \boldsymbol{s\theta}_y) p(\boldsymbol{sz}|\boldsymbol{s\theta}_z) \pi(\boldsymbol{s\theta}).$$

The marginalization in (5) requires high-dimensional integration. The integrated nested Laplace approximation (INLA), proposed by Rue *et al.* [12], enables fast computation without posterior sampling, using the Laplace approximation of the posterior distribution of the sparse latent variables, which is normally approximated, and using numerical integration on the distribution of hyper-parameters, which is far from normal. It has been shown that the INLA has comparable estimation performance to the Markov chain Monte Carlo [12].

An R package `INLA` (`http://www.r-inla.org/`) [9] for Bayesian analysis with the latent GMRF models and the INLA is available, and we used it through our study.

3 Cluster detection by echelon scan

Spatial scan statistic

Consider the null hypothesis $\mathrm{H}_0 : \lambda_C = \lambda_N$ versus the alternative hypothesis $\mathrm{H}_1 : \lambda_C > \lambda_N$ in the standard model for the relative risk (2). For a candidate cluster C, the spatial scan statistic based on the likelihood ratio [6] is defined as follows:

$$L(C) = \begin{cases} \dfrac{\hat{\lambda}_C^{y_C} \hat{\lambda}_N^{y_N}}{\tilde{\lambda}_T^{y_T}} & \text{if } \hat{\lambda}_C > \hat{\lambda}_N, \\ 1 & \text{otherwise,} \end{cases} \tag{6}$$

where

$$\hat{\lambda}_C = \frac{\sum_{i\in C} y_i}{\sum_{i \in C} e_i}, \quad \hat{\lambda}_N = \frac{\sum_{i\notin C} y_i}{\sum_{i \notin C} e_i}, \quad \text{and} \quad \tilde{\lambda}_T = \frac{\sum_{i=1}^n y_i}{\sum_{i=1}^n e_i}.$$

The scan statistic (6) is computed for each candidate cluster, and clusters for which the statistic (6) is (locally) maximized are detected. The circular scan by Kulldorff [6] restricts candidate clusters to be circular-shaped. The flexibly-shaped scan by Tango and Takahashi [14] allows non-circular-shaped clusters, but it requires much computational cost.

Echelon scan based on spatial scan statistic

Echelons represent geographic structure of variation in spatial data. It consists of some peaks, foundations, foundations of foundations and so on, which are stacked on the study area. The peaks are separated collections of adjacent regions,of higher-level values, and the foundations are collections of adjacent regions of lower-level values under some peaks. The echelon dendrogram is a graph to indicate location relationship between such peaks and foundations.

We need to know observed values (or their ranks) in each region and information on neighboring relationship between regions to construct echelons. However, echelons constructed from raw data (y_i/e_i) tend to include too many peaks because of their large variability, which may lead to complicated echelon structure. An alternative method is to construct echelons from empirical Bayes estimates of the relative risk λ_i.

The echelon scan [3, 4, 7] is a method of scanning regions based on the echelon dendrogram, and detecting clusters of high prevalence. The procedure is described as follows:

1. Choose the top peak of echelons, and we take the highest-level region in the top peak to form an initial cluster, say C. Then add to C a region that is a neighbor of C in the same peak, subsequently in descending order of the level.
2. Choose the second highest peak, and form clusters subsequently in descending order of the level as in Step 1.
3. Choose each peak or foundation in descending order, and form clusters as in the previous steps. In scanning on a foundation, any regions in higher-level peaks and/or foundations are always included in the initial cluster.

The scan statistic (6) is computed for each candidate cluster obtained in the above procedure, and clusters for which the statistic (6) is maximized are detected.

The echelon scan enables us to deal with clusters of non-circular shape, and to detect significant clusters more efficiently than standard spatial scan methods that need to search all possible candidate clusters.

Echelon scan based on latent GMRF models

We propose a model-based echelon scan as described below:

1. Fit the Poisson model (1) with the latent GMRF (3) to spatial data, and estimate the posterior means, say $\hat{f}_i$, of f_i, $i = 1, \ldots, n$.
2. Construct an echelon dendrogram from the posterior means $\{\hat{f}_1, \ldots, \hat{f}_n\}$.
3. For each cluster C scanned on the echelon dendrogram, fit the following Poisson model with a two-phase latent variable and random effect:

$$\log \lambda_i = \begin{cases} f_C + u_i, & i \in C, \\ f_N + u_i, & i \notin C, \end{cases} \tag{7}$$

   and obtain the following marginal likelihood:

$$p(\boldsymbol{sy}|C) = \int p(\boldsymbol{sy}|\boldsymbol{sz}, \boldsymbol{s\theta}_y)p(\boldsymbol{sz}|\boldsymbol{s\theta}_z)\pi(\boldsymbol{s\theta})d\boldsymbol{sz}\, d\boldsymbol{s\theta}. \tag{8}$$

4. Detect clusters C for which the marginal likelihood (8) is maximized.

Fitting the latent model(3) and (7) can be conducted by using the function `inla` in the R-package `INLA`, and the log of the marginal log-likelihood (8) can be obtained as `inla.object$mlik` from the object output by `inla`.

A justification of using the marginal likelihood (8) can be explained by the Bayes factor, the ratio of the marginal likelihood. For two candidate clusters C_1 and C_2, the Bayes factor is defined as

$$\frac{p(\boldsymbol{sy}|C_2)}{p(\boldsymbol{sy}|C_1)} = \frac{P(C_2|\boldsymbol{sy})}{P(C_1|\boldsymbol{sy})} \Big/ \frac{P(C_2)}{P(C_1)}. \tag{9}$$

So, if we assume that $P(C_1) = P(C_2)$, then a cluster of larger marginal likelihood should have higher posterior probability.

The posterior means of $\{f_1, \ldots, f_n\}$ are spatially smooth under the prior of GMRF with the sparse precision matrix, so the resulting echelons and clusters are expected to have more aggregated shape than using raw data or empirical Bayes estimates of the relative risks.

An extention to the case of covariate adjustment is straightforward. In step 1, we consider a model with adding the covariate effect, and in step 3, we replace the smooth spatial effect with the two-phase latent variable.

Model selection

In the model (7), the two cases of clusters that $C = \{\text{All regions}\}$ and $C = \{\text{No region}\}$ are unidentifiable because both correspond to the null model of no significant cluster: $f_C = f_N$. So we need some model selection criterion to choose between the null model and the alternative model with significant clusters detected.

Here we use the deviance information criterion [13, 2]

$$\text{DIC}(C) = -2 \log p(\boldsymbol{s}y | \hat{\boldsymbol{s}z}_B, \hat{\boldsymbol{s}\theta}_B) + 2d_{\text{DIC}}, \tag{10}$$

where $\hat{\boldsymbol{s}z}_B$ and $\hat{\boldsymbol{s}\theta}_B$ are Bayes estimates of $\boldsymbol{s}z$ and $\boldsymbol{s}\theta$, respectively, and d_{DIC} is the effective number of parameters. According to the website of R-INLA, the R function `inla` computes the DIC value using posterior mean of $\boldsymbol{s}z$ and the posterior mode of $\boldsymbol{s}\theta$. We compare the DIC values for the cluster detected model with the one for the null model.

4 Example

We applied the proposed method to the Germany oral cavity cancer data set (`Germany` in R-INLA) [11]. For each district $i = 1, \ldots, n(= 544)$ (district names not indicated in the R data set), let y_i be number of cases, e_i be the offset values proportional to the number of people, and x_i be the covariate measuring smoking consumption.

We fitted Poisson models (1) with the following three formulas suggested in [9]:

a) $\log \lambda_i = \mu + f_i + u_i$,

b) $\log \lambda_i = \mu + f_i + \beta x_i + u_i$,

c) $\log \lambda_i = \mu + f_i + g(x_i) + u_i$,

where f_i is the spatial effect, for which the intrinsic GMRF (4) is assumed, μ and β are unknown constants, $g(x_i)$ is a smooth function of x_i, on which the second-order random walk prior is imposed, and u_i is the independent random region effect. For each of the three models a)–c), clusters were scanned on the echelons constructed from the posterior means of f_i, as described in Section 3. Disjoint clusters giving positive values of the marginal log likelihood ratio $\log\{p(\boldsymbol{s}y|C)/p(\boldsymbol{s}y|\text{H}_0)\}$ were detected. The computation were conducted with R 3.2.5, on which the package INLA was used for fitting the intrinsic GMRF models. Programs for constructing echelons and scanning clusters on the echelons, not implemented in R packages, were built by the author. The largest cluster size was restricted to 50 regions.

For the purpose of comparison, we also fitted a Poisson model (1) in which each λ_i is assumed to follow the gamma distribution with the parameters estimated by the empirical Bayes method. Then clusters were scanned on the echelons constructed from the empirical Bayes estimates of λ_i. Disjoint clusters giving small (less than 0.05) Monte Carlo p-values (based on 10000 replications) for the spatial scan statistic (6) were detected. Furthermore we applied the circular scan and the flexibly-shaped scan using the software FleXScan [15] (`https://sites.google.com/site/flexscansoftware/`). Note that these methods based on the spatial scan statistic (6) does not incorporate either covariate effect or random region effect.

Table 1 shows the number of peaks and foundations in the echelons constructed from the empirical Bayes estimates of λ_i, and those from the posterior means of f_i in the latent GMRF models a)–c). The posterior means for the three models generated less numbers of peaks and foundations, which suggested that the resulting clusters are expected to be more aggregated.

Figure 1 shows the cluster with the largest spatial scan statistic (6) detected by the circular scan, the flexible-shaped scan and the echelon scan on the empirical Bayes estimates of λ_i, and the cluster with the largest marginal log likelihood ratio detected by the echelon scan on the posterior means of f_i in the models a), b) and c). The regions belonging to the detected cluster were filled in gray. The model a), taking account of the random effect between regions, detected a more aggregated cluster than that for the empirical Bayes methods. The model b) and c), taking account of the covariate effect, detected clusters including different regions from those for the model a).

Table 2 shows clusters with p-value less than 0.05 detected by the circular scan and the flexibly-shaped scan, and five clusters with largest statistic values(6) detected by echelon scan on the empirical Bayes estimates of λ_i. Table 3 shows the clusters with the positive marginal log likelihood ratio detected by the echelon scan on the posterior means of f_i in the models a), b) and c). For the detected clusters, the DIC values (10) were computed. For readers' reference, the spatial scan statistic (6) were also computed. For the models a)–c), the cluster with the largest marginal log-likelihood also provided the smallest DIC. The result suggested that the clusters scanned on the echelons from latent GMRF model were reasonable.

5 Simulation Study

We evaluated the performance of the proposed method by a simple simulation study. Consider the square region $[0,1] \times [0,1]$ on $\mathbf{R}^2$, which was divided into 10×10 square areas $a_i = [j/10, (j+1)/10] \times [k/10, (k+1)/10]$ $(j = 0, 1, \ldots, 9;\ k = 0, 1, \ldots, 9)$, where $i = 10k + j + 1$ is the area code.

We generated 100 data sets of size 100 (y_i, x_i) $(i = 1, \ldots, 100)$, according to the following distribution:

$$y_i \sim \mathrm{Po}(\lambda_i), \quad \lambda_i = 1_{\{a_i \in C_0\}} + x_i + u_i,$$

where $1_{\{a_i \in C_0\}}$ is the indicator function, $x_i = j + 0.5$, and $u_i \sim \mathrm{N}(0, 0.5^2)$. The true cluster C_0 that contains high-risk regions were set as $C_0 = [0.3, 0.6] \times [0.3, 0.6]$.

We fitted a Poisson model (1) in which each λ_i is assumed to follow the gamma distribution with the parameters estimated by the empirical Bayes method. A cluster giving the largest spatial scan statistic were detected by scanning on the echelons constructed from the empirical Bayes estimates of λ_i. We also fitted Poisson models with the following two formulas:

a) $\log \lambda_i = \mu + f_i + u_i$,

b) $\log \lambda_i = \mu + f_i + \beta x_i + u_i$,

where f_i is the spatial effect, for which the intrinsic GMRF (4) is assumed, μ and β are unknown constants, and u_i is the independent random region effect. For each of the models a) and b), a cluster giving the largest marginal log likelihood ratio were detected by scanning on the echelons constructed from the posterior means of f_i.

Table 1. The number of peaks and foundations in the echelons

Method/Model		Peaks	Foundations	Total
Empirical Bayes		112	104	216
Latent GMRF	a)	68	63	131
	b)	71	69	140
	c)	70	67	137

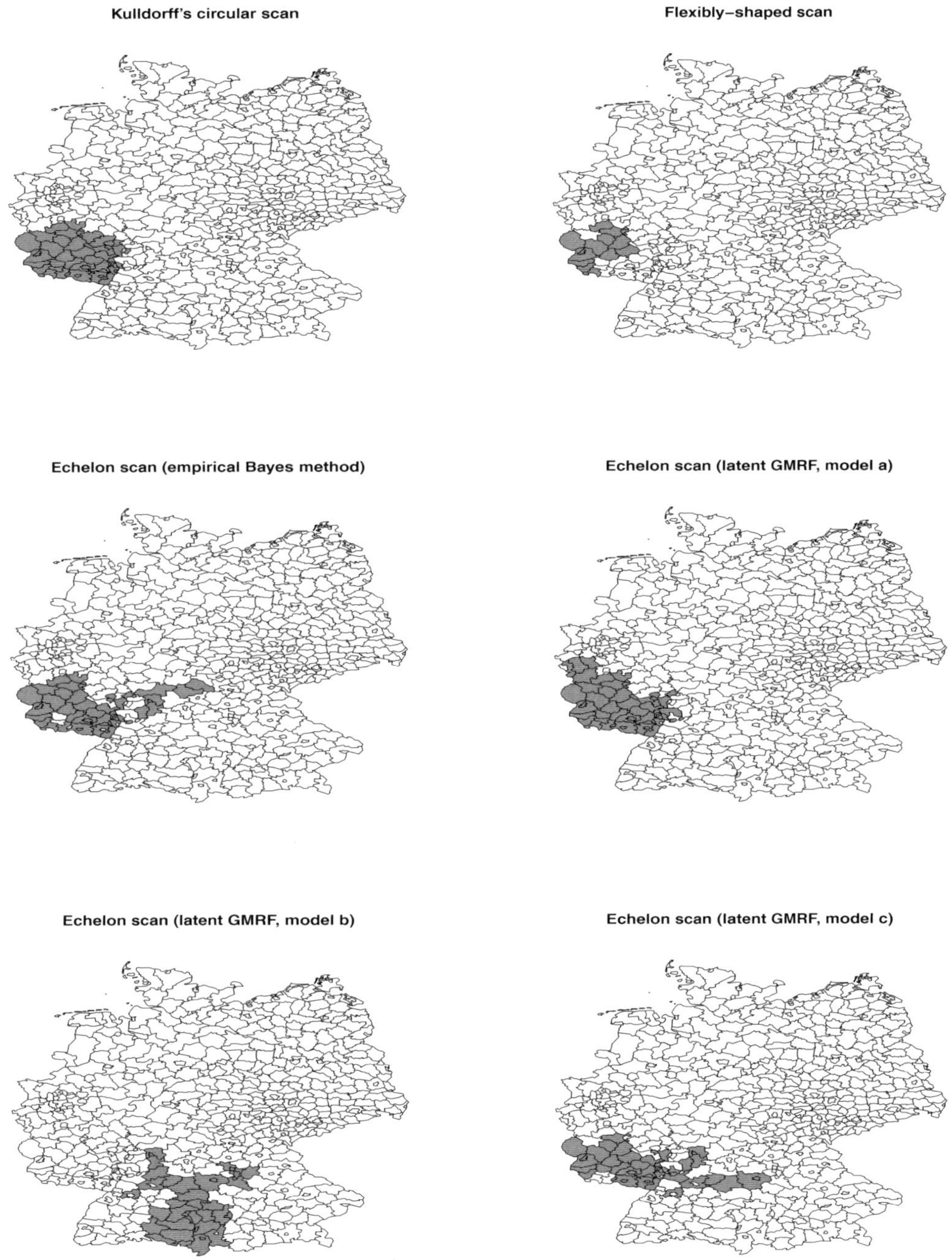

Figure 1. The detected clusters with the largest scan statistic (6) or marginal log likelihood ratio

Table 2. Clusters detected by the spatial scan statistic

Method	Cluster	$\log L(C)$	p-value	Region numbers included
Circular	1	31.57	0.001	120, 123, 126, 129, 132, ..., 327 (46 regions)
	2	22.61	0.001	68, 71, 72, 73, 75, ..., 119 (15 regions)
Flexible	1	31.13	0.001	146, 147, 149, 150, 151, ..., 325 (14 regions)
	2	24.94	0.001	68, 72, 73, 75, 77, 93, 94, 108, 109, 112
Echelon	1	44.57	0.000	121, 122, 123, 126, 128, ..., 326 (46 regions)
	2	27.95	0.000	66, 67, 68, 70, 71, ..., 327 (35 regions)
	3	13.73	0.000	234, 244, 252, 253, 257, ..., 292 (15 regions)
	4	11.43	0.000	17, 23, 26, 27, 28, ..., 504 (21 regions)
	5	7.74	0.000	328, 333, 338, 341, 345, ..., 469 (19 regions)

$L(C)$: spatial scan statistic (6)

Table 3. Latent GMRF models: detected clusters with positive marginal log likelihood ratio

Model	Cluster	MLLR	DIC	$\log L(C)$	Region numbers included
a)	1	23.09	−43.00	40.59	85, 86, 87, 88, 121, ..., 327 (48 regions)
	2	6.57	−13.18	24.74	66, 67, 68, 71, 72, ..., 116 (18 regions)
	3	0.06	−6.45	6.32	253, 257
	4	0.03	−6.51	9.84	28, 30, 31, 34, 35, 36, 43, 106
	5	0.03	−8.35	8.23	328, 333, 335, 336, 337, ..., 407 (24 regions)
b)	1	19.16	−36.06	10.33	185, 189, 190, 191, 193, ..., 321 (38 regions)
	2	13.40	−29.17	31.90	126, 149, 150, 151, 154, ..., 327 (29 regions)
	3	1.49	−4.52	21.22	94
	4	0.40	−7.23	9.84	28, 30, 31, 34, 35, 36, 43, 106
c)	1	22.79	−44.36	36.86	126, 130, 149, 150, 151, ..., 327 (44 regions)
	2	6.69	−19.07	6.47	120, 121, 123, 124, 125, ..., 326 (50 regions)
	3	2.86	−10.70	9.84	28, 30, 31, 34, 35, 36, 43, 106
	4	0.17	−5.01	6.71	285, 287, 288, 292

MLLR: the marginal log likelihood ratio between the cluster-detected model and the null model
DIC: the difference between the cluster-detected model and the null model

The generation of random numbers and the computation were conducted with R 3.2.5, on which the package INLA was used for fitting the intrinsic GMRF models. Programs for constructing echelons and scanning clusters on the echelons, not implemented in R packages, were built by the author. The largest cluster size was restricted to 50 regions.

Table 4 shows percentage that each region was contained in the detected cluster. The boldface numbers correspond to the true high-risk regions. The model b), taking account of the covariate effect, detected the true high-risk regions appropriately. The other two methods have more tendency to detect clusters outside the true high-risk regions.

6 Concluding remarks

We proposed to detect significant clusters based on the latent GMRF models using the echelon scanning. We illustrated that the proposed method, which used the posterior means of the smooth spatial effect, constructed more aggregated echelons and clusters than the echelon scanning based on empirical Bayes estimates of relative risk. Moreover, detected clusters by the proposed method provided smallest DIC values.

For future research, we should conduct simulation studies to evaluate performance of the proposed method and compare with existing cluster detection methods.

Currently we have used the R package `INLA`. Fitting the GMRF models and computing the marginal log-likelihood were conducted for each candidate cluster. We are studying more computationally effective methods.

Acknowledgement

The author would like to thank reviewers for their useful comments. Some serious errors in the computation program and the results could be found.

This research was supported by JSPS KAKENHI Grant No. 26330042.

Table 4. Percentage that each region was contained in the detected cluster

Echelon scan with empirical Bayes estimates

9	0	1	1	1	3	4	3	9	7	6
8	0	0	3	2	9	10	11	16	16	14
7	0	2	6	9	15	17	19	18	21	25
6	0	4	7	23	24	30	19	15	24	36
5	3	7	12	**65**	**66**	**75**	28	33	37	32
4	5	12	23	**66**	**71**	**77**	32	47	40	40
3	1	8	17	**69**	**72**	**80**	45	40	44	41
2	4	5	12	26	33	41	42	40	48	38
1	3	3	6	11	22	33	35	38	44	39
0	1	1	1	6	14	25	31	36	37	40
(j,k)	0	1	2	3	4	5	6	7	8	9

Latent GMRF model a)

9	0	1	1	5	8	6	5	8	9	3
8	0	0	1	6	13	16	16	14	10	6
7	0	2	5	10	16	19	16	13	14	10
6	2	4	8	19	23	28	17	10	9	14
5	3	3	12	**47**	**52**	**63**	17	13	16	12
4	5	8	16	**52**	**59**	**64**	15	19	14	10
3	2	4	11	**55**	**56**	**60**	23	18	18	11
2	4	3	6	16	16	19	19	16	16	8
1	2	3	3	6	10	14	16	13	15	7
0	1	0	1	1	5	7	11	11	8	5
(j,k)	0	1	2	3	4	5	6	7	8	9

Latent GMRF model b)

9	3	4	3	7	7	9	5	7	6	3
8	0	2	4	7	13	17	10	11	10	6
7	1	4	11	14	20	20	12	9	11	11
6	5	10	18	30	30	30	15	8	7	9
5	8	7	18	**63**	**68**	**71**	17	12	12	5
4	10	17	29	**71**	**76**	**75**	19	17	12	8
3	4	12	18	**68**	**70**	**66**	29	14	10	4
2	6	11	14	23	27	26	27	12	10	3
1	3	7	9	12	20	22	19	13	11	6
0	3	2	4	5	11	15	10	7	4	3
(j,k)	0	1	2	3	4	5	6	7	8	9

Bibliography

[1] Besag, J., York, J. and Mollie, A. (1991) Bayesian image restoration with two applications in spatial statistics (with discussion). *Annals of the Institute of Statistical Mathematics*, **43**, 1–59.

[2] Gelman, A., Hwang, J. and Vehtari, A. (2014). Understanding predictive information criteria for Bayesian models. *Statistics and Computing*, **24**, 997–1016.

[3] Ishioka, F. and Kurihara, K. (2006) Spatial structure for multidimensonal spatial lattice data. *COMPSTAT 2006: Proceedings in Computational Statistics* (eds: Rizzi, A. and Vichi, M.), Springer, CD-ROM, pp. 1209–1216.

[4] Ishioka, F. and Kurihara, K. (2013) Evaluation of hotspot detection method based on echelon structure. *Proceedings of the 59th ISI World Statistics Congress*, pp. 5366–5371.

[5] Jung, I. (2009) A generalized linear models approach to spatial scan statistics for covariate adjustment. *Statistics in Medicine*, **28**, 1131–1143.

[6] Kulldorff, M. (1997) A spatial scan statistic. *Communications in Statistics, Theory and Methods*, **26**, 1481–1496.

[7] Kurihara, K. (2003) The detection of hotspots based on the hierarchical spatial structure. *Bulletin of the Computational Statistics of Japan*, **15**, 171–183 (in Japanese).

[8] Kurihara, K. (2004) Classification of geospatial lattice data and their graphical representation. *Classification, Clustering and Data Mining Applications* (eds: D. Banks *et al.*), pp. 251–258, Springer.

[9] Martino, S. and Rue, H. (2010) Implementing approximate Bayesian inference using integrated nested Laplace approximation: a manual for the inla program. Department of Mathematical Sciences, NTNU, Norway.

[10] Myers, W., Patil, G. P. and Joly, L. (1997) Echelon approach to areas of concern in synoptic regional monitoring. *Environmental and Ecological Statistics*, **4**, 131–152.

[11] Rue, H. and Held, L. (2005) *Gaussian Markov Random Fields: Theory and Applications.* Chapman and Hall.

[12] Rue, H., Martino, S. and Chopin, N. (2009) Approximate Bayesian inference for latent Gaussian models by using integrated nested Laplace approximations (with discussion). *Journal of the Royal Statistical Society, Series B*, **71**, 319–392.

[13] Spiegelhalter, D. J., Best, N. G., Carlin, B. P. and van der Linde, A. (2002) Bayesian measures of model complexity and fit (with discussion). *Journal of the Royal Statistical Society, Series B*, **64**, 583–639.

[14] Tango, T. and Takahashi, K. (2005) A flexibly shaped spatial scan statistic for detecting clusters. *International Journal of Health Geographics*, **4**, 11.

[15] Takahashi, K., Yokoyama, T. and Tango, T. (2010) FleXScan User Guide for Version 3.1. National Institute of Public Health, Japan.

[16] Wakefield, J. and Kim, A. (2013) A Bayesian model for cluster detection. *Biostatistics*, **14**, 752–765.

[17] Zhang, T., Zhang, Z. and Lin, G. (2011) Spatial scan statistics with overdispersion. *Statistics in Medicine*, **31**, 762–774.

Non-reduced versus reduced-bias estimators of the extreme value index —efficiency and robustness

M. Ivette Gomes, *CEAUL and DEIO, FCUL, Universidade de Lisboa*, ivette.gomes@fc.ul.pt
Helena Penalva, *ESCE - IPS and CEAUL, Universidade de Lisboa*, helena.penalva@esce.ips.pt
Frederico Caeiro, *CMA and DM, Universidade Nova de Lisboa*, fac@fct.unl.pt
M. Manuela Neves, *ISA and CEAUL, Universidade de Lisboa*, manela@isa.ulisboa.pt

Abstract. The *extreme value index* (EVI) is the primary parameter of extreme events. The EVI is used to characterize the tail behavior of a distribution, and it helps to indicate the size and frequency of certain extreme events under a given probability model: for large events, the bigger the EVI is, the heavier is the right-tail of the underlying parent distribution. The Lehmer mean of order p of the k log-excesses over the $k+1$-th upper order statistic has been recently considered in the literature for the estimation of a positive EVI, associated with large extreme events. Such a Lehmer mean of order p generalizes the arithmetic mean ($p = 1$), the classical Hill estimator of a positive EVI, and for $p > 1$ has revealed to be very competitive for small values of the EVI, comparing favorably with one the simplest classes of reduced-bias EVI-estimators, a corrected-Hill estimator. Now, the comparison to other EVI-estimators is performed, and some information on the robustness of such a general class is provided, including its resistance to possible contamination by outliers.

Keywords. Efficiency, extreme values, heavy tails, Monte-Carlo, resistance to outliers, robustness, semiparametric statistics

1 Introduction and scope of the article

Given a random sample, $(X_1, \ldots, X_n)$, of independent, identically distributed *random variables* (RVs) from a *cumulative distribution function* (CDF) F, let us denote by $(X_{1:n} \le \cdots \le X_{n:n})$ the sample of associated ascending *order statistics* (OSs). Let us further assume that there exist sequences of real constants $\{a_n > 0\}$ and $\{b_n \in \mathbb{R}\}$ such that the maximum, linearly normalised, i.e. $(X_{n:n} - b_n)/a_n$, converges in distribution to a non-degenerate RV, as usual in a framework of statistical *extreme value theory* (EVT). Then, the limit distribution is necessarily of the type of the general *extreme value* (EV) CDF ([15]), given by

$$\mathrm{EV}_\xi(x) = \begin{cases} \exp(-(1+\xi x)^{-1/\xi}),\ 1+\xi x > 0, & \text{if } \ \xi \neq 0, \\ \exp(-\exp(-x)),\ x \in \mathbb{R}, & \text{if } \ \xi = 0. \end{cases} \tag{1}$$

The parameter ξ is the *extreme value index* (EVI), the primary parameter of extreme and large events, and the CDF F is said to belong to the *max-domain of attraction* of EV_ξ, in (1). We then use the notation $F \in \mathcal{D}_{\mathcal{M}}(\mathrm{EV}_\xi)$. Such a limiting result is robust to changes in the structure of the available data, which can come from any weakly dependent and stationary process, like the AR and ARCH processes, among others. The EVI is used to characterize the right tail behavior of the distribution underlying any set of stationary data. Indeed, the EVI measures the heaviness of the *right-tail function* (RTF),

$$\overline{F}(x) := 1 - F(x), \tag{2}$$

and helps to indicate the size and frequency of large extreme events under a given CDF, in the sense that the larger ξ is, the heavier is the RTF underlying the available data.

Let us denote by $\mathcal{R}_a$ the class of regularly varying functions at infinity with an index of regular variation equal to $a \in \mathbb{R}$ (see [6], among others). In this article we work with Pareto-type underlying CDFs, with a positive EVI, i.e. models such that $\overline{F}(x) = x^{-1/\xi}L(x)$, $\xi > 0$, with $L \in \mathcal{R}_0$, a slowly varying function at infinity, i.e. a regularly varying function with an index of regular variation equal to zero. These heavy-tailed models, in $\mathcal{D}_{\mathcal{M}}^{+} := \mathcal{D}_{\mathcal{M}}(EV_{\xi>0})$, are quite common in a large variety of fields of application, like biostatistics, insurance, finance, social sciences and telecommunications, among others.

For these Pareto-type models, the most prominent classical EVI-estimators are the *Hill* (H) estimators ([27]), which are constructed on the the basis of *maximum-likelihood* (ML) estimation, and are the averages of the log-excesses, i.e.

$$\mathrm{H}(k) := \frac{1}{k}\sum_{i=1}^{k} V_{ik}, \quad V_{ik} := \ln X_{n-i+1:n} - \ln X_{n-k:n}, \quad 1 \le i \le k < n, \tag{3}$$

with k an adequately chosen tuning parameter. Indeed, whenever $k = k_n$, $1 \le k < n$, is an intermediate sequence of integers, i.e.

$$k = k_n \to \infty \quad \text{and} \quad k_n = o(n), \quad \text{as } n \to \infty, \tag{4}$$

the log-excesses, V_{ik}, $1 \le i \le k$, in (3), are approximately the k top OSs of a sample of size k from an exponential parent with mean value ξ. This justifies the H EVI-estimator, also in (3), as the average of the k log-excesses. As usually happens with ML-estimators, the H EVI-estimators are not quite robust. And even in EVT, where the most extreme data usually receive the greatest attention, this can be a problem. It thus looks sensible to introduce an extra tuning parameter, which will enable playing with both robustness and efficiency.

Note next that a simple generalization of the mean is Lehmer's mean of order p, as can be seen in [26]. Given a set of positive numbers $\underline{\mathbf{a}} = (a_1, \ldots, a_k)$, such a mean generalizes both the arithmetic mean ($p = 1$) and the harmonic mean ($p = 0$), being defined as

$$L_p(\underline{\mathbf{a}}) := \frac{\sum_{i=1}^{k} a_i^p}{\sum_{i=1}^{k} a_i^{p-1}}, \quad p \in \mathbb{R}.$$

The H EVI-estimator can thus be considered as the Lehmer mean of order $p = 1$ of the k log-excesses $\underline{\mathbf{V}} := (V_{ik},\ 1 \le i \le k)$, in (3). We now more generally consider the Lehmer mean of order p of those statistics. With $\{E_i\}_{i \ge 1}$ denoting a sequence of independent unit exponential RVs, since under the validity of (4),

$$V_{ik}^p \approx \xi^p E_{k-i+1:k}^p, \quad 1 \le i \le k,$$

and $\mathbb{E}(E^p) = \Gamma(p+1)$ for any real $p > -1$, where $\Gamma(\cdot)$ stands for the Gamma function, the law of large numbers enables us to say that

$$\frac{1}{k}\sum_{i=1}^{k} V_{ik}^p \underset{n\to\infty}{\overset{p}{\longrightarrow}} \Gamma(p+1)\xi^p.$$

Hence the reason for the class of *Lehmer* (L) EVI-estimators, introduced and studied in [29],

$$\mathrm{L}_p(k) := \frac{\mathrm{L}_p(\underline{\mathbf{V}})}{p} = \frac{1}{p}\,\frac{\frac{1}{k}\sum_{i=1}^{k} V_{ik}^{p}}{\frac{1}{k}\sum_{i=1}^{k} V_{ik}^{p-1}} =: \frac{M_{k,n}^{(p)}}{pM_{k,n}^{(p-1)}} \qquad \Big[\mathrm{L}_1(k) \equiv \mathrm{H}(k),\ \text{ in } (3)\Big], \tag{5}$$

consistent for all $\xi \geq 0$ and real $p > 0$. The class in (5) is a particular case of the possibly *reduced-bias* (RB) class of EVI-estimators in [8] (see also [7] and [9]),

$$\mathrm{CG}_{p,\delta}(k) := \frac{\Gamma(p)}{M_{k,n}^{(p-1)}}\left(\frac{M_{k,n}^{(\delta p)}}{\Gamma(\delta p+1)}\right)^{1/\delta}, \quad \delta > 0,\ p \geq 1 \quad \Big[\ \mathrm{CG}_{p,1}(k) \equiv \mathrm{L}_p(k)\ \text{ in } (5)\Big]. \tag{6}$$

For $\delta = 2$ in (6), we obtain a class studied in [7], which can be RB, provided that we choose $p \equiv p_0 = -\ln\big(1-\rho-\sqrt{(1-\rho)^2-1}\big)/\ln(1-\rho)$, where ρ is a shape second-order parameter, to be defined in Section 2. Then, for an adequate estimator, $\hat\rho$, of such a shape second-order parameter, ρ, and with $\mathrm{CG}_{p,\delta}(k)$ given in (6), we get the RB EVI-estimators,

$$\mathrm{CG}_0(k) := \mathrm{CG}_{\hat p_0,2}(k), \quad \hat p_0 = -\frac{\ln\Big(1-\hat\rho-\sqrt{(1-\hat\rho)^2-1}\Big)}{\ln(1-\hat\rho)}. \tag{7}$$

We shall further consider the simplest class of RB EVI-estimators, the one introduced in [10]. With $(\hat\beta,\hat\rho)$ an adequate estimator of the vector second-order parameters (β,ρ), to be defined in Section 2, we shall thus consider the class of *corrected-Hill* (CH) EVI-estimators defined by

$$\mathrm{CH}(k) := \mathrm{H}(k)\left(1-\frac{\hat\beta(n/k)^{\hat\rho}}{1-\hat\rho}\right). \tag{8}$$

The estimators in (8) can be second-order *minimum-variance reduced-bias* (MVRB) estimators, for adequate levels k and an adequate external estimation of the vector of second-order parameters, (β,ρ), i.e. the use of $\mathrm{CH}(k)$ enables us to eliminate the dominant component of bias of the H EVI-estimator, $\mathrm{H}(k)$, keeping its asymptotic variance. The class of EVI-estimators in (8), as well as other RB EVI-estimators, like the weighted-Hill EVI-estimators in [20], which involve a linear combination of the log-excesses with a lighter weight of largest values, are expected to be more resistant than the H EVI-estimators in (3) to data contamination.

More generally than the class in (8), we shall now also introduce and play with the direct reduction of the dominant bias component of $\mathrm{L}_p(k)$, in (5), working with the RB Lehmer's EVI-estimators,

$$\mathrm{L}_p^{\mathrm{RB}}(k) := \left(1-\frac{\hat\beta(n/k)^{\hat\rho}}{(1-\hat\rho)^p}\right)\mathrm{L}_p(k) \quad [\mathrm{L}_1^{\mathrm{RB}} \equiv \mathrm{CH}\ \text{ in } (8)]. \tag{9}$$

In this article, after the introduction, in Section 2, of a few technical details in the field of EVT, we deal in Section 3 with the finite sample comparison of some of the aforementioned classes of EVI-estimators, done through a Monte-Carlo simulation study, developed for small positive ξ, the values which usually appear in practice. Some information on the robustness of the L-class of EVI-estimators is also provided, including its resistance to possible contamination by outliers. A few conclusions are provided in Section 4.

2 A few technical details in the field of EVT

After a reference to the most common first and second-order frameworks for heavy-tailed models, and second-order parameters' estimation, we briefly review the asymptotic behaviour of the aforementioned EVI-estimators. Recent reviews on the topic of statistical univariate EVT can be found in [2] and [16].

A brief review of first and second-order conditions. Second-order parameters' estimation

In the area of statistical EVT and whenever working with large values, a model F is usually said to be heavy-tailed whenever the right tail function $\overline{F}$, in (2), is a regularly varying function with a negative index of regular variation equal to $-1/\xi$, $\xi > 0$ ([15]), or equivalently, the reciprocal *right tail quantile function* (RTQF), $U(t) := F^{\leftarrow}(1 - 1/t)$, is of regular variation with an index ξ ([21]), i.e. for all $x > 0$,

$$F \in \mathcal{D}_{\mathcal{M}}^{+} = \mathcal{D}_{\mathcal{M}}\left(\mathrm{EV}_\xi\right)_{\xi>0} \iff \overline{F} \in RV_{-1/\xi} \iff U \in RV_\xi. \tag{10}$$

The second-order parameter ρ (≤ 0) rules the rate of convergence in any of the first-order conditions, in (10), and can be defined as the non-positive parameter appearing in the limiting relation

$$\lim_{t\to\infty} \frac{\ln U(tx) - \ln U(t) - \xi \ln x}{A(t)} = \begin{cases} (x^\rho - 1)/\rho, & \text{if} \quad \rho < 0, \\ \ln x, & \text{if} \quad \rho = 0, \end{cases} \tag{11}$$

which is assumed to hold for every $x > 0$, and where $|A|$ must then be of regular variation with index ρ ([14]). This condition has been widely accepted as an appropriate condition to specify the RTF of a Pareto-type distribution in a semi-parametric way. For technical simplicity, we often assume that we are working in Hall-Welsh class of models ([24]), with an RTQF

$$U(t) = C\, t^\xi \Big(1 + \xi\ \beta\ t^\rho/\rho + o(t^\rho)\Big), \quad \text{as } t \to \infty, \tag{12}$$

$C > 0$, $\beta \neq 0$ and $\rho < 0$. Models like the log-Gamma, associated with $\rho = 0$, in (11), are thus excluded from this class. The standard Pareto ($\rho = -\infty$) is also excluded. But most heavy-tailed models used in applications, like the EV_ξ, in (1), the Fréchet, $F_\xi(x) = \exp(-x^{-1/\xi})$, $x \geq 0$, and the Student's t_ν CDFs, among others, belong to Hall-Welsh class. Further details on first and second-order conditions can be found in [1], [22] and [13], among others. For details on algorithms for the (β, ρ)-estimation, see [19] and [20], among others. We have so far suggested the use of the ρ-estimators in [12] and the β-estimators in [17]. Overviews on reduced-bias EVI and second-order parameters estimation can be found in Chapter 6 of [30], [2] and [16].

Asymptotic behaviour of the EVI-estimators

Under the validity of the second-order condition in (12), and with intermediate k such that $\lambda_A := \lim_{n\to\infty} \sqrt{k}\ A(n/k)$ is finite, trivial adaptations of the results in [23], [10] and [29], respectively for the H, CH and L_p EVI-estimators, enable us to guarantee the asymptotic normality of all the aforementioned estimators. More precisely, for adequate regions of the tuning parameters, any of the classes L_p and $\mathrm{CG}_{p,\delta}$, respectively given in (5) and (6), generally denoted $\hat{\xi}^\bullet(k)$, are asymptotically normal, i.e.

$$\sqrt{k}\left(\hat{\xi}^\bullet(k) - \xi\right) \underset{n\to\infty}{\overset{d}{\longrightarrow}} \mathcal{N}\left(\lambda_A b_\bullet, \sigma_\bullet^2\right),$$

where $\mathcal{N}\left(\mu, \sigma^2\right)$ stands for a normal RV with mean value μ and variance σ^2. Moreover, if β and ρ are consistently estimated through $\hat{\beta}$ and $\hat{\rho}$, with $\hat{\rho} - \rho = o_p(1/\ln n)$, we get a null dominant component of bias for $\mathrm{CG}_0 = \mathrm{CG}_{\hat{\rho}_0,2}$ and for $\mathrm{L}_p^{\mathrm{RB}}$, respectively given in (7) and (9), i.e. $b_{\mathrm{L}_p^{\mathrm{RB}}} = b_{\mathrm{CG}_0} (= b_{\mathrm{CH}}) = 0$. The variance is kept at the same level of the associated non-RB EVI-estimators.

3 Finite sample properties of the EVI-estimators

We have implemented multi-sample Monte-Carlo simulation experiments of size 1000×10, essentially for the class of L_p EVI-estimators, in (5), but comparatively with the RB EVI-estimators, in (7), (8), and more generally in (9), for sample sizes $n = 50, 100(100)1000$. The results will be essentially illustrated with the EV CDF, $F(x) = \mathrm{EV}_\xi(x)$, with $\mathrm{EV}_\xi(x)$ given in (1), $\xi = 0.1$ ($\rho = -\xi = -0.1$), but also with the Student-$t_{\nu=10}$ CDF ($\xi = 1/\nu = 0.1, \rho = -2/\nu = -0.2$). However, other values of ξ and other models have been included in the simulation. For details on multi-sample simulation, see [18], among others.

Mean values and mean square error patterns

For each value of n and for each of the above-mentioned models, we have first simulated the mean value (E) and the RMSE of the aforementioned estimators, as functions of the sample fraction, k/n, for $p = 1, 2, 3, 6, 10(10)40$. As can be seen in Figure 1, based on the first replicate with a size 1000, and at optimal levels in the sense of minimal RMSE, L_3 (and indeed, even L_2, not included in the picture for sake of simplicity) beats the MVRB EVI-estimators, CH, in (8), also pictured in Figure 1. But L_{10} has the smallest RMSE, among the estimators considered.

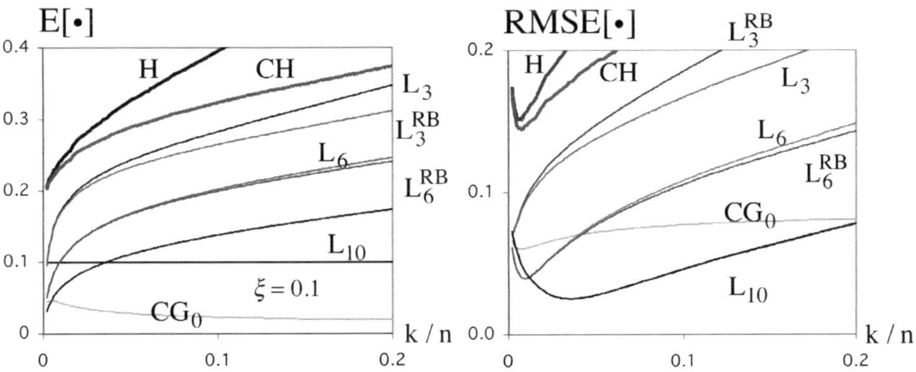

Figure 1. Mean values (left) and RMSEs (right) of the EVI-estimators under study for an EV_ξ CDF with $\xi = 0.1$

We can always find an optimal p, clear from these pictures in what concerns RMSE, but often also valid for mean values at optimal levels, as can be seen in [29].

Mean values, RMSEs and relative efficiency indicators of the EVI-estimators at optimal levels

Tables 1 and 2 are respectively related to simulated EV_ξ and Student t_{10} parents both with $\xi = 0.1$. We there present, for $n =$ 100, 200, 500 and 1000, the simulated mean values at optimal levels (levels where RMSE are minima as functions of k) of the EVI-estimators under study. Information on 95% confidence intervals, computed on the basis of the 10 replicates with 1000 runs each, is also provided. In the simulation experiments, we have included the values $p = 2, 3, 6, 10(10)40$, but we make explicit only the values up to the one leading to the minimal squared bias. Bias reduction in (9) is not at all visible for large p, as expected, and we have shown only the values associated with an overall increase in the REFF-indicator of $\mathrm{L}_p^{\mathrm{RB}}$ comparatively to L_p, with the REFF-indicator to be defined in (13). Among the estimators considered, and distinguishing 3 regions, a first one with (H, CH, CG_0), a second one with L_p, and a third one with $\mathrm{L}_p^{\mathrm{RB}}$, the one providing the smallest squared bias is written in **bold** whenever there is an out-performance of the behaviour achieved in the previous region. The overall smallest squared bias is also underlined.

Note that in all cases, the mean value of $L_p(k)$ is decreasing in p, and we thus can always find a value of p associated with minimal squared bias, often not visible in the tables.

Table 1. Simulated mean values, at optimal levels, of the EVI-estimators under study, for EV_ξ underlying parents with $\xi = 0.1$, together with 95% confidence intervals

n	100	200	500	1000
H	0.335 ± 0.0026	0.285 ± 0.0031	0.241 ± 0.0022	0.222 ± 0.0036
CH	0.284 ± 0.0037	0.260 ± 0.0034	0.232 ± 0.0042	0.219 ± 0.0037
CG_0	$\mathbf{0.055} \pm 0.0011$	$\mathbf{0.053} \pm 0.0007$	$\mathbf{0.048} \pm 0.0006$	$\mathbf{0.045} \pm 0.0007$
L_2	0.254 ± 0.0022	0.219 ± 0.0028	0.189 ± 0.0020	0.174 ± 0.0019
L_3	0.199 ± 0.0019	0.172 ± 0.0021	0.149 ± 0.0016	0.137 ± 0.0015
L_6	0.112 ± 0.0010	$\underline{\mathbf{0.099}} \pm 0.0011$	$\underline{\mathbf{0.099}} \pm 0.0014$	$\underline{\mathbf{0.100}} \pm 0.0011$
L_{10}	$\underline{\mathbf{0.099}} \pm 0.0015$	0.097 ± 0.0009	0.098 ± 0.0007	0.098 ± 0.005
L_2^{RB}	0.228 ± 0.0019	0.208 ± 0.0027	0.184 ± 0.0020	0.171 ± 0.0019
L_3^{RB}	0.185 ± 0.0016	0.166 ± 0.0020	0.146 ± 0.0016	0.136 ± 0.0015
L_6^{RB}	0.109 ± 0.0009	0.096 ± 0.0011	0.098 ± 0.0016	0.099 ± 0.0012

Table 2. Simulated mean values, at optimal levels, of the EVI-estimators under study, for Student t_ν underlying parents with $\nu = 10$ ($\xi = 1/10 = 0.1$), together with 95% confidence intervals

n	100	200	500	1000
H	0.273 ± 0.0034	0.223 ± 0.0023	0.186 ± 0.0026	0.168 ± 0.0029
CH	0.214 ± 0.0054	0.198 ± 0.0026	0.179 ± 0.0030	$\mathbf{0.167} \pm 0.0033$
CG_0	$\mathbf{0.046} \pm 0.0017$	$\mathbf{0.039} \pm 0.0011$	$\mathbf{0.036} \pm 0.0004$	0.034 ± 0.0005
L_2	0.204 ± 0.0025	0.172 ± 0.0018	0.146 ± 0.0013	0.132 ± 0.0015
L_3	0.159 ± 0.0020	0.135 ± 0.0014	0.115 ± 0.0011	0.104 ± 0.0013
L_6	0.098 ± 0.0020	0.099 ± 0.0011	0.099 ± 0.0013	0.100 ± 0.0010
L_{10}	$\underline{\mathbf{0.098}} \pm 0.0006$	$\underline{\mathbf{0.099}} \pm 0.0008$	$\underline{\mathbf{0.100}} \pm 0.0005$	$\underline{\mathbf{0.100}} \pm 0.0003$
L_2^{RB}	0.174 ± 0.0037	0.159 ± 0.0018	0.141 ± 0.0012	0.130 ± 0.0017
L_3^{RB}	0.143 ± 0.0028	0.128 ± 0.0015	0.112 ± 0.0010	0.104 ± 0.0017

We have further computed the Hill estimator, in (5) whenever $p = 1$, at the simulated value of $k_{0|1} := \arg\min_k \text{RMSE}\big(L_1(k)\big)$, the simulated optimal k in the sense of minimum RMSE. Such an estimator is denoted $L_{0|1}$. We have also compute $L_{0|p}$, the estimator L_p computed at the simulated value of $k_{0|p} := \arg\min_k \text{RMSE}\big(L_p(k)\big)$. The simulated *relative efficiency* (REFF)-indicators are

$$\text{REFF}_{p|1} := \frac{\text{RMSE}\left(L_{0|1}\right)}{\text{RMSE}\left(L_{0|p}\right)}, \tag{13}$$

and again the values $p = 2, 3, 6, 10(10)40$ were included in the simulation experiment. Similar indicators have also been simulated for all other EVI-estimators under study, and the higher these indicators are, the better the associated EVI-estimators perform, comparatively to $H_0 = L_{0|1}$.

Again as an illustration of the results obtained, we present Tables 3–4. In the first row, we provide the RMSE of $L_{0|1}$, so that we can easily recover the RMSE of all other estimators. The following rows provide the REFF-indicators of CH, CG_0, and $\text{REFF}_{p|1}$ in (13), for the same L_p and L_p^{RB} EVI-estimators considered in the Tables 1-2. Just as before, similar marks are used.

For a better visualization of the results associated with Table 1 and Table 3 , Figure 2 is presented.

Table 3. Simulated RMSE of H (first row) and REFF-indicators of other EVI-estimators comparatively to H, for EV_ξ underlying parents, $\xi = 0.1$, together with 95% confidence intervals

n	100	200	500	1000
$RMSE_0(H)$	0.270 ± 0.2529	0.217 ± 0.2439	0.171 ± 0.2297	0.149 ± 0.2207
CH	1.226 ± 0.0065	1.130 ± 0.0037	1.068 ± 0.0045	1.044 ± 0.0046
CG_0	**5.666** ± 0.1647	**4.063** ± 0.0913	**2.955** ± 0.0621	**2.477** ± 0.0413
L_2	1.447 ± 0.0179	1.430 ± 0.0170	1.401 ± 0.0239	1.383 ± 0.0239
L_3	2.063 ± 0.0327	2.055 ± 0.0315	2.024 ± 0.0485	1.995 ± 0.0466
L_6	5.447 ± 0.0918	5.027 ± 0.0915	4.351 ± 0.0952	4.019 ± 0.0688
L_{10}	**8.776** ± 0.1247	**7.772** ± 0.1213	**6.816** ± 0.1490	**6.301** ± 0.1003
L_2^{RB}	1.649 ± 0.0218	1.547 ± 0.0197	1.459 ± 0.0260	1.416 ± 0.0250
L_3^{RB}	2.250 ± 0.0368	2.161 ± 0.0352	2.076 ± 0.0510	2.023 ± 0.0477
L_6^{RB}	5.505 ± 0.0888	5.043 ± 0.0921	4.364 ± 0.0964	4.028 ± 0.0686

Table 4. Simulated RMSE of H (first row) and REFF-indicators of other EVI-estimators comparatively to H, for Student-t_ν underlying parents, $\nu = 10$ ($\xi = 1/\nu = 0.1$), together with 95% confidence intervals

n	100	200	500	1000
$RMSE_0(H)$	0.203 ± 0.2290	0.152 ± 0.2197	0.113 ± 0.2024	0.095 ± 0.1913
CH	1.348 ± 0.0216	1.197 ± 0.0058	1.114 ± 0.0064	1.073 ± 0.0070
CG_0	**3.640** ± 0.1405	**2.454** ± 0.0488	**1.702** ± 0.0274	**1.379** ± 0.0240
L_2	1.532 ± 0.0125	1.491 ± 0.0223	1.435 ± 0.0186	1.400 ± 0.0355
L_3	2.269 ± 0.0319	2.198 ± 0.0423	2.046 ± 0.0296	1.900 ± 0.0610
L_6	5.276 ± 0.0920	4.377 ± 0.0648	3.657 ± 0.0482	3.278 ± 0.0938
L_{10}	**7.894** ± 0.1563	**6.737** ± 0.1218	**5.669** ± 0.0662	**5.089** ± 0.1419
L_2^{RB}	1.859 ± 0.0369	1.674 ± 0.0307	1.524 ± 0.0189	1.448 ± 0.0366
L_3^{RB}	2.563 ± 0.0560	2.354 ± 0.0502	2.108 ± 0.0292	1.927 ± 0.0614

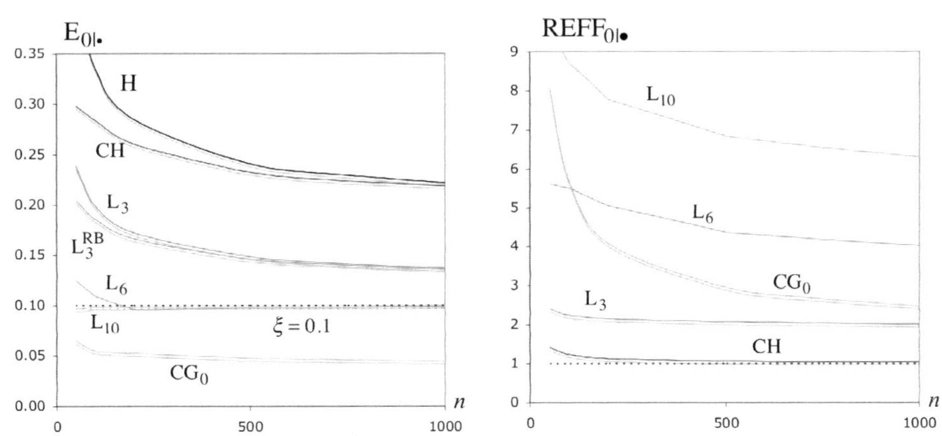

Figure 2. Mean values at optimal levels (*left*) and REFF-indicators (*right*) of the EVI-estimators under study for an EV_ξ CDF with $\xi = 0.1$

Robustness of Lehmer's EVI-estimator towards contamination

The terminology robustness is used here not in the strict sense of Huber's ([28]) and Hampel's ([25]) work, but in a more wide sense, also contemplated by these authors, of insensitivity (or weak sensitivity) of our semi-parametric estimators to changes of the underlying model F in $\mathcal{D}^+_{\mathcal{M}}$. Just as in [11], among others, for all generated samples, we have concomitantly generated corresponding contaminated samples, obtained by multiplying by 10, 5% of randomly chosen observations. As an illustration, we present now the results attained for such a contaminated $EV_{0.1}$ underlying parent. Tables 5 and 6 are thus respectively similar to Tables 1 and 3, without repeating the mean values and RMSE of the H EVI-estimators for $EV_{0.1}$ samples. Any of the EVI-estimators (generally denoted $\bullet$), associated with the contaminated sample, are denoted $\bullet|c$. The resistance of CG_0 to this type of contamination is amazing, and was not at all expected. But, sometimes going up to $p = 40$, we can always find a value of p such that L_p beats CG_0, both regarding squared bias and RMSE.

Table 5. Simulated mean values, at optimal levels, of the EVI-estimators under study, for contaminated EV_ξ underlying parents with $\xi = 0.1$, together with 95% confidence intervals

n	100	200	500	1000
H|c	0.563 ± 0.0047	0.565 ± 0.0031	0.553 ± 0.0019	0.548 ± 0.0015
CH|c	0.628 ± 0.0190	0.642 ± 0.0023	0.645 ± 0.0021	0.554 ± 0.0075
CG_0|c	**0.089** ± 0.0007	**0.088** ± 0.0006	**0.088** ± 0.0006	**0.085** ± 0.0007
L_2|c	0.671 ± 0.0056	0.629 ± 0.0177	0.501 ± 0.0052	0.466 ± 0.0064
L_3|c	0.509 ± 0.0006	0.484 ± 0.0075	0.396 ± 0.0046	0.373 ± 0.0053
L_6|c	0.281 ± 0.0034	0.271 ± 0.0042	0.222 ± 0.0025	0.208 ± 0.0029
L_{10}|c	0.172 ± 0.0021	0.167 ± 0.0025	0.137 ± 0.0015	0.127 ± 0.0018
L_{20}|c	0.099 ± 0.0008	0.100 ± 0.0013	0.100 ± 0.0011	0.099 ± 0.0014
L_{30}|c	<u>**0.099**</u> ± 0.0004	<u>**0.100**</u> ± 0.0006	0.100 ± 0.0004	0.100 ± 0.0002
L_{40}|c	0.097 ± 0.0023	0.099 ± 0.0005	<u>**0.100**</u> ± 0.0002	<u>**0.100**</u> ± 0.0002
L_2^{RB}|c	0.597 ± 0.0046	0.607 ± 0.0047	0.497 ± 0.0052	0.464 ± 0.0064
L_3^{RB}|c	0.496 ± 0.0064	0.480 ± 0.0076	0.394 ± 0.0046	0.372 ± 0.0053
L_6^{RB}|c	0.279 ± 0.0034	0.270 ± 0.0042	0.222 ± 0.0025	0.208 ± 0.0029

Table 6. REFF-indicators (comparatively to the H for an $EV_{0.1}$ model) of EVI-estimators, for contaminated EV_ξ underlying parents, $\xi = 0.1$, together with 95% confidence intervals

n	100	200	500	1000
H|c	0.548 ± 0.0060	0.452 ± 0.0082	0.378 ± 0.0053	0.332 ± 0.0027
CH|c	0.446 ± 0.0835	0.390 ± 0.0075	0.315 ± 0.0047	0.2889 ± 0.0047
CG_0|c	**9.055** ± 0.4137	**7.956** ± 0.1390	**6.459** ± 0.1227	**5.270** ± 0.1101
L_2|c	0.448 ± 0.0056	0.370 ± 0.0106	0.364 ± 0.0075	0.326 ± 0.0072
L_3|c	0.572 ± 0.0097	0.493 ± 0.0148	0.471 ± 0.0108	0.418 ± 0.0096
L_6|c	1.220 ± 0.0219	1.041 ± 0.0339	1.007 ± 0.0249	0.904 ± 0.0232
L_{10}|c	2.571 ± 0.0521	2.210 ± 0.0831	2.128 ± 0.0586	1.863 ± 0.0555
L_{20}|c	7.033 ± 0.1301	6.028 ± 0.1891	4.630 ± 0.1328	3.772 ± 0.1019
L_{30}|c	10.342 ± 0.1668	8.986 ± 0.2691	7.005 ± 0.2006	5.634 ± 0.1570
L_{40}|c	<u>**12.321**</u> ± 0.4677	<u>**11.717**</u> ± 0.3507	<u>**9.312**</u> ± 0.2537	<u>**7.496**</u> ± 0.2051
L_2^{RB}|c	0.509 ± 0.0062	0.413 ± 0.0080	0.366 ± 0.0076	0.328 ± 0.0072
L_3^{RB}|c	0.581 ± 0.0098	0.497 ± 0.0150	0.473 ± 0.0109	0.419 ± 0.0096
L_6^{RB}|c	1.226 ± 0.0216	1.042 ± 0.0340	1.008 ± 0.0249	0.905 ± 0.0232

To have a better visualization of the results above, we next present Figure 3, similar to Figure 2, but related essentially to the contaminated $EV_{0.1}$ model. We have kept the same scale in both figures. This is the reason why bot H|c and CH|c mean values do not appear, since they are above 0.5, another unexpected result relatively to the expected resistance of the CH EVI-estimators to changes in the underlying parent.

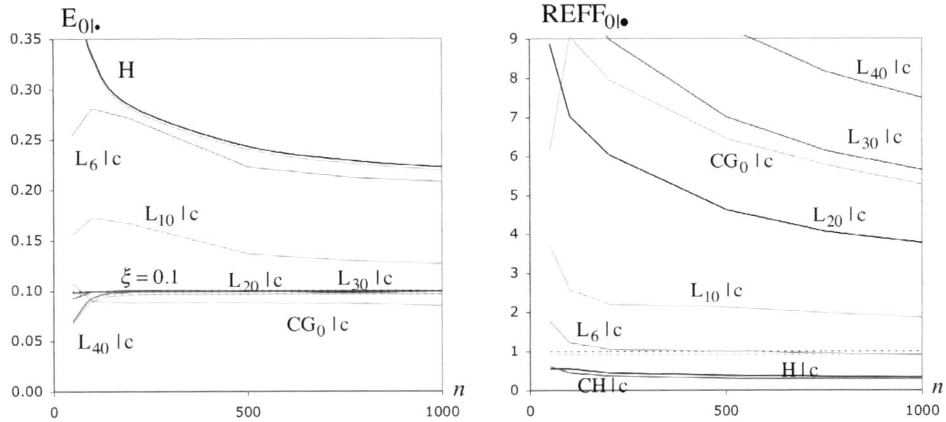

Figure 3. Mean values at optimal levels (*left*) and REFF-indicators (*right*) of the EVI-estimators under study for the $EV_{0.1}$ contaminated CDF

4 Conclusions

- The results obtained for the L_p EVI-estimators, defined in (5), were slight astonishing from a theoretical point of view. Indeed, the highest efficiency was obtained for large p, a long way from the asymptotically optimal p, which lies between 1 and 2 (see [29], for details). However, further note that at such an asymptotically optimal p, not difficult to estimate in practice, the estimators in (5) outperform the asymptotically optimal H EVI-estimator in the whole (ξ, ρ)-plane, an unusual property among 'classical' EVI-estimators.
- For small values of the EVI, say $\xi < 1$, The L_p EVI-estimators outperform the MVRB EVI-estimators, exhibiting reasonably stable sample paths, as a function of k, and small RMSEs for adequate large values of p. But for any real or simulated heavy-tailed data set, with $\xi < 1$, the highest stability of sample paths is associated with CG_0, despite of with a systematic underestimation of ξ.
- As intuitively expected, $L_{0|p}$ are decreasing in p until a value p_{min}, approaching the true value of ξ, for all simulated models. But we cannot forget that as p increases to $+\infty$, $L_{0|p}$ approaches zero, being no longer consistent. Consequently, if p is further increased in $L_p(k) \equiv CG_{p,1}(k)$, the sample paths approach zero and the RMSEs increase. The choice of p needs thus to be carefully done, not on the basis of sample path stability, but on the basis of reliable estimated values of the RMSE, as a function of k and p. For the choice of (k, p) we thus think sensible the use of a non-parametric double-bootstrap algorithm, like the one suggested in [4]. The 'parametric' double-bootstrap methodology, used in in [5], among others, leads to a p close to the one that maximizes the asymptotic relative efficiency. Then they beat the H and CH EVI-estimators, but not so strongly as happens with the large values of p here considered.
- The resistance of the CG_0 RB EVI-estimators to the type of contaminations under consideration, together with the fact that, also under contamination, we can always find a value of p such that L_p beats CG_0, both regarding squared bias and RMSE, deserves a theoretical study of the robustness of these EVI-estimators in the lines of [3]. Indeed, for contaminated samples the value of p needs to be highly increased in order to obtain the smallest possible RMSE.

Acknowledgement

Research partially supported by National Funds through **FCT** — Fundação para a Ciência e a Tecnologia, projects UID/MAT/00006/2013 (CEA/UL), UID/MAT/0297/2013 (CMA/UNL), and COST Action IC1408.

Bibliography

[1] Beirlant, J., Goegebeur, Y., Segers, J. and Teugels, J. (2004). *Statistics of Extremes. Theory and Applications.* Wiley.

[2] Beirlant, J., Caeiro, F. and Gomes, M.I. (2012). *An overview and open researh topics in statistics of univariate extremes.* Revstat, *10*:1, 1–31.

[3] Beran, J., Schell, D. and Stehlik, M. (2014). *The harmonic moment tail index estimator: asymptotic distribution and robustness.* Ann. Inst. Statist. Math., *66*, 193–220.

[4] Brilhante, M.F., Gomes, M.I. and Pestana, D. (2012). *Non-parametric double-bootstrap method for an adaptive MOP EVI-estimation.* In T.E. Simos *et al.* (eds.), Numerical Analysis and Applied Mathematics ICNAAM 2012, AIP Conferecence Proceedings, 1708-1711.

[5] Brilhante, M.F., Gomes, M.I. and Pestana, D. (2013). *A simple generalization of the Hill estimator.* Computational Statistics and Data Analysis, *57*:1, 518–535.

[6] Bingham, N., Goldie, C.M. and Teugels, J.L. (1987). *Regular Variation.* Cambridge Univ. Press, Cambridge.

[7] Caeiro, F. and Gomes, M.I. (2002a). *A class of asymptotically unbiased semi-parametric estimators of the tail index.* Test, *11*:2, 345–364.

[8] Caeiro, F. and Gomes, M.I. (2002b). *Bias reduction in the estimation of parameters of rare events.* Theory of Stochastic Processes, *8*:24, 67–76.

[9] Caeiro, F. and Gomes, M.I. (2014). *Comparison of asymptotically unbiased extreme value index estimators: a Monte Carlo simulation study.* AIP Conference Proceedings *1618*, 551–554.

[10] Caeiro, F., Gomes, M.I. and Pestana, D.D. (2005). *Direct reduction of bias of the classical Hill estimator.* Revstat, *3*:2, 111–136.

[11] Cantoni, E. and Ronchetti, E. (2006). *A robust approach for skewed and heavy-tailed outcomes in the analysis of health care expenditures.* Journal of Health Economics, *25*:2, 198–213.

[12] Fraga Alves, M.I., Gomes, M.I. and de Haan, L. (2003). *A new class of semi-parametric estimators of the second order parameter.* Portugaliae Mathematica, *60*:1, 193–213.

[13] Fraga Alves, M.I., Gomes, M.I., de Haan, L. and Neves, C. (2007). *A note on second order conditions in extreme value theory: linking general and heavy tails conditions.* Revstat, 5:3, 285–305.

[14] Geluk, J. and de Haan, L. (1987). *Regular Variation, Extensions and Tauberian Theorems.* CWI Tract 40, Center for Mathematics and Computer Science, Amsterdam, Netherlands.

[15] Gnedenko, B.V. (1943). *Sur la distribution limite du terme maximum d'une série aléatoire.* Annals of Mathematics, *44*:6, 423–453.

[16] Gomes, M.I. and Guillou, A. (2015). *Extreme value theory and statistics of univariate extremes: a review.* International Statistical Review, *83*:2, 263–292.

[17] Gomes, M.I. and Martins, M.J. (2002). *"Asymptotically unbiased" estimators of the extreme value index based on external estimation of the second order parameter.* Extremes, *5*:1, 5–31.

[18] Gomes, M.I. and Oliveira, O. (2001). *The bootstrap methodology in Statistics of Extremes: choice of the optimal sample fraction.* Extremes, *4*:4, 331–358.

[19] Gomes, M.I. and Pestana, D.D. (2007). *A sturdy reduced-bias extreme quantile (VaR) estimator.* J. American Statistical Association, *102*:477, 280–292.

[20] Gomes, M.I., de Haan, L. and Henriques-Rodrigues, L. (2008). *Tail index estimation for heavy-tailed models: accommodation of bias in weighted log-excesses.* J. Royal Statistical Society B, *70*:1, 31–52.

[21] de Haan, L. (1984) *Slow variation and characterization of domains of attraction.* In J. Tiago de Oliveira, ed., *Statistical Extremes and Applications.* D. Reidel, Dordrecht, 31–48.

[22] de Haan, L. and Ferreira, A. (2006). *Extreme Value Theory: An Introduction.* Springer Science+Business Media, LLC, New York.

[23] de Haan, L. and Peng, L. (1998). *Comparison of extreme value index estimators.* Statistica Neerlandica, *52*, 60–70.

[24] Hall, P. and Welsh, A.W. (1985). *Adaptive estimates of parameters of regular variation.* Ann. Statist., *13*, 331–341.

[25] Hampel, F.R., Ronchetti, E.M., Rousseeuw, P.J. and Stahel, W.A. (1986). *Robust Statistics: The Approach Based on Infuence Functions.* Wiley, New York.

[26] Havil, J. (2003). *Gamma: Exploring Euler's Constant.* Princeton, NJ: Princeton Univ. Press.

[27] Hill, B.M. (1975). *A simple general approach to inference about the tail of a distribution.* Ann. Statist., *3*, 1163–1174.

[28] Huber, P.J. (1981). *Robust Statistics.* Wiley, New York.

[29] Penalva, H., Caeiro, F., Gomes, M.I. and Neves, M.M. (2016). *An efficient naive generalisation of the Hill estimator—discrepancy between asymptotic and finite sample behaviour.* Notas e Comunicações CEAUL 02/2016.

[30] Reiss, R.-D. and Thomas, M. (2007). *Statistical Analysis of Extreme Values, with Application to Insurance, Finance, Hydrology and Other Fields*, 2nd edition; 3rd edition, Birkhäuser Verlag.

Risk profiles for severe mental health difficulty: classification and regression tree analysis

Yoshitake Takebayashi, *Institute of Statistical Mathematics*, ytake2@ism.ac.jp
Takafumi Kubota, *Tama University*, kubota@tama.ac.jp
Hiroe Tsubaki, *Institute of Statistical Mathematics*, tsubaki@ism.ac.jp

Abstract. Severe mental health difficulty is a leading cause of suicide. Its development has been associated with several risk factors. Comprehensive approaches must be used to examine the degree to which these factors co-act and interact to develop severe mental health difficulty risk profiles. Our literature review reveals no report of a study exploring interactions among severe mental health condition factors in subsets of people with high suicidal risk (mental disorders, unemployment, and caregivers of relatives) in Japan. This study aims to use a classification and regression tree (CART) approach to establish risk profiles and examine their performance for diagnostic accuracy. Data were obtained from the National Comprehensive Survey of Living Conditions. Outcome measures (K6) were categorized into low, moderate, and high, applying the recommended cut-off values. Socio-demographic status, financial status, and subjective stress were included as predictors in the CART. CART analysis results indicate that subjective stress in daily life is the strongest predictor for severe mental health difficulties in the high-suicide-risk group. Additionally, results show that all high-suicide-risk group divided into several sub-groups that reflect interactions among predictors.

Keywords. machine learning, ordinal data, regression trees

1 Introduction

Severe mental health difficulty (e.g., major depression) is a leading cause of suicide [1]. Its development has been associated with several risk factors. Evidence from earlier studies has indicated that, in addition to the daily stress itself, socio-demographic, financial, and health status might interact with daily stress, resulting in severe mental health difficulty[2, 3]. Although multiple studies evaluating risk factors individually to predict severe mental health are useful, more comprehensive approaches are necessary to examine the degree to which these factors interact to develop severe mental health difficulty risk profiles. The Japanese National Comprehensive Survey of Living Conditions [4] is administered to obtain basic data related to health, medical care, welfare, pension, and income, which have been required for health, labor and welfare administration planning and management. Since 2007, Kessler 6 (K6), a well-established and widely used measure to screen mental health difficulties, has been included in this national survey to infer means of suicide prevention planning and management. The survey is apparently extremely useful because it incorporates variables that might be related to severe mental health, such as demographic, financial, and health status factors. Psychiatric disorder, unemployment, and caregivers of relatives are well-established groups associated with high suicide risk. However, our literature search revealed no report of a study exploring complex interactions among factors related to severe mental health difficulties in subsets of people with high suicidal risk in Japan. This study uses a classification and regression tree (CART) approach to establish severe mental health difficulty risk profiles in those populations.

2 Classification and Regression Tree for ordered response outcome

Independent observations n to be classified are characterized by a p-dimensional vector of predictors $x_i = (x_{i1}, x_{i2} ..., x_{ip})$. Each observation x_i falls into one of J classes. All observations start together in root node t to derive a classification tree. Then, the optimal split is ascertained for predictors 1, 2...,p. The optimal split is defined as the split resulting in the largest decrease in node impurity. For node t, the optimal split divides the observations into left and right descendent nodes, t_L dan t_R, respectively. The proportion of cases in each of the J classes within these nodes are called the node proportions, i.e., $p(\omega_j|t)$ for $j = 1, 2..., J$ such that $p(\omega_1|t) + p(\omega_2|t) + ... + p(\omega_j|t) = 1$. For nominal response classification, the within-node impurity measure most commonly used is the Gini criterion [5], defined as shown below.

$$i(t) = \sum_k \sum_{k \neq l} p(\omega_k|t)p(\omega_l|t) \tag{1}$$

One impurity function that is useful for ordinal response prediction is the generalized Gini impurity[5], defined as

$$i_{GG}(t) = \sum_k \sum_{k \neq l} C(\omega_k|\omega_l)(\omega_k|t)p(\omega_l|t) \tag{2}$$

which incorporates $C(\omega_k|\omega_l)$ as a misclassification cost in class observations as belonging to class k. Presuming that a set of increasing scores $s_1 < s_2 < s_j$ is assigned to the ordered categories of the response Y, we will use the quadratic misclassification cost, where $C(\omega_k|\omega_l) = (S_k - S_l)^2$.

We use two pruning functions for the ordinal CART model to control the tree size. The pruning functions are included in the rpartScore package in R [6]. The pruning approach used here is the cost-complexity measure in equation (2), which combines a measure of predictive performance and a measure of the complexity of the tree, usually the number of leaves or terminal nodes. Specifically, the cost-complexity measure is

$$R_\alpha(T) = R(T) + \alpha \times card(T), \tag{3}$$

where $R(T)$ is the predictive performance, $card(T)$ is the size of the tree T measured by the number of terminal nodes. Also, α is a tuning parameter that controls the tradeoff between predictive performance and model complexity. The rpartScore MC sets the predictive performance measure to the total misclassification cost, whereas rpartScore MR sets it to the total number of misclassified observations. The latter is a sum of all observations that are classified incorrectly, whereas the former is a sum over all observations of the absolute difference between the observed score and the predicted score.

To build and prune the nominal and ordinal CART model, we selected the best model for each tree type using the 1-SE rule [5], where SE was estimated using ten-fold cross-validation in the testing set and a complexity parameter (in equation (2)) 0.001, a small minimum bound to allow for pruning. The 1-SE rule selects the tree that has the maximum predictive error that is within 1 SE of the minimum predictive error.

3 Samples and measures

Sample Data were obtained from the Japanese National Comprehensive Survey of Living Conditions[4], which was conducted to obtain basic data related to health, medical care, welfare, pension, and income, which have been required to the planning and management for the Health, Labour, and Welfare Ministry. We extracted three groups of high suicidal risk: people with psychiatric disorders ($n = 259$), unemployed people ($n = 428$), and caregivers for relatives ($n = 505$). We excluded observations with missing outcomes from the analytical dataset. Consequently, the sample sizes were reduced: psychiatric disorders, $n = 198$; unemployment, $n = 550$; caregivers for relatives, $n = 304$).

Outcome Severe psychological distress was assessed by Kesseler 6 (K6). Scores were divided into three mental health difficulty severities of low (0–4), moderate (5–9), severe (>9) based on recommended cut-off values[8]. The frequencies of outcome categories by group are shown in Table 1.

Table 1. Distribution of each group by K6 category.

K6 rank	Psychiatric disorders	Caregiver	No job
low (0)	47	149	232
moderate (1)	59	78	106
severe (2)	92	77	90

Predictors We were unable to extract many variables related to socio-demographic, financial, and health conditions. We generated 63 variables and used them as predictors to classify the outcome categories. These predictors are shown in Table 2. Factor scores of subjective stress were obtained from graded response IRT model using several indicators (subjective rating of global health condition (very good (5) to poor (1)), perceived physical symptoms (yes or no), days having trouble carrying out daily activities (1–24), and presence of stress (yes or no)).

CART analysis We used the packages rpart [7] and rpartScore [6] of the R language and environment for statistical computing and graphics to conduct CART analysis.

4 Results

Diagnostic accuracy We evaluated the agreement between the tree predictions and responses from the original dataset using Somers' d, which measures differences in terms of ordinal association between predicted and observed scores and which can be interpreted similarly to a correlation coefficient [6]. Based on Somers' d in the test dataset, the nominal CART model was more accurate than other models among the psychiatric disorders group and the unemployed group. The ordered CART model with CR was more accurate than other models for the test dataset in the caregivers group.

Variable Importance The variable importance among groups is shown in Figure 1. Results show that subjective stress was the most important predictor for severe mental health difficulty in all high-suicide-risk groups. It is particularly interesting that predictors related to financial status were also important, especially among the caregiver group. Details of variables included in final CART model are shown in Table 4.

Table 2. Predictors

Demographic/Financial status		Medical status (yes/no)/Subjective stress (continuous)			
1	gender	22	diabetes	43	dental
2	age	23	obesity	44	atopic dermatitis
3	marital status	24	hyperlipidemia	45	dermatitis
4	education level	25	thyroid disease	46	gout
5	number of household members	26	psychiatric	47	rheumatosis
6	type of household	27	dementia	48	arthritis
7	total household expenditure	28	Parkinson	49	stiff shoulder
8	total income	29	neurologic disease	50	lower back pain
9	amount of contributions	30	eye	51	osteoporosis
10	premium payment	31	ear	52	kidney disease
11	amount of savings	32	hypertension	53	prostatomegaly
12	loss of savings	33	stroke	54	menopausal disorder
13	amount of loans	34	cardiovascular disease	55	fracture
14	medical expenses	35	circulatory organ	56	injury
15	job	36	cold	57	anemia
16	childcare expenses	37	rhinitis	58	malignant neoplasm
17	person requiring care	38	asthma	59	pregnant
18	living with person requiring care	39	respiratory disease	60	infertility treatment
19	type of person requiring care	40	gastroduodenal disease	61	other
20	smoking	41	liver or gallbladder disease	62	unknown
21	number of smoking	42	gastrointestinal disease	63	subjective stress

Table 3. Diagnostic accuracy: Somers' d values based on the validation set comparing the predicted estimates from three CART models.

	psychiatric disorders		unemployment		caregivers	
	training set	test set	training set	test set	training set	test set
nominal	0.58	0.47	0.38	0.25	0.55	0.44
ordered/CM	0.55	0.46	0.55	0.24	0.44	0.51
ordered/CR	0.45	0.38	0.54	0.20	0.58	0.49

Psychiatric Disorders The nominal CART tree in psychiatric disorders is shown in Figure 2. In the end, six groups are distinguished. Three groups were categorized as having severe mental health difficulties. These groups are defined based on interactions between subjective stress, household type, amount of saving, and total income. People who have high subjective stress tend to be have more severe mental health difficulty when they live in one-person households in a dormitory (categorized as 1), single-parent family (categorized as 5), three-generation household (categorized 6). Even though there are other types of households, they tend to have severe mental health difficulty when they have fewer savings. It is particularly interesting that few patients reported severe mental health difficulties when their total income is high.

Unemployment An ordered CART tree in unemployment group is shown in Figure 3. In the end, eight groups were distinguished and two groups are categorized as having severe mental health difficulties. These groups are defined based on interactions between subjective stress, age, type of household, amount of household expenditure, and lower back pain. People with high subjective stress tended to report severe mental health difficulties when they experienced lower back pain. People who have medium subjective

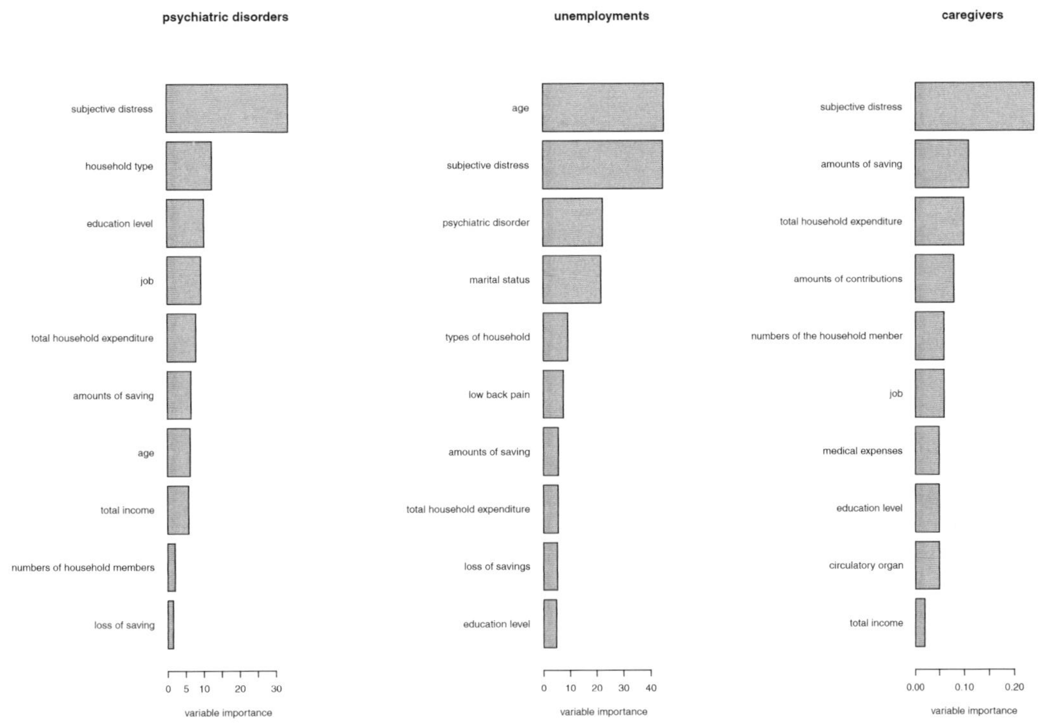

Figure 1. Top ten important variables identified using the CART model.

stress and high household expenditure are classified as having severe mental health difficulty in their thirties.

Caregivers of relatives Nominal CART tree in the caregiver group is shown in Figure 4. Eventually, seven groups were distinguished; two groups were categorized as having severe mental health difficulty. These groups are defined based on interactions among subjective stress, type of a person requiring care, total household expenditure, medical expenses, and education level. Total household expenditures play an important role in distinguishing moderate from severe mental health difficulty. People who have higher subjective stress tend to show severe mental health difficulty when their household expenditure is low.

5 Discussion and Conclusion

Results of current application of nominal and ordinal CART model showed that the ordinal CART model is not always superior to the nominal CART model. These results are consistent with findings in Nindahayati et al. (2015), which revealed no significant difference in predictive accuracy between the nominal and ordinal CART model. However, Wheeler et al. (2015) revealed that ordinal CART is superior to nominal CART in predictive accuracy. Further research is necessary to clarify the conditions that are suitable for using an ordinal or nominal CART model.

Table 4. Details of predictors included in the final CART model.

predictor	contents
subjective stress	factor score from IRT model (mean=0, variance=1)
total income, amount of saving	1=10,000 yen
household expenditures, medical expenses	1=1,000 yen
lower-back	1=yes, 0=no
household type	
1	one person household (loadings)
2	one person household (other)
3	household with married couple only
4	household with married couple and unmarried child
5	household with one parent and unmarried child
6	three-generation household
7	other
education level	
1	junior high school or less
2	high school
3	vocational college
4	junior college or technical college
5	university
6	graduate school
type of person requiring care	
1	spouse
2	child
3	child's spouse
4	parents
5	other relatives
6	nursing care service
7	other

CART analysis revealed several important findings. Although subjective stress factors are a commonly important factor in all three high risk groups, patterns of interaction between subjective stress and other factors to predict severe mental health difficulties differ greatly among groups. In the psychiatric disorder group, severe mental health difficulty was diagnosed by the interaction between subjective stress, household type, and financial parameters (savings and income). Subjective stress and financial problems are well-established risk factors for severe mental health, although types of households get little attention in this research area. Further confirmatory research must be conducted to clarify the relation between the type of household and severe mental health difficulty. In the unemployment group, severe mental health difficulty was diagnosed mainly by the interaction between subjective stress and lower back pain. There is plenty of evidence related to chronic pain disease and depression. Mental health difficulties might tend to be worse because of limited daily activity. In the caregiver group, severe mental health difficulty was diagnosed mainly through interaction between subjective stress and household expenditures.

In summary, this exploratory application suggests that it might be useful to consider financial status

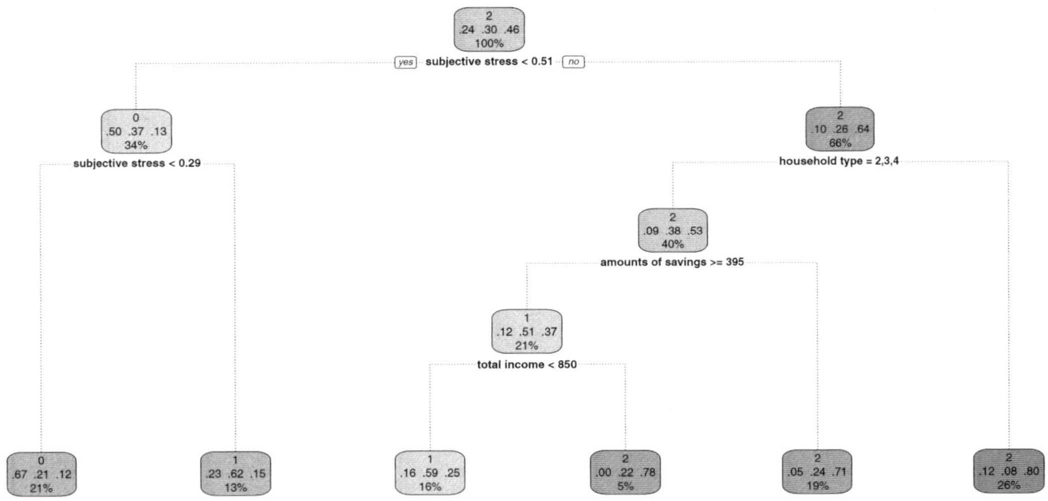

Figure 2. Classification tree representing severe mental health difficulty predictors of psychiatric disorders.

in addition to the subjective stress level during planning and management for severe mental health difficulty. The findings of this study were limited by its cross sectional design. To examine predictive performance more precisely, the study results should be replicated in a prospective cohort study.

6 Acknowledgement

This study was supported by a Health Labour Sciences Research Grant (H26-Seishin-Ippan-003) from the Ministry of Health, Labour and Welfare of Japan.

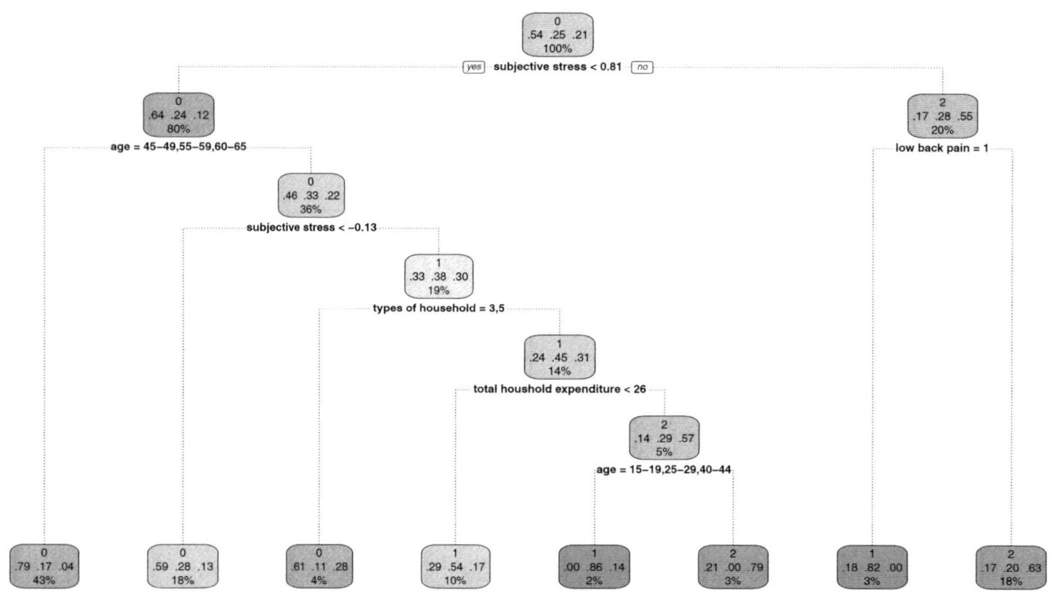

Figure 3. Classification tree representing the predictors of severe mental health difficulty in unemployment.

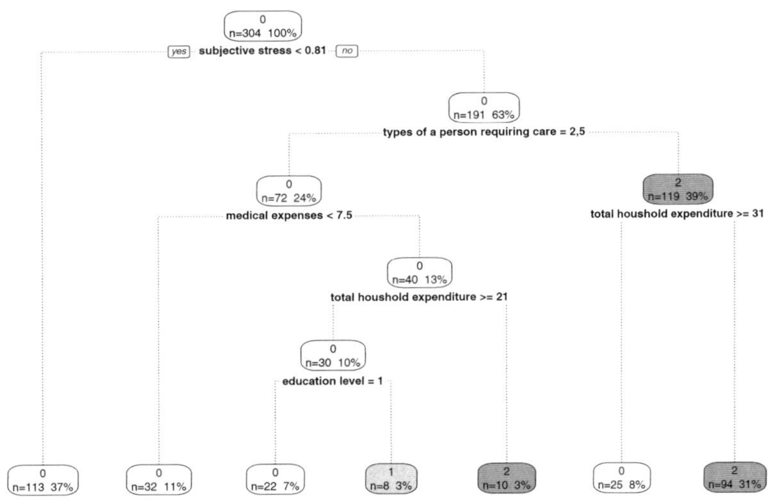

Figure 4. Classification tree representing the predictors of severe mental health difficulty in caregivers.

Bibliography

[1] World Health Organization (WHO) *Depression. Media Centre; 2012* http://www.who.int/mediacentre/factsheets/fs269/en/

[2] Belle, D., and Doucet, J. (2003) *Poverty, inequality, and discrimination as sources of depression among U.S. women.* Psychology of Women Quarterly, **27**, 101-113.

[3] Spearing, R., and Bailey, J. (2012) *Depression and chronic physical illness: its prevalence and diagnosis, and implications for therapeutic practice.* International Journal of Therapy & Rehabilitation, **19(7)**, 394-404.

[4] Ministry of Health and Welfare. (2010) *Comprehensive Survey of the Living Conditions of People on Health and Welfare.* Statistics and Information Department Minister's Secretariat. R package ver. 1.0-1. Retrieved from http://cran.rproject.org/package=rpartScore.

[5] Breiman, L., Friedman, J. H., Olshen, R. A., and Stone, C. J. (1984) *Classification and regression trees.* New York: Chapman & Hall.

[6] Galimberti G, Soffritti G, and Di Maso M. (2012) *Classification trees for ordinal responses in R: The rpartScore package.* Journal of Statistical Software, **47**, 1-25.

[7] Therneau, T., Atkinson, B., and Ripley, B. (2015) *rpart: Recursive partitioning and regression trees.* R package ver. 4.1-9. Retrieved from http://cran.r-project.org/package=rpart.

[8] Sakurai, K., Nishi, A., Kondo, K., Yanagida, K., and Kawakami, N. (2011). *Screening performance of K6/K10 and other screening instruments for mood and anxiety disorders in Japan.* Psychiatry and Clinical Neurosciences, **65(5)**, 434-441.

[9] Nindahayati. Y., Wijayanto, H., and Sartono, B. (2015) *Building a model to predict school accreddiation rank using boosted classification tree.* Proceeding of International Conference on Research, Implementation and Education of Mathematics and Sciences 2015 (ICRIEMS, 2015), 17-19.

[10] Wheeler, D. C., Archer, K. J., Burstyn, I., Yu, K., Stewart, P. A., Colt, J. S., Baris, D., Karagas, M. R., Schwenn, M., Johnson, A., Armenti, K., Silverman, D. T, and Friesen, M. C. (2015) *Comparison of ordinal and nominal classification trees to predict ordinal expert-based occupational exposure estimates in a case-control study.* Annals of Occupational Hygiene, **59(3)**, 324-335.

Forecasting financial time big data using interval time series

Carlos G. Maté, *Comillas Pontifical University*, cmate@icai.comillas.edu
Javier Redondo, *Comillas Pontifical University*, javier@javierredondo

Abstract. An interval time series (ITS) assigns to each period an interval covering the values taken by the variable. Each interval has four characteristic attributes, since it can be defined in terms of lower and upper boundaries, center and radius. The analysis and forecasting of ITS is a very young research area, dating back less than 15 years, and still presents a wide array of open issues. One main issue with time series in a big data context consists of deciding if to handle it as classic time series (CTS) or to proceed with some kind of aggregation in order to get a time series of symbolic data like ITS. Using the k-Nearest Neighbours (kNN) method, in this paper both approaches are applied to forecast exchange rates. Based on usual distances for interval-valued data such as Haussdorff, Ichino-Yaguchi and so on; the reduction in mean distance error using ITS instead of CTS suggests that the ITS approach could be a better way to forecast exchange rates using large data or data streaming. Some interesting conclusions about monthly and daily aggregation horizons are obtained and further research issues are proposed.

Keywords. exchange rates, interval analysis, interval-valued data, kNN, symbolic data analysis

1 Introduction

In CTS analysis, the variable is represented by a single point for each period. This approach was built in a world where collecting and summarizing information was a difficult and time consuming task. Nowadays, we face the opposite situation. New fields of research, such as symbolic data analysis (SDA) or functional data analysis (FDA), have given birth to innovative approaches to the analysis of information. Variables describing entities are valued by complex elements such as intervals or histograms in SDA, and by functions in FDA. SDA considers that, compared to point-valued variables, symbolic variables (such as lists, intervals or histograms) are more accurate representations of the magnitudes of complex real-life situations.

Figure 1 shows the framework to handle CTS and ITS in a Big Data context. As part of this framework, it has been included the information about exchange rates forecasts in CTS contexts we can find on the Internet. Additional information about this issue can be seen in [13]. In the middle of the right part of Figure 1 we can see the monthly ITS of the EURUSD values in the period 2003-April to 2013-March versus the monthly CTS obtained by the centre of intervals taking the last values of every day. This ITS is obtained aggregating the CTS of daily close values of such rate in that period.

[5] reviewed the first advances in the field of SDA. The study of the knowledge available in symbolic data time series (such as ITS or histogram time series (HTS)), whose development had started in [2], has

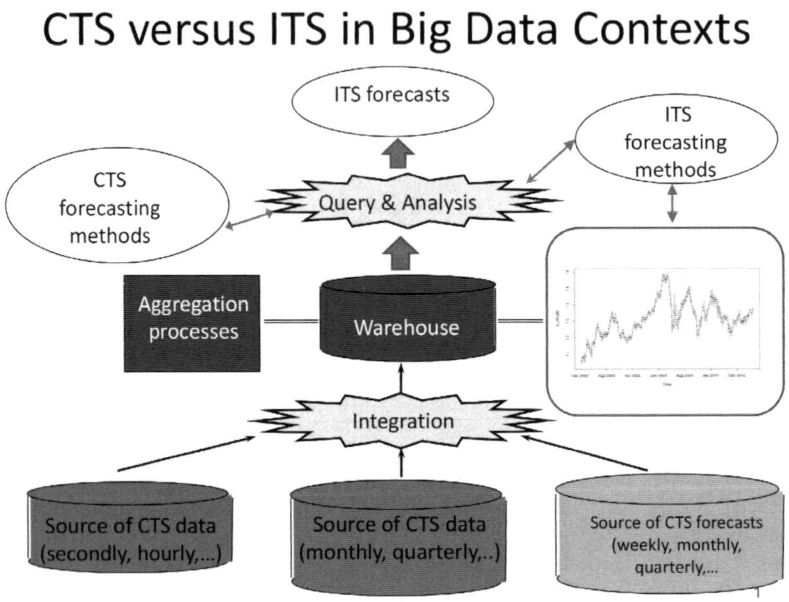

Figure 1. EURUSD as an ITS reflecting monthly range values in a Big Data context

been reviewed more recently in a new survey of SDA concepts and methods in [15]. On the other hand, FDA considers that data can be described by smooth curves rather than point-valued observations.

The temporal aspects of big data management has been reviewed recently in [7]. In addition, [8] has indicated that the big data paradigm raises new tasks for prediction models, suggesting that nonlinearity becomes important and considering several cases from the FX market. Between the proposed methods is not included the kNN. However, in other statistical approaches used to deliver solutions for problems in a Big Data context such as clustering, the kNN method has been considered very adequate.

One main issue with time series forecasting in a big data context consists of deciding if to handle it as classic time series (CTS), or to proceed with some kind of aggregation in order to get a time series of symbolic data like ITS or HTS, evaluating the performance of both approaches. This issue was not considered in [2] neither in [4] or in [3]. Using the k-Nearest Neighbours (kNN) method, in this paper both approaches are applied to forecast exchange rates. Based on usual distances for interval-valued data such as Haussdorff, Ichino-Yaguchi and so on; the reduction in mean distance error using ITS instead of CTS is analyzed. The domain of forecasting in the FOREX market has been reviewed by [17] and very recently [6], among many others. A recent paper about modeling, forecasting and trading the EUR exchange rates is [18].

The organization of this paper is as follows. Section 2 introduces essential concepts about ITS. Section 3 shows the kNN forecasting method for ITS and provides a brief review of applications using this key method with economic and financial time data. Using some case studies, Section 4 discusses data, methodology and empirical results. Finally, Section 5 collects some concluding remarks and outlines some open lines of research.

2 Interval time series

Since 1960 Interval analysis (IA) has become an active research area, and an impressive development in the field of methods and applications of interval data has followed. However, the analysis and forecasting of ITS is a very young research area, dating back less than 15 years, and still presents a wide array of open issues. Among them are the problems with asserting the stationarity of the ITS, as well as the lack of guidance concerning the analysis of cointegration between several ITS. In the following, some important definitions and facts about ITS will be provided.

Definition 1. An interval $[x]$ over the base set $(E, \leq)$ is an ordered pair $[x] = [x_L, x_U]$ where $x_L, x_U \in E$ are the endpoints or boundaries of the interval, such that $x_L \leq x_U$.

An equivalent representation of an interval is given by the center (midpoint) and radius (half-range) of the interval, namely $[x] = \langle x_C, x_R \rangle$ where $x_C = \frac{1}{2}(x_L + x_U)$ and $x_R = \frac{1}{2}(x_U - x_L)$. An interval $[x] = [x_L, x_U]$ may also be written as $[x_1, x_2]$.

Let $\mathcal{K}_c(\mathbb{R})$ denote the class of nonempty compact intervals.

Simple operations with intervals Let $[x_1, x_2]$ and $[y_1, y_2]$ be two intervals. The sum between two intervals is defined by:

$$[x_1, x_2] + [y_1, y_2] = [x_1 + y_1, x_2 + y_2]$$

Since $-[x_1, x_2] = [-x_2, -x_1]$, the difference between two intervals is defined as:

$$[x_1, x_2] - [y_1, y_2] = [x_1 - y_2, x_2 - y_1]$$

Some stochastic issues with interval time series

In this section, we consider the stochastic issues that arise when using interval-valued time series (ITS). Following [3], [12] and [19]; we first define the important concepts of interval random variable and interval stochastic process, and consequently define the concept of ITS.

Let $(\Omega, \mathcal{F}, P)$ be a probability space, where Ω is the set of elementary events, $\mathcal{F}$ is the σ-field of events and $P : \mathcal{F} \to [0, 1]$ the σ-additive probability measure. We define a partition of Ω into sets $A(x)$ such that $A_X(x) = \{\omega \in \Omega | X(\omega) = x\}$, where $x \in [x_L, x_U]$,.

Definition 2. A mapping $[X] : \mathcal{F} \to \mathcal{K}_c(\mathbb{R}) \subset \mathbb{R}$ such that for each $x \in [x_L, x_U]$, there exists a set $A_X(x) \in \mathcal{F}$, is called an interval random variable.

Definition 3. An interval-valued stochastic process is a collection of interval random variables that are indexed by time, that is $\{[X]_t\}$ for $t \in T \subset \mathbb{R}$ with each $[X]_t$ following Definition 2.

Definition 4. An interval-valued time series is a realization of an interval-valued stochastic process. It may be equivalently denoted as $\{[x]_t\} = \{[x_L, x_U]_t\} = \{\langle x_C, x_R \rangle_t\}$ for $t = 1, 2, \ldots, T$.

Differencing an ITS

In this paper, the following definition is proposed for interval differencing:

$$\frac{d\langle y_c, y_r \rangle}{d(t-1, t]} = \left\langle y_{c,t} - y_{c,t-1}, \frac{y_{r,t-1} + y_{r,t}}{2} \right\rangle$$

This definition works particularly well if the radius describes a stationary process, which is very often the case in the FOREX market.

Choosing a distance measure with interval valued-data and ITS

Forecasting accuracy requires a measure of distance between the forecast and the actual value. For CTS, the absolute or the squared value of the difference between the observation and the forecast is considered as the distance. What remains is to find a definition of distance that can be applied to intervals. Subtraction can be defined using the principles of interval arithmetic as it was mentioned above. However, the arithmetic difference is not a suitable measure of error; for example, consider the case when the forecast exactly matches the actual observation [1]. In such case, one would expect the distance to evaluate to zero. Following the definition, for $[x_1, x_2] = [y_1, y_2]$ the result is [-2r, 2r]. This implies that the distance will only evaluate to $[0, 0]$ in the case of degenerate intervals ($r = 0$). Therefore, the arithmetic difference is not a representative measure of distance in the general case. [19] obtain a similar conclusion because the space $mathcalK_c(\mathbb{R})$ is only semilinear.

Several distance measures for intervals have been defined in the literature. Table 1, adapted from [11], lists the most frequently used ones. Each distance is expressed both in terms of lower and upper bound, and in terms of centre and radius. Other interesting distances have been proposed in the literature. See, for example, [10].

Distance	Formula	
	$d_X([x_L, x_U], [y_L, y_U])$	$d_X(\langle x_C, x_R\rangle, \langle y_C, y_R\rangle)$
Hausdorff $d_H([x], [y])$	$max(\lvert x_L - y_L\rvert, \lvert x_U - y_U\rvert)$	$\lvert x_C - y_C\rvert + \lvert x_R - y_R\rvert$
Ichino-Yaguchi $d_{IY}([x], [y])$	$w([x] \cup [y]) - w([x] \cap [y])$ $+ \gamma(2w([x] \cap [y]) - w([x]) - w([y]))$, where $w(I) = I_U - I_L$	$w([x] \cup [y]) - w([x] \cap [y])$ $+ \gamma(2w([x] \cap [y]) - w([x]) - w([y]))$, where $w(I) = 2I_R$
De Carvalho $d_{DC}([x], [y])$	$\frac{d_{IY}([x],[y])}{w([x] \cup [y])}$, where $w(I) = I_U - I_L$	$\frac{d_{IY}([x],[y])}{w([x] \cup [y])}$, where $w(I) = 2I_R$
Euclidean $d_E([x], [y])$	$\frac{1}{\sqrt{2}}\sqrt{(x_L - y_L)^2 + (x_U - y_U)^2}$	$\sqrt{(x_C - y_C)^2 + (x_R - y_R)^2}$
Weighted Euclidean $d_{WE}([x], [y])$	$\frac{1}{\sqrt{2}}\sqrt{\beta(x_L - y_L)^2 + (1-\beta)(x_U - y_U)^2}$ where $0 \leq \beta \leq 1$	$\sqrt{\beta(x_C - y_C)^2 + (1-\beta)(x_R - y_R)^2}$ where $0 \leq \beta \leq 1$
Bertoluzza $d_\theta([x], [y])$	‐ ‐ ‐ ‐ ‐ ‐ ‐ ‐ ‐ ‐ ‐ ‐ ‐ ‐ ‐ ‐ ‐	$\sqrt{(x_C - y_C)^2 + \theta(x_R - y_R)^2}$ where $0 < \theta < \infty$

Table 1. Some distance measures for interval data

Given a realized and a forecast ITS $\{[x]_t\}$ and $\{[\hat{x}]_t\}$, respectively, with $t = 1, ..., T$ the Mean Distance Error to quantify the accuracy of the forecast will be given by

$$MDE_J^q = \left(\frac{\sum_{t=1}^{T}(d_J([x]_t, [\hat{x}]_t))^q}{T}\right)^{\frac{1}{q}}, J = \{H, IY, ...\} \tag{1}$$

where d_J is a particular distance measure and q is the order of the distance, such that for $q = 1$ the MDE is the mean absolute error (MAE) loss function and for $q = 2$ the root mean squared error (RMSE) loss function. We consider $q = 1$ in this paper. Hence we will compare MAE values.

3 The k-nearest neighbors forecasting method for CTS and ITS

The k-nearest neighbors forecasting method for CTS

As described in [2], the (k-NN) method determines forecast values by identifying k sequences in the past that resemble the last observations. Those sequences are known as the k nearest neighbors. Therefore, this procedure takes two steps:

1. a search for the k nearest neighbors in the data,
2. a prediction based on the next observation for those k nearest neighbors.

Choice of the nearest neighbors The k-NN method takes several parameters to determine which neighbors to base the forecast on:

- the number of neighbors k. Higher values gives a smoother but not necessarily more accurate forecast.
- the length of the query vector l. It depends on the track of historic data, where more information could possibly make a larger value's forecast more accurate, although it may also often be set to 1.
- the transformation f applied to the past data.

Let $\left(y_{t_i-(l-1)}, \ldots, y_{t_i}\right)$ with $1 \leq i \leq k$ be the query vectors for the k nearest neighbors and let $\left(y_{T-(l-1)}, \ldots, y_T\right)$ be the last l observations available.

For scalar values the euclidean distance is often used in the minimization test, and then the norms of the following k vectors are minimal:

$$\left\| f\left(y_{t_i-(l-1)}, \ldots, y_{t_i}\right) - \left(y_{T-(l-1)}, \ldots, y_T\right) \right\| \mid 1 \leq i \leq k \tag{2}$$

Forecast generation The forecast is generated by averaging the transformed values of the next observations of all k nearest neighbors. As shown in [2], it is possible to weigh the different neighbors, so that every neighbor y_{t_i} is assigned a weight w_i.

The forecast is finally given by:

$$\hat{y}_{T+h} = \frac{\sum_{i=1}^{k} w_i\, f\left(y_{t_i+h}\right)}{\sum_{i=1}^{k} w_i} \tag{3}$$

Parameter choice The choices required for this method are:

- a value for k and a value for l.
- a transformation $f\colon \mathbb{R}^l \to \mathbb{R}^l$, which can often be linear and therefore optimized with the least squares method, or may even be set to be the identity function so that $f\colon x \mapsto x$;
- a sequence of k weights to generate the forecast, which is often set to $w_i = 1$ for all $1 \leq i \leq k$, so that the forecast is an arithmetic mean.

Parameter sensitivity As discussed in [2], the main problem with the k-NN method is the high sensitivity to the parameter setup. The choice of the number of neighbors is particularly interesting in that the results may not only be different in form but also in characteristics. When a forecast combines too many neighbors, it tends to be biased. On the other hand, forecasts that combine too few neighbors tend to exhibit a very large variance. More details can be seen in [18].

The k-nearest neighbors forecasting method for ITS

Due to its simplicity, the k-NN forecasting method can easily be adapted to interval time series as presented in [2]. We will denote this method by $k - \mathrm{NN}_I$

Choice of the nearest neighbors The interval k-NN forecasting method takes the same parameters as the classic time series approach, namely:

- the number of interval neighbors k,
- the length of the query interval list l,
- the transformation f applied to the past interval data.

Let d be any measure of distance between intervals, so that $d\colon \mathcal{K}_c(\mathbb{R}) \times \mathcal{K}_c(\mathbb{R}) \mapsto \mathbb{R}_+$ assigns the value $d([x],[y])$ to the couple of intervals $[x]$ and $[y]$. Let the application of d to a list of intervals be defined so that:

$$d\left(\left([x]_1, \ldots, [x]_n\right), \left([y]_1, \ldots, [y]_n\right)\right) = \left(d\left([x]_1, [y]_1\right), \ldots, d\left([x]_n, [y]_n\right)\right)$$

Let $\left([y]_{t_i-(l-1)}, \ldots, [y]_{t_i}\right)$ with $1 \leq i \leq k$ be the query vectors for the k nearest neighbors and let $\left([y]_{T-(l-1)}, \ldots, [y]_T\right)$ be the last l observations available.

Then the norm of the following k vectors should be minimized:

$$\left\| d\left(f\left([y]_{t_i-(l-1)}, \ldots, [y]_{t_i}\right), \left([y]_{T-(l-1)}, \ldots, [y]_T\right)\right)\right\| \mid 1 \leq i \leq k \tag{4}$$

Forecast generation Like in the case of CTS, a weighted average of the nearest neighbors' next observation provides the forecast. Applied to intervals, the formulation yields:

$$[\hat{y}]_{T+1} = \frac{\sum_{i=1}^{k} w_i\, f\left([y]_{t_i+1}\right)}{\sum_{i=1}^{k} w_i}$$

Typical parameters According to [2], the following parametric choices are made:

- l is set to 1,
- f is set to be the identity transformation,
- w_i are set to 1 for all $1 \leq i \leq k$.

In this case, there are simply k distances to minimize:

$$d\left([y]_{t_i}, [y]_T\right) \mid 1 \leq i \leq k$$

The forecast in this case is given by:

$$[\hat{y}]_{T+1} = \frac{1}{k}\sum_{i=1}^{k} [y]_{t_i+1} \tag{5}$$

The k-nearest neighbors forecasting method for financial CTS and ITS

The kNN method has a not very long tradition in forecasting economic and financial time series. It has been used to forecast interest rates and to establish automatic stock trading combining technical analysis and nearest neighbor classification.

One of the first attempts to review the kNN method with exchange rates is [14]. Following the abstract of this paper, forecasts were generated by a linear AR-GARCH model and four non-linear methods, including three nearest neighbour methods and locally weighted regression. Five data frequencies were used: daily, four-hourly, two-hourly, hourly and half-hourly. Using root mean square error as a measure, significantly greater accuracy than a no-change forecast was achieved for two-hourly and higher frequency data sets. Using a test by Peseran and Timmerman, significant predictive directional accuracy was found for four-hourly and higher frequency data sets. These results were supported by simulated trading based on forecast direction. No evidence was found that the FX rate behaviour is better represented by a non-linear generating process than by a linear model.

Concerning time series of symbolic data, the first use of the k-NN algorithm is implemented in [2] to forecast the so called histogram time series (HTS). The forecasting ability of the k-NN adaptation is illustrated with meteorological and financial data, and promising results are obtained. [3] is a review of ITS and HTS forecasting methods. They conclude concerning the forecast of the low/high SP500 index that a VEC model and the k-NN methods have the best forecasting performance. In addition, they state that economic and financial questions will benefit greatly from this new approach (symbolic data) to the analysis of large data sets.

4 Case studies

Setup

Following [16] the goal is to assess the performance of different forecasting methods (Naïve, ETS, ARIMA and k-NN) in mostly stochastic environments. The tool developed is based in different functions described in [9] such as ets, arima and so on. Comparisons will be made between:

- classic and interval time series forecasting,
- monthly and daily aggregation,
- linear and nonlinear methods based on one time series,
- linear and nonlinear methods based on multiple time series.

The source data contains:

- EUR/USD bid prices (September 2012 to March 2013), hourly close;
- EUR/USD bid prices (April 2003 to March 2013), daily close;
- BRL/USD bid prices (April 2003 to March 2013), daily close;

The data was extracted from `http://www.oanda.com/`OANDA. The data in each analysis is split into two sets. In the case of single time series analysis, the first set will be used for fitting and the second one for testing. This was done as follows. For hourly data, the last two months were predicted; for daily data, the last two years were predicted. In the case of multiple time series models, 80% of the data was used for training.

Analysis

Aggregated data will be produced as follows:

- for hourly close raw data, the daily range will constitute the interval for the day;

- for daily close raw data, the monthly range will constitute the interval for the month.

The following prices will be compared.

- BRL/USD and EUR/USD from raw data using the monthly aggregate of univariate daily CTS forecasts using several methods,
- BRL/USD and EUR/USD from monthly aggregated data using ITS forecasts by several methods,
- EUR/USD from raw data using the daily aggregate of univariate hourly CTS forecasts using several methods,
- EUR/USD from daily aggregated data using ITS forecasting methods,

All forecasts are compared to the real values using the distance measures described in previous Sections. The Ichino-Yaguchi distance uses the suggested value $\gamma = 0.5$ and the general $\mathcal{L}^2$ metric uses $\theta = 1$. Normalized distances are produced by calculating distances from normalized values. If the time series has (not) been differentiated, a normalized value $\dot{y}_t$ is:

$$\dot{y}_t = \frac{y_t}{\frac{1}{T}\sum_{i=1}^{T}|y_i|} \left(\dot{y}_t = \frac{y_t}{y_0} - 1 \right) \tag{6}$$

Monthly BRL/USD bid

Table 2 presents the normalized distances that were observed after forecasting with CTS methods and aggregating the daily results into monthly intervals. Between brackets the normalized distances based on the forecasting of monthly intervals using ITS forecasting methods. This notation will remain in the rest of the section. In the fit period no clear conclusion is obtained. However, in the test period the $k-\mathrm{NN}_I$ outperforms k-NN with CTS and some of the other methods.

		Naïve	ETS	ARIMA	k-NN
Fit	Hausdorff	**0.0030** [0.0523]	0.0031 [0.0506]	0.0042 [**0.0493**]	0.3345 [0.2780]
	Ichino-Yaguchi	1.0738 [0.9942]	1.0740 [1.0370]	1.0757 [1.0169]	**0.4608** [**0.4121**]
	De Carvalho	0.9265 [0.9262]	0.9266 [0.9263]	0.9274 [0.9283]	**0.9042** [**0.8542**]
	General $\mathcal{L}^2$	**0.0015** [0.0301]	0.0016 [0.0208]	0.0022 [**0.0287**]	0.2193 [0.1844]
Test	Hausdorff	0.3015 [0.2416]	0.3015 [0.2551]	0.2981 [0.2170]	**0.1326** [**0.1360**]
	Ichino-Yaguchi	**0.5665** [0.5074]	**0.5665** [**0.4712**]	0.5666 [0.4829]	0.5799 [0.5653]
	De Carvalho	0.9239 [0.8935]	0.9239 [**0.8298**]	0.9240 [0.8503]	**0.8999** [0.8485]
	General $\mathcal{L}^2$	0.1883 [0.1568]	0.1883 [0.1591]	0.1859 [0.1396]	**0.0820** [**0.0805**]

Table 2. Normalized distances for classic [interval] TS forecasting of monthly BRL/USD bid

Monthly EUR/USD bid

Table 3 presents the normalized distances that were observed after forecasting with CTS methods and aggregating the daily results into monthly intervals. In order to simplify only results of the test period are shown in the following.

Daily EUR/USD bid

Table 4 presents the normalized distances that were observed after forecasting with classic time series methods and aggregating the results into daily intervals.

		Naïve	ETS	ARIMA	k-NN
Test	Hausdorff	0.1069 [0.0823]	0.1085[0.0819]	0.1075 [0.0832]	**0.0910** [**0.0562**]
	Ichino-Yaguchi	0.2269 [0.2078]	0.2268 [0.2045]	0.2268 [0.2019]	**0.2219** [**0.20019**]
	De Carvalho	0.9038 [0.8254]	0.9036 [0.8120]	0.9037 [0.8017]	**0.8841** [**0.7979**]
	General $\mathcal{L}^2$	0.0633 [0.0536]	0.0645 [0.0534]	0.0637 [0.0537]	**0.0540** [**0.0343**]

Table 3. Normalized distances for classic [interval] TS forecasting of monthly EUR/USD bid

		Naïve	ETS	ARIMA	k-NN
Test	Hausdorff	0.0372 [0.0359]	0.0372 [0.0440]	0.0523 [0.0338]	**0.0201** [**0.0200**]
	Ichino-Yaguchi	0.0716 [0.0708]	0.0716 [0.0736]	0.0928 [**0.0702**]	**0.0709** [0.0711]
	De Carvalho	0.9628 [0.9279]	0.9628 [**0.9195**]	0.9685 [0.9199]	**0.9535** [0.9204]
	General $\mathcal{L}^2$	0.0246 [0.0239]	0.0246 [0.0304]	0.0353 [0.0230]	**0.0130** [**0.0130**]

Table 4. Normalized distances for classic [interval] TS forecasting of daily EUR/USD bid

Conclusions: Classic and interval time series forecasting

Table 5 shows the reduction in error resulting from the use of ITS forecasting methods on aggregated periods, when compared to aggregated CTS forecasts. The values have been obtained according to the following equation

$$RedindisJmethodX\frac{ITS}{CTS} = \left(1 - \frac{NDJITS(M_X)}{NDJCTS(M_X)}\right) * 100\% \tag{7}$$

where J belongs to the set {H,IY,DC,General $\mathcal{L}^2$} and X belongs to the set {Naïve, ETS, ARIMA, k-NN}. For example, the fifth number in the last column of Table 5, 38.2%, is obtained using 0.0562 for ITS and 0.0910 for CTS, the first couple of values in the last column in Table 3. Negative values indicate an increase in error.

		Naïve	ETS	ARIMA	k-NN
Monthly BRL/USD	Hausdorff	19.9%	15.4%	**27.2%**	-2.6%
	Ichino-Yaguchi	10.4%	**16.8%**	14.8%	2.5%
	De Carvalho	3.3%	**10.2%**	8.0%	5.7%
	General $\mathcal{L}^2$	16.7%	15.5%	**24.9%**	1.8%
Monthly EUR/USD	Hausdorff	23.0%	24.5%	22.6%	**38.2%**
	Ichino-Yaguchi	8.4%	9.8%	**11.0%**	9.5%
	De Carvalho	8.7%	10.1%	**11.3%**	9.8%
	General $\mathcal{L}^2$	15.3%	17.2%	15.7%	**36.5%**
Daily EUR/USD	Hausdorff	3.5%	-18.3%	**35.4%**	0.5%
	Ichino-Yaguchi	1.1%	-2.8%	**24.4%**	-0.3%
	De Carvalho	3.6%	4.5%	**5.0%**	3.5%
	General $\mathcal{L}^2$	2.8%	-23.6%	**34.8%**	0.0%

Table 5. Reduction in error in the test period when applying ITS forescasting methods versus CTS forecasting methods and then aggregating

It appears that ITS forecasting methods are better at predicting than CTS methods and k-NN is one of the best forecasting strategies when aggregating and handling CTS and ITS. The analyzed time series look like random walks, then linear methods output forecasts which are mostly flat. In the case of CTS methods, the aggregated forecasts have a consequently low radius, which results in a large departure from the actual values. However, since interval methods produce take a radius close to the last observation's, it ends up being larger, and closer to the actual values. In the case of the k-NN methods, since a much larger number of neighbors is necessary to produce a forecast because the observations are not aggregated, the output forecast tends to present less variations. This in turn increases departures from actual values, because the radii shrink and the centers do not follow the real centers closely enough. In the examples examined, aggregation seems to improve results the most when dealing with monthly time series. As mentioned earlier, interval models improved results from classic methods the most in volatile environments. Since variations tend to be larger in the scale of months than in the scale of days, this could explain the phenomenon, particularly in the case of linear forecasts.

5 Concluding remarks

This paper has proposed the SDA approach to tackle Financial Time Big Data (FTBD) by the use of ITS. In order to forecast ITS extracted from FTBD the kNN method has been proposed. Two alternatives have been considered with FTBD of two exchange rates (EURUSD and BRLUSD) and two frequencies (daily and monthly). The first one consists of handling FTBD as CTS and the second one to proceed with some kind of aggregation in order to get a time series of symbolic data like ITS. The reduction in error using ITS instead of CTS suggests that the ITS approach could be a better way to forecast exchange rates. Further research is required with high frequency trading values and other currencies in different frequencies. An interesting topic for future research is the performance of other forecasting methods based on machine learning such as neural networks with FTBD.

Acknowledgement

Our thanks to referees for the time they have dedicated to read the paper and write some suggestions to improve it.

Bibliography

[1] Arroyo, J. and Maté, C. (2006) *Introducing interval time series: accuracy measures.* Compstat, proceedings in computational statistics, Heidelberg: Physica-Verlag, 1139–1146.

[2] Arroyo, J. and Maté, C. (2009)*Forecasting histogram time series with k-nearest neighbours methods.* International journal of forecasting,**25**,1,192–207.

[3] Arroyo, J., González-Rivera, G. and Maté, C. (2011a) *Forecasting with interval and histogram data: some financial applications.* Handbook of empirical economics and finance, Chapman & Hall/CRC, 247–279.

[4] Arroyo, J., Espínola, R. and Maté, C. (2011b) *Different approaches to forecast interval time series: A comparison in finance.* Computational economics, **2**,169-191.

[5] Billard, L. and Diday, E. (2003)*From the statistics of data to the statistics of knowledge: Symbolic data analysis.* Journal of the american statistical association,**98**,462,470–487.

[6] Byrne, J. P., Korobilis, D. and Ribeiro, P. J. (2016) *Exchange rate predictability in a changing world.* Journal of international money and finance, **62**,1–24.

[7] Cuzzocrea, A. (2015)*Temporal aspects of big data management: state-of-the-art analysis and future research directions.* Temporal representation and reasoning (TIME), 2015 22nd international symposium on, IEEE, 180–185.

[8] Dietz, S. (2013) *Big data impacts on stochastic forecast models: Evidence from fx time series.* Pakistan journal of statistics and operation research,**9**,3,277–291.

[9] Hyndman, R. and Khandakar, Y. (2008) *Automatic time series forecasting: the forecast package for R.* Journal of statistical software, **27**,1, 1–22.

[10] Irpino, A. and Verde, R. (2008) *Dynamic clustering of interval data using a Wasserstein-based distance.* Pattern recognition letters,**29**,11,1648–1658.

[11] Kao, C. H., Nakano, J., Shieh, S. H., Tien, Y.J., Wu, H.M., Yang, C. K. and Chen, C.H. (2014) *Exploratory data analysis of interval-valued symbolic data with matrix visualization.* Computational statistics and data analysis, **79**, 14–29.

[12] Kubica, B. G. and Malinowski, K. (2006) *Interval random variables and their application in queueing systems with long–tailed service times.* Soft methods for integrated uncertainty modelling, Springer Berlin Heidelberg,393–403.

[13] Maté, C. G. (2011) *A multivariate analysis approach to forecasts combination: application to foreign exchange markets.* Revista colombiana de estadística,**34**,2,347–375.

[14] Meade, N. (2002) *A comparison of the accuracy of short term foreign exchange forecasting methods.* International journal of forecasting,**18**, 1,67–83.

[15] Noirhomme-Fraiture, M. and Brito, P. (2011) *Far beyond the classical data models: symbolic data analysis.* Statistical analysis and data mining,**4**,2,157–170.

[16] Redondo, J. (2013) *Interval time series: analysis and forecasting.* Master's thesis, Universidad ponficia comillas.

[17] Rossi, B.(2013) *Exchange rate predictability.* Journal of economic literature,**51**,4,1063–1119.

[18] Sermpinis, G., Stasinakis, C., Theofilatos, K. and Karathanasopoulos, A. (2015) *Modeling, forecasting and trading the EUR exchange rates with hybrid rolling genetic algorithms Support vector regression forecast combinations.* European journal of operational research,**247**,3,831–846.

[19] Sinova, B. and Van Aelst, S. (2015) *On the consistency of a spatial-type interval-valued median for random intervals.* Statistics & probability letters, **100**, 130–136.

A toolkit for stability assessment of tree-based learners

Michel Philipp, *University of Zurich*, Michel.Philipp@psychologie.uzh.ch
Achim Zeileis, *Universität Innsbruck*, Achim.Zeileis@R-project.org
Carolin Strobl, *University of Zurich*, Carolin.Strobl@psychologie.uzh.ch

Abstract. Recursive partitioning techniques are established and frequently applied for exploring unknown structures in complex and possibly high-dimensional data sets. The methods can be used to detect interactions and nonlinear structures in a data-driven way by recursively splitting the predictor space to form homogeneous groups of observations. However, while the resulting trees are easy to interpret, they are also known to be potentially unstable. Altering the data slightly can change either the variables and/or the cutpoints selected for splitting. Moreover, the methods do not provide measures of confidence for the selected splits and therefore users cannot assess the uncertainty of a given fitted tree. We present a toolkit of descriptive measures and graphical illustrations based on resampling, that can be used to assess the stability of the variable and cutpoint selection in recursive partitioning. The summary measures and graphics available in the toolkit are illustrated using a real world data set and implemented in the R package **stablelearner**.

Keywords. Stability, Recursive partitioning, Variable selection, Cutpoint selection, Decision trees

1 Introduction

Recursive partitioning approaches, such as classification and regression trees (CART, [2]), conditional inference trees [5] or model-based recursive partitioning [7], are widely used for modelling complex and possibly high-dimensional data sets [11]. The methods are able to detect high-degree interactions and nonlinear structures in a data-driven way. Therefore, these methods have been frequently applied in many scientific disciplines, as well as in many industries for predictive modeling purposes [8].

Nowadays, more complex and more flexible methods exist for predictive learning, that often achieve a better prediction accuracy (e.g., random forests, boosting, support vector machines, neural networks). Recursive partitioning, however, is still a popular method in situations where the aim is to infer and interpret the structure of the underlying process that has generated the data. For this purpose, recursive partitioning is often favoured over other methods, since the results can be illustrated in the form of decision trees, which are relatively easy to interpret. Therefore tree-based methods are widely used as exploratory modeling techniques in many fields, such as social and behavioral sciences (see e.g., [7]).

Recursive partitioning algorithms recursively split the predictor space $\mathcal{X} \in \mathbb{R}_p$ to form homogenous groups of observations. The various algorithms, that have been proposed in the literature, mainly differ with respect to the criteria for selecting the split variable, choosing the cutpoint and stopping the recursion (see [5]). CART, for example, selects the variable and the cutpoint that best unmixes the classes in case of a classification problem, or that most reduces the squared error loss in case of a regression problem. Conditional inference trees, on the other hand, perform splitting and stopping based on a statistical inference procedure.

Despite their popularity, a major drawback of recursive partitioning methods is their instability. By studying the predictive loss of different regularization techniques, Breiman [3] identified recursive partitioning (among others) as unstable. It is well known that small changes in the training data can affect the selection of the split variable and the choice of the cutpoint at any stage in the recursive procedure, such that the resulting tree can take a very different form [8, 11, 12]. Moreover, recursive partitioning methods do not provide measures of confidence for the results. Therefore, users cannot assess the degree of certainty for selected variables and cutpoints. Hence, the question remains to what extend one can rely on the splits in a single tree to draw conclusions.

Previous research has already focussed on assessing the stability of trees from different perspectives and with different goals (see e.g., [1, 4, 9]). Their methods are commonly based on a measure that is used to compare the distance (or similarity) between pairs of trees. In [4], for example, a measure of similarity between trees is proposed to stabilize the selection of the splits in a specific tree algorithm. And more recently, measures and theory were proposed to detect observations that influence the prediction or the pruning in CART [1]. While in these approaches the prediction, partitioning and the structure of the trees are considered separately, they may also be combined in one measure [9]. Thus, while previous research has focussed on reducing instability, measuring the influence of individual observations or assessing the distance between trees, we focus on assessing and visualizing two important aspects that reveal the stability of a tree resulting for a given data set: the variable and the cutpoint selection.

In this paper, we first discuss instability of results from recursive partitioning using a practical example. In the second part, we present a computational procedure and a number of graphical tools that support users for assessing the stability of the variable and the cutpoint selection. The proposed methods are implemented in the software package **stablelearner** (currently available from `https://R-Forge.R-project.org/projects/stablelearner/`) for the free R system for statistical computing [10]. By using a real world data set, the package will be used throughout the article for illustrating the proposed methods.

2 Instability of trees

To illustrate the instability of trees we have used recursive partitioning to predict the survival of the passengers during the sinking of the RMS Titanic in 1912 by several passenger characteristics. A complete passenger list is available online on `http://www.encyclopedia-titanica.org/` (accessed on 2016-04-05). According to the list, 1317 passengers (excluding crew members) were aboard from which 500 survived the sinking. The passenger information, that was used for training the tree, was gender, age, fare, class (1st, 2nd or 3rd), place of embarkment (B = Belfast, C = Cherbourg, Q = Queenstown, S = Southampton), number of siblings/spouses aboard (abbreviated as sibsp) and number of parents/children aboard (abbreviated as parch). The last two features were obtained from an overlapping data set available on
`http://biostat.mc.vanderbilt.edu/wiki/Main/DataSets`. The tree was generated using the function `ctree` from the **partykit** package [6] that performs recursive partitioning in a conditional inference framework in R and is illustrated in the form of a tree in the upper panel of Figure 1. In the following, we will refer to this result as the original tree, since the partitioning was performed on the original passenger data (as opposed to random samples drawn from the original data set employed subsequently).

Based on a bootstrap sample taken from the original passenger data, we generated a second tree, which is illustrated in the lower panel of Figure 1. The structures of the trees look quite different at first sight, which suggests a large instability of the tree. However, when looking closer one can identify variables that were selected in both trees and split at the same or a similar cutpoint. For example, the numerical variable `age` was split at 4 and 54 in the original tree and at the values 4 and 36 in the boostrap

(a) Tree based on the original data set:

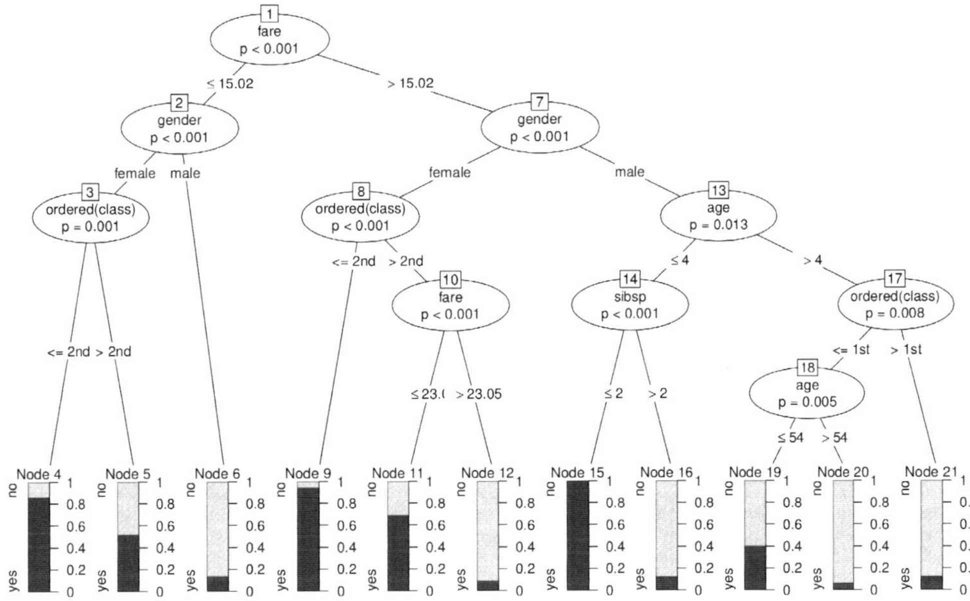

(b) Tree based on a bootstrap sample drawn from the original data set:

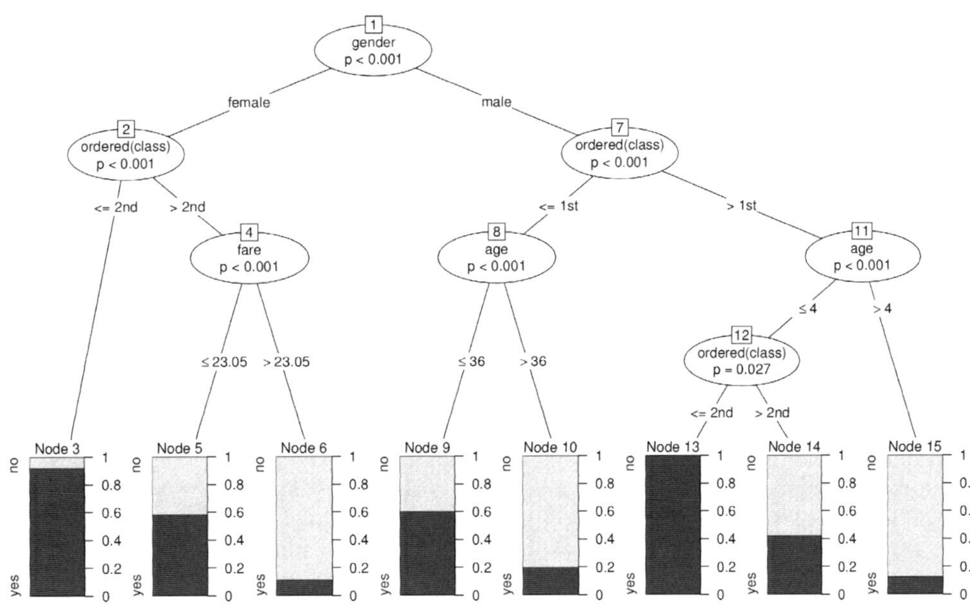

Figure 1. Tree representation of results from recursive partitioning for the RMS Titanic passenger data.

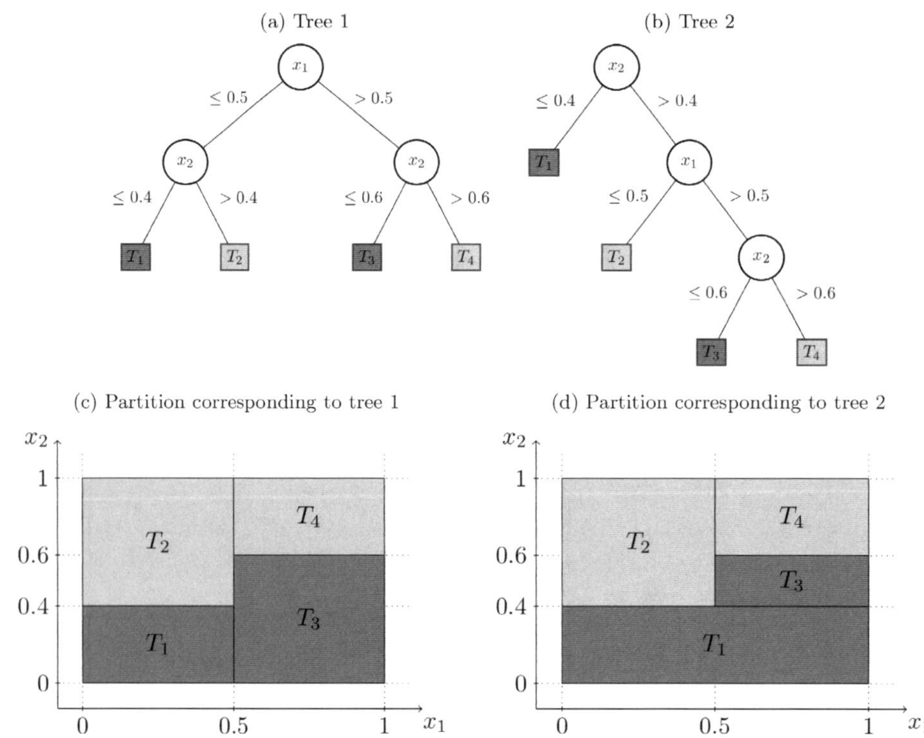

Figure 2. Examples of different tree structures, but equivalent partitions and interpretations.

tree, or the numerical variable `fare` was split twice in the original tree at 15.02 and 23.05 and at 23.05 in the bootstrap tree. Thus, many splits appeared in both trees, only the order and the cutpoints for numerical variables were slightly different.

As Turney [12] elaborates in his work, two trees that are structurally different can be logically equivalent. This means that two trees can lead to very similar or even the same interpretation although their structures (in particular the order of the splits) look very different. To illustrate this principle, we consider two hypothetic trees for a simple classification problem with two classes and a two-dimensional predictor space. Note however, that the statement also holds for any type of response variable and also for predictor spaces with more dimensions. Figure 2 shows two trees (upper row) and representations of the corresponding partitioning of the feature space (bottom row). In the illustration the predicted class in each terminal node is indicated by the colors red and green. According to the figures in panel (a) and (b), the tree structures differ by the split variable in their root node, their path structure and the sequence of split variables in the paths between the root node and the leafs. Yet, though the two trees are structurally different, the predictions are equivalent for any point in the predictor space. By mentally merging the two partitioning representations, it becomes evident that the two trees have identical splits, that only appear in a different order in the tree representation.

To assess whether a tree is stable or not, it is therefore principally important to investigate the stability of the splits, rather than the stability of the entire tree structure. From a single tree representation it is not possible to identify which splits are stable. It is possible, however from an ensemble of trees, e.g., generated by resampling from the original data. From the ensemble, the stability of the splits can be assessed by investigating the variable selection frequency and the cutpoint variability.

3 Measuring variable selection and cutpoint stability

In the following we will outline what steps are necessary from a conceptual point of view to assess the stability of variable and cutpoint selection in trees. Subsequently, these steps will be illustrated for a binary classification tree modeling survival vs. non-survival on the RMS Titanic.

The first step to assess stability is to draw several samples from the original data. The second step is to compute the descriptive measures and graphics provided in our toolkit over all samples. The options implemented in the package for generating samples in the first step are bootstrap sampling (sampling with replacement), subsampling (sampling without replacement), k-fold sample splitting (partitioning the original data into k equally sized samples), leave-k-out jackknife sampling, or further user-defined strategies. Since each option has its specific peculiarities, they will likely generate different results. For the further illustration we will focus on bootstrap sampling, which is most widely used and was chosen as the default option in the function `stabletree()` that performs the resampling and extracts the required information from the ensemble for further analysis:

```
R> library("stablelearner")
R> data("titanic", package = "stablelearner")
R> m <- ctree(survived ~ gender + age + fare + ordered(class) + embarked +
+    sibsp + parch, data = subset(titanic, class %in% c("1st", "2nd", "3rd")))
R> s <- stabletree(m, B = 500)
```

The function `stabletree()` requires a tree-based model object that either inherits from class `party` (like, e.g., the result of `ctree()` or `glmtree()`) or can be coerced to it (like, e.g., the results of `rpart()` or `J48()`). Additionally, parallelization can easily be utilized with a convenience option for multicore computation based on **parallel** (for platforms that support this).

In the remaining part of this section, descriptive measures and graphical illustrations are introduced for investigating the stability of the splits, specifically for the variable and the cutpoint selection. First, the measures will be briefly discussed and then illustrated for the Titanic example.

Variable selection analysis

The aim of the variable selection analysis is to investigate *whether* variables that are selected for splitting in the original tree are also consistently selected for splitting in the resampled data sets. Furthermore, it can be compared *how often* (on average) a variable is selected within the original tree and the repetitions, respectively.

The first descriptive measure is simply the relative frequency of selecting variable x_j for splitting, computed over all repetitions in the procedure. Let $b = 1, \ldots, B$ denote the index for the repetitions and $j = 1, \ldots, p$ the index of the variables considered for partitioning. Further, let $\mathbf{S} = \{s_{bj}\}$ be a binary matrix, where $s_{bj} = 1$ if variable x_j was selected for splitting in repetition b and 0 otherwise. Then, the relative *variable selection frequency* is computed by $100 \cdot \frac{1}{B} \sum_{b=1}^{B} s_{bj}$ and is expected to be large (i.e., close to 100%) for those variables selected in the original tree, if the result is stable. The variable selection frequency can be illustrated graphically using a `barplot()` method that generates the barplot depicted in the left panel of Figure 3. The variables depicted on the x-axis are sorted in decreasing order with respect to their variable selection frequencies (here and in all the following graphical tools). The bars of variables selected in the original tree are colored in dark gray and the corresponding labels are underlined. Thus, from the plot we can infer that the variables `gender`, `class`, `age`, `fare` and `sibsp` were selected for splitting in the original tree. The height of the bars corresponds to the variable selection frequency depicted on the y-axis. The first two bars reach the upper limit of 100%, which means that the variables `gender` and `class` were selected for splitting in each repetition. The variable `age`, represented by the third bar, was selected slightly less than 100% (but still very often) over the repetitions. The variables `fare` and `sibsp`, represented by the fourth and the fifth bar, were selected in the original tree, but not as frequently over all repetitions. This indicates that the splits in those variables in the original tree must

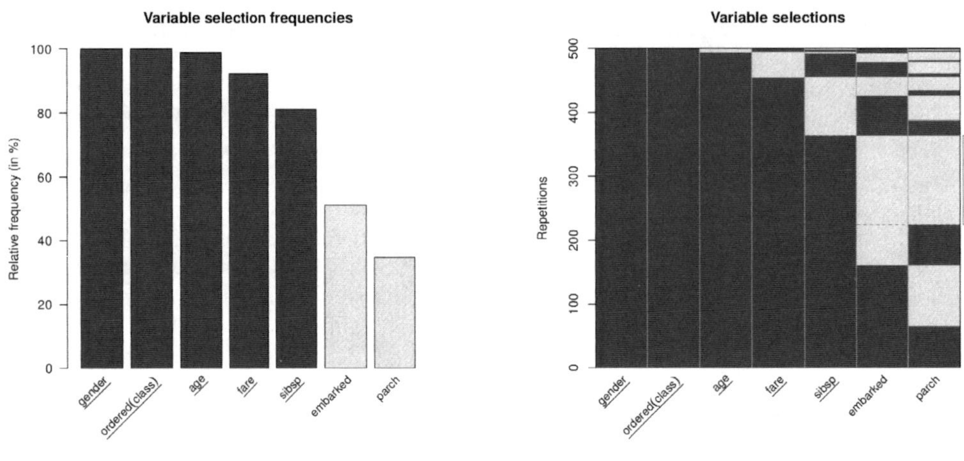

Figure 3. Graphical variable selection analysis.

be considered less reliable compared to the splits of the variable `gender`, `class` and `age`. The last two bars represent the variables `embarked` and `parch`, which were not selected in the original tree. They were selected for splitting in less than 50% of the repetitions. This indicates that although those variables seem to carry some information that is useful for predicting survival, they are not predominant. From a content perspective one may assume for this example that, over the repetitions, the variables `embarked` and `parch` occasionally acted as a proxy for the other variables in the data set.

The `summary()` method prints the corresponding table with the variable selection frequency (termed `freq`) in the first column for each variable. The second column (headed by an asterisk) indicates whether the variable was selected for splitting in the original tree:

```
                 freq * mean *
gender          1.000 1 1.644 2
ordered(class)  1.000 1 2.578 3
age             0.988 1 2.316 2
fare            0.922 1 1.676 2
sibsp           0.810 1 1.132 1
embarked        0.510 0 0.638 0
parch           0.348 0 0.444 0
(* = original tree)
```

The third column in the table (termed `mean`) contains the values of another descriptive measure and denotes the average count splitting in variable x_j per tree. Let $\mathbf{C} = \{c_{bj}\}$ be an integer matrix, where c_{bj} equals the number of times x_j was used for splitting in the tree for repetition b. Note that this number can be greater than one, because the same variable may be used for splitting in different parts of the tree. The *average variable split count* is computed by $\frac{1}{B}\sum_{b=1}^{B} c_{bj}$ and is expected to be close to the count of splitting in variable x_j in the original tree. The last column in the table (also headed by an asterisk) indicates how many times the variable was selected for splitting in the original tree. For example, the variable `gender`, was used on average 1.644 times over all repetitions and twice in the original tree. It is possible, that the variable `gender` was often split on a higher level (and thus less often used for splitting) in the repetitions, as compared to the original tree. The reverse may be assumed for the variable `age`, which was on average more often used for splitting over the repetitions than it was used for splitting in the original tree. Similar interpretations follow from the information for the other variables.

Furthermore, we can investigate the combinations of variables selected in the various trees over the repetitions. This can be illustrated using the function `image()`. The resulting plot, that is illustrated in the right panel of Figure 3, is a graphical illustration of the binary matrix **S** that contains the variable selections over the repetitions. A fine grid of rectangles is drawn for each element in **S**, which are colored dark gray if $s_{bj} = 1$ and light gray if $s_{bj} = 0$. The repetitions (illustrated in the y direction) are ordered such that similar combinations of selected variables are grouped together. The combination of variables used for splitting in the original tree is marked on the right side of the plot using a thin solid red line. The area representing the combination is additionally enclosed by two dashed red lines. Notice that this is also the most frequent combination of variables selected over all repetitions. Repetitions that included additional variables beyond the combination in the original tree are illustrated below the marked area. Hence, we can deduce from the illustration that the variables `embarked` and `parch` were sometimes additionally used for splitting. In the replications above the marked area some splitting variables from the original tree were substituted with other variables.

Cutpoint analysis

The variable selection analysis showed that there are some variables which are consistently used for splitting, indicating that those variables are more relevant in predicting survival than others. However, even when the same variables are selected, the splits may still vary with respect to the cutpoints chosen for splitting. Therefore a further important step in assessing the stability of the splits is the analysis of the cutpoints, which provides more detailed information about the variability of the splits.

We suggest different graphical illustrations for analyzing the variability of the cutpoints for numerical, unordered categorical and ordered categorical variables. Using the function `plot()` these illustrations can be generated for all variables specified in the model. According to the type of variable the correct illustration is chosen automatically and the variable names are underlined if the variable was selected for splitting in the original tree. Figure 4 illustrates these plots for the variables in the Titanic passenger data set.

To analyze the cutpoints for ordered categorical variables, we suggest to use a barplot that shows the frequency of all possible cutpoints. Those are sorted on the x-axis by their natural order that arises from the ordering of the categories of the variables. Examples are given for the variables `class`, `sibsp` and `parch` in Figure 4. Additionally, the cutpoints chosen in the original tree are marked using a vertical dashed red line. The number above each line indicates at which level the split occured in the original tree. For example, the cutpoint between the first and the second class is selected more than 500 times (the number of repetitions in this example). This means that for some repetitions the split appeared several times in different positions in the same tree (for example in parallel branches). However, the passengers were split even more often between the second and the third class. The illustration indicates that the observations were consistently split by their class affiliation over the repetitions to predict survival of the passengers. The cutpoint in the variable `sibsp`, on the other hand, was less stable. Although the variable was quite frequently selected for splitting, the variable was often split between lower categories over the repetitions as compared to the original tree. The variable `parch`, which was not used in the original tree, was split only few times between the lower categories and can thus be considered as not very relevant.

To analyze the partition for unordered categorical variables (avoiding ambiguities by using the term "partition" rather than "cutpoint" here), we suggest to use image plots, as illustrated for the variables `gender` and `embarked` in Figure 4. When using binary splits, observations with the same categories are partitioned into the left or the right daughter node. Thus, the categories are assigned to the left or to the right branch of the split, respectively. For visualizing the partitions over the repetitions, categories that are directed to the same daughter node are illustrated by the same color. For the variable `gender`, there is only one possible split between the two categories `Female` and `Male`. The plot illustrates, however, that this split occurs many times (more than 500) over all repetitions, which underscores the relevance of the split. The combination of categories that represent a partition as it occurred in the original tree, is marked on the right side of the plot using a thin solid red line. The area representing the partition is additionally enclosed by two dashed lines (this is a little hard to see here, because the binary variable `gender` only offers one possible partition). Furthermore, the number(s) on the right side of the marking also represent(s) the level(s) of the corresponding split(s) in the structure of the original tree. The two

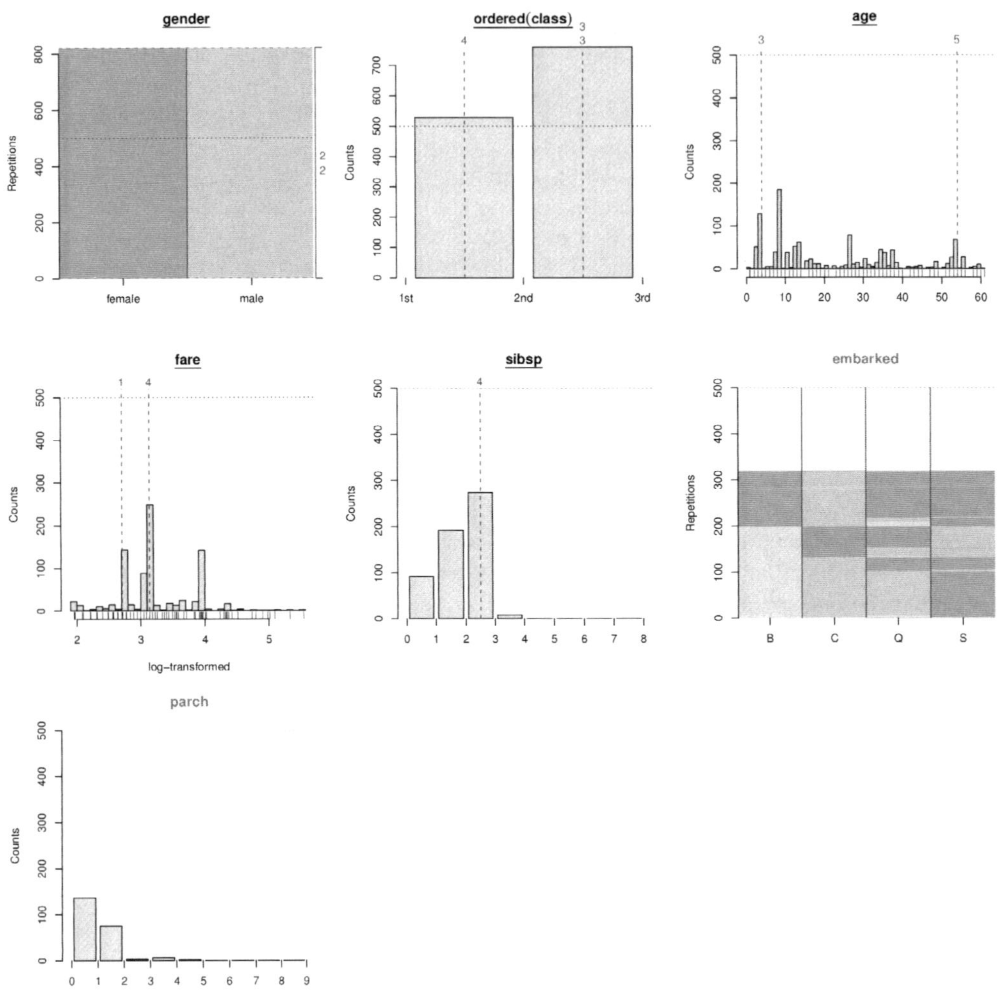

Figure 4. Graphical cutpoint analysis.

numbers on the right side of the illustration for the variable `gender` in Figure 4 indicate that `gender` was split twice on the second level in the original tree.

The plot becomes more detailed for variables with more than two categories such as the variable `embarked`. This variable, however, was not used for splitting in the original tree. Nevertheless it was used relatively often for splitting over all repetitions. In this illustration the additional color light gray is used when a category was no more represented by the observations left for partitioning in the particular node. The partitions over all repetitions are ordered such that equal partitions are grouped together. The most frequent partitions are $[C, Q]$ versus $[S]$ and $[C]$ versus $[B, Q, S]$. Since passengers from the different classes tended to embark in different cities (e.g., most third class passengers embarked in Southampton), the variable `embarked` may in some repetitions (but not in the original tree) have been used as a proxy for the variable `class` in parts of the tree.

To analyze the cutpoints for numerical variables, we suggest to use a histogram, as illustrated for the variables `age` and `fare`. According to the distribution illustrated for the variable `age`, the cutpoints selected over the repetitions spread over the complete range of possible cutpoints. Although some cutpoints were selected more frequently than others, there were no unique cutpoints that were selected over

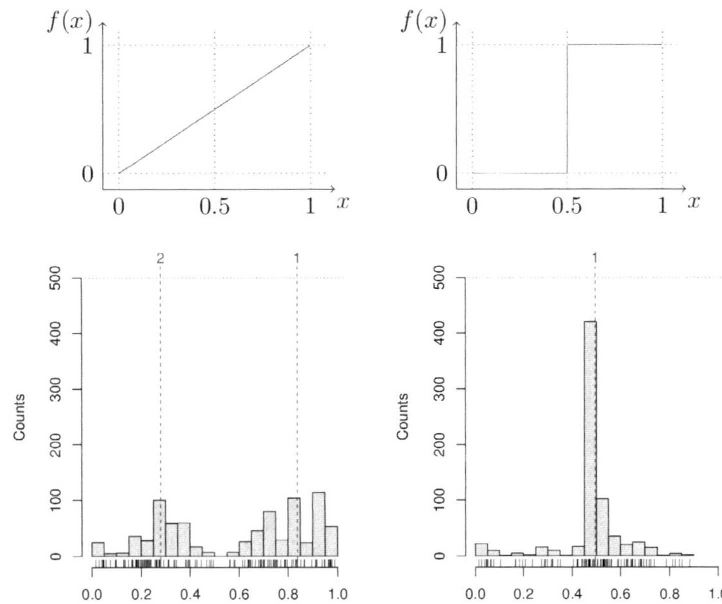

Figure 5. Cutpoint analysis for artificial regression problem.

most repetitions. The selected cutpoints of the variable `fare` are illustrated on a logarithmic scale in Figure 4, as it makes the picture easier to read. Again, the cutpoints selected over the repetitions spread over the complete range of possible cutpoints. However, the cutpoints selected in the original tree match two of three distinct peaks in the histogram and can be considered slightly more stable as compared to the cutpoints within the variable `age`.

From a conceptual point of view, the cutpoint pattern reflects the underlying functional shape. Due to the recursive nature of trees, smooth functions need to be approximated by several splits while piecewise constant step functions can be described straightforwardly by individual splits (see also [11]). This is illustrated in Figure 5. The upper left panel illustrates a linear relationship. To approximate this functional form, a tree algorithm will split the variable several times at different cutpoints. Altering the data would thus very likely lead to a different cutpoint. For a piecewise constant function like the one illustrated in the upper right panel of Figure 5, on the other hand, the functional form is captured by a single split, that is relatively easy to detect and will not change much if the data are altered.

To further demonstrate how the cutpoint stability plots can reflect the underlying functional form, we have simulated 500 observations from the model $y = f(x) + \varepsilon$ for each of the two functions displayed in the top row of Figure 5. The variable x was sampled from a uniform distribution $\in [0, 1]$ and ε was sampled from a standard normal distribution. In the bottom row of Figure 5 the stability of the cutpoints for the variable x is illustrated for the two artificial examples. As expected, the identification of a stable cutpoint failed for the example with the linear relationship (see lower-left panel). In the example with the piecewise constant relationship, however, the cutpoint at 0.5 was correctly recovered over most repetitions (see lower-right panel). For the Titanic example illustrated in Figure 4 this means that the cutpoints selected in the original tree for the variable `age` are rather unstable and should not be overinterpreted because the underlying functions seems to be smooth rather than piecewise constant. The cutpoints selected for the variable `fare` are slightly more stable.

To sum up, the stability analysis of the binary classification tree fitted for the Titanic data revealed that many splits in the original tree illustrated in Figure 1 were rather stable, but some parts were quite variable. First, the splits of the variables `gender` and `class` can be considered as important and stable. Second, the splits of the variable `age` are ambiguous, although the variable is definitely relevant for predicting survival of the passengers. Furthermore, the splits of the variable `fare` are fairly stable, but the variable was a few times not selected for splitting over the repetitions. Thus, if the data were altered slightly, the variable might also had been omitted for splitting in the original tree. And finally, the split of the variable `sibsp` is least stable and should not be overinterpreted.

4 Discussion

In this paper we have presented a toolkit of descriptive measures and graphical illustrations that can be used to investigate the stability of the variable and cutpoint selection in models resulting from recursive partitioning. It was demonstrated how the tools are used and illustrated how intuitive they are by a real world data set. The analysis revealed that many aspects of the fitted tree were rather stable, but some parts were quite variable. Notice that the toolkit is not limited to classification trees, but can also be used to investigate the stability of regression trees or model-based trees. It was further illustrated that clear cutpoints from piecewise constant functions in the underlying data generating process, can be identified using the proposed graphics for the cutpoint analysis.

To acknowledge some limitations associated with the tools it should be mentioned, that they produce less meaningful results for very large trees with many splits. If the structure of the underlying data generating process is complex, the sample size or the number of predictors is large, it can become tedious to interpret a tree. Assessing the variable selection and cutpoint stability of such trees is computationally very intensive and the result might be unclear. However, the complexity of a tree can be reduced by modifying the settings (i.e., the pruning rule or the stopping criteria) of the recursive partitioning algorithm. Furthermore one should always be aware that any resampling scheme can only mimic what would happen if a new sample could be drawn from the population. And finally, the proposed tools do not assess the predictive stability of trees, which is another important aspect for their interpretation, as we briefly saw in Section 2. This aspect will be addressed in future research.

Bibliography

[1] Bar-hen, A., Gey, S., and Poggi, J.-M. (2015). Influence Measures for CART Classification Trees. Journal of Classification, **32(1)**, 21–45.

[2] Breiman, L., Friedman, J., Stone, C. J., and Olshen, R. A. (1984). *Classification and Regression Trees.* Belmont: Wadsworth.

[3] Breiman, L. (1996). *Heuristics of Instability and Stabilization in Model Selection.* The Annals of Statistics, **24(6)**, 2350–2383.

[4] Briand, B., Ducharme, G. R., Parache, V., and Mercat-Rommens, C. (2009). A Similarity Measure to Assess the Stability of Classification Trees. Computational Statistics & Data Analysis, **53(4)**, 12081217.

[5] Hothorn, T., Hornik, K., and Zeileis, A. (2006). *Unbiased Recursive Partitioning: A Conditional Inference Framework.* Journal of Computational and Graphical Statistics, **15(3)**, 651–674.

[6] Hothorn, T. and Zeileis, A. (2015). ***partykit**: A Modular Toolkit for Recursive Partytioning in R.* Journal of Machine Learning Research, **16**, 3905–3909.

[7] Kopf, J., Augustin, T., and Strobl, C. (2013). *The Potential of Model-Based Recursive Partitioning in the Social Sciences – Revisiting Ockham's Razor.* In McArdle, J. J. and Ritschard, G. (Ed.), Contemporary Issues in Exploratory Data Mining in the Behavioral Sciences (pp. 75–95). New York: Routledge.

[8] Kuhn, M. and Johnson, K. (2013). *Applied Predictive Modeling.* New York: Springer-Verlag.

[9] Miglio, R. and Soffritti, G. (2004). The Comparison Between Classification Trees Through Proximity Measures. Computational Statistics & Data Analysis, **45(3)**, 577–593.

[10] R Core Team (2016). *R: A Language and Environment for Statistical Computing.* R Foundation for Statistical Computing, Vienna, Austria. URL `https://www.R-project.org/`.

[11] Strobl, C., Malley, J., and Tutz, G. (2009). *An Introduction to Recursive Partitioning: Rationale, Application, and Characteristics of Classification and Regression Trees, Bagging, and Random Forests.* Psychological Methods, **14(4)**, 323–348.

[12] Turney, P. (1995). *Technical Note: Bias and the Quantification of Stability.* Machine Learning, **20(1–2)**, 23–33.

Variable Importance in Clustering Using Binary Decision Trees

Pierre Michel, *Aix Marseille University*, pierre.michel@univ-amu.fr
Badih Ghattas, *Aix Marseille University*, badih.ghattas@univ-amu.fr

Abstract. We consider different approaches for assessing variable importance in clustering. We focus on clustering using binary decision trees, which is a non-parametric top-down hierarchical clustering method designed for both continuous and nominal data. We suggest a measure of variable importance for this method similar to the one used in Breiman's classification and regression trees. We analyze the efficiency of this score on different data simulation models in presence of noise, and compare it to other classical variable importance measures.

Keywords. Clustering, decision trees, CUBT, deviance, variable importance, variables ranking

1 Introduction

In most statistical modeling and data analysis tasks, scoring variables is essential. Its most frequent use is for dimension reduction or feature selection [12], in order to reduce the complexity of the models, to reduce the noise in the data and hence to gain in model accuracy and interpretability. Variables are generally scored with respect to a model or a specific task.

In supervised learning, variable importance is often related to its correlation or dependence with a target variable Y. It may be assessed within the model learning process, or once the model is estimated. The most known approaches are classification and regression trees (CART [3]), random forests (RF, [5]) and support vector machines (SVM, [10, 15]), where variable importance is strongly related to the model structure and accuracy.

Unsupervised learning concerns data clustering and density estimation. We consider here only clustering methods which aim to construct a partition of a set of n observations in k clusters, where k is specified a priori or determined by the method. In clustering, there is a need to determine which variables are the most important to get an output partition. For example, it can be useful to detect noisy variables or irrelevant variables for the partition.

Clustering using binary trees (CUBT, [7]) is a non-parametric top-down hierarchical method designed for both continuous and nominal data. A decision tree constructed with CUBT can be used to identify which variables of the dataset are active and take part directly in the growing stage of the tree. However, although some input variables may be irrelevant for the tree construction, they may be competitive with the active variables at different splits of the tree. Identifying these variables may be very useful for many applications.

An example of CUBT output obtained for the Iris dataset is given in Figure 1. Table 1 gives the score of variable importance for each input variable. The only active variable (*Petal.Length*) has the highest score. The other variables (*Sepal.Length*, *Sepal.Width*, *Petal.Width*) are not active, however two of them have high scores.

The goal of this paper is to define variable importance for clustering using unsupervised binary trees. The role of important variables is analyzed and some results about the efficiency of the score of variable importance are provided. The paper is structured as follows: Section 2 presents some classical methods to assess variable importance in clustering. Section 3 gives a brief description of CUBT and presents the method. Section 4 presents the new methodology to assess variable importance in clustering, based on CUBT. Finally, section 5 presents some experiments and results with different data simulation models and different clustering methods.

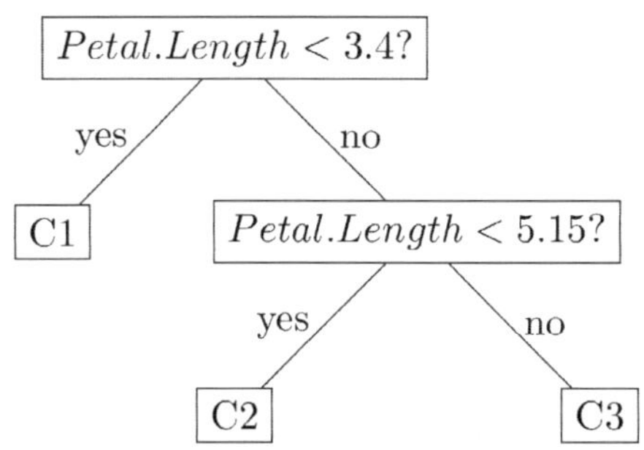

Figure 1. Tree structure obtained by CUBT with the Iris dataset. The only active variable is *Petal.Length*.

Sepal.Length	*Sepal.Width*	*Petal.Length*	*Petal.Width*
3.24	0.14	**3.97**	3.82

Table 1. The four input variables of the iris dataset with their corresponding importance score.

2 Variable importance in clustering

In this section we present some scoring approaches for variables in the context of clustering. Two of them are strongly related to the clustering approach, unsupervised random forests (URF) and two-level variable weighting clustering algorithm TW-k-means (TWKM). A third scoring approach, Laplacian score (LS), computes a score of variable importance independently from any partitioning model or algorithm. Finally, we propose an intuitive scoring method, leave-one-variable out (LOVO) adapted for all clustering methods.

Unsupervised random forests

RF and their variable importance measure are one of the most known and used supervised methods both for regression and classification. Breiman [4, 5] adapted RF and suggested transforming the clustering problem into a binary classification problem: The unlabeled data are assigned to the first class. The second class contains observations generated by independent sampling from the one-dimensional marginal distributions of the first class data. The clustering task is thus transformed into a binary classification problem and variable importance is computed accordingly. The importance of a variable corresponds to the mean decrease of the Gini criterion at any node split using this variable within all the trees of the forest. URF are adapted to both continuous and nominal data.

Two-level variable weighting clustering algorithm

The two-level variable weighting clustering algorithm TW-k-means (TWKM) [6] is a weighted subspace clustering algorithm which can compute weights for the variables in a dataset, that can be used to assess variable importance. This algorithm is an extension of the entropy weighted k-means clustering algorithm, which itself is an extension of the k-means algorithm [13]. First, a set of k starting centers are randomly selected. Observations are assigned to the nearest centers using a dissimilarity measure. Then, new centers are updated, based on these new clusters. Weights are computed for each variable within each cluster, based on the current clustering. The variable weight is inversely proportional to the sum of the within-cluster variances of the variable in the clusters. Weights are taken into account in the computation of the dissimilarity measure. The process continues in this way until convergence.

Laplacian score

The Laplacian score (LS) [1] is a popular variable ranking index for unsupervised learning. LS assesses variable importance by evaluating the power of locality preserving of a variable. For each variable, this score is computed according to proximities between observations in a k-nearest neighbor graph, with the assumption that two observations are related if they are connected in the graph. A similarity matrix S between observations is first computed. For each observation, similarities are put to zero for each observation which is not in the k-nearest neighborhood. Let $D = diag(S\mathbf{1})$ (where $\mathbf{1}$ is an identity vector) the diagonal matrix containing for each observation the sum of the similarities to its k-nearest neighbors. Then Laplacian matrix is defined to be $L = D - S$, each variable $X_.j$, $j \in \{1, ...p\}$, is "centered" getting:

$$\tilde{X}_.j = X_.j - \frac{X'_.jD\mathbf{1}}{\mathbf{1}'D\mathbf{1}}\mathbf{1}$$

Finally, the LS of the jth variable is computed as follows:

$$Imp(X_{.j}) = \frac{\tilde{X}'_.jL\tilde{X}_.j}{\tilde{X}'_.jD\tilde{X}_.j}$$

LS assigns highest scores to variables that best respect the graph structure of the nearest neighbor. It is often used in filter methods for feature selection. In comparison with other classical scores such as data variance or Fisher score, LS showed best results in terms of efficiency [18].

Leave-one-variable-out score

We propose an intuitive scoring method that is adapted to any clustering method. In this approach, we use the output partition obtained with CUBT, k-means or k-modes to rank the variables according to a within-cluster heterogeneity based criterion. Each variable is removed one-by-one from the data. For each removed variable the criterion is computed. The removed variable providing the highest amount of within-cluster heterogeneity is defined as the most important, and so on. Denote $P^{(j)}$ the partition obtained when omitting variable j from the data and $R(P^{(j)})$ the within-cluster heterogeneity measure of $P^{(j)}$. The importance of variable j is defined as follows:

$$Imp(X_{.j}) = R(P^{(j)})$$

We denote this method LOVO for leave one variable out. The actual R package CUBT [8] has been updated to take into account these computations.

3 Clustering using unsupervised binary trees

CUBT is a clustering method inspired from CART. It defines a set of clusters (a partition) using binary decision rules over the original variables. The obtained partition is thus interpretable in terms of the original variables, the obtained tree may be used to simply assign a cluster to new observations. The algorithm is parallelizable, thus usable for large data sets.

A clustering tree is obtained in three stages. In the growing stage the sample is split recursively into two subsamples reducing the heterogeneity of the data within the new subsamples according to an heterogeneity measure. In the second and third stages the maximal tree obtained from the growing stage is pruned using two different criteria, one for the sibling nodes and the other one for all the terminal nodes.

In recent works, an extension of CUBT was proposed for both ordinal and nominal data [9], which uses the Shannon entropy and mutual information. Comparisons with other classical methods using several data simulation models and real data applications showed that CUBT outperforms most of the other methods, especially in terms of prediction. Some heuristics were proposed for a fine tuning of the parameters used in the method.

Notations

Let $X \in E = \prod_{j=1}^{p} supp(j)$, be a random p-dimensional vector with coordinates $X_{.j}$, $j \in \{1, ..., p\}$, and $supp(j)$ is the support of the jth variable, i.e the set of values it can take. We have a set S of n random observations of X, denoted as X_i with $i \in \{1, ..., n\}$ and X_{ij} is the ith observation of the jth component of X. Similar notations are used with small letters to denote the realizations of these variables: x, x_i, $x_{.j}$ and x_{ij}.

Heterogeneity Criteria

For any node t (a subset of S), let n_t be the number of observations in t, let $X^{(t)}$ be the restriction of X to node t, i.e $X^{(t)} = \{X | X \in t\}$. We define $R(t)$, the heterogeneity measure of t for continuous data as follows:

$$
\begin{aligned}
R(t) &= \frac{n_t}{n}\, trace(cov(X^{(t)})) \\
&= \frac{\sum_{X_i \in t} \| X_i - \bar{X}_t \|^2}{n}
\end{aligned}
$$

where $cov(X^{(t)})$ is the covariance matrix of $X^{(t)}$ and $\bar{X}_t = \frac{\sum_{X_i \in t} X_i}{n_t}$.
For nominal data, the heterogeneity measure is defined as follows:

$$
\begin{aligned}
R(t) &= trace(\mathbf{MI}(X^{(t)})) \\
&= -\sum_{j=1}^{p} \sum_{k \in supp(j)} p_{kj}^{(t)} \log_2 p_{kj}^{(t)}
\end{aligned}
$$

where $\mathbf{MI}(X^{(t)})$ is the mutual information matrix of $X^{(t)}$ and $p_{kj}^{(t)}$ is the probability for the component j of X to take value k within node t.

Growing Stage

Initially, the root node of the tree contains all the observations of S. The sample is split recursively into two disjoint subsamples using binary splits of the form $x_{.j} \in \mathcal{A}_j$, where $j \in \{1, ..., p\}$ and $\mathcal{A}_j$ is a subset of $supp(j)$. For continuous data, $\mathcal{A}_j$ will be a real interval of the form $\mathcal{A}_j =]\inf(supp(j)); a_j]$, with $a_j \in supp(j)$.

Thus, a split of a node t into two subnodes t_l and t_r is defined by a pair $(j, \mathcal{A}_j)$ as follows:

$$t_l = \{x \in E : x_{.j} \in \mathcal{A}_j\} \text{ and } t_r = \{x \in E : x_{.j} \notin \mathcal{A}_j\}$$

The best split of t into two sibling nodes t_l and t_r is defined by:

$$argmax_{(j,\mathcal{A}_j)}\{\Delta(t, j, \mathcal{A}_j)\}$$
$$\text{with } \Delta(t, j, \mathcal{A}_j) = R(t) - R(t_l) - R(t_r)$$

The new subnodes are recursively split, and the growing process may stop when at least one of the two following criteria is satisfied:

$$n_t < minsize \text{ or } \Delta(t, j, \mathcal{A}_j) < mindev \times R(S)$$

where *minsize* and *mindev* are fixed thresholds, and $R(S)$ is the deviance of the entire sample. Once the algorithm stops, a class label is assigned to each leaf of the maximal tree. A partition of the initial dataset is obtained, and each leaf from the tree corresponds to a cluster.

Details of both pruning stages of CUBT are described in previous works [7, 9].

4 Assessing variable importance in CUBT

To define variable importance in CUBT we follow similar ideas used in CART. For that we first define the surrogate splits for each variable within each node of a tree.

Surrogate splits

Let s be the best split of a node t in the tree T, based on variable j_0 which splits t into t_L and t_R. Let $s_j = s_j(t)$ be any split of t using any variable $j \neq j_0$ splitting t into t'_L and t'_R.

The probability for an observation to be sent to the left node for both splits is defined as follows:

$$p(t_L \cap t'_L) = \frac{Card\{t_L \cap t'_L\}}{n_t}$$

Then, the probability that both splits send an observation to the left node is:

$$p_{LL}(s, s_j) = \frac{p(t_L \cap t'_L)}{p(t)}$$

where $p(t)$ can be estimated by $\frac{n_t}{n}$, $p_{RR}(s, s_j)$ is defined equivalently.

The probability that s_j predicts well s is:

$$p(s, s_j) = p_{LL}(s, s_j) + p_{RR}(s, s_j)$$

The *surrogate split* for s over variable j in node t, denoted $\tilde{s_j}(t)$ is defined by:

$$p(s, \tilde{s_j}(t)) = \max_{s_j \in S_j} p(s, s_j)$$

where S_j is the set of all splits over variable j.

Surrogate splits are used to compute variable importance. They may also be used when predicting new observations with missing data.

Variable importance

We define the importance of variable j as follows:

$$Imp(X_{.j}) = \sum_{t \in T} \Delta(t, \tilde{s_j}(t))$$

The score of importance is the total loss of deviance induced if each split in the tree T is replaced by the surrogate split over $X_{.j}$.

5 Experiments

To test the efficiency of variable importance scoring based on CUBT we use some data simulation models where the important variables defining the clusters are known. We choose the same simulation models previously defined in [7, 9]. For each of these models we add irrelevant variables using this configuration: $p' = p$ irrelevant variables are added.

Simulation models

We consider nine data simulation models. Four models are designed for generating continuous data while the five remaining are designed for generating nominal data.

M1: 2D-model In this model, we fix $k = 4$ and $X \in \mathbb{R}^2$ following a multivariate normal distribution $N(\mu_l, \Sigma)$, $l \in \{1, ..., k\}$, where $\mu_1 = (-1, 0)$, $\mu_2 = (1, 0)$, $\mu_3 = (0, -1)$ and $\mu_4 = (0, 1)$, and the covariance matrix $\Sigma = diag(\sigma^2 \mathbf{1})$, with $\sigma = 0.1$.

M2: 5D-model This model generates $k = 10$ clusters of observations in $\mathbb{R}^5$, having different multivariate normal distributions $N(\mu_l, \Sigma)$, $l \in 1 : k$, with $\mu_1 = (1, 0, 0, 0, 0)$, $\mu_2 = (0, 1, 0, 0, 0)$, $\mu_3 = (0, 0, 1, 0, 0)$, $\mu_4 = (0, 0, 0, 1, 0)$, $\mu_5 = (0, 0, 0, 0, 1)$ and $\mu_i = -\mu_{i-5}$ for $i \in \{6, ..., 10\}$. The covariance matrix is $\Sigma = diag(\sigma^2 \mathbf{1})$, with $\sigma = 0.1$.

M3: Cocentric cluster Model In this model, two cocentric clusters in $\mathbb{R}^2$; each cluster has observations distributed uniformly between two cocentric circles centered in $(0, 0)$. The first cluster is delimited by circles having radius between 50 and 80, the second by circles of radius from 200 to 230.

M4: High-dimensional Model In this model, three clusters in $\mathbb{R}^{50}$ normally distributed $\mu_1 = (-1, ..., -1)$, $\mu_2 = (0, ..., 0)$ and $\mu_3 = (1, ..., 1)$, and the covariance matrix is $\Sigma = diag(\sigma^2 \mathbf{1})$, with $\sigma = 0.01$.

M5: Linear combination (LC) Model In this model, each variable $X_{.j}$, $j \in \{1, ..., p = 9\}$ has $m = 5$ levels. We define $k = 3$ clusters, each characterized by a high frequency of one level. For observations from cluster 1, $P(X_{.j} = 1) = q$, and a uniform probability is used for the other levels i.e, $P(X_{.j} = l) = \frac{1-q}{m-1}$ for $l \neq 1$. For clusters 2 and 3, the frequent levels are 3 and 5, respectively, using the same probabilities. We fix $q = 0.8$.

M6: Tree Model 1 We use here a tree structure model. We fix the dimension $p = 3$ and the number of groups $k = 4$. Each variable $X_{.j}$, $j \in \{1, ..., p\}$, has $m = 4$ levels, $X_{.j} \in \{1, 2, 3, 4\}$. Clusters are defined as follows:

- $C1$: x_1 and x_2 have odd levels, and x_3 is arbitrary
- $C2$: x_1 has odd levels, x_2 has even levels, and x_3 is arbitrary

- $C3$: x_1 has even levels, x_3 has odd levels, and x_2 is arbitrary
- $C4$: x_1 and x_3 have even levels, and x_2 is arbitrary

This model produces clusters that are expected to be easily found by CUBT, as each cluster is characterized by some levels of each variable. However, as these levels are uniformly distributed, their contribution to the entropy is high, making the optimal splits difficult to retrieve.

M7: Tree Model 2 We use the same tree structure model. As in the previous case, we fix $p = 3$ and the number of groups $k = 4$. Here, each variable $X_{.j}$, $j \in \{1, ..., p\}$, has $m = 4$ levels. The only difference is that variable levels are not uniformly distributed in each cluster. Here, we consider a parameter p_0 that controls the non-uniformity of the distribution of levels. In our experiments, we fix $p_0 = 0.8$. Clusters are defined as follows:

- $C1$: x_1 and x_2 have odd levels with $P(x_1 = 1) = P(x_2 = 1) = p_0$, and x_3 is arbitrary
- $C2$: x_1 has odd levels, x_2 has even levels with $P(x_1 = 1) = P(x_2 = 2) = p_0$, and x_3 is arbitrary
- $C3$: x_1 has even levels, x_3 has odd levels with $P(x_1 = 2) = P(x_3 = 1) = p_0$, and x_2 is arbitrary
- $C4$: x_1 and x_3 have even levels with $P(x_1 = 2) = P(x_3 = 2) = p_0$, and x_2 is arbitrary

M8: Nominal IRT-based model We use here an item response theory (IRT) model designed for nominal data. We fix the dimension $p = 9$ and the number of groups $k = 3$. Each variable has $m_j = 5$ levels. We now suppose that variables are representing multiple-choice items. The nominal response model [2] (NRM) can address nominal data. It is a specialization of the general model for multinomial response relations and is defined as follows:

Let θ be a level of latent ability underlying the response to the items. The probability that a subject of ability θ responds category k for item j is given by

$$\Psi_{jk_j}(\theta) = \frac{\exp[z_{jk_j}(\theta)]}{\sum_{h=1}^{m_j} \exp(z_{jh}(\theta))}$$

where $z_{jh}(\theta) = c_{jh} + a_{jh}\theta$ with $h = 1, 2, ..., k_j, ..., m_j$, θ is a latent trait, and c_{jh} and a_{jh} are item parameters associated with the h-th category of item j. We generate random datasets using the NRM by simulating latent trait values for the four groups. For $c \in \{1, 2, 3\}$, we simulate a vector of latent trait values for each group c using $N(\mu_c, \sigma^2)$, $\mu = (-3, -1, 1, 3)$ and $\sigma^2 = 0.2$. For $j \in \{1, ..., p\}$, the values of c_{jh} range uniformly between -2 and 2 while c_{jh} are distributed as $N(1, 0.1)$. Simulations are performed using the *NRM.sim* function of the *mcIRT* package [16] with **R**.

M9: IRT-based Model We use again IRT models. These models allow us to assess the probability of observing a level for each variable given a latent trait level. The latent trait is an unobservable continuous variable that defines the individual's ability, measured by the observed variables. In the IRT framework, the variables called items are ordinal. The observations can be either binary or polytomous. Here, we introduce a polytomous IRT model to generate data in a probabilistic way. The generalized partial credit model [14] (GPCM) is an IRT model that can address ordinal data. It is an extension of the 2-parameter logistic model for dichotomous data. The model is defined as follows:

$$p_{jx}(\theta) = P(X_{ij} = x|\theta) = \frac{\exp \sum_{k=0}^{x} \alpha_j(\theta_i - \beta_{jk})}{\sum_{r=0}^{m_j} \exp \sum_{k=0}^{r} \alpha_j(\theta_i - \beta_{jk})}$$

where θ is the latent trait and θ_i represents the latent trait level of individual i. β_{jk} is a difficulty threshold parameter for the category k of the item j. For $j \in \{1, ..., p\}$, β_j is a vector of dimension $m - 1$. α_j is a discrimination parameter represented by a scalar. We generate random datasets using the GPCM by simulating latent trait values for the three groups. For $c \in \{1, 2, 3\}$, we simulate a vector of latent trait values for each class c using $N(\mu_c, \sigma^2)$, $\mu = (-3, 0, 3)$ and $\sigma^2 = 0.2$. For $j \in \{1, ..., p\}$, α_j is distributed as $N(1, 0.1)$, and β_j is a vector of ordered values that range uniformly between -2 and 2. Simulations are performed using the *rmvordlogis* function of the *ltm* package [17] with **R**.

Adding noise to the simulated datasets

We propose here a way to add some noise variables to our different datasets generated by models presented above. Let p be the initial number of variables for each model. We define p' the number of noise variables, we fix $p' = p$.

For continuous models M1 to M4, we simulate $\frac{p'}{2}$ variables following the normal distribution $N(0, \frac{\sigma_0}{2})$, where $\sigma_0 = \min_{j\in\{1,\dots,p\}} \sigma_j$ is the minimum standard deviation observed in the initial set of important variables. The remaining $\frac{p'}{2}$ variables follow the uniform distribution $U(-\sigma_0\sqrt{\frac{3}{4}}, \sigma_0\sqrt{\frac{3}{4}})$.

For nominal simulation models M5 to M9, for each of the p relevant variables in the model, we generate r corresponding irrelevant variables as follows: if the original variable has m levels, its corresponding noise variable has the distribution $(p_1, ..., p_m)$ over the same support where $p_l = 0.8$ and $p_j = \frac{0.2}{m-1}$ for $j \neq l$ and level l is chosen arbitrarily.

Results

For each simulation model, we vary the sample size using $n = 100$, 300 and 500, and we run 100 replicates computing importance scores from the following methods: CUBT, URF, LOVO-cubt, LOVO-km, TWKM and LS. For both continuous and nominal models, clusters are equally sized. For CUBT, the score of variable importance is computed from the optimal clustering tree obtained after both the pruning stages of the method.

To assess the efficiency of each scoring method, we compute the proportion of true important variables appearing in top p highest score variables. This index is similar to a true positive rate (TPR). Highest values correspond to a highest ability to detect important variables. When it is far from 100% it may be interesting to see how are scored the "undetected" true important variables. We report also the highest rank (HR) among the ranks of the p true important variables. These two indices are averaged over the 100 replicates.

Table 5 provides the results for all the models and all the methods. It gives the TPR together with the HR (between parentheses). For continuous data (models M1 to M4), LOVO-km uses k-means, and for the other models it uses k-modes.

In terms of top p ranking, CUBT, with a TPR greater than 90%, is always one of the two best performing methods, for all models, for all sample sizes. For continuous models (M1 to M4), URF shows poor results, that are worse when increasing the sample size. This is not the case for TWKM and LS, whose results remain stable or are better when increasing this parameter, but they are still unsatisfactory. CUBT is often placed equal with both LOVO approaches in model M1 (and model 4 for LOVO-cubt) and with TWKM in model M4 with $n = 300$. For nominal models (M5 to M9), CUBT is often placed equal with URF. The two methods have a TPR greater than 90%, and outperform the LOVO approaches. The results according to the HR are complementary with those already exposed (higher proportions induce lower ranks).

6 Conclusion

We have presented a new method to assess variable importance within Clustering using binary decision trees. We have compared this approach to other classical scoring methods over several data simulation models, in presence of noise. CUBT Variable Importance score was the best performing method in these experiments. More experiments may be undertaken to analyze the efficiency of the presented scores with respect to correlation or redundancy between the variables.

Model	n	CUBT	URF	LOVO-cubt	LOVO-km	TWKM	LS
M1	100	**100(2)**	28(4)	**100(2)**	**100(2)**	23(4)	0(4)
$p = 2$	300	**100(2)**	19(4)	**100(2)**	**100(2)**	56(3)	0(4)
	500	**100(2)**	14(4)	**100(2)**	**100(2)**	56(3)	0(4)
M2	100	**97(5)**	57(8)	89(6)	94(6)	35(9)	20(9)
$p = 5$	300	**100(5)**	58(8)	86(6)	96(6)	72(8)	20(9)
	500	**100(5)**	57(8)	85(6)	95(6)	90(6)	20(9)
M3	100	**98(2)**	34(4)	40(3)	78(3)	50(3)	58(3)
$p = 2$	300	**100(2)**	13(4)	39(3)	74(3)	50(3)	60(3)
	500	**100(2)**	4(4)	86(2)	75(3)	50(4)	50(3)
M4	100	**100(50)**	1(100)	**100(50)**	71(98)	69(70)	0(100)
$p = 50$	300	**100(50)**	0(100)	**100(50)**	70(98)	**100(50)**	0(100)
	500	**100(50)**	0(100)	**100(50)**	74(98)	81(60)	0(100)
M5	100	**100(9)**	98(9)	56(16)	84(14)	-	-
$p = 9$	300	**100(9)**	**100(9)**	44(18)	89(13)	-	-
	500	**100(9)**	**100(9)**	49(17)	82(13)	-	-
M6	100	**100(3)**	**100(3)**	33(6)	60(4)	-	-
$p = 3$	300	**100(3)**	**100(3)**	27(6)	20(6)	-	-
	500	**100(3)**	**100(3)**	27(6)	20(6)	-	-
M7	100	**100(3)**	**100(3)**	20(6)	53(5)	-	-
$p = 3$	300	**100(3)**	**100(3)**	60(5)	67(4)	-	-
	500	**100(3)**	**100(3)**	60(5)	47(6)	-	-
M8	100	91(11)	**98(9)**	62(14)	47(17)	-	-
$p = 9$	300	**98(9)**	**98(9)**	56(18)	44(18)	-	-
	500	96(10)	**100(9)**	47(17)	49(16)	-	-
M9	100	**100(9)**	96(9)	53(14)	67(14)	-	-
$p = 9$	300	**100(9)**	98(9)	49(16)	58(17)	-	-
	500	**100(9)**	93(10)	56(16)	62(15)	-	-

Table 2. True positive rate (TPR) for each simulation model and each method. Values in parentheses represent the highest rank (HR) among the ranks of the p true important variables. Bold values correspond to the best performing method(s). p is the number of true important variables.

Future work will analyze the stability of this score with respect to the data and to the tuning parameters of CUBT. The use of our score for feature selection in clustering is also under study.

Bibliography

[1] Belkin, M. and Niyogi, P. (2001) *Laplacian Eigenmaps and Spectral Techniques for Embedding and Clustering.* Advances in Neural Information Processing Systems 14, 585–591.

[2] Bock, R.D. (1972) *Estimating item parameters and latent ability when responses are scored in two or more nominal categories.* Psychometrika, **37**, 29–51.

[3] Breiman, L., Friedman, J., Olshen, R. and Stone, C. (1984) *Classification And Regression Trees.* Wadsworth and Brooks.

[4] Breiman, L. (2001) *Looking inside the black box.* Wald Lecture 2, Berkeley University.

[5] Breiman, L. (2001) *Random forests.* Machine Learning, **45**, 1, 5–32.

[6] Chen, X., Xu X., Huang J.Z. and Ye Y. (2013) *TW-k-means: Automated Two-Level Variable Weighting Clustering Algorithm for Multiview Data.* IEEE Transactions on Knowledge and Data Engineering, **25**, 4, 932–944.

[7] Fraiman, R., Ghattas, B. and Svarc, M. (2013) *Interpretable clustering using unsupervised binary trees.* Advances in data analysis and classification, **7**, 125–145.

[8] Ghattas, B., Svarc, M., Fraiman, R. and Michel, P. (2016) ***R**-package for interpretable clustering using binary trees.* version 3.0, http://lumimath.univ-mrs.fr/ ghattas/CUBT.html.

[9] Ghattas, B., Michel, P. and Boyer, L. (2016) *Clustering nominal data using Unsupervised Binary decision Trees: Comparisons with the state of the art methods.* Pattern Recognition. (in revision).

[10] Guyon, I., Weston, J., Barnhill, S. and Vapnik, V.N. (2002) *Gene selection for cancer classification using support vector machines.* Machine Learning, **46**, (1–3), 389–422.

[11] Kaufman, L. and Rousseeuw, P.J. (1987) *Clustering by means of Medoids.*, in Statistical Data Analysis Based on the L_1Norm and Related Methods, edited by Y. Dodge, North-Holland, 405416.

[12] Liu, H., and Yu, L. (2005) *Toward integrating feature selection algorithms for classifcation and clustering.* IEEE TKDE, 17, 491-502.

[13] MacQueen, J. (1967) *Some methods for classication and analysis of multivariate observations.* Proceedings of the Fifth Berkeley Symposium on Mathematical Statistics and Probability, Editions L. M. Le Cam & J. Neyman, 1, 281297.

[14] Muraki, E. (1992) *A generalized partial credit model: application of an EM algorithm.* Applied Psychological Measurement, **16**, 159–176.

[15] Rakotomamonjy, A. (2003) *Variable selection using SVM-based criteria.* Journal of Machine Learning Research. **3**, 1357–1370.

[16] Reif, M. (2014) *mcIRT: IRT models for multiple choice items.* R package version 0.41. https://github.com/manuelreif/mcIRT.

[17] Rizopoulos, D. (2006) *ltm: An R package for Latent Variable Modelling and Item Response Theory Analyses* Journal of Statistical Software, **17**, 5, 1–25. http://www.jstatsoft.org/v17/i05/.

[18] Zhu, L., Miao, L., Zhang, D. (2012) *Iterative Laplacian Score for Feature Selection.* in Pattern Recognition. Volume 321 of the series Communications in Computer and Information Science, 80–87.

Time series changes of the categorical data using the text data regarding radiation

Takafumi Kubota, *Tama University*, kubota@tama.ac.jp
Hitoshi Fujimiya, *DYNACOM Co.,Ltd.*, h.fujimiya@dynacom.co.jp
Hiroyuki A. Torii, *University of Tokyo*, torii@radphys4.c.u-tokyo.ac.jp

Abstract. This paper is a report about the research which was gathered on how the classified class of the questions keywords included in the text-based is changing in terms of time. As for the research background, after the accident of TEPCO's Fukushima Daiichi Nuclear Power Station in March 2011, a lot of information about the keyword relating to the radiation has been taken up in the media, and a group of people who receives the media has been becoming uneasy and dissatisfied about the information. To solve this issue, the Japan Health Physics Society started a website of questions and answers which related radiation (radiation QA). Within this, the experts have been answering towards the questions that the people asked (mainly the people in metropolitan area and the Fukushima prefecture). The goal for this research is to hold a comparative verification by checking against one another of the various happenings that occurred after the incident with the related keywords of radiation, and also the groups the keyword belongs to, of time alongside with how the transitions were made, using the text data which is open to public on the radiation QA.

Keywords. Classification,Text mining,Time series

1 Introduction

Recently, the applied study on text data is carried out in various fields. This study focused on the text data which was related radiation. We tried to find out how the changes of the results of text mining are. In this paper we target the text data of web site of questions and answers which related radiation (radiation QA).

As for the research background, after the accident of TEPCO's Fukushima Daiichi Nuclear Power Station in March 2011, a lot of information about the keyword relating to the radiation has been taken up in the media, and a group of people who receives the media has been becoming uneasy and dissatisfied about the information.

To solve this issue, the Japan Health Physics Society (JHPS) [1] started a website of questions and answers which related radiation. Within this, the experts have been answering towards the questions that the people asked (mainly the people in metropolitan area and the Fukushima prefecture).

Furthermore, to apply for visualization for the result of text data, there are some previous studies. For example, for the application of association rules [2] used a three-dimensional display with interactions. However, it has not been fully discussed for the interactions for time series changes. Therefore, we tried the application using the shiny packages of R to apply the results of text mining and check the time series changes of the results.

2 Data and keywords of groups

Data

Radiation QA has 1870 records that include radiation related questions from private citizens and answers of these questions from expert. The records also include variables; published date of asked questions from March of 2011 to March of 2013, types of questions and characteristics of questioner such as address, age groups, occupations and sex. In this study we only used the date and the text of questions to text mining. For example of question there is a text as follows:

"Please tell me about the prevention of the radiation exposure."

Keywords of groups

We set 93 keywords and divide into four groups; Administration, Physics, Medical care and Health and Environment. The groups of the keywords are as follows

- Administration (Gyosei in Japanese) (26 words):
  - Evacuation, conduct, caution, plan, instructions, warning, message, guidance, decontamination, temporary housing, regulation (value), standard (value), tentativeness, inspection, Ministry of Health, Labor and Welfare, Ministry of Economy, Trade and Industry, Ministry of Agriculture, Forestry and Fisheries, administration, prefectures, cabinet, Ministry of Internal Affairs and Communications, the Fire and Disaster Management Agency, Center(a principal place).
- Physics (Butsuri in Japanese)(14 words):
  - Radiation (quantity), Sievert (Sv), Becquerel (Bq), estimate, collapse, the measurement (quantity of measurement, measuring instrument), iodine, cesium, gamma ray, density, nuclide, effective, standard (value), material
- Medical care, Health (Iryo in Japanese)(30 words):
  - Thing (something to eat, food), internal exposure, eating and drinking, pregnancy, pregnant woman, sterility, health damage, disease, treatment, medicine, water, marine products, mushroom, edible wild plant, milk, water purifier, thyroid gland, examination, fish, nursery school, school, mother (child), the toxicity, bath, vegetables, faucet, sight, hearing, the sense of touch, taste, sense of smell
- Environment (Kankyo in Japanese)(23 words):
  - Water, timber, tree (wood), the outskirts, the outdoors, wall (surface), the sea, the forest, domestic animals (cow, pig), wild bird, car, the atmosphere, rain, wind, dust, mud, acorn, camping, the soil, composure, pesticide, air, radon

3 Methods

Text mining

In Japanese text words are not separated by blanks, therefore we have to split words from text data. As a method of the analysis, we used the Text Mining Studio of the NTT DATA Mathematical Systems Inc. We used all text data to summarize monthly by all words not only the keywords that we set in section two but also other every words included in all text.

Figure 1 shows area chart that indicate monthly frequencies of top ten appearing words; radioactive, material, influence, radiation, value, dose, think, amount, cesium and high. In the figure there are two peaks; from April to June of 2011 and from November of 2011 to January of 2012. Frequent words of these peaks are radioactive, material, cesium and high.

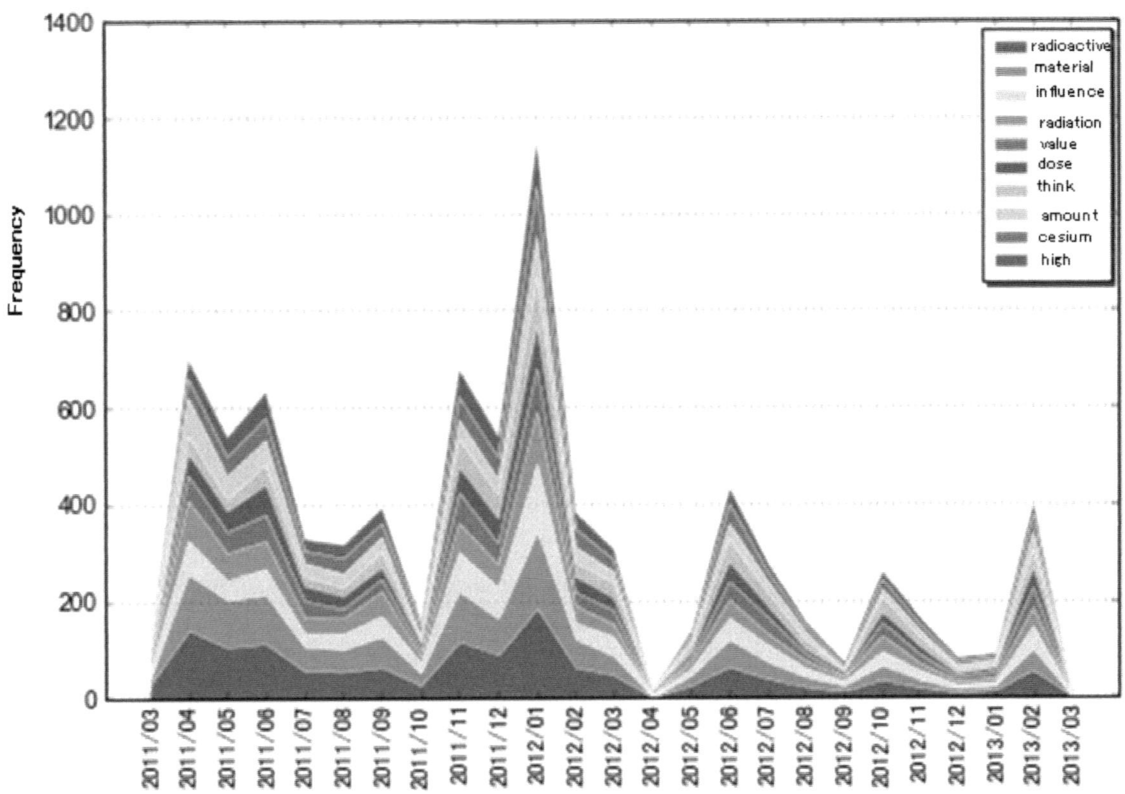

Figure 1. Frequencies of top ten appearing words; radioactive, material, influence, radiation, value, dose, think, amount, cesium and high

Figure 2 shows area chart that indicate degree of increases of top 20 appearing words; reference 2, reference 1, 2011, state, new ref. value, H. 24, order, food, ingredient, report, receive, 100Bq/kg, Tochigi Pref., hide, gene, 3 times, 1 Bq dose, boundary, ref. 1 and rare. In figure 2 there are three peaks; from December of 2011 to February of 2012, from June to August of 2012 and from October to November of 2012. The increased words in the first peaks are meal, food and ingredient that correspond to concerns of food. While the increased words in the second peaks are 100 Bq/kg and boundary that correspond to the period when new standard of dose of radiation were started. The third peaks are not important because these frequencies are not so high.

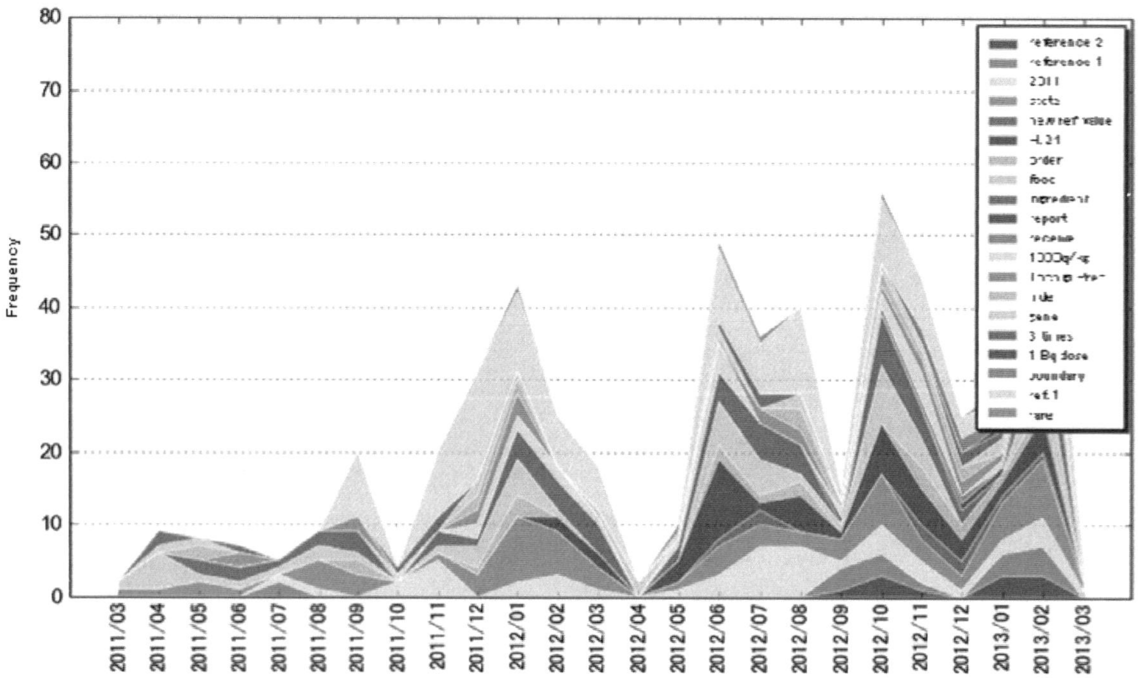

Figure 2. Degree of increases of top 20 appearing words

Shiny application

We also used the text data of radiation QA to restructure the data to text-word matrix with elements a_{dmwg} where if the text of day d of month m have a word w of group g then $a_{dmwg} = 1$, else $a_{dmwg} = 0$. Then we calculate the mean values within groups of keywords as follows.

$$b_{dmg} = \frac{1}{n_g} \sum_{w=1}^{n_g} a_{dmwg}$$

where n_g is the number of keywords in group g.

Then we calculate the mean values of months as follows.

$$c_{mg} = \frac{1}{n_m} \sum_{d=1}^{n_m} b_{dmg}$$

where n_m is the number of days in month m.

For the total frequencies of all groups, we also calculate count values of a_{dmwg} as b'_{dmg} and c'_{mg} as follows.

$$b'_{dmg} = \sum_{w=1}^{n_g} a_{dmwg}$$

Then we calculate the count values of months as follows.

$$c'_{mg} = \sum_{d=1}^{n_m} b_{dmg}$$

To visualise c_{mg} and c'_{mg}, we draw line chart of $c_{mg}(g = 1, 2, 3, 4)$ and bar chart of c'_{mg} for the interest group g.

To add interactive control we apply R package *shiny* for the graphs. Figure 3 shows the application of the graphs. The top figure shows four line charts which corresponds to four groups. The bottom figure shows bar chart of which user select the group from top left select menu.

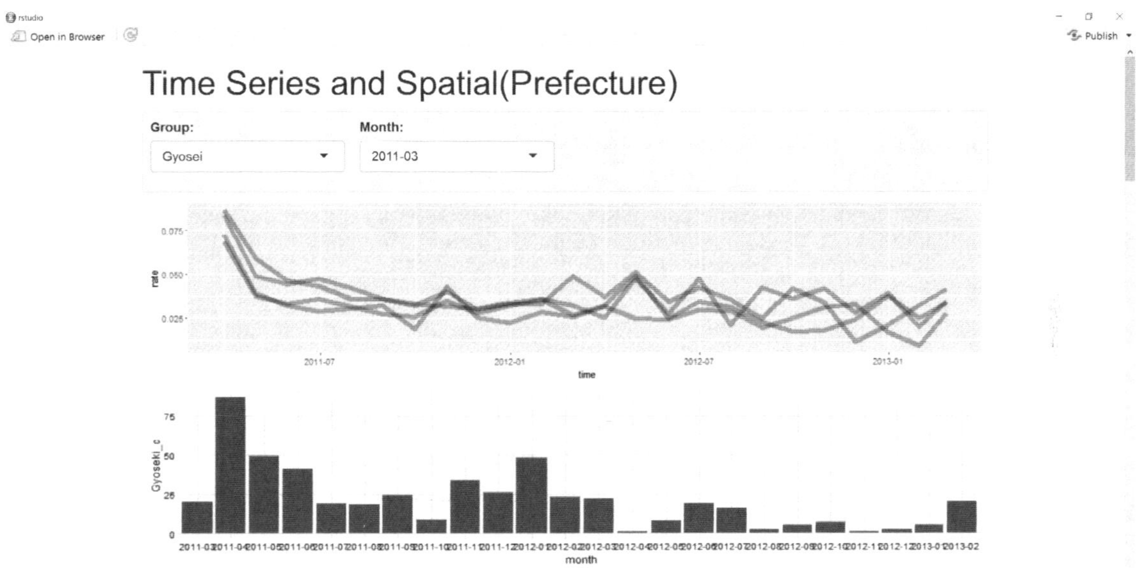

Figure 3. Shiny application of the results of text mining

To detect spatial transition, we also add the variable of address (prefecture) where the question were asked, and then we also apply shiny applications as well as the line charts and bar chart.

Figure 4 shows application window which user are selecting month from menu. Figure 5 corresponds choropleth map of selected month with selected group from the menu of figure 4. In figure 5 there is also the table of the mean rate value of every month with every group.

4 Concluding remarks and future studies

Looking over the monthly results that were analyzed for every group of the keyword for the first 3 months after the radiation QA was explained, every groups questions and keyword frequency was high. The groups frequency of the keyword became low in the next three months. However, half a year later in the environmental group, a year later in the administrative group and also the physics group had spiked (the part that frequency increased rapidly), and the frequency of the health group rose at the end of the period.

We also applied shiny package to interactive control to graphs. This is for the user especially an ordinal person who interest in the result of text ming.

JHPS tweeted in twitter at the same time of replying the questions, and many people reacted to the tweets. As for the future studies, the tweets will be retrieved, and we are planning to analyze about the

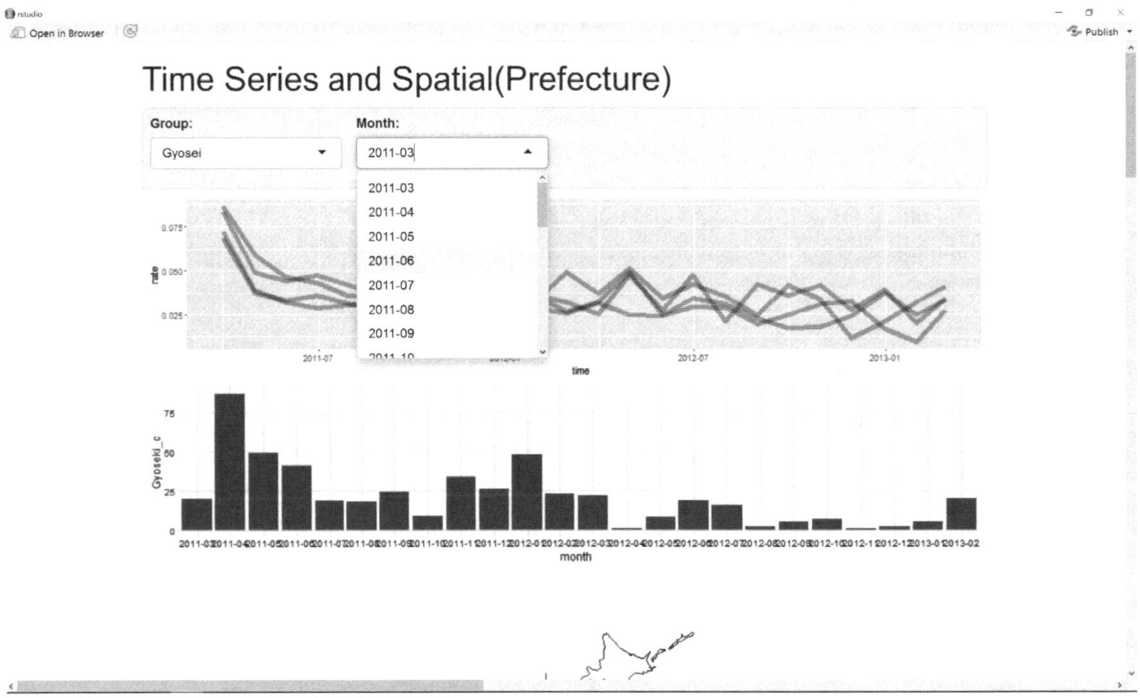

Figure 4. Shiny application which user are selecting Month from menu

expanse including the number of retweeting and also to examine the connection with the text data of radiation QA.

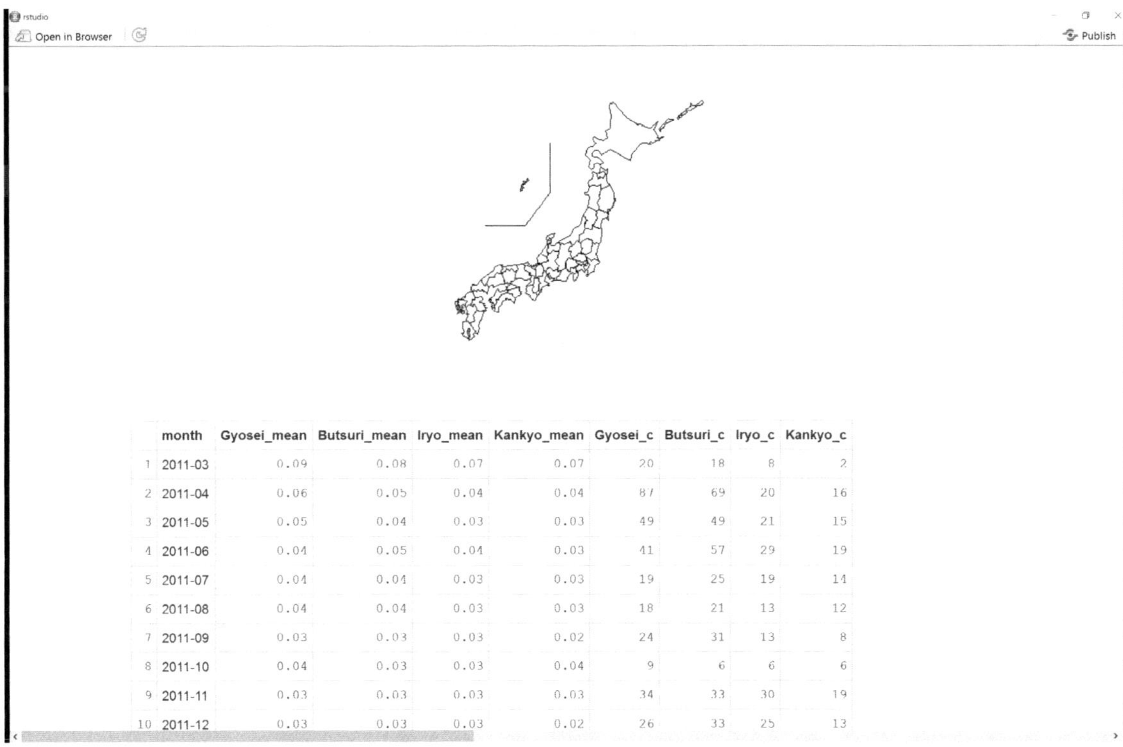

	month	Gyosei_mean	Butsuri_mean	Iryo_mean	Kankyo_mean	Gyosei_c	Butsuri_c	Iryo_c	Kankyo_c
1	2011-03	0.09	0.08	0.07	0.07	20	18	8	2
2	2011-04	0.06	0.05	0.04	0.04	87	69	20	16
3	2011-05	0.05	0.04	0.03	0.03	49	49	21	15
4	2011-06	0.04	0.05	0.04	0.03	41	57	29	19
5	2011-07	0.04	0.04	0.03	0.03	19	25	19	14
6	2011-08	0.04	0.04	0.03	0.03	18	21	13	12
7	2011-09	0.03	0.03	0.03	0.02	24	31	13	8
8	2011-10	0.04	0.03	0.03	0.04	9	6	6	6
9	2011-11	0.03	0.03	0.03	0.03	34	33	30	19
10	2011-12	0.03	0.03	0.03	0.02	26	33	25	13

Figure 5. The results of choropleth map and table

Bibliography

[1] The Japan Health Physics Society (2014), *Expert's Answers: Living of radiation Q and A* (In Japanese), URL: *http://warp.da.ndl.go.jp/info:ndljp/pid/8699165/radi-info.com/* (2016-04-30)

[2] Pak Chung Wong, Paul Whitney, Jim Thomas (1999), *Visualizing Association Rules for Text Mining*,Proceedings of Information Visualization, 1999. (Info Vis '99).

A Multimomental ARMA model: initial formulation and a case study

Thomas Michael Bartlett, *School of Electrical and Computer Engineering, University of Campinas*, th.m.bartlett@gmail.com
Levy Boccato, *School of Electrical and Computer Engineering, University of Campinas*, lboccato@dca.fee.unicamp.br

Abstract. Recently, non-normal distribution functions have been developed to model more precisely and richly the behavior of time series of data. Here, we aim at developing a times series model that describes, by mixing Gaussian distributions, the evolution of each statistical moment up to the third order - mean, variance and skewness. To each instant of time and to each moment an ARMA law of evolution is applied and the estimation of the model is done by optimizing a quasi likelihood function that depends on the ARMA coefficients of the three moments. Thus, the method of moments for gaussian mixtures is used to obtain a probability density function which has the desired instantaneous moments. Employing Newton's Method and Nelder-Mead optimization procedures to estimate the model, an empirical analysis is done studying the model's performance and consistency using a synthetic time series.

Keywords. Time Series, Skewness, Method of Moments, Gaussian Mixture, GARCH

1 Introduction

The observation of a system evolving over time led to the development of various methods for modeling, analyzing and forecasting such set of samples, called a time series. One of the most widely used techniques exploits the natural intuition that the future can be explained based on the present value of the system and the values observed in the recent past. The formalization of this idea led to the Box-Jenkins method [4], which establishes the use of autoregressive linear models and moving average (ARMA) for modeling and prediction of the behavior of the time series associated with the system. Currently, this approach is used in many areas of science, such as economics, ecology, meteorology, speech recognition and tomography [21].

The classical linear models, viz., AR, MA and ARMA, are based on the assumption that the errors (ϵ_t) involved are i.i.d. and have a Gaussian distribution with zero mean and a constant variance. However, this hypothesis, concentrated only on the average value of the system, can be very restrictive and less suitable for modeling real time series, such as those related to financial assets [7]. Therefore, it was necessary to consider also the possibility of dynamical variations in the probability density function (PDF) underlying the observations. This perspective has led to more flexible models, such as the heteroskedastic models, in which the variance of the PDF that generates the sample changes with time (GARCH, [3]). More specifically, the GARCH [3] considers that the variance of the errors has a ARMA evolution itself, given by $\sigma_t^2 = \alpha_0 + \sum_{i=1}^{p} \omega_i \sigma_{t-i}^2 + \sum_{i=1}^{q} \gamma_i \epsilon_{t-i}^2$ and $\epsilon_t = \sigma_t z_t$, where $z_t \sim N(0,1)$.

Although the GARCH [3] model certainly represents an improvement towards a more adequate modelling of the considered random phenomenon, the use of additional higher-order moments can potentially bring additional flexibility to the statistical model. Thus, Harvey et al. [12] analyzed the performance of a model that not only considers an ARMA process for the variance, but also for the third moment of the distribution, the so-called skewness $E((x-\mu)^3)$. The approach of Brooks et al. [5] goes further and assumes that the fourth moment, i.e., the kurtosis $E((x-\mu)^4)$, is also time-variant and follows an ARMA model. Finally, Jondeau et al. [15] encompassed both ideas and considered an ARMA evolution for all moments up to the fourth order.

Another perspective that aims at making the GARCH model [3] more precise when dealing with non-normal time series consists in using mixtures of distributions. One example that exploits this idea is known as Finite Mixtures Markov Switching Model [9], [6]. Additionally, Haas [8, 1, 10] proposed a model in which the error PDF is generated by a mixture of normal distributions, and the variance of each distribution evolves according to a different ARMA process.

The strategy of modeling time series by resorting to higher-order moments and/or by using non-normal distributions is particularly relevant in the context of exchange rate finance [14, 13] and risk analysis VaR [16]. Within the framework of risk analysis, an accurate estimate of the underlying distribution time series is absolutely crucial, e.g., in calculating the maximum loss when one of the worst 5% cases occurs, that is, calculating the VaR_5, which emphasizes the benefits of models associated with more flexible PDFs, such as those based on mixtures of distributions [16].

This paper explores the idea of analyzing the samples of a time series as taken from a time-variant non-normal distribution. In particular, we propose to describe the evolution of the underlying PDF in terms of the variation of its first three moments, each of them being represented by an ARMA model, whose optimum parameters are obtained with the aid of a quasi-maximum likelihood estimator.

Additionally, the PDF of the samples shall be represented as a mixture of Gaussian functions, whose parameters – the individual means, variances and weights – are selected via the method of moments, assuring that the PDF attains the desired values for the first three moments.

In this work, we initially study the characteristics and the performance of the proposed model in the context of a case study involving a synthetic non-stationary time series. The obtained results, albeit preliminary, indicate the flexibility of the model as well as motivate further investigations concerning the approximation capability of the model and/or extensions to the proposed method.

This paper is organized as follows. Section 2 describes the proposed method, indicating how to select the parameters of the mixture of distributions (Gaussians) in order that the first three moments present specific values. Then, in section 2, based on an approach of quasi-maximum likelihood, we present a strategy to estimate the optimum parameters of the ARMA model associated with each moment. Then, in Section 3, the proposed model is analyzed in the context of a synthetic non-stationary time series and some preliminary observations are presented. Finally, Section 6 concludes the paper and provides perspectives for further research.

2 Proposed Model

In this work, our aim is to conceive a mathematical model that may be able to follow the variation of the probability distribution associated with the observations of a random dynamical system. In this sense, we propose to extend the works of Harvey, Brooks and Jondeau [12, 5, 15] by defining a general model that explicitly considers the dynamic variations of higher-order moments of the error distribution.

The proposed model represents the time evolution of each statistical moment according to an ARMA process and describes the underlying PDF of the data observed as a mixture of Gaussians, whose parameters – mean, variance and weights – vary with the time and are obtained so that the moments of the actual PDF reach the exact values generated by the ARMA process. In the following sections, we present the main aspects of the proposed model, called Multimomental ARMA model. Our initial effort shall consider only the variation of the moments up to the third order, but we intend to extend this method for more higher-order moments in future works.

Method of Moments Applied to Gaussian Mixtures

In order to explore the information present in the statistical moments up to the third order, it is necessary to deal with a class of distributions whose mathematical expression of the PDF takes into account the knowledge of such moments. In this work, we shall use the Gaussian mixture model to represent the PDF and employ the approach based on moments [18] to obtain the parameters of the mixture. According to this method, if it is established a set of centralized moments $m_1,\ m_2,\ m_3$ of a distribution F, then it is possible to calculate the values of $\pi_1,\ \mu_1,\ \mu_2$ of the mixture such that

$$\begin{aligned}
&F(x) = \pi_1 G(x, \mu_1, \sigma) + (1-\pi_1) G(x, \mu_2, \sigma)\\
&G(x, \mu, \sigma) = \frac{1}{\sqrt{2\pi\sigma}} \exp\left(\frac{-(x-\mu)^2}{2\sigma^2}\right)\\
&E_F(X^1) = m_1\\
&E_F((X-m_1)^2) = m_2\\
&E_F((X-m_1)^3) = m_3
\end{aligned}$$

Making use of the definition of $F(x)$, we determine the desired moments from those of the mixture distribution:

$$\begin{aligned}
m_1 &= \pi_1\mu_1 + (1-\pi_1)\mu_2\\
m_2 &= \pi_1\mu_1^2 + (1-\pi_1)\mu_2^2 - m_1^2 + \sigma^2\\
m_3 &= \pi_1\mu_1^3 + (1-\pi_1)\mu_2^3 - 3m_1m_2 - m_1^3 + 3\sigma^2 m_1
\end{aligned}$$

As demonstrated by [18], we can define three variables $Q_1,\ Q_2,\ Q_3$ that will help the deduction of parameters μ_i.

$$\begin{aligned}
Q_1 &= \pi_1\mu_1 + (1-\pi_1)\mu_2 = m_1\\
Q_2 &= \pi_1\mu_1^2 + (1-\pi_1)\mu_2^2 = m_2 + m_1^2 - \sigma^2\\
Q_3 &= \pi_1\mu_1^3 + (1-\pi_1)\mu_2^3 = m_3 + 3m_1m_2 + m_1^3 - 3\sigma^2 m_1
\end{aligned}$$

Then, as shown in [18], Equation (1) can be used to calculate the value of $\mu_i,\ i = 1,2$.

$$\begin{vmatrix} 1 & Q_1 & 1\\ Q_1 & Q_2 & \mu_i\\ Q_2 & Q_3 & \mu_i^2 \end{vmatrix} = 0 \iff (Q_2 - Q_1^2)\mu_i^2 + (Q_3 - Q_1Q_2)\mu_i + (Q_1Q_3 - Q_2^2) = 0 \tag{1}$$

Although it is not trivial to deduce this equation, we can check its validity by replacing the values of Q_i in (1), confirming that, in fact, solutions are $\mu_i,\ i = 1,2$. We also emphasize that the above equation has an exact solution by the formula of Bhaskara. However, to ensure the existence of two roots, we need that $(Q_2 - Q_1^2) > 0 \iff m_2 - \sigma^2 > 0$.

Therefore, we define a new fixed parameter $\lambda \in (0.1)$ such that $\sigma^2 = \lambda m_2$, thereby ensuring the aforementioned inequality. Hence, after calculating the values $\mu_i,\ i = 1,2$, we can determine the parameters π_i by the following expression:

$$\begin{bmatrix} 1 & 1\\ \mu_1 & \mu_2 \end{bmatrix} \begin{bmatrix} \pi_1\\ \pi_2 \end{bmatrix} = \begin{bmatrix} 1\\ Q_1 \end{bmatrix}$$

So, we obtained the set of PDF parameters, namely $\pi_1,\ \mu_1,\ \mu_2,\ \sigma$, which guarantee the equality between the moments of $F(x)$ and the desired values m_1, m_2 and m_3. Futhermore the mixture distribution with the parameters $(\pi_1, \mu_1, \mu_2, \sigma)$ is uniquely determined using the moments (m_1, m_2, m_3) (Theorem 2A(b) from [17]).

Multimomental ARMA Model

In this paper, the proposed model assumes that the system value x_t was generated by non-stationary distribution which is a mixture of gaussian distributions and depends on the time varying moments up to the third order.

$$x_t \sim F\left(\cdot; m_1(t), m_2(t), m_3(t)\right) = \pi_1(t)G(\cdot, \mu_0(t), \sigma(t)) + (1 - \pi_1(t))G(\cdot, \mu_1(t), \sigma(t)) \quad (2)$$

And to determine $\pi_1(t)$, $\mu_0(t)$, $\mu_1(t)$ and $\sigma(t)$ we use the method of moments described in the section 2 above. And although it is possible to adopt a nonlinear law to describe the evolution of the moments, let us assume, for simplicity, linearity of the evolution of each moment and error, as shown by the following equations:

$$\epsilon(t) = x_t - m_1(t) \quad (3)$$

$$m_1(t) = a_1 + \sum_{i=1}^{r_1} b_i^{(1)} m_1(t-i) + \sum_{j=1}^{s_1} c_j^{(1)} \epsilon(t-j) + \epsilon(t),\ m_1(t) = E(x_t|\mathcal{F}_{t-1}) \quad (4)$$

$$m_2(t) = a_2 + \sum_{i=1}^{r_2} b_i^{(2)} m_2(t-i) + \sum_{j=1}^{s_2} c_j^{(2)} \epsilon(t-j)^2,\ m_2(t) = E(\epsilon_t^2|\mathcal{F}_{t-1}) \quad (5)$$

$$m_3(t) = a_3 + \sum_{i=1}^{r_3} b_i^{(3)} m_3(t-i) + \sum_{j=1}^{s_3} c_j^{(3)} \epsilon(t-j)^3,\ m_3(t) = E(\epsilon_t^3|\mathcal{F}_{t-1}) \quad (6)$$

$$\theta = \left(a_1, b_1^{(1)}, ..., b_{r_1}^{(1)}, c_1^{(1)}, ..., c_{s_1}^{(1)}, a_2, b_1^{(2)}, ..., b_{r_2}^{(2)}, c_1^{(2)}, ..., c_{s_2}^{(2)}, a_3, b_1^{(3)}, ..., b_{r_3}^{(3)}, c_1^{(3)}, ..., c_{s_3}^{(3)}\right) \quad (7)$$

At each time $t = 1, \ldots, T$, the error of the model is represented by $\epsilon(t)$. For orders $h = 1, 2, 3$, the coefficient a_h is the constant parameter around which the h-order moment $m_h(t)$ oscilates. The parameters $b_1^{(h)}, \ldots, b_{r_h}^{(h)}$ are the ARMA parameters which relates the value of the moment of order h at time t, represented as $m_h(t)$, with its values $m_h(t-i)$ up to r_h instants in the past. Similarly, the parameters $c_1^{(h)}, \ldots, c_{s_h}^{(h)}$ determine how the h-moment at time t, $m_h(t)$, depends on the value of the error raised to the power h, $\epsilon(t-j)^h$, up to s_h instants in the past.

The estimation of the parameter vector θ will be based on the principle of quasi-maximum likelihood (QML). The method of the original maximum likelihood considers that the vector of optimal parameters is the one for which the joint PDF of the observations x_t, $t = 1, \ldots, T$, given θ, is maximum. Intuitively, the higher the value of this joint PDF using a parameter vector θ, the greater the probability that θ is, in fact, the parameter vector giving rise to the observed data.

When the sequence of observations is i.i.d., the joint conditional PDF can be factored as the product of the conditional distributions. However, in our model, there is a clear dependency within the sampled variables $(x_t, x_{t-1}, \ldots)$, which means that the joint distribution cannot be factored. Nevertheless, we will use the product of conditional distributions as an approximation of the joint PDF, resulting in an approximate likelihood function for estimating the optimum parameter vector. This approach is known as quasi-maximum likelihood estimation, which, under some assumptions, is consistent and asymptotically normal.

At each time t this conditional PDF, represented as $f(t, \theta, \lambda)$, is the distribution 2 applied to value of the system x_t.

$$f(t, \theta, \lambda) = F\left(x_t; m_1(t), m_2(t), m_3(t)\right) \quad (8)$$

This conditional PDF $f(t, \theta, \lambda)$ measures how probable the value x_t of the system was generated by a PDF whose moments are $m_1(t)$, $m_2(t)$, $m_3(t)$. The bigger is the probability $f(t, \theta, \lambda)$ then the bigger is the probability that the system parameter vector is θ. Thus, the quasi-likelihood function of our model is given by:

$$\mathcal{L}(\theta, \lambda) = \prod_{t=1}^{T} f(t, \theta, \lambda)$$

More specifically, we will deal with the logarithm of the quasi-likelihood function:

$$l(\theta, \lambda) = \sum_{t=1}^{T} \log(f(t, \theta, \lambda)) \tag{9}$$

Optimization Methods

The optimum coefficient values of the Multimomental ARMA model are obtained by maximizing the quasi-likelihood function. To solve this optimization problem, we shall use the Newton's method with random initial parameters, which explores the information from the first and second-order derivatives of the function to be optimized to search the optimum point.

The parameter vector that maximizes the function $l(\theta, \lambda)$ corresponds to the limit of θ_k with $k \to \infty$ where θ_k is calculated by:

$$\theta_{k+1} = \theta_k - \left[\frac{\partial^2 l}{\partial\theta\partial\theta}(\theta_k)\right]^{-1} \frac{\partial l}{\partial\theta}(\theta_k) \tag{10}$$

Generally, the quasi-likelihood function is multimodal, which means that the Newton's method, depending on the initial condition, can converge to one of the local optima. Therefore, it is common to consider a set of several different initializations to Newton's method, being the final solution the one that reaches the maximum value of the objective function. The main steps of Newton's method are described in 2.1.

Algorithm 2.1 QML Estimation by Newton's Method

Input: time series x_t, initial parameters θ_0, finalization tolerance tol and maximum number of iterations max_it

Step 1 $k = 0$

Step 2.1 WHILE $\{k < max\_it$ and $size\_step > tol\}$ DO

Step 2.2 Compute θ_{k+1} from θ_k by eq. 10

Step 2.3 $size\_step = ||\theta_{k+1} - \theta_k||^2$

Step 2.4 $k = k + 1$

Step 3 $\theta_{final} = \theta_{k_*}$

The possibility of obtaining a set of parameters suboptimal for the Multimomental ARMA model may compromise the analysis of its behavior, since, if we observe an imprecise approximation of the time series considered, this may be a consequence of both the limitation of the model – i.e., it is not flexible enough to model that type of random phenomenon – and of the solution found to its parameters. In the latter case, although there was sufficient approximation capability, it has not been adequately explored due to the fact that the optimization process was not capable of reaching the correct optimum solution for the model.

Therefore, we will also study the behavior and performance of the Multimomental ARMA model when its parameters are adjusted, in the sense of quasi-maximum likelihood, with the aid of Nelder-Mead search heuristic [19].

3 Empirical Analysis: a Case Study

In order to assess the behavior of the proposed model, we shall study its performance in the context of a synthetic time series. The set of observations is constructed from the use of the first three moments with an ARMA model. In other words, we selected three specific values for the initial moments $m_1(0)$, $m_2(0)$, $m_3(0)$ and we sampled a value of the system having the following distribution $\pi_1(0)G(x,\mu_0(0),\sigma(0)) + (1-\pi_1(0))G(x,\mu_1(0)),\sigma(0))$ with $\pi_1(0),\mu_0(0),\mu_1(0)$ obtained from the initial values $m_1(0)$, $m_2(0)$, $m_3(0)$. Then, for each subsequent time instant, $m_1(t)$, $m_2(t)$ and $m_3(t)$ are generated via expressions (3-7) and, finally, the value of the system is sampled from the corresponding PDF: $x_t \sim \pi_1(t)G(\cdot,\mu_0(t),\sigma(t)) + (1-\pi_1(t))G(\cdot,\mu_1(t),\sigma(t))$.

After the data set is generated, we estimated the optimum parameter vector θ using both the Newton's method and the Nelder-Mead search heuristic [19] considering special initial values. To generate these initial values, we sampled 1000 vectors of random parameters with a uniform distribution on the set $[-1,1]^3 \times [0,1]^3 \times [-1,1]^3$, selecting the 5 vectors with the highest values of quasi-likelihood $l(\theta)$. The optimum solution is considered to be the vector θ that achieves the highest value of the quasi-likelihood function after the optimization algorithm is executed.

For the set of synthetic data generated from a vector of initial parameters, we present in Table 1 the optimum solution found by the methods of Newton, Nelder-Mead and standard ARMA-GARCH [3].

Parameter	Original Vector	Newton	Nelder-Mead	ARMA-GARCH
logL	-13205.1	-1686.7	-1590.7	-
a_1	0.1	0.0903	0.1991	0.7098
$b_1^{(1)}$	0.8	-0.1503	0.6681	0.8288
$c_1^{(1)}$	0.4	0.3003	0.6290	0.3900
a_2	0.1	0.9765	0.8312	0.0583
$b_1^{(2)}$	0.6	0.3206	4.6768	0.7142
$c_1^{(2)}$	0.3	0.8312	0.7903	0.2059
a_3	0.2	0.1110	0.3661	-
$b_1^{(3)}$	0.5	-0.6143	5.7900	-
$c_1^{(3)}$	0.3	-0.8189	0.8313	-
MSE	-	0.8271	0.1291	1.2494

Table 1. 1: QML Estimation results

Three pertinent points can be observed in Table 1: (i) the optimum solutions obtained by Newton's method and by Nelder-Mead heuristic are quite different from the original parameter vector;(ii) the log-likelihood value associated with the original parameter vector is significantly smaller than those related to the solutions found by the algorithms. This means that the actual parameter vector does not represent an attractive solution in the quasi-maximum likelihood sense. This undesired characteristic has been studied [20] and modifications have been proposed to recover the consistency of the quasi-ML estimator, which we intend to incorporate in the design of the Multimomental ARMA model in future works; (iii) the models considering three moments present a smaller Mean Squared Error (MSE) than the ARMA-GARCH [3] approach, which shows that the multimomental model is indeed more precise.

Notwithstanding, it is interesting to notice that parameters quite distinct from the original can still lead to an adequate approximation of the time series. This is confirmed by Figure 1, which exhibits the original time series (red) as well as the sequence of values generated by the Multimomental ARMA model whose parameters are those offered by Newton's method (blue) and by Nelder-Mead heuristic (black). Additionally, we show in Figures 2 and 3 the time evolution of the original and estimated conditional variance and of the conditional skewness, respectively.

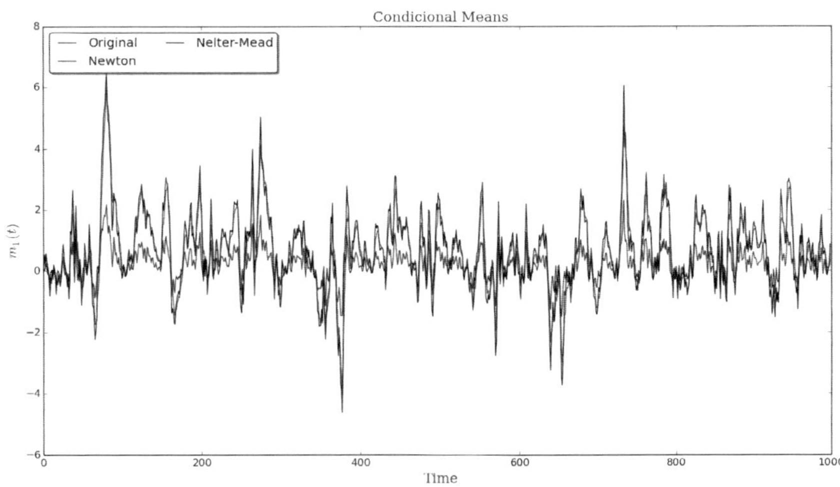

Figure 1. Evolution of conditional means series

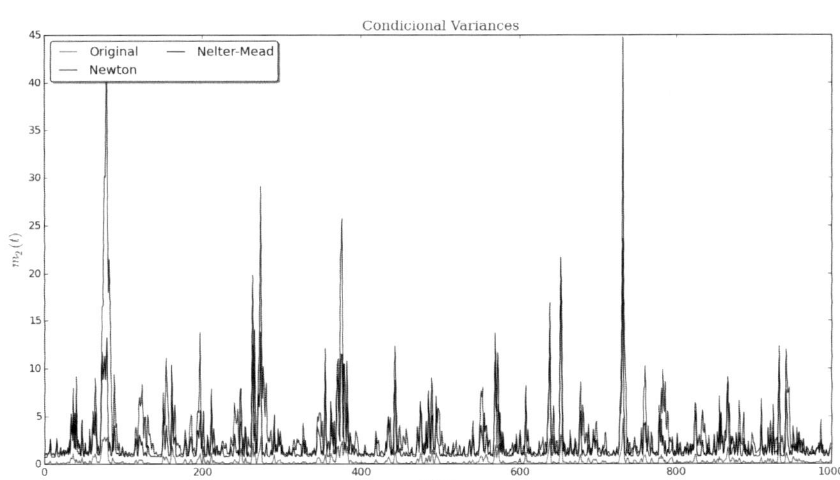

Figure 2. Evolution of conditional variances series

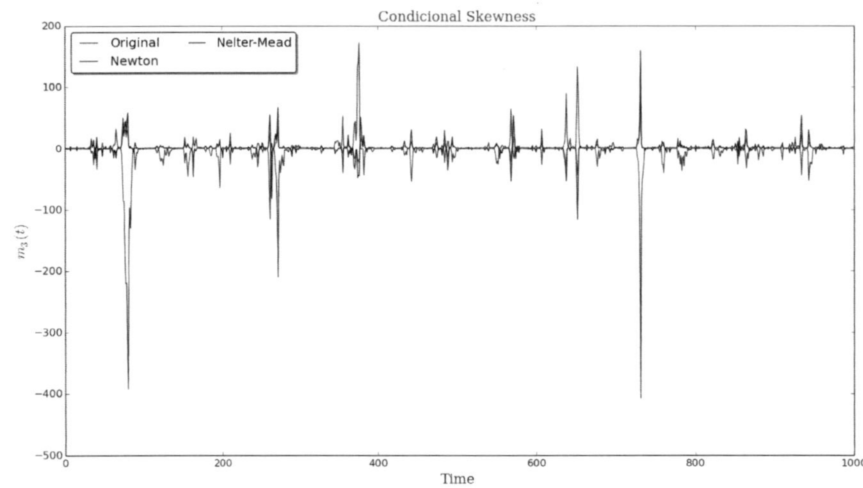

Figure 3. Evolution of conditional skewnesses series

As we can notice, the Multimomental ARMA model tends to deviate from the original variance, presenting increased values for its variance. Moreover, with respect to the conditional skewness, it tends to reduce the amount of skewness when compared with the original skewness. Both behaviors supports the case that the quasi-maximum likelihood is inconsistent as appeared in [20].

Illustrating the estimation methodology, we show below the implied mixture parameters generated by Nelder-Mead estimation. In the figure 4, the left plot shows the evolution of the mixture centers $\mu_1(t)$ (in blue) and $\mu_2(t)$ (in red), and the right plot shows the logarithm of the mixture's weights $\log(\pi_1(t))$ (in blue) and $\log(\pi_2(t))$ (in red). In both plots, we can notice a noisy behaviour of the mixture's centers and weights meaning that in some time instants the conditional PDF has considerable fat tail and non normal behaviour.

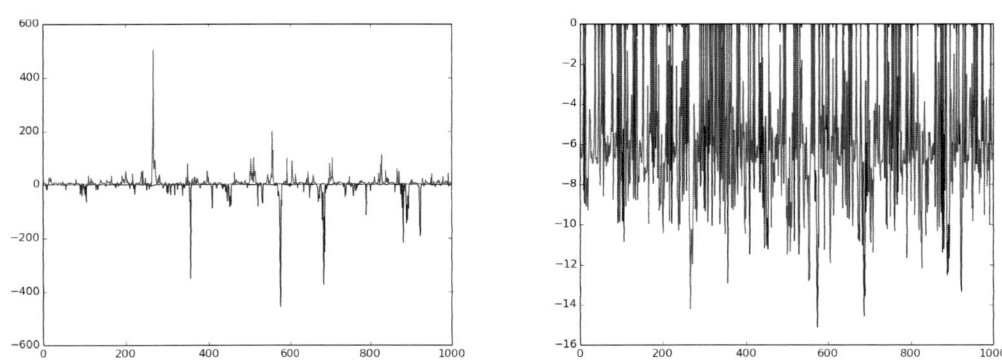

Figure 4. Evolution of mixture's centers (left) and logarithm of the mixture's weights (right)

The histograms of $\epsilon(t)$ resulting from each vector parameters are shown below. The red curve represents the normal PDF with the same mean and variance of the set $\{\epsilon(t)\}$. The histograms show that to each case - original (left), Newton's Method (center) and Nelder-Mead (right) - the distribution is not normal because the histogram's shapes differ from the red normal curve indicating that distribution of $\epsilon(t)$ have heavy tails. This fact justify the use of non normal approach to model the series.

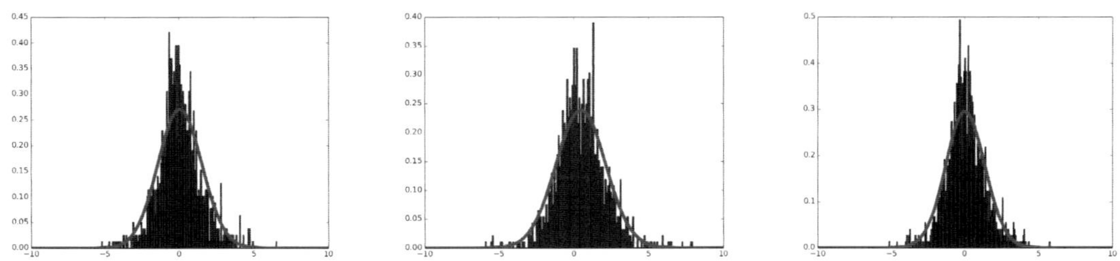

Figure 5. Histogram of $\epsilon(t)$ to each case (original (left), Newton Method (center) and Nelder-Mead (right))

4 Conclusion

In this paper, we have proposed a new model for approximating non-stationary time series that explicitly deals with the variations of the first three statistical moments, which are modeled as ARMA processes. Additionally, the PDF of the observations is represented by a mixture of two gaussian functions, whose parameters are adjusted in order to ensure that the statistical moments reach the desired values (i.e., those obtained from the corresponding ARMA evolution). We believe that an analogous approach to model the fifth and seventh moments is possible following the same steps as in this paper and fulfilling the moment matrices conditions to existence of the corresponding mixture distribution (Theorem 2A(b) from [17]).

The parameters of the Multimomental ARMA model were adjusted according to the approach based on the concept of quasi-maximum likelihood estimation. Finally, two optimization algorithms have been applied to search the optimum solution for the parameters: the Newton's method and the Nelder-Mead search heuristic.

In the context of a synthetic time series, we observed that the quasi-ML estimator was not consistent, since the actual parameter vector (i.e., that which effectively was used to generate the sequence of observations) did not represent an attractive solution. Still, the solutions found by the search algorithms were capable of producing a sequence of samples quite similar to the original time series and they estimated a model more accurate than the standard ARMA-GARCH model[3].

In order to improve the estimation performance, it is certainly relevant to study the application of different optimization strategies for determining the optimum parameters- with quasi-maximum likelihood - of the developed Multimomental ARMA model, particularly some search methods, such as the *simulated annealing* and populational metaheuristics belonging to the evolutionary algorithm class [2], that possess mechanisms to escape from local optima.

The consistency of the quasi-ML estimator represents a topic that also deserves future investigations. Finally, the behavior and the performance of the Multimomental ARMA model needs to be further analyzed, considering other synthetic multimomental time series, like that treated in Section 3, and real time series, such as those related to financial measures and to monthly streamflow of rivers.

Bibliography

[1] Alexander, C., and Lazar, E. (2006). *Normal mixture GARCH (1, 1): Applications to exchange rate modelling.* Journal of Applied Econometrics, **21(3)**, 307-336.

[2] Back, T., Fogel, D. B., and Michalewicz, Z. (Eds.). (2000). *Evolutionary computation 1: Basic algorithms and operators* (Vol. 1). CRC Press.

[3] Bollerslev, T. (1986). *Generalized autoregressive conditional heteroskedasticity.* Journal of econometrics, **31(3)**, 307-327.

[4] Box, G. E., and Jenkins, G. M. (1976). *Time series analysis: Forecasting and control.* In Holden-Day series in time series analysis. Holden-Day.

[5] Brooks, C., Burke, S. P., Heravi, S., and Persand, G. (2005). *Autoregressive conditional kurtosis.* Journal of Financial Econometrics, **3(3)**, 399-421.

[6] Cheng, X., Yu, P. L., and Li, W. K. (2009). *On a dynamic mixture GARCH model.* Journal of Forecasting, **28(3)**, 247-265.

[7] Engle, R. F. (1982). *Autoregressive conditional heteroscedasticity with estimates of the variance of United Kingdom inflation.* Econometrica: Journal of the Econometric Society, 987-1007.

[8] Haas, M., Mittnik, S., and Paolella, M. S. (2004). *Mixed normal conditional heteroskedasticity.* Journal of Financial Econometrics, **2(2)**, 211-250.

[9] Haas, M., and Paolella, M. S. (2012). *Mixture and RegimeSwitching GARCH Models.* Handbook of volatility models and their applications, 71-102.

[10] Haas, M., Krause, J., Paolella, M. S., and Steude, S. C. (2013). *Time-varying mixture GARCH models and asymmetric volatility.* The North American Journal of Economics and Finance, **26**, 602-623.

[11] Hansen, B. E. (1994). *Autoregressive conditional density estimation.* International Economic Review, 705-730.

[12] Harvey, C. R., and Siddique, A. (1999). *Autoregressive conditional skewness.* Journal of financial and quantitative analysis, **34(04)**, 465-487.

[13] Huisman, R., Koedijk, K., Kool, C., and Palm, F. (2002). *Notes and communications: the tail-fatness of fx returns reconsidered.* De Economist, **150(3)**, 299-312.

[14] Johnston, K., and Scott, E. (2000). *GARCH models and the stochastic process underlying exchange rate price changes.* Journal of Financial and Strategic Decisions, **13(2)**, 13-24.

[15] Jondeau, E., and Rockinger, M. (2003). *Conditional volatility, skewness, and kurtosis: existence, persistence, and comovements.* Journal of Economic Dynamics and Control, **27(10)**, 1699-1737.

[16] Kuester, K., Mittnik, S., and Paolella, M. S. (2006). *Value-at-risk prediction: A comparison of alternative strategies.* Journal of Financial Econometrics, **4(1)**, 53-89.

[17] Lindsay, B. G. (1989). *Moment matrices: applications in mixtures.* The Annals of Statistics, 722-740.

[18] Lindsay, B. G., Pilla, R. S., and Basak, P. (2000). *Moment-based approximations of distributions using mixtures: Theory and applications.* Annals of the Institute of Statistical Mathematics, **52(2)**, 215-230.

[19] Nelder, J. A., and Mead, R. (1965). *A simplex method for function minimization.* The computer journal, **7(4)**, 308-313.

[20] Newey, W. K., and Steigerwald, D. G. (1997). *Asymptotic bias for quasi-maximum-likelihood estimators in conditional heteroskedasticity models.* Econometrica: Journal of the Econometric Society, 587-599.

[21] Shumway, R. H., and Stoffer, D. S. (2010). *Time series analysis and its applications: with R examples.* Springer Science & Business Media.

[22] Tsay, R. S. (2000). *Time series and forecasting: Brief history and future research.* Journal of the American Statistical Association, **95(450)**, 638-643.

[23] Yule, G. U. (1927). *On a method of investigating periodicities in disturbed series, with special reference to Wolfer's sunspot numbers.* Philosophical Transactions of the Royal Society of London. Series A, Containing Papers of a Mathematical or Physical Character, **226**, 267-298.

Joint modeling of inflation and real interest rate dynamics with application to equity-linked investment

Tommi Salminen, *University of Tampere*, tommi.salminen@nordea.com
Lasse Koskinen, *University of Tampere*, lasse.koskinen@uta.fi
Arto Luoma, *University of Tampere*, arto.luoma@uta.fi

Abstract. This paper introduces a realistic model for the joint dynamics of real interest rate and inflation so that it could be used for various prediction purposes, for example to analyze the role of inflation in the pricing and hedging of financial derivatives. In a combined auto-regressive process, normal or more stable inflation periods are explained by a standard AR-process, while unanticipated peaks are captured by an additional process following Student's t-distribution. Next, the effect of inflation on the pricing of an equity index linked insurance product is studied. The models are estimated using Bayesian methods with US data.

Keywords. regression method, equity-linked contract, option valuation, insurance product, normal-inverse Gaussian distribution, Bayesian estimation

1 Introduction

Understanding inflation dynamics is one of the key challenges for economic agents. Even so that virtually all currency areas have authorized a specific public body, namely a central bank, to monitor and also control the development of inflation. Traditionally, high inflation has been seen as a more severe risk than too slow inflation (deflation). This is probably due to the history of OECD countries, where the periods of positive/high inflation seem to have taken place more often than deflation [4]. However, this is not to say that high inflation is more serious a problem than deflation. Actually, due to the lessons in history, we are probably better equipped to face the challenges of high inflation than deflation. Our intention is to model inflation and real rate dynamics.

The most recent years have moved the risk pendulum towards too low inflation or even deflation. As an outcome, many central banks have adopted a very expansive monetary policy stance, by cutting their key rates to practically zero and simultaneously entering so called quantitative easing [5]. This has taken place e.g. in the US and more recently in the euro area.

This recent development creates an additional challenge to inflation and real rate modeling; we have witnessed the periods of very high inflation, even so called hyper-inflation, but how do we expect the inflation to behave when it becomes negative? High inflation can persist relatively long periods of time, but we know from experience that (monetary) policy can eventually find means to moderate it. With deflation we are probably not that certain about the dynamics. However, it seems natural to assume that inflation does not stay at the double digit negative territory for extensive periods of time. Therefore we have decided to incorporate such a deflation limiting factor to our combined inflation process.

As mentioned above, inflation risk is present in the decision making of practically all economic agents. This has created the need to investigate different methods to hedge against inflation risk. An obvious method is to apply inflation clauses in various types of contracts. On the other hand, it has been suggested that investing in certain asset classes, e.g. real estate, would provide a natural shelter from inflation risk. The evidence supporting this approach has been controversial [13]. The performance of equity markets, also often claimed to be providing a natural hedge against inflation, is not trivial either, as suggested by e.g. [14] and [9].

We model inflation and real rate dynamics with a combined auto-regressive process. Normal or more stable inflation periods are explained by a standard AR-process, while unanticipated peaks are captured by an additional process that follows Student's t-distribution. As an outcome, the real rate process also follows such a combined process. Alternative approaches to model time varying inflation and real rate process exist, e.g. the regime shift model presented in [10].

An interesting angle to inflation risk in relation to equity markets comes via certain types of life insurance products, more specifically index linked (or unit-linked) products. These have become more popular during the recent years as insurance companies have moved away from traditional guarantee yield towards products where the investment risk is at the policy holder's end. This is often done by offering equity linked yield, probably combined with a relatively low guarantee yield. Again, these products are marketed as natural hedges against inflation. Various types of structures have been investigated in this manner [17].

At the same time these new types of savings products expose the insurer to inflation risk, when hedging the product portfolio. A key characteristic of such insurance contracts is often the policy holder's right to early exercise, before the final maturity date. According to the financial markets vocabulary, the policy holder has the put option with multiple exercise rights. In our example the policy holder can exercise the option any day, so it is of American type [16]. Given the non-trivial value of such exercise right, we need to implement a pricing method in order to value the option and eventually the insurance product. Our choice is the so-called regression method.

We shall investigate the effect of the introduced inflation process on the pricing of an equity index linked life insurance product. The log returns of the equity index are assumed to follow the normal-inverse Gaussian distribution, which has certain beneficial features, e.g. infinite divisibility, as we shall witness later.

Bayesian methodologies in model estimation have expanded in popularity during the past ten to twenty years, in tandem with the increasing availability of the processing capacity for computation. The immediate benefit of the Bayesian approach is the possibility of taking parameter uncertainty into account in the estimation process in a coherent way.

This paper is organized as follows: Chapter 2 presents the inflation and real rate models we use and Chapter 3 describes the estimation methodology. Then, Chapter 4 introduces the actual data set and summarizes the results. Finally, Chapter 5 concludes our findings.

2 Model

Our purpose has been to build a realistic model for the joint dynamics of real interest rate and inflation so that it could be used for various prediction purposes, for example to analyze the role of inflation in the pricing and hedging of financial derivatives. We found the following model to be adequately consistent with empirical data.

Inflation and real interest rate

We assume that

$$\begin{aligned} i_t &= y_t + w_t, \\ y_t &= e^{x_t} - g, \\ x_t &= \phi_0 + \phi_1 x_{t-1} + \epsilon_t^{(x)}, \end{aligned} \tag{1}$$

where i_t is inflation at time t, and $w_t \sim t_\nu(0, \sigma_w^2)$ and $\epsilon_t^{(x)} \sim N(0, \sigma_x^2)$ are independent and identically distributed (iid), mutually independent error processes with zero mean. By allowing w_t to be Student-t distributed with a small degrees-of-freedom parameter ν we can account for the jumps in the inflation process. We can think of the process y_t as 'intrinsic inflation' from which transitory peaks have been filtered. If we assume that this intrinsic inflation can be sometimes negative, so that there is deflation, we can assign a positive value for g. Note also that the transformed inflation process x_t is autoregressive of order 1, i.e. AR(1); we only assume that $|\phi_1| < 0$ to guarantee its stationarity.

We also assume that the real interest rate r_t follows an AR(1) process disturbed by the peak process w_t:

$$r_t = \psi_0 + \psi_1 r_{t-1} + \zeta_0 w_t + \zeta_1 w_{t-1} + \epsilon_t^{(r)}, \tag{2}$$

where $\epsilon_t^{(r)} \sim iid\ N(0, \sigma_r^2)$ and $|\psi_1| < 0$. Further, the process $\epsilon_t^{(r)}$ is assumed to be independent of the other error processes w_t and $\epsilon_t^{(x)}$. Thus, r_t is also assumed to be mean-reverting (and stationary), but no restrictions for its range are set. According to our estimation results $\zeta_0 \approx -1$ and $\zeta_1 \approx 1$, which indicates that although the inflation peaks are not anticipated in the nominal interest rate, they are fully accounted for in the next time period.

Initial time series analysis indicated significant cross-correlation between r_t and i_t, but this was fully explained by the shared process w_t. The process w_t also helps explain autocorrelation in r_t and i_t, so that the AR(1) structure turned out to suffice in both cases.

Figure 1 shows the real interest rate and inflation series, continued with simulations paths. Details of the used data will be provided in Section 4.

Equity index

Since our goal is to study the effect of inflation on equity-linked contracts, we also need a model for equity returns. We decided to model the returns separately from the real interest rate and inflation processes because our 3-month data did not show significant cross-correlations. Specifically, we assume that the distribution of log returns is normal-inverse Gaussian (NIG), which is a special case of the generalized hypergeometric distribution, introduced by [2, 3]. The density is

$$f(x) = \frac{\alpha\delta K_1(\alpha\sqrt{\delta^2 + (x-\mu)^2})}{\pi\sqrt{\delta^2 + (x-\mu)^2}} e^{\delta\gamma + \beta(x-\mu)},$$

where $K_\nu(\cdot)$ denotes a modified Bessel function of the third kind, and $\gamma = \sqrt{\alpha^2 - \beta^2}$. Here, μ is the location parameter and δ the scale parameter, and α represents tail heaviness, and β skewness.

This model has the advantage that while being fairly simple it can fit the skewness and kurtosis apparent in return distributions. Further, it is infinitely divisible, meaning that it can be constructed as a sum of *iid* components belonging to the same family of distributions. This is a useful property, since we can estimate the model using quarterly data, and simulate daily returns using the same distribution but with different parameters.

The moment generating function

$$\mu(z) = \mathbb{E}e^{zX} = e^{\mu z + \delta(\gamma - \sqrt{\alpha^2 - (\beta + z)^2})}$$

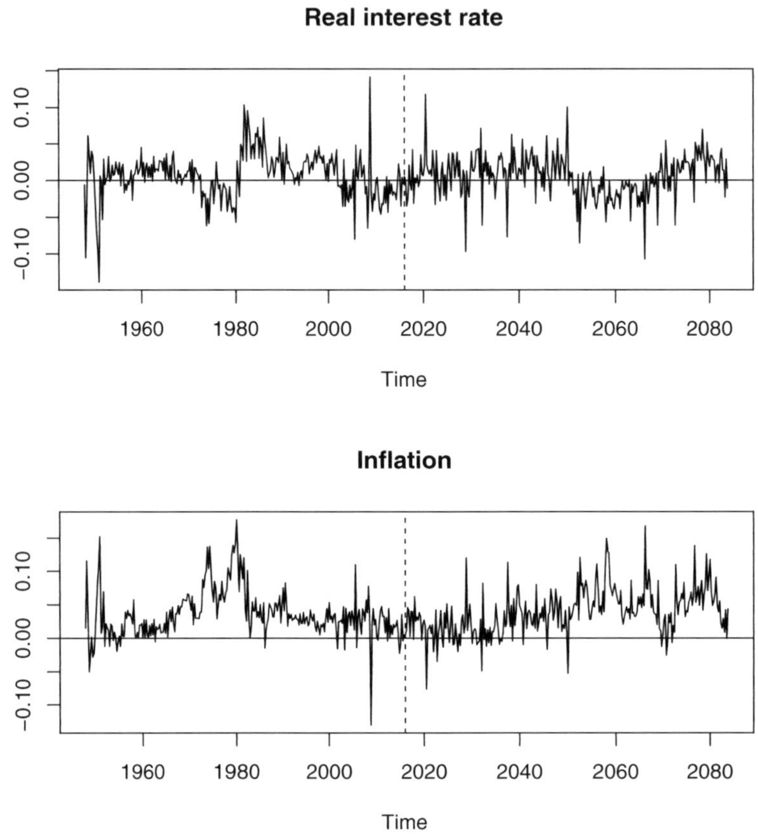

Figure 1. Real interest rate and inflation series, based on quarterly data from 1948 to 2015. The inflation series represents 3 months annualized inflation, calculated from the US Consumer Price Index, all urban consumers. The real interest rate is obtained by subtracting inflation from the nominal three months T-bill yield. The parts on the left are realized series and on the right simulations.

can be used to show that if $X \sim \mathrm{NIG}(\alpha, \beta, \mu, \delta)$ and $X = X_1 + ... + X_n$, where the X_is are *iid*, then $X_i \sim \mathrm{NIG}(\alpha, \beta, \mu/n, \delta/n)$. Further, if X is a log-return over a time interval Δt and r an instantaneous riskless interest rate, and we set

$$\mu = r\Delta t - \delta(\gamma - \sqrt{\alpha^2 - (\beta + 1)^2}),$$

then $\mathbb{E}e^X = \mathbb{E}e^{r\Delta t}$, so that the return follows a risk-neutral probability measure. Note, however, that the real-world probability measure has no unique equivalent risk-neutral probability measure in this case; the markets are incomplete with this model.

3 Estimation

In the estimation of our financial models, we employ Bayesian techniques. This helps us to take parameter uncertainty into account at the prediction stage. More specifically, we use Markov Chain Monte Carlo (MCMC) simulation.

Inflation and real interest rate

In the estimation of the joint model of inflation and real interest rate, we use a single-component Metropolis-Hastings algorithm (see Chapter 1 of [7]). We denote the data vectors as $r = (r_1, ..., r_T)$ and $i = (i_1, ...i_T)$ for the real interest rate and inflation, respectively. The components, which are updated one by one, include $\theta_1 = (\psi_0, \psi_1, \zeta_0, \zeta_1, \sigma_r^2)$, $\theta_2 = (\phi_0, \phi_1, \sigma_x^2)$, $w = (w_1, ..., w_T)$, $u = (u_1, ..., u_T)$, g, ν, τ^2 and η^2.

The components u, τ^2 and η^2 do not appear explicitly in Equations (1) and (2), but they play an auxiliary role in estimation. The Student-t distribution is constructed as an inverse-χ^2 mixture of normal distributions with varying variances. Thus, we have an auxiliary model

$$\begin{aligned} w_t &\sim N(0, \eta^2 u_t), \\ u_t &\sim \text{Inv-}\chi^2(\nu, \tau^2), \end{aligned}$$

where $\eta > 0$ is an additional scale parameter. This parameter is not absolutely necessary but is helpful by providing an additional dimension so that the algorithm does not get stuck so easily (see [6]).

The unknown parameter vectors can be updated in a straightforward way. Gibbs steps with conjugate distributions can be used (see e.g. [6]), except for g and ν, for which Metropolis steps can be applied. We used uniform priors for g and ν but restricted g to be nonnegative to improve MCMC mixing. However, updating the elements of w is somewhat more involved. Suppressing the parameters in notation, the full conditional of w_t is

$$\begin{aligned} p(w_t|w_{-t}, i, r, u) &\propto p(w_t|w_{-t}, i_{-t}, u)p(i_t, r|w_t, w_{-t}, i_{-t}, u) \\ &\propto p(w_t|u_t)p(i_t|w_t, x_{-t})p(r_t|r_{t-1}, w_t, w_{t-1})p(r_{t+1}|r_t, w_{t+1}, w_t). \end{aligned} \quad (3)$$

All factors on the right side of (3) are normal densities except $p(i_t|w_t, x_{-t})$, which is log-normal. If all of them were normal, the full conditional would also be normal. We use a normal approximation for $p(i_t|w_t, x_{-t})$ in order to obtain an approximation for $p(w_t|w_{-t}, i, r, u)$ and to generate a proposal for w_t. This proposal is then either accepted or rejected according to the Metropolis-Hastings acceptance rule. The details are provided in the Appendix.

Equity index

We first compute the maximum likelihood (ML) estimate and its approximate covariance matrix for the parameters of the NIG model. Then we use the Metropolis algorithm to simulate the posterior distribution. The proposals can be generated from a multivariate normal distribution with covariance matrix $2.4^2/4 \cdot \Sigma$, where Σ is the covariance matrix of the ML estimator. For optimization of the Metropolis algorithm, see Section 11.9 of [6].

4 Application

Data

In order to test our modeling framework, we looked for relatively long and consistent time series. Soon it became natural to apply the US data. Not only is there a long history available, but also one can assume that the pricing of tradeable assets is efficient in the sense that the connection between inflation, rates and equity markets could exist in the way our model predicts. Another interesting currency area would of course be the euro area. However, the data history of euro currency, extending not further than the start of the new millennium, makes it challenging for estimation purposes. This is especially so as one cannot operate with daily data here.

As the nominal interest rate time series we chose the end-of-month, three months T-bill yield. The raw data were monthly observations, of which we created a chain of quarterly yields i.e. we picked every third data point of the monthly series. The inflation time series applied is the US Consumer Price Index, all urban consumers, which is also known as the headline CPI figure. We then calculated the quarterly inflation figures by annualizing the three month change of the CPI. Now, the corresponding three month real rate was achieved simply by subtracting the quarterly inflation from the three month nominal rate. The equity index series is S&P500, end-of-month, price return index. We then calculated the three month log return of the index.

Thus, the whole data set was in quarter-by-quarter format. The time span was from 1948Q1 to 2015Q4. We approached this period as one time series. One could of course share arguments why it would be interesting to investigate certain sub-periods, e.g. 'before and after the oil crises of the 1970s', as the inflation process could be different ([15]). Here we see potential for future research.

Equity-linked investment

As an example case we define an equity-linked savings contract and consider its valuation in the presence of inflation. We assume that the yield of the contract is a fixed percentage (participation rate) of the growth of an equity index. This yield is added yearly to accumulated savings. However, negative returns of the index do not decrease the savings. Further, the investor has a surrender (early exercise) option which gives him the right to reclaim the contract at any time before the final expiration date without extra expenses. In the case of early exercise, the investor will obtain the savings accumulated thus far.

Thus, after n years, the value of the savings is given by

$$P\prod_{t=1}^{n}\left\{1+\max\left(\lambda\left(\frac{S_t}{S_{t-1}}-1\right),0\right)\right\},$$

where P is the premium, S_t the value of the index at time t and λ the participation rate. This contract is similar to the compound annual ratchet (CAR) contract defined in [8]. However, it is different in that it does not provide guaranteed interest but includes a surrender option instead.

We introduced a more general version of the contract (e.g. it could include guaranteed interest rate, or the savings could be evaluated daily) in [12]. We also tackled the inverse prediction problem of determining a break-even participation rate λ, such that the value of the contract equals the premium. Here, we do not deal with this problem but use a fixed value for λ. Note, however, that in order to such a solution exist, one must make the further assumption that the contract is not reclaimed immediately.

Regression method

Since the underlying processes are complicated and the contract includes a surrender option, it is not possible to determine a closed-form formula for its price. Instead, we use the regression (or least squares) method, in which paths of the underlying processes are simulated and the continuation value of the contract is determined at each time point with linear regression. The version of the algorithm we use was introduced by [18]; a slightly different version is provided by [11]. In our Bayesian approach, the simulations are generated from the posterior predictive distribution. However, the equity index process is adjusted so that it follows a risk-neutral probability measure. This is done by adjusting the drift parameter μ as described in Section 2.

In order to obtain the regression variables, we first define the variables $x_1 = S_t/S_0$, $x_2 = 100(i_t + r_t)$, and $x_3 = 100(i_t - r_t)$. Here, x_2 is the nominal interest rate, so x_3 represents the additional effect of inflation. Note that x_2 and x_3 are approximately uncorrelated. Then, we use the first two Laguerre polynomials,

$$\begin{aligned} L_0(X) &= \exp(-X/2) \\ L_1(X) &= \exp(-X/2)(1-X) \end{aligned}$$

of x_1, x_2 and x_3 to form the regression variables. Using Laguerre polynomials as basis functions is suggested by [11]. In addition, we use the cross-products $L_0(X_1)L_0(X_2)$, $L_0(X_1)L_0(X_3)$, $L_0(X_2)L_0(X_3)$, $L_1(X_1)L_0(X_2)$ and $L_0(X_1)L_1(X_2)$. Thus, we have altogether 11 explanatory variables in the regression. We also tried leaving out the regressors depending on x_3 or adding lagged values of inflation, but these modifications did not considerably affect the price estimates.

Results

Table 1 shows that the posterior distributions of ψ_1 and ϕ_1 are concentrated on values close to 1. Thus the real interest rate process r_t and the transformed inflation process x_t are almost non-mean-reverting. One could assume that the real rate process is mean-reverting due to the often stated link between real rates and potential GDP growth rate [1]. However, here one must take into account the long observation horizon in this study; it is a fair assumption that the potential GDP growth rate is not constant over the long run.

Moreover, the posteriors of ζ_0 and ζ_1 are concentrated on the neighborhoods of -1 and 1, respectively, which suggests that unexpected inflation peaks do not have any long-term effect on the real interest rate. This makes sense in the way that it is expected to be the nominal interest rate which balances out the changes in inflation, leaving the real rates untouched. One could expect this neutrality to hold especially for the 'intrinsic inflation' (y_t), whereas short-term peaks in inflation could have a short-term effect, both to the real rate and potential growth. It could be interesting to investigate if applying long-term rates, say, 10-year T-bond rate, would bring alternative results, due to the 'averaging-out' effect of long rates. However, the central bank policy rules are defined in terms of short rates, and, therefore, the 3-month T-bill rate is a natural choice.

Table 2 shows the estimation results of the NIG model. One can see that the estimates are not very accurate with this number of observations and that the posterior distributions are skewed. However, it is obvious that the fitted distribution is left-skewed (skewness < 0) and heavy-tailed (excess kurtosis > 0), and thus differs substantially from normality. Therefore, the use of the normal distribution might lead to crucial pricing errors.

Table 3 shows the results of a sensitivity test, where the effect of inflation mean reversion on the price of the savings contract is studied for various initial inflation levels. In all cases it is assumed that the initial nominal interest rate is 5%. The participation rate λ is set at 0.33, in order that the price of the contract would be approximately equal to the premium 100 (euros).

In the first price column original posterior simulations have been used. In this case the autoregressive parameter ϕ_1 is about 0.975 and the median level of inflation 2.5%. In the other cases, the median level is set at 2%. We see that when original parameter simulations are used, the initial inflation level does not significantly affect the price estimate. But when ϕ_1 is decreased, the price estimate becomes larger when the initial inflation level is above its median level. This finding is due to the mean reversion of inflation. Decreasing inflation implies decreasing nominal interest rate, which affects the price through discounting.

One can also note that in two cases the price for $i_0 = 0$ is larger than that for $i_0 = 0.01$. This might be due to the slight mean-reverting behavior of the real interest rate, for which the median level is about 0.7% and the auto-regressive parameter ψ_1 about 0.98.

As the second sensitivity test we study the effect of the innovation variance σ_x^2 of the transformed inflation process x_t on the precision of the price estimates. Table 4 shows the standard errors for various initial inflation levels. It is obvious that the results are accurate if σ_x^2 is at most 0.15, which is about five times as large as the value estimated from the data. For larger values of σ_x^2 the results explode so that it is no longer possible to determine the price (at least with the regression method).

	Mean	Median	SD	Lower (95%)	Upper (95%)
g	0.00111	0.00081	0.00106	1.56e-07	0.00317
ν	2.92	2.81	0.652	1.91	4.26
ψ_0	0.000144	0.000145	0.000222	-0.000318	0.000521
ψ_1	0.979	0.98	0.0121	0.958	1
ζ_0	-1.01	-1.01	0.0122	-1.04	-0.989
ζ_1	0.997	0.997	0.0179	0.964	1.03
σ_r^2	9.71e-06	9.58e-06	2.21e-06	5.7e-06	1.41e-05
ϕ_0	-0.0927	-0.0901	0.0478	-0.187	-0.011
ϕ_1	0.975	0.975	0.0132	0.951	0.999
σ_x^2	0.0268	0.0265	0.0041	0.0189	0.0348
σ_w^2	0.000223	0.00022	3.96e-05	0.000149	0.000298

Table 1. Estimation results of the joint model of real interest rate and inflation, based on 100000 iterations of a cyclic Metropolis-Hastings algorithm. The 95% HPD (highest posterior density) intervals are provided. The data were calculated from the 3-month T-bill yield and consumer price index from 1948 to 2015.

	Mean	Median	SD	Lower (95%)	Upper (95%)
α	35	31	16	16	78
β	-19	-15	12	-51	-3.2
μ	0.092	0.085	0.034	0.042	0.17
δ	0.12	0.12	0.032	0.068	0.19
skewness	-0.75	-0.72	0.23	-1.2	-0.33
ex. kurt.	1.7	1.5	1.1	0.4	3.9

Table 2. Estimation results of the NIG model based on 100000 iterations of the Metropolis algorithm. The 95% HPD (highest posterior density) intervals are provided. The data are quartely returns of the S&P500 equity index from 1948 to 2015.

5 Conclusions

We have introduced a parsimonious, yet flexible approach to the joint modeling of inflation and real interest rate processes. Our model replicates the most prominent features apparent in empirical data. We have also shown how to estimate the model using the Bayesian approach.

An important application is to study the significance of inflation risk in the pricing of financial derivatives. As an example we considered an equity-linked savings contract. We found that inflation did not have a considerable effect on its pricing under current circumstances. However, sensitivity analysis showed that changing some of the model parameters made inflation risk significant.

One can expect that inflation will play an even greater role in the hedging of derivatives, and this will be a natural next research topic. We did not explicitly model the term structure of interest rate and inflation expectations, and it would be possible to extend the analysis to that direction also.

		ϕ_1				
r_0	i_0	orig.	0.95	0.9	0.8	0.7
0.00	0.05	100.1	100.5	100.9	101.5	102.0
0.01	0.04	100.1	100.3	100.5	100.9	101.1
0.02	0.03	100.0	100.1	100.3	100.4	100.5
0.03	0.02	100.0	100.1	100.1	100.1	100.1
0.04	0.01	100.0	100.0	100.0	99.9	99.9
0.05	0.00	100.1	100.1	100.0	99.9	99.8

Table 3. Price estimates of the contract for various initial inflation levels and ϕ_1.

		σ_x^2					
r_0	i_0	orig.	0.05	0.1	0.15	0.2	0.25
0.00	0.05	0.02	0.02	0.03	0.10	115.55	777770.98
0.01	0.04	0.02	0.02	0.02	0.02	22.01	31372.48
0.02	0.03	0.02	0.02	0.02	0.02	3.21	1199.07
0.03	0.02	0.02	0.02	0.02	0.01	0.06	41.99
0.04	0.01	0.02	0.02	0.02	0.02	0.01	0.52
0.05	0.00	0.03	0.03	0.03	0.02	0.02	0.02

Table 4. Standard errors for the price estimates of the contract for various initial inflation levels and σ_x^2. The results are based on 10 repetitions of 1000 simulation paths for each case.

Appendix

We can approximate the full conditional of w_t in (3) by

$$\hat{p}(w_t|w_{-t}, i, r, u) \propto p(w_t|u_t)\hat{p}(i_t|w_t, x_{-t})p(r_t|r_{t-1}, w_t, w_{t-1})p(r_{t+1}|r_t, w_{t+1}, w_t), \tag{4}$$

were $\hat{p}(i_t|w_t, x_{-t})$ is a normal approximation for $p(i_t|w_t, x_{-t})$. We can list the densities in (4) as follows:

$$\begin{aligned}
p(w_t|u_t) &\propto \frac{1}{\alpha\sqrt{u_t}} \exp\left\{-\frac{1}{2\alpha^2 u_t} w_t^2\right\}, \\
\hat{p}(i_t|w_t, x_{-t}) &\propto \frac{1}{\sigma_{i.p}} \exp\left\{-\frac{1}{2\sigma_{i.p}^2} \left(i_t - (\hat{y}_t + w_t)\right)^2\right\}, \\
p(r_t|r_{t-1}, w_t, w_{t-1}) &\propto \frac{1}{\sigma_r} \exp\left\{-\frac{1}{2\sigma_r^2} \left(r_t - \psi_0 - \psi_1 r_{t-1} - \zeta_0 w_t - \zeta_1 w_{t-1}\right)^2\right\}, \\
p(r_{t+1}|r_t, w_{t+1}, w_t) &\propto \frac{1}{\sigma_r} \exp\left\{-\frac{1}{2\sigma_r^2} \left(r_{t+1} - \psi_0 - \psi_1 r_t - \zeta_0 w_{t+1} - \zeta_1 w_t\right)^2\right\}.
\end{aligned}$$

Collecting the terms involving w_t we obtain that $\hat{p}(w_t|w_{-t}, i, r, u)$ is a normal density $N(w_t|\hat{w}_t, \sigma_{w.p}^2)$, with

$$\begin{aligned}
\hat{w}_t &= \sigma_{w.p}^2 \left(\frac{1}{\sigma_{i.p}^2}(i_t - \hat{y}_t) + \frac{\zeta_0^2}{\sigma_r^2}(w_t - \hat{w}_{t1}) + \frac{\zeta_1^2}{\sigma_r^2}(w_t - \hat{w}_{t2})\right), \\
\sigma_{w.p}^2 &= \left(\frac{1}{\alpha^2 u_t} + \frac{1}{\sigma_{i.p}^2} + \frac{\zeta_0^2}{\sigma_r^2} + \frac{\zeta_1^2}{\sigma_r^2}\right)^{-1},
\end{aligned}$$

where

$$\hat{w}_{t1} = \frac{1}{\zeta_0}(r_t - \psi_0 - \psi_1 r_{t-1} - \zeta_1 w_{t-1}),$$
$$\hat{w}_{t2} = \frac{1}{\zeta_1}(r_{t+1} - \psi_0 - \psi_1 r_t - \zeta_0 w_{t+1}).$$

These formulas apply for $t = 2, ..., T-1$. The terms from $p(r_t|r_{t-1}, w_t, w_{t-1})$ or $p(r_{t+1}|r_t, w_{t+1}, w_t)$ are left out for $t = 1$ or $t = T$, respectively.

Now, w_t can be updated by generating a proposal w_t^* from $\hat{p}(w_t|w_{-t}, i, r, u)$. The proposal is accepted with probability

$$\min\left(1, \frac{p(i_t|w_t^*, x_{-t})/\hat{p}(i_t|w_t^*, x_{-t})}{p(i_t|w_t, x_{-t})/\hat{p}(i_t|w_t, x_{-t})}\right).$$

Next, let us have a closer look at $p(i_t|w_t, x_{-t})$. Since x_t is an AR(1) process, its predictive distribution is $x_t|x_{-t} \sim N(\hat{x}_t, \sigma^2_{x.p})$, with

$$\hat{x}_t = \begin{cases} \mu_x + \frac{\phi_1}{1+\phi_1^2} + [(x_{t-1} - \mu_x) + (x_{t+1} - \mu_x)] & \text{for } t = 2, ..., T-1, \\ \mu_x + \phi_1(x_2 - \mu_x) & \text{for } t = 1, \\ \mu_x + \phi_1(x_{T-1} - \mu_x) & \text{for } t = T, \end{cases}$$

and

$$\sigma^2_{x.p} = \begin{cases} \frac{\sigma_x^2}{1+\phi_1^2} & \text{for } t = 2, ..., T-1, \\ \sigma_x^2 & \text{for } t = 1, T, \end{cases}$$

where $\mu_x = \phi_0/(1 - \phi_1)$ is the unconditional mean of x_t.

Since $i_t = \exp(x_t) - g + w_t$, its predictive density is

$$p(i_t|w_t, x_{-t}) = \frac{1}{[i_t - (w_t - g)]\sigma_{x.p}\sqrt{2\pi}} \exp\left\{-\frac{\log[i_t - (w_t - g)] - \hat{x}_t]^2}{2\sigma^2_{x.p}}\right\},$$

which can be approximated by a normal density $\hat{p}(i_t|w_t, x_{-t}) = N(i_t|\hat{i}_t, \sigma^2_{i.p})$ with

$$\hat{i}_t = e^{\hat{x}+\sigma^2_{x.p}/2} + (w_t - g),$$
$$\sigma^2_{i.p} = (e^{\sigma^2_{x.p}} - 1)e^{2\hat{x}_t+\sigma^2_{x.p}}.$$

Bibliography

[1] Anderson, Richard G., Buol, Jason J., and Rasche, Robert H. (2004) *A Neutral Federal Funds Rate?* Economic Synopses #28, Federal Reserve Bank of St. Louis.

[2] Barndorff-Nielsen, Ole (1977) *Exponentially decreasing distributions for the logarithm of particle size.* Proceedings of the Royal Society of London. Series A, Mathematical and Physical Sciences (The Royal Society), **353(1674)**, 401–409.

[3] Barndorff-Nielsen, Ole (1978) *Hyperbolic Distributions and Distributions on Hyperbolae.* Scandinavian Journal of Statistics, **5**, 151–157.

[4] Boschen, John F., and Weise, Charles L. (2003) *What Starts Inflation: Evidence from the OECD Countries.* Journal of Money, Credit and Banking **35(3)**, 323–349.

[5] Fawley, Brett W., and Neely, Christopher J.(2013) *Four Stories of Quantitative Easing.* Federal Reserve Bank of St. Louis Review, January/February 2013, **95(1)**, 51–88.

[6] Gelman, A., Carlin, J.B., Stern, H.S., Rubin, D.B., (2004) *Bayesian data analysis.* Chapman & Hall/CRC.

[7] Gilks, W. R., Richardson, S., and Spiegelhalter, D., eds. (1996) *Markov chain Monte Carlo in practice.* Chapman & Hall.

[8] Hardy, M.R. (2003) *Investment guaranteed – Modeling and risk management for equity-linked life insurance.* Wiley.

[9] Kaliva, Kasimir, and Koskinen, Lasse(2008/09) *The Long-Term Risk Caused By The Stock Market Bubble.* The Journal of Risk, **11(2)**, 65–77.

[10] Kanas, Angelos (2008) *On Real Interest Rate Dynamics and Regime Switching.* Journal of Banking & Finance, **32**, 2089–2098.

[11] Longstaff, F.A., and Schwartz, E.S. (2001) *Valuing American options by simulation: A simple least-squares approach.* Review of Financial Studies, **14**, 113–148.

[12] Luoma, A., Puustelli, A., Koskinen, L. (2014) *Bayesian analysis of equity-linked savings contracts with American-style options.* Quantitative Finance, **14**, 343-356.

[13] Sebastian, Steffen P., and Maurer, Raimond (2002) *Inflation Risk Analysis of European Real Estate Securities.* Journal of Real Estate Research, **24(1)**, 47–77.

[14] Koller, Tim, Goedhart, Marc, and Wessels, David (2010) *Valuation: Measuring and Managing the Value of Companies.* John Wiley & Sons, 605–619.

[15] Koskinen, Lasse, and Pukkila, Tarmo (1996) *An Application of the Vector Autoregressive Model with a Markov Regime to Inflation Rates.* Aktuarielle Anstze fr Finanzrisken Band I / Volume I, 1095–1108.

[16] Rebonato, Riccardo (2002) *Interest-Rate Option Models.* John Wiley & Sons, 36–41.

[17] Tiong, Serena (2013) *Pricing inflation-linked variable annuities under stochastic interest rates.* Insurance: Mathematics and Economics, **52**, 77–86.

[18] Tsitsiklis, J., and Van Roy, B. (2001) *Regression methods for pricing complex American-style options.* IEEE Transactions on Neural Networks, **12**, 694–703.

Visualization of cross tabulation by the Association rules by using the Correspondence analysis

Yoshiro Yamamoto, *School of Science, Tokai University*, yama@tokai-u.jp
Sanetoshi Yamada, *Graduate School of Science and Technology, Tokai University*, S.Yamada@star.tokai-u.jp

Abstract. When comparing the response in the survey by gender and age, we make the cross-tabulation tables then visualize them by such like mosaic plot. For many answers to multiple-choice item, we want to find the item that the reaction of a particular layer (gender and age) is different from the others. Association rule analysis are suitable for the kind of analysis. By using the coordinates by correspondence analysis it is possible to plot the relationship between items and media layers. In this visualization, correspondence analysis and association rules analysis is the function of the mutually complementary. In addition, by showing the percentage of respondents each item and each layer, it become possible to understand the trend between items and layers. The visualization is constructed by using RStudio Shiny. It is possible to change the various parameters of the association rule analysis interactively, and It is also possible to change the plot style to facilitate visualization it.

Keywords. Visualization of the Questionnaire, Association Rule Analysis, Correspondence Analysis

1 Introduction

The information that customer data usually provides is the personal profile information including gender, age and so on. But, we can obtain the personal internal information by analyzing questionnaire data. This paper propose the visualization of the multiple choice questionnaire to find the difference of the internal characteristic, about 6 ($= 2 \times 3$) layers we call the media layers. The media layers are M1 (male from 20 years old to 34 years old), M2 (male from 35 years old to 49 years old), M3 (male over 50 years old), F1 (female from 20 years old to 34 years old), F2 (female from 35 years old to 49 years old) and F3 (female over 50 years old).

Offered data have 31 multiple choice questionnaires. This time, we think about this question "Please check all appropriate items about your health worries". Figure 1-1 shows total results of people that checked "catch a cold easily" at this question. We could understand that total results are similar at all media layers. However, because the numbers of people of media layers are different from each other, if we look at each ratios of media layers like Figure 1-2, we understand that young groups have relatively at their health worries. About other health worries, M2 layer and M3 layer mind "body odor of old people", F1 layer and F2 layer mind "period pains" (Figure 1-3, 1-4, 1-5, 1-6). However, it is difficult to find out the difference of media layers about all health worries by this method becauce we should find out all question items.

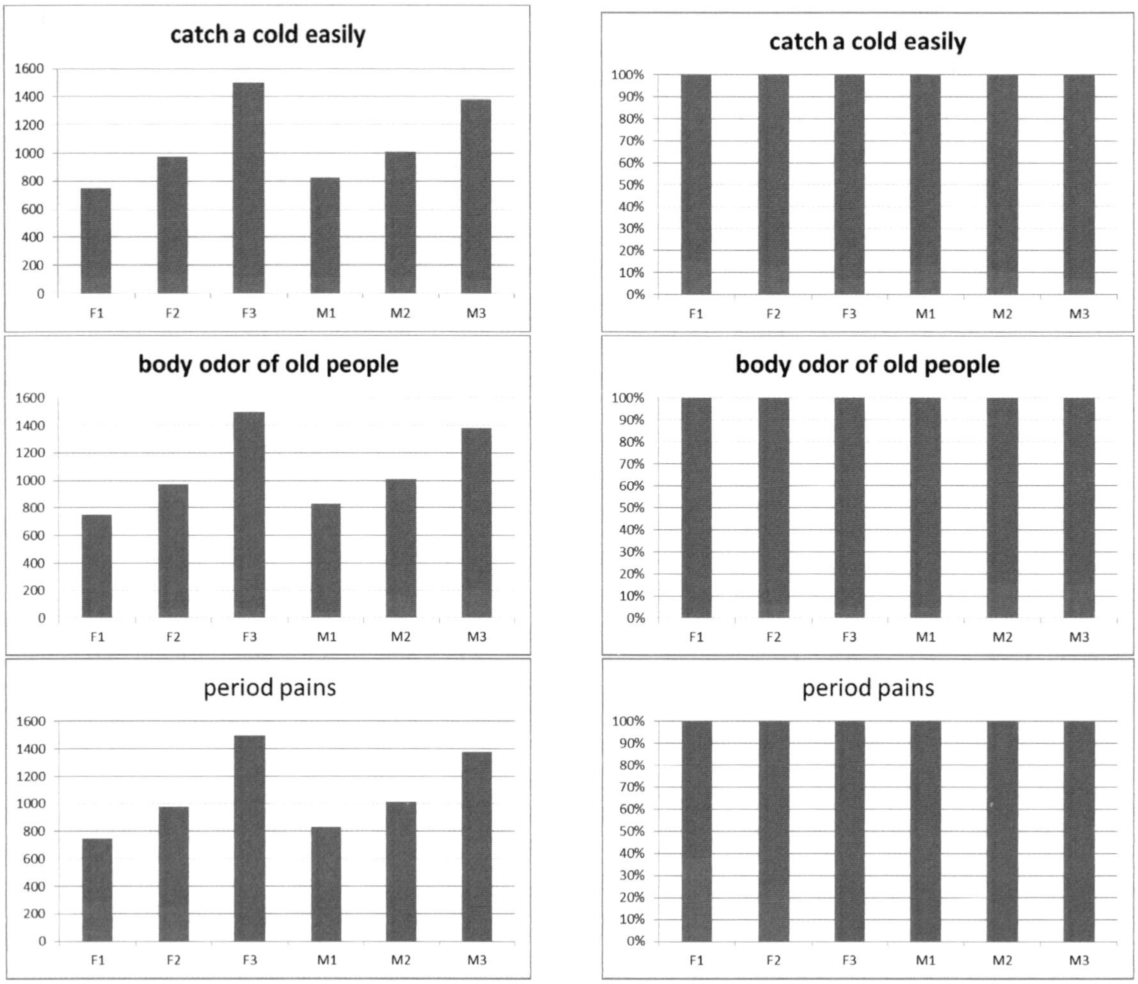

Figure 1. Total and Ratio of the questionnaire result by media layers
(Top left; 1-1, top right; 1-2, center left; 1-3, center right; 1-4, bottom left; 1-5, bottom right; 1-6.)

Figure 2 shows ratios of people that answered that there is a worry about each item of 40 question items. We see that body worries like “flab” and “stay fat” answer rate is high. Thus, we saw that mdia worries like “body odor of old people” and “period pain” answer rate is low.

2 Extraction of strong relationship between attribute and item by association rule analysis

We looked at the tendency of reaction of the media layer in Figure 1 about three question items, but we should find out all question items. There, we used association rule analysis (see [1,2,3]) as a method to find difference of tendency of answer by a group about all question items. It is basket data that

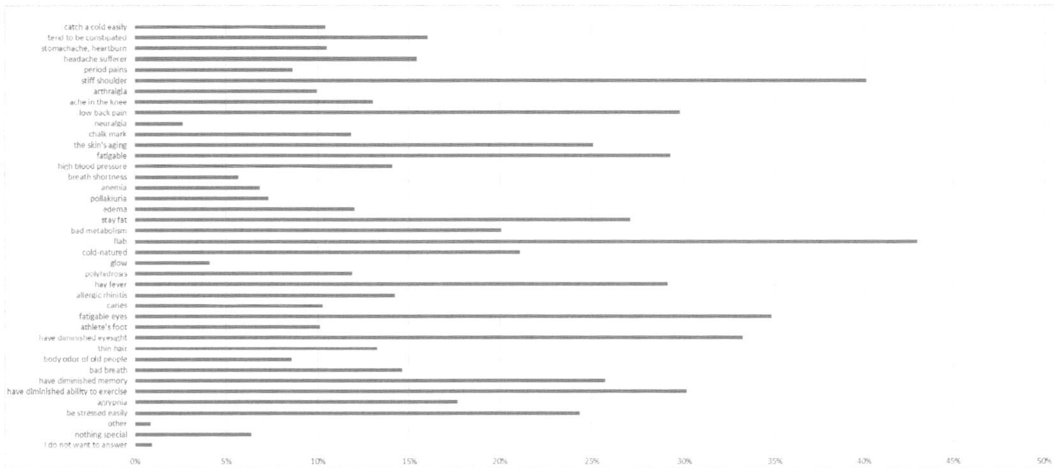

Figure 2. Ratios of questions that replied "yes"

association rule analyses are used well, but questionnaire data can be thought of 0-1 data as basket data. We also treated 0-1 data about media layer (Table 1).

Table 1. Data structure to treat

	questionnaires data				media layers					
ID	**Q1_1**	**Q1_2**	**...**	**Q1_40**	**F1**	**F2**	**F3**	**M1**	**M2**	**M3**
1	1	0	...	0	1	0	0	0	0	0
2	0	0	...	1	0	0	0	0	1	0
3	1	1	...	0	0	1	0	0	0	0
⋮	⋮	⋮	⋮	⋮	⋮	⋮	⋮	⋮	⋮	⋮
6438	0	1	...	0	0	0	0	1	0	0

To find out about question items that reacted the media layer, condition part extracted only the rule that was the media layer. We used *apriori* function for association rule analysis in statistical analysis software R. In addition, we used *subset* function to limit the association rule. When we set support more than 0.01, confidence more than 0.1 and lift more than 1.5 to find out strong rules, we could extract 27 rules.

We understood about ratio of health worry that "period pains" of F1 and F2 layers and "high blood" of M3 layer is high by Table 2. We could find the class of media and the relations of the health worry.

However, it is difficult to grasp the whole 27 rules. Therefore we visualized association rules. Figure 3 is association rules visualized by using `arulesViz` package.

If support of rule is high, the thickness of the arrow of this visualization is big, and if lift of rule is high, the depth of the arrow of this visualization is deep. Then, the variable randomly places so that we can see all rules. We understood also about ratios of health worry that "period pains" of F1 and F2 layers and "body odor of old people" of M2 and M3 layers is high. Figure 3 is easily understandable than Table 2 because similar rules were extracted in similar age and same gender.

We could see the rule very well by this placement method, but the position between variables is unrelated to strength of relationship. Therefore we tried improvement of visualization to reflect the position relationship of variables.

Table 2. Association rule of the health worry in the media layer

lhs		rhs	*Supp*	*Conf*	*Lift*
F1	⇒	period pains	0.044	0.381	4.408
F2	⇒	period pains	0.039	0.255	2.946
F1	⇒	chalk mark	0.038	0.323	2.726
F1	⇒	anemia	0.019	0.164	2.389
F1	⇒	edema	0.033	0.283	2.354
F1	⇒	feeling of cold	0.055	0.471	2.230
F2	⇒	anemia	0.023	0.151	2.201
M3	⇒	high blood pressure	0.066	0.309	2.196
M3	⇒	athlete's foot	0.045	0.211	2.085
M1	⇒	nothing special	0.017	0.131	2.060
F2	⇒	edema	0.037	0.243	2.020
F3	⇒	the skin's aging	0.116	0.500	1.993
F1	⇒	tend to be constipated	0.036	0.309	1.932
F3	⇒	ache in the knee	0.058	0.250	1.918
F2	⇒	feeling of cold	0.060	0.400	1.894
F1	⇒	headache sufferer	0.033	0.285	1.850
M2	⇒	body odor of old people	0.025	0.156	1.825
F2	⇒	the skin's aging	0.068	0.451	1.799
F3	⇒	aching joint	0.041	0.178	1.783
F2	⇒	bat metabolism	0.054	0.358	1.779
M3	⇒	pollakiuria	0.028	0.130	1.773
M3	⇒	body odor of old people	0.032	0.151	1.758
F2	⇒	headache sufferer	0.040	0.263	1.706
M3	⇒	thinning hair	0.048	0.226	1.704
F2	⇒	chalk mark	0.028	0.186	1.572
F3	⇒	bat metabolism	0.072	0.311	1.550
F1	⇒	bat metabolism	0.036	0.307	1.526

3 Visualization of association rules using the correspondence analysis

We used the correspondence analysis (see [4,8]) to get the position relationship of variables. Correspondence analysis uses cross tabulation. And so that the correlation of the element of the row and the element of the column becomes biggest, correspondence analysis calculate row score and column score and plot them. First, the correspondence matrix Z that becomes basic of the correspondence analysis is expressed in the following expressions,

$$z_{ij} = \frac{f_{ij} - f_{i.} \times f_{.j}/n}{\sqrt{f_{i.} \times f_{.j}}} \tag{1}$$

where f_{ij} is each ingredient of the cross tabulation, $f_{i.}$ each row sum of the cross tabulation, $f_{.j}$ each column sum of the cross tabulation, n Total of the cross tabulation. And row score X and column score

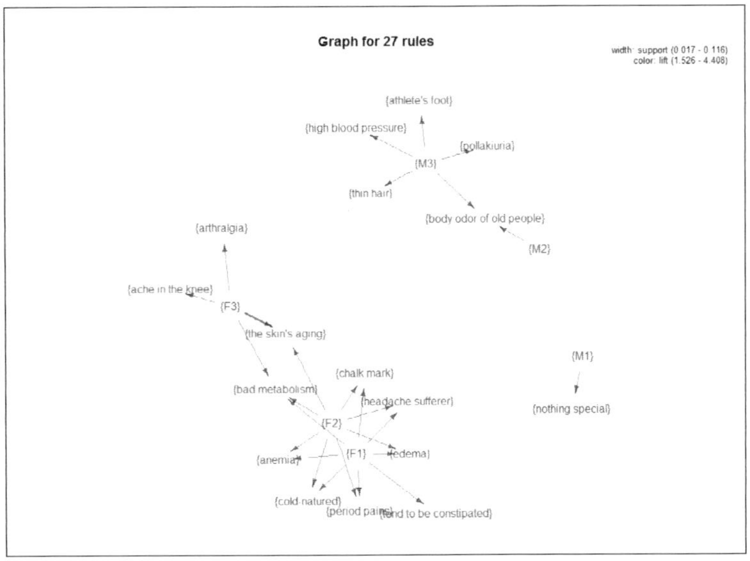

Figure 3. Visualization of association rules by using `arulesViz` package

Y are expressed in the following expressions,

$$X = D_r^{-\frac{1}{2}} V \tag{2}$$

$$Y = D_c^{-\frac{1}{2}} U \tag{3}$$

where D_r is the diagonal matrix which assumed $p_{i.}$ element ($p_{i.} = \frac{f_{i.}}{n}$), V is characteristic vector of ZZ^t, D_c is the diagonal matrix which assumed $p_{.j}$ element ($p_{i.} = \frac{f_{.j}}{n}$), U is characteristic vector of $Z^t Z$.

We explain how to make of the visualization. First, we performed the correspondence analysis of the result that performed cross tabulation by questionnaire items and media layers to set the position of each item, and we displayed the second axis of the correspondence analysis (Figure 4).

Next, we displayed circles that size depended on the number of the check and we displayed squares that size depended on the number of each media layer. Then, if its media layer is male, color of squares are blue, if its media layer is female, color of squares are red, and if its media layer is old, color of squares are deep (Figure 5).

Last, we added association rules (Figure 6). Then, if support of rule is high, the thickness of the arrow of this visualization is big, and if lift of rule is high, the depth of the arrow of this visualization is deep. In this way, we made the visualization of the questionnaire result that reflected the position relationship of items.

By Figure 6, about the question of the worry of the health, we understand that M2 layer and M3 layer are worried in "body odor of old people" and F1 layer and F2 layer are worried in "period pains". And, because there is "overweight" midmost and there is much answer number, we understand many people are worried in "fatness". But, because association rules are not shown, we understand that the relations with the specific media layer are not accepted on "fatness". In this way, we can do much consideration in one plot.

Figure 4. Plot of correspondence analysis

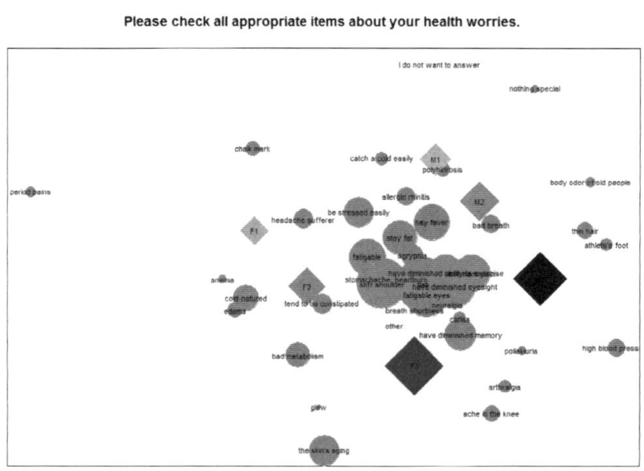

Figure 5. Improved version of the plan of the correspondence analysis

4 Interactive plot on RStudio and Shiny application

About the association rule, by rules extracted by setting of the lower limit of the support, confidence and lift changes, differences of plot that is obtained are seen. To display rules that are easy to characterize with the media layer, so that we can interactively coordinate the parameter of association rules about this plot, we built the indication system using the *manipulate* function on RStudio (see [7]) and Shiny Application (Figure 7 and Figure 8). Then, for the improvement of indication and plots of other

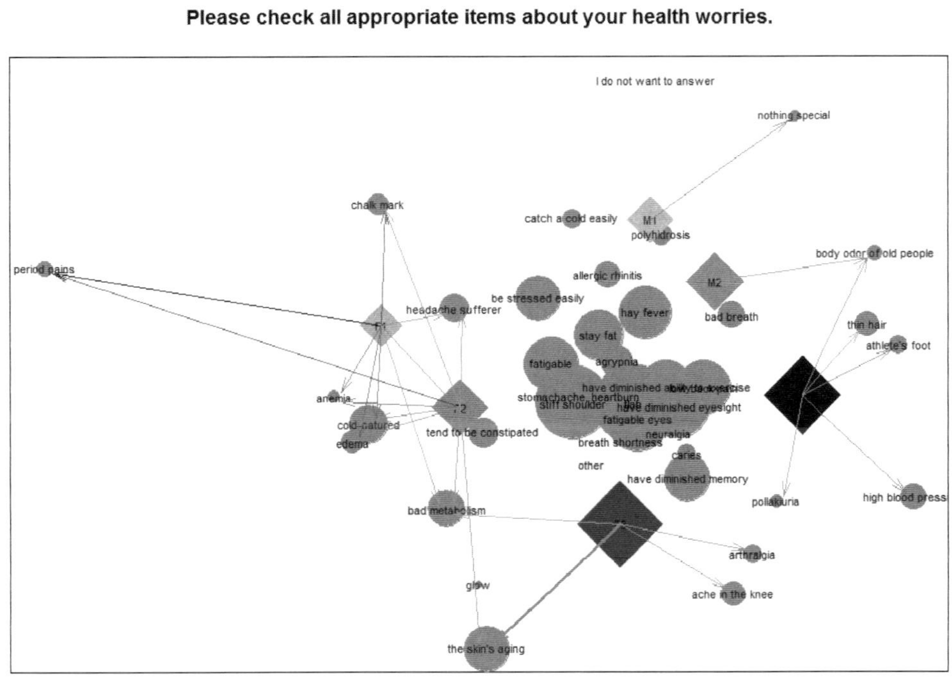

Figure 6. Visualization of association rules using the correspondence analysis

questions, We enabled change and adjustment of question contents, size of fonts, size of circles and size of squares.

5 Conclusion

By the visualization that combined association analysis with the correspondence analysis, we enabled grasping a reaction of the media layer about multiple choice questionnaires. We recognized the health worries for many people and health worries in a particular layers. In the proposed visualization, correspondence analysis and association rules analysis had a relationship to assist each other. It is possible to adjust the parameters of the association rules by *manipulate* function on RStudio and Shiny Application.

In the future, we would like to realize the visualization of association rules of over two items.

Acknowledgement

This work was supported by JSPS KAKENHI Grant Number 15K11997. The data for this research was provided by the data analysis competition 2013, Joint Associate Study Group of Manegement Science. The data source was provided by Macromill, Inc.

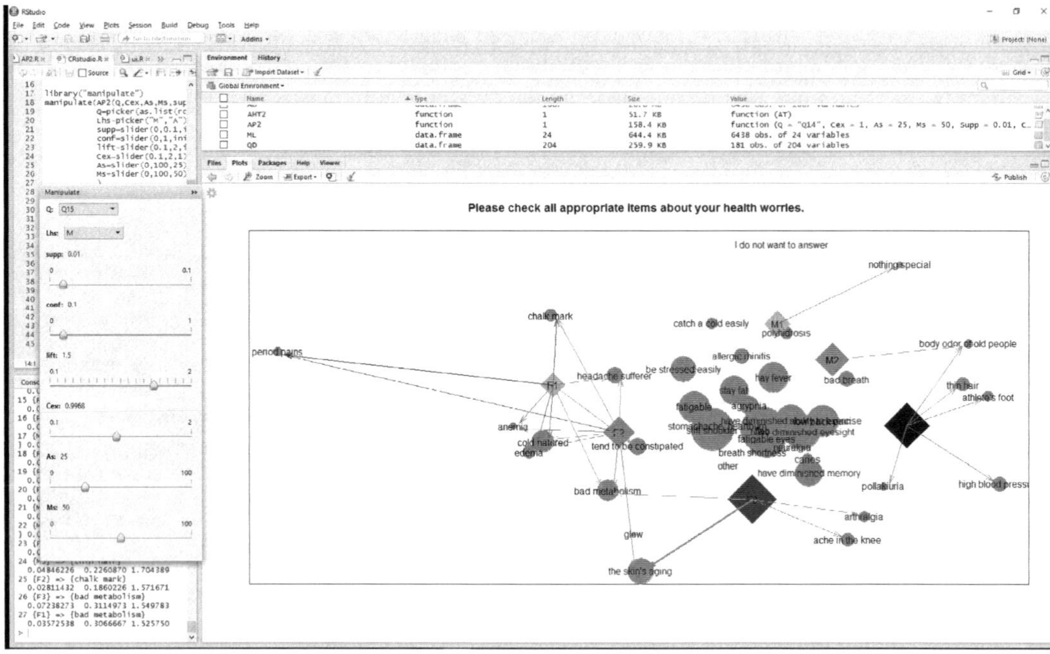

Figure 7. Interactive Plot on *manipulate* function on RStudio

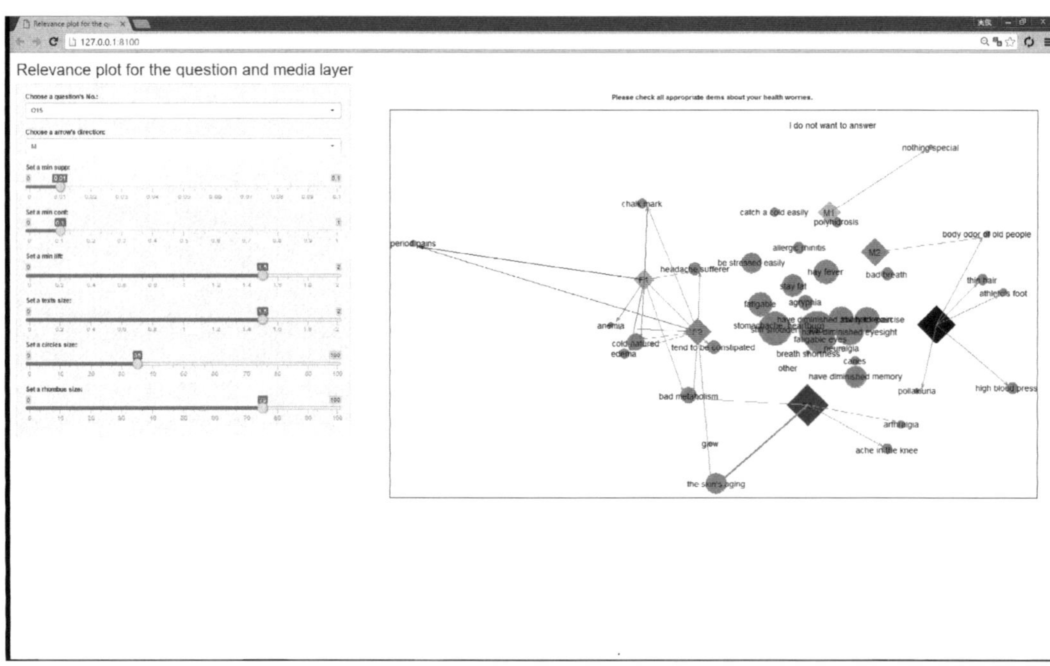

Figure 8. Interactive Plot on Shiny Application

Bibliography

[1] Agrawal,R., Imielinski, T. and Swami, A. (1993) *Mining association rules between sets of items in large databases.* In Proceedings of the ACM SIGMOD International Conference on Management of Data, 207–216.

[2] Agrawal, R. and Srikant, R.(1994) *Fast Algorithms for Mining Association Rules*, Proc. 20th int. conf. very large data bases, VLDB 1215, 487–499

[3] Ito, A., Yoshikawa, T., Furuhashi, T., Ikeda, R. and Kato, T.(2010) *Search of an interesting rule using the visualization in the association analysis*, 26th Fuzzy System Symposium, 684–689,

[4] Gower, J.C. and Hand, D.J.(1996) *Biplots.* Chapman & Hall.

[5] Lantz, B.(2015) *Machine Learning with R. 2nd Edition*, Packt Publishing.

[6] Linoff,G.S. and Berry, M.J.A.(2011) *Data Mining Techniques: For Marketing, Sales, and Customer Relationship Management. 3rd Edition.* Wiley.

[7] *RStudio*, http://www.rstudio.com/

[8] Venables, W.N. and Ripley, B.D.(2002) *Modern Applied Statistics with S. 4th Edition.* Springer.

[9] Wong, PC., Whitney, P. and Thomas, J.(1999) *Visualizing Association Rules for Text Mining*, Pacific North-west National Laboratory.

Inference in nonlinear systems with unscented Kalman filters

Diana Giurghita, *University of Glasgow*, D.Giurghita.1@research.gla.ac.uk
Dirk Husmeier, *University of Glasgow*, Dirk.Husmeier@glasgow.ac.uk

Abstract. An increasing number of scientific disciplines, most notably the life sciences and health care, have become more quantitative, describing complex systems with coupled nonlinear differential equations. While powerful algorithms for numerical simulations from these systems have been developed, statistical inference of the system parameters is still a challenging problem. A promising approach is based on the unscented Kalman filter (UKF), which has seen a variety of recent applications, from soft tissue mechanics to chemical kinetics. The present study investigates the dependence of the accuracy of parameter estimation on the initialisation. Based on three toy systems that capture typical features of real-world complex systems: limit cycles, chaotic attractors and intrinsic stochasticity, we carry out repeated simulations on a large range of independent data instantiations. Our study allows a quantification of the accuracy of inference, measured in terms of two alternative distance measures in function and parameter space, in dependence on the initial deviation from the ground truth.

Keywords. Unscented Kalman Filter, Parameter estimation, Nonlinear models, Ordinary differential equations, Stochastic differential equations.

1 Introduction

Mathematics is transforming biology in the same way it shaped physics in the previous centuries [2]. The underlying paradigm shift that distinguishes modern quantitative systems biology from more traditional non-quantitative approaches is based on a conceptualisation of elementary processes as a complex network of interactions, and its representation with an adequate mathematical description, typically in terms of coupled differential equations. While the intrinsic nonlinearities typically defy analytical tractability, advances in high-performance computing provide the hardware for fast numerical solutions. This allows an in silico exploration of complex biological systems under varying experimental conditions and in different environmental contexts. However, this forward modelling approach (simulation via numerical solution of a given mathematical description) assumes that the system under investigation is known, i.e. that the parameters defining the kinetics and dynamics of the interactions are given. Such detailed knowledge is rarely available in practice. What is needed is a solution of the so-called inverse problem, i.e. the rigour of statistical inference to systematically infer the kinetic parameters from given data.

The direct approach to parameter estimation is to minimise a divergence measure between the predictions from the model and the data, like in [3]. This approach is computationally expensive and suffers from susceptibility to local optima. Gradient matching bypasses the computationally expensive numerical solution of the differential equations and thereby allows a more exhaustive exploration of the parameter space. However, this comes at the price of information loss inherent in gradient matching, which is the subject of current methodological research [5]. A promising idea is based on Bayesian filtering, and in particular the unscented Kalman filter (UKF). The idea of the UKF, as proposed in [7], is to relax the linearity constraint of the underlying dynamics without incurring an explosion of the computational complexity (as opposed e.g. to particle filters [6]), and to include the unknown system parameters in an augmented state vector, so as to automatically track them with established Bayesian filtering techniques. In more detail, the mathematical description of the biological system leads to iterative nonlinear equations relating an augmented state vector at the present time point to a nonlinear function of the state vector at the previous time point perturbed by additive noise. The noise distribution is assumed to be multivariate normal. Starting from an initial distribution of state vectors, drawn from a multivariate normal distribution with an initial covariance matrix, state vectors are iteratively subjected to the nonlinear dynamics of the state equations, which can easily be parallelized. From these sampled vectors, the so- called sigma points, the new covariance matrix is computed. By iterative assimilation of new measurements and application of established and computationally efficient Kalman filtering techniques [6], we can obtain the posterior distribution of the model parameters. This procedure was successfully applied to inference in soft tissue mechanics of the heart [9] and chemical kinetics [1].

A potential drawback of the UKF is the dependence of the posterior distribution on the initial state, as pointed out in [8]. Given the Markovian nature of the process, the initialisation should not matter if the system is ergodic, but there is no guarantee that this condition is met in practice. The performance presented in the seminal article of [7] is impressive, but a closer inspection reveals that all results were obtained from highly informative initialisations, which started from values close to the true parameters. The objective of the present article is to investigate the dependence on the initialisation more systematically, based on an extensive range of numerical simulations. To this end, we choose the systems in [7], which are representative of what we typically encounter in systems biology: (1) a deterministic dynamical system with periodic attractor, (2) a deterministic dynamical system with chaotic attractor, and (3) a stochastic system. We systematically quantify the accuracy of parameter inference based on two alternative divergence measures: mean square error in function space, and relative bias in parameter space. Our study provides practical guidelines about the robustness of inference with UKF, and indications of when complementary techniques for informed initialisation, e.g. [8], are needed.

2 Methods

This section provides a short summary of important UKF concepts, as provided in [6], where the reader will find more details, discussion and derivations on the topic. The general form of the state space model consists of the following state and observation equations: $\mathbf{x}_t = g(\mathbf{x}_{t-1}, \mathbf{u}_t, \boldsymbol{\epsilon}_t)$, $\mathbf{y}_t = h(\mathbf{x}_{t-1}, \mathbf{u}_t, \boldsymbol{\eta}_t)$, where $\mathbf{x}_t$ is the hidden state, $\mathbf{u}_t$ is an optional input or control signal, $\mathbf{y}_t$ is the observation, g is the transition model describing the system dynamics, h is the observation model, and $\boldsymbol{\epsilon}_t$ and $\boldsymbol{\eta}_t$ representing the system respectively observation noises at time t. The Kalman filter algorithm performs exact Bayesian filtering for linear-Gaussian state space models (where g and h are linear and $\boldsymbol{\epsilon}_t$ and $\boldsymbol{\eta}_t$ are Gaussian). The advantage of the method comes from the fact that the probability distribution of the predictor step: $p(\mathbf{x}_t|\mathbf{y}_{1:t-1})$ and the probability distribution of the updating step $p(\mathbf{x}_t|\mathbf{y}_t, \mathbf{y}_{1:t-1})$ can be obtained in closed form because of the Gaussian assumption (see [6] for derivations).

However, in practice, models are often nonlinear and the Gaussian assumption does not hold and the UKF is one of the methods that can accommodate such scenarios. The UKF, proposed by Julier and Uhlmann [4] is based on the idea that it is easier to approximate a Gaussian than to linearise a function. First, a set of points, called sigma points, are chosen in a deterministic way, and then passed through a nonlinear function. Second, a Gaussian distribution is fitted to the resulting transformed sigma points. This is called the unscented transform and it is performed as follows. Given $p(\mathbf{x}) = \mathcal{N}(\mathbf{x}, \boldsymbol{\mu}, \boldsymbol{\Sigma})$, suppose we want to estimate $p(\mathbf{y})$ where $\mathbf{y} = \mathbf{f}(\mathbf{x})$ and $\mathbf{f}$ is a nonlinear function. A set of $2d + 1$ deterministically chosen points, the sigma-points are obtained as follows: $\mathbf{x} =$

$(\boldsymbol{\mu}, \{\boldsymbol{\mu} + (\sqrt{(d+\lambda)\boldsymbol{\Sigma}})_{:i}\}_{i=1}^{d}, \{\boldsymbol{\mu} - (\sqrt{(d+\lambda)\boldsymbol{\Sigma}})_{:i}\}_{i=1}^{d})$, where λ is a scaling parameter, d is the dimension of $\mathbf{x}$, $\boldsymbol{\mu}$ is the mean vector and $\boldsymbol{\Sigma}_{:i}$ represents the i'th column of the covariance matrix. The sigma points are propagated through f to obtain $\mathbf{y}_i$, and the mean and covariance of $\mathbf{y}$ are calculated based on the weighted $\mathbf{y}_i$'s. See [6] for more details about the unscented transform. The UKF uses the unscented transform twice. First, the sigma points are transformed using the system dynamics model g to compute the prediction distribution: $p(\mathbf{x}_t|\mathbf{y}_{1:t-1})$ and, second, the sigma points are transformed using the measurement model function h to compute the update distribution: $p(\mathbf{x}_t|\mathbf{y}_{1:t})$.

Parameter estimation can be accomplished with the UKF [7] by considering the parameter vector $\boldsymbol{\lambda}$ as a dynamical variable that follows the dynamical model: $\boldsymbol{\lambda}_t = \boldsymbol{\lambda}_{t-1}$. Similarly, the process noise, $\boldsymbol{\epsilon}_t$, and the measurement noise, $\boldsymbol{\eta}_t$, are added to the state vector, resulting in the joint state vector $\mathbf{j}_t$ that has the following state and observation equations:

$$\mathbf{j}_t = \begin{bmatrix} \mathbf{x}_t \\ \boldsymbol{\lambda}_t \\ \boldsymbol{\epsilon}_t \\ \boldsymbol{\eta}_t \end{bmatrix} = \begin{bmatrix} \mathbf{g}(\mathbf{x}_{t-1}, \boldsymbol{\lambda}_{t-1}) + \boldsymbol{\epsilon}_{t-1} \\ \boldsymbol{\lambda}_{t-1} \\ \boldsymbol{\epsilon}_{t-1} \\ \boldsymbol{\eta}_{t-1} \end{bmatrix} = \mathbf{g}^j(\mathbf{j}_{t-1}), \qquad \mathbf{y}_t = \mathbf{h}^j(\mathbf{j}_t) = \mathbf{h}(\mathbf{x}_t) + \mathbf{h}^{\eta}(\boldsymbol{\eta}_t) \tag{1}$$

In the models considered in this paper $\mathbf{h}^{\eta}$ is the identity matrix, but in scenarios with correlated noise it will be a non-diagonal matrix.

3 Data

This section presents the three models used in [7]: the Lotka-Volterra system, the chaotic Lorenz system, the stochastic van der Pol system, and the results of the augmented UKF (as described in Section 2) for signal tracking and parameter estimation.

Lotka-Volterra system

The Lotka-Volterra (LV) system is structured as a system of two ordinary differential equations (ODEs) as follows:

$$\frac{dx_{1t}}{dt} = \lambda_{1t}x_{1t} - \lambda_{2t}x_{1t}x_{2t} \qquad \frac{dx_{2t}}{dt} = \lambda_{2t}x_{1t}x_{2t} - \lambda_{3t}x_{2t} \tag{2}$$

The model describes the interaction of prey and predator populations which are represented as concentrations by the variables (x_2) respectively (x_1). The parameters: $\lambda_{1t}, \lambda_{2t}, \lambda_{3t}$ representing the growth rate of prey population, death rate of predator population, respectively growth rate of the prey population are considered unknown and augmented to the state variables for the joint estimation using the UKF, as discussed in Section 2 . Data was generated by numerical integration of the equations in (2), using a Runge-Kutta method implemented in `Matlab` by function `ode45`. The sampling step size is $\Delta t = 0.1$, the parameters are $\lambda_{1t} = 1, \lambda_{2t} = 1.5, \lambda_{3t} = 2$ and the initial values for the numerical integration are $\mathbf{x}_0 = (0.5, 1)^T$. To obtain the measurements, the concentration of prey population x_{1t} has been corrupted with additive Gaussian noise with standard deviation 0.1: $\boldsymbol{\eta}_t \sim \mathcal{N}(0, \mathbf{R})$ to ensure a signal to noise ratio (SNR) of 8.26. Thus, the joint state space vector for the UKF contains the population densities $\mathbf{x}_t = (x_{1t}, x_{2t})^T$, the unknown parameters $\boldsymbol{\lambda} = (\lambda_{1t}, \lambda_{2t}, \lambda_{3t})^T$, as well as the measurement noise $\boldsymbol{\eta}_t$ which is assumed to be known, as detailed in Section 2. The initialisation for the UKF is the first observation for the states: $\hat{\mathbf{x}}_0 = (y_0, y_0)^T$, and for the parameters twice the true values: $\hat{\boldsymbol{\lambda}}_0 = (2, 3, 4)$. The covariance matrix is initialised as the identity matrix: $\hat{\mathbf{P}}_0 = \mathbb{I}_5$.

The resulting limit cycle for the chosen set of parameters displays periodic oscillations which can be observed in Figure 1. The UKF algorithm tracks the real signal closely, with the estimates overlapping the clean solution towards the end of the observation time window. This is due to the fact that the UKF initialisation at y_0 is very close to the actual value given the small amount of measurement noise in the data. Furthermore, the unobserved state x_2, and the parameters are estimated with high precision using the UKF, as suggested by the small standard errors reported in Table 1.

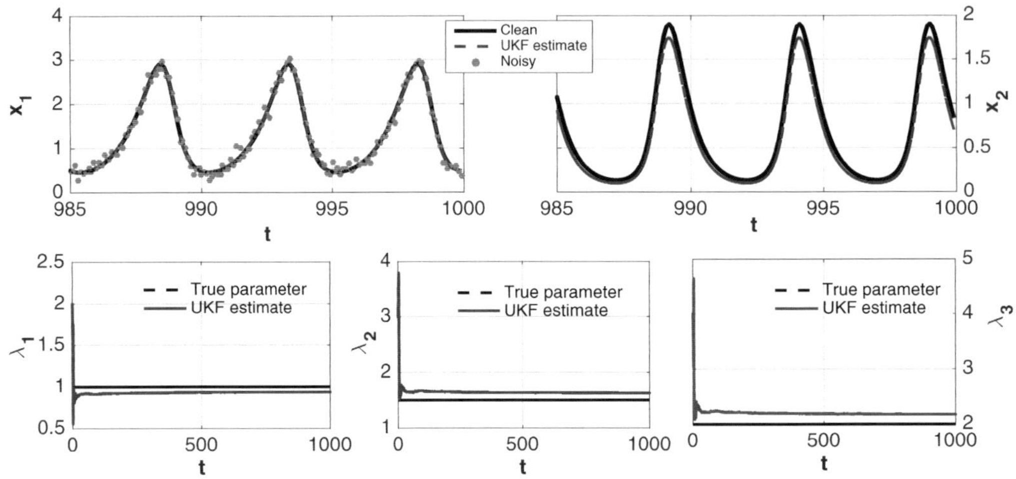

Figure 1. **Lotka-Volterra model** with parameters $\lambda_1 = 1, \lambda_2 = 1.5, \lambda_3 = 2$. *Top*: UKF tracking of prey and predator population concentrations for $t \in [998, 1000]$. *Bottom*: UKF estimation for Lotka-Volterra parameters.

Lorenz system

The Lorenz system has been extensively studied in relation to models of Earth's atmospheric convection. The system's chaotic behaviour poses an interesting challenge for the UKF in terms of convergence, especially since the system dynamics can drastically change depending on initial conditions. The mathematical form of the Lorenz system consists of the following three ODEs:

$$\frac{dx_{1t}}{dt} = -\lambda_{1t}x_{1t} + \lambda_{1t}x_{2t}; \quad \frac{dx_{2t}}{dt} = \lambda_{2t}x_{1t} - x_{2t} - x_{1t}x_{3t}; \quad \frac{dx_{3t}}{dt} = -\lambda_{3t}x_{3t} + x_{1t}x_{2t} \tag{3}$$

The ODEs in (3) with parameters $\lambda_{1t} = 10, \lambda_{2t} = 46, \lambda_{3t} = 8/3$, and initial conditions $\mathbf{x}_0 = (1, 1, 1)$ are numerically integrated using a Runge-Kutta method. $N = 10000$ data samples are obtained at equidistant intervals, $\Delta t = 0.01$, and the measurements are obtained by adding Gaussian noise $\boldsymbol{\eta}_t \sim \mathcal{N}(\mathbf{0},\ \mathbf{R})$ to the x_{1t} component only. To reproduce the results in Sitz et al. [7], the noise variance is chosen as: $\mathbf{R} = 4$. The initialisation for the UKF is the first observation for the states: $\hat{\mathbf{x}}_0 = (y_0, y_0, y_0)^T$ and for the parameters it is half the true values: $\hat{\boldsymbol{\lambda}}_0 = (5, 23, 4/3)^T$. The covariance matrix is initialised as a diagonal matrix: $\hat{\mathbf{P}}_0 = 10\mathbb{I}_6$.

In Figure 2(*Left*), the complexity of the system dynamics can be recognised by looking at the nonlinearities in 2D projections and non-periodic oscillations of the $\mathbf{x}_1$, $\mathbf{x}_2$ and $\mathbf{x}_3$ components in time. The UKF proves it is able to reconstruct the original signal (Figure 2, *Right*) providing highly accurate parameter tracking with bias within three decimal points and standard errors less than 1%, as reported in Table 1. As before, this is a consequence of the initialisation for the observed state that is very close to the true values due to the low level of measurement noise.

Van der Pol system

Consider the following equations describing the van der Pol oscillator:

$$\frac{dx_{1t}}{dt} = x_{2t}, \quad \frac{dx_{2t}}{dt} = \lambda_{1t}(1 - x_{1t}^2)x_{2t} - x_{1t} + \epsilon_t \tag{4}$$

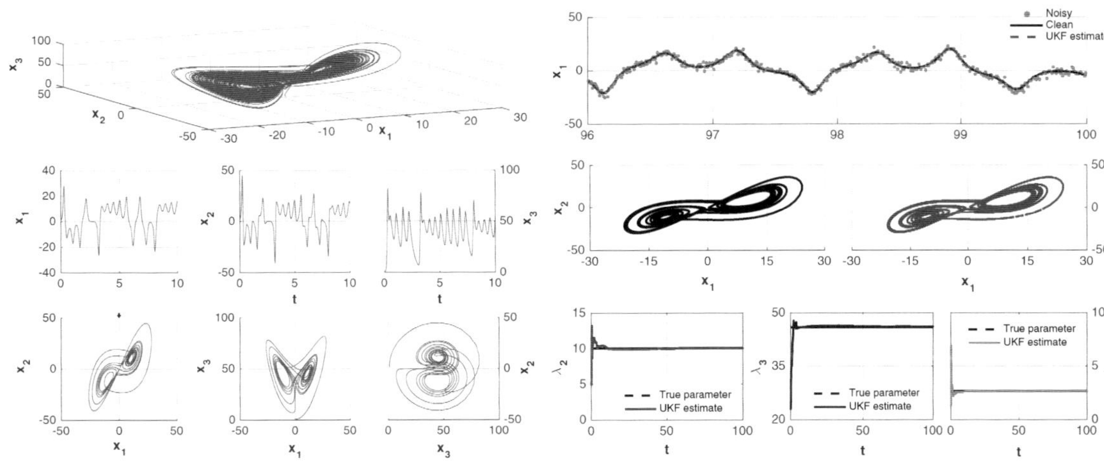

Figure 2. **Lorenz attractor** with parameters $\lambda_1 = 10, \lambda_2 = 46, \lambda_3 = 8/3$. *Left:* System dynamics and projections in 2D and 1D. *Right:* Signal reconstruction for $t = [98, 100]$, projection on 2D using actual and UKF estimated components x_1 and x_2, and parameter estimation using the UKF.

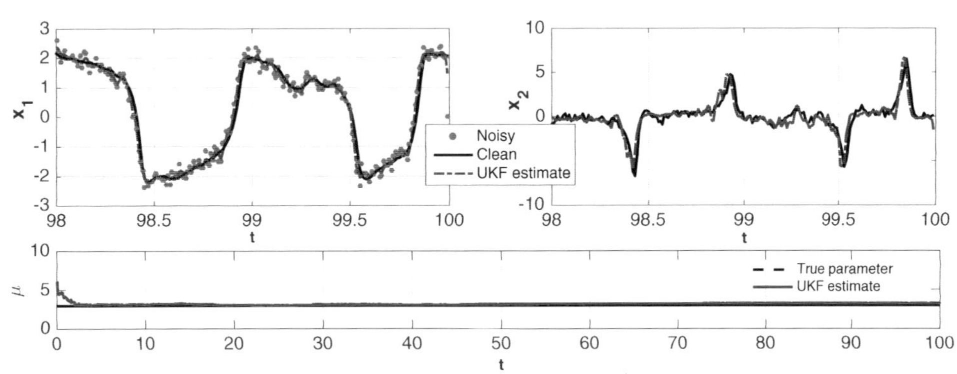

Figure 3. **Stochastic van der Pol system** with parameter $\lambda_1 = 3$: signal reconstruction (*Top*) and parameter estimation (*Bottom*) using the UKF.

The stochasticity of the system comes from the second component which contains an uncorrelated noise term ϵ_t. For the augmented UKF, the process noise is considered fixed and known: $\epsilon \sim N(0,1)$. The noise propagates through both states x_1 and x_2, which means the system will display stochastic oscillations throughout the observation time window. See Figure 3 for a solution of the stochastic van der Pol system, with parameter $\lambda_{1t} = 3$ and $[1,0]$ as the initial conditions for the numerical integration. Measurements are obtained from the first component, $\mathbf{x}_{1t}$ by adding Gaussian noise: $\boldsymbol{\eta}_t \sim \mathcal{N}(\mathbf{0}, \mathbf{R})$ with variance $R = 0.2^2$. This ensures a SNR of around 10%. As with all stochastic systems, the numerical integration has to be performed using a very small step size, $\delta t = 0.001$, and a low order integration method such as the Euler method [7].

The initialisation for the UKF is the first observation for the states: $\hat{\mathbf{x}}_0 = (y_0, y_0)^T$, and for the parameter twice the true value: $\hat{\boldsymbol{\lambda}}_1 = 6$. The covariance matrix is initialised as a diagonal matrix: $\hat{\mathbf{P}}_0 = 2\mathbb{I}_3$. In Figure 3 and Table 1 we report the results of the UKF estimation for the van der Pol system. Since the UKF initialisation for the observed system state is relatively close to the true one (SNR is 10%), the estimated path of the signal is very close to the real signal. However, the tracking for the second component is not as precise, which is also reflected in the higher standard error for $\mathbf{x}_2$. Nevertheless, this is not unusual given the stochastic nature of the system, and the fact that this state is not observed.

To conclude this section, we report that our results from the UKF estimation for the three considered systems are consistent with those reported in [7].

Table 1. UKF estimation results for the Lotka-Volterra, Lorenz and van der Pol systems. Point estimates and standard errors (in brackets) are reported using the last prediction step of UKF.

	States and parameter estimates using the UKF					
	$\hat{\mathbf{x}}_1$	$\hat{\mathbf{x}}_2$	$\hat{\mathbf{x}}_3$	$\hat{\lambda}_1$	$\hat{\lambda}_2$	$\hat{\lambda}_3$
Lotka-Volterra	0.474 (0.001)	0.837 (0.004)	-	1 (0.001)	1.5 (0.003)	2 (0.004)
Lorenz	-4.836(0.855)	-10.518(1.808)	17.665(0.635)	10(0.019)	46 (0.101)	2.667 (0.009)
Van der Pol	2.075 (0.111)	0.157 (0.627)	-	3 (0.093)	-	-

4 Simulations

The main aim of this study is to systematically investigate the influence of the UKF initialisation, given that this is an important practical aspect that has not been addressed in the literature in great depth. We complement the results presented in [7] with a series of simulation studies to provide a more complete and informative picture of UKF estimation for nonlinear models. Hence, we preserve the modelling choices reported in [7]: sampling step size, integration step size, sampling time interval, system parameters, model errors etc. Note that the covariance matrix initialisation for the UKF will most likely be different from the one chosen by Sitz et al. [7], as it was not reported by the authors.

Initialisation

The impact of the initialisation has been assessed by considering a range of offsets for the states and model parameters: $\{0\%, 25\%, 50\%, 100\%, 150\%\}$, which include the choices in [7] across the 3 models included. For consistency, the offsets have been calculated as percentages from the true values, with the choice of a positive or negative offset being decided at random for each UKF instantiation. The offset has been considered for the states $\mathbf{x}_0$ as well, even though the most common scenario would be to initialise using the first observed states $\mathbf{y}_0$. The reasoning behind this choice is that, in practice, it is also plausible to consider a different initialisation other than the first observation, which would be based on prior knowledge of the system dynamics.

Number of observations

A secondary aim of this study is to investigate the impact of reducing the number of observations, given the results reported in Section 3 indicate that comparable UKF estimates could be obtained using a smaller number of observed samples. reaches Two scenarios have been compared, see Table 2: the 'Same frequency' scenario which relies on fixing the same sampling step size, Δt, and reducing the time window, T, to acquire the desired number of observations, N, and the 'Less frequency' scenario which relies on increasing the sampling step size until a set number of observations is achieved in a certain time window. Given that originally the results have been produced using $N = 10000$ observation, which is an excessive number for most applications in practice, we opted for a more realistic $N = 3000$ observations, which is a 70% decrease in the number of observations. These original settings, denoted as 'Original' in Table 2 have been used as a benchmark for the other two scenarios.

Table 2. Details on the simulation scenarios for investigating the impact of sampling frequency on the UKF results.

	Lotka-Volterra system	Lorenz system	Van der Pol system
Original	$\Delta t = 0.1, N = 10000$ $T = [0, 1000]$	$\Delta t = 0.01, N = 10000$ $T = [0, 100]$	$\Delta t = 0.01, N = 10000$ $T = [0, 1000]$
Same frequency	$\Delta t = 0.1, N = 3000$ $T = [0, 300]$	$\Delta t = 0.01, N = 3000$ $T = [0, 30]$	$\Delta t = 0.1, N = 3000$ $T = [0, 300]$
Less frequency	$\Delta t = 0.33, N = 3000$ $T = [0, 1000]$	$\Delta t = 0.033, N = 3000$ $T = [0, 100]$	$\Delta t = 0.33, N = 3000,$ $T = [0, 1000]$

Reporting results

Two measures have been calculated for each scenario to asses differences in the performance in parameter space and in function space: relative bias $RelBias_t = \frac{\hat{\mathbf{x}}_t - \mathbf{x}_t}{\mathbf{x}_t}$ and mean squared error $MSE = \sum_{t=\frac{N}{2}}^{N} (\hat{\mathbf{x}}_t - \mathbf{x}_t)^2$. For the calculation of the MSE, the first half of the UKF estimation is discarded across the simulations, to ensure the UKF estimation has enough time to converge to the true value, which will also ensure a fairer comparison across data sets with very different initialisations. Although in recursive methods such as the UKF, the last step (after all the data has been seen) provides the best estimate for the state, in stochastic models the bias will often exhibit small fluctuations, in which case the bias reported has been calculated as an average of the biases for the last 100 estimates.

5 Results

This section contains the results for the simulations carried out to investigate the effect of the initialisation on parameter estimation using the augmented UKF, as well as the effect of reducing the sample size, as discussed in Section 4 and summarized in Table 2. Across the scenarios considered, 10 simulation have been run for each situation e.g.: 10 sets of 'Original', 10 sets of 'Same frequency', 10 set of 'Less frequency' etc. Without loss of generality, the results for the unobserved states have been excluded from the plots since they exhibit similar patterns as the observed system state.

Lotka Volterra system

The simulation results in Figure 4 suggest that decreasing the observed sampling time either by increasing the sampling frequency time step, or by reducing the time interval for the system measurements will result in a deterioration of the results, in terms of bias and MSE. Although the range of the bias and MSE suggests keeping the sampling step size constant, the median indicates that increasing the sampling size will increase the bias by on average 5%.

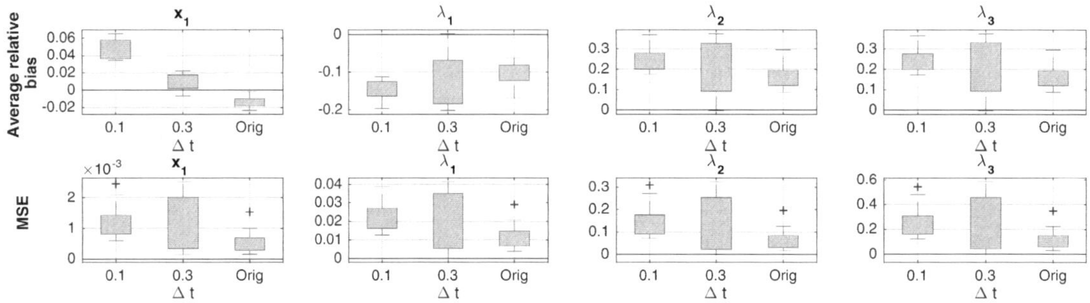

Figure 4. **Lotka-Volterra system**: observation sample size. Boxplots of relative bias (*Top*) and MSE (*Bottom*) over 10 data sets for the three scenarios considered in Table 2.

Figure 5 shows the effect of the initialisation offset on the UKF estimation for the three models considered, by looking at box plots of the relative bias from 10 data sets. For the Lotka Volterra model, as the relative offset increases to 150% the estimation deteriorates by, on average, 7 % for the parameters bias, and 0.7 % for the state bias. Notice that the bias range increases by a factor of between 4 and 7 for the parameters, which means the estimation becomes considerably uncertain for 150% offset. The MSE at 150% offset is, on average, about 50 times higher for signal and up to 10 times higher for parameters compared to 0% offset, indicating the UKF estimation is considerably worse at higher offset.

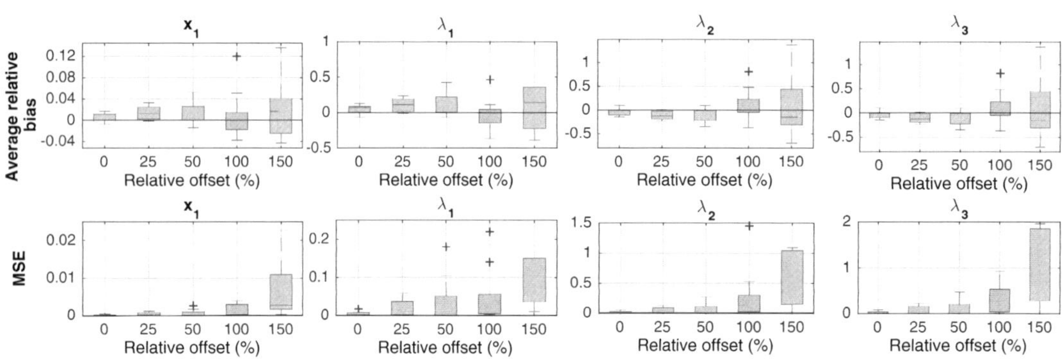

Figure 5. **Lotka-Volterra system**: initialisation. Boxplots of relative bias (*Top*) and MSE (*Bottom*) over 10 data sets with relative offsets:$\{0\%, 25\%, 50\%, 100\%, 150\%\}$.

Lorenz system

In Figure 6, the bias indicates that maintaining the sampling size at $\Delta t = 0.01$ produces slightly better results compared to decreasing the sampling frequency, although a difference of around 1-2% in parameter space is likely to be irrelevant in practice. However, the range of the bias and MSE suggests that UKF is more likely to produce estimates that are more inaccurate if the number of observations is reduced (by either scenario). Furthermore, the presence of outliers in reduced sample size scenarios means that the algorithm can get stuck in local optima of the likelihood landscape, and one way to overcome this would be a larger sample size. This is supported by the MSE which is, on average, up to 10 times higher for parameter estimates for two scenarios compared to the 'Original' scenario.

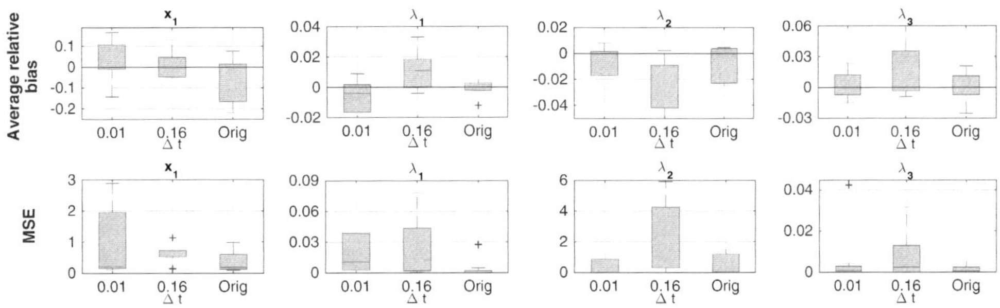

Figure 6. **Lorenz system**: observation sample size. Boxplots of relative bias (*Top*) and MSE (*Bottom*) over 10 data sets for the three scenarios considered in Table 2.

Figure 7 shows results for only 4 of the 5 offsets considered in the other scenarios due to the fact that in the chaotic Lorenz system a large initial offset is very likely to produce very different system dynamics. For example, with an initial offset of 150% the UKF performed very poorly due to numerical instabilities which caused the algorithm to stop before the end, as such results are not available for the comparison. Note also that the estimation degrades for the parameter bias, on average, up to 50% and up to a factor of 150 for the parameter MSE at 100% offset.

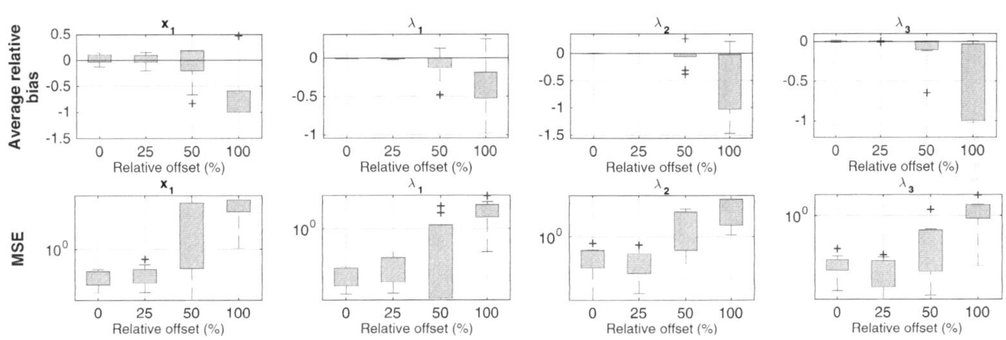

Figure 7. **Lorenz system**: initialisation. Boxplots of relative bias, zoomed-in to a more informative scale (*Top*), and MSE (*Bottom*), on a log scale due to high range of values. Estimates obtained over 10 data sets with relative offsets:$\{0\%, 25\%, 50\%, 100\%\}$.

Van der Pol system

In Figure 8, the relative bias and the MSE indicate that reducing the sampling frequency reduces the quality of the estimation in terms of parameter bias by, on average, 140%, and in terms of MSE by a factor of 2 on the $\log_{10}$ scale. The stochasticity of the system poses a greater challenge for the UKF, and the increase in sampling frequency is leading to a significant loss of information that is required to provide more accurate estimates. Given that the difference between the bias of the 'Original' and '$\Delta t = 0.1$', is less than 1% means the UKF achieves convergence really fast, even before the first 30% of the observations. This makes sense given the periodic oscillation in the observed system. When dealing with similar periodic systems, having data that covers several cycles of the observed signal would be sufficient for the UKF to provide reliable estimates. In Figure 9 we present the results of UKF estimation

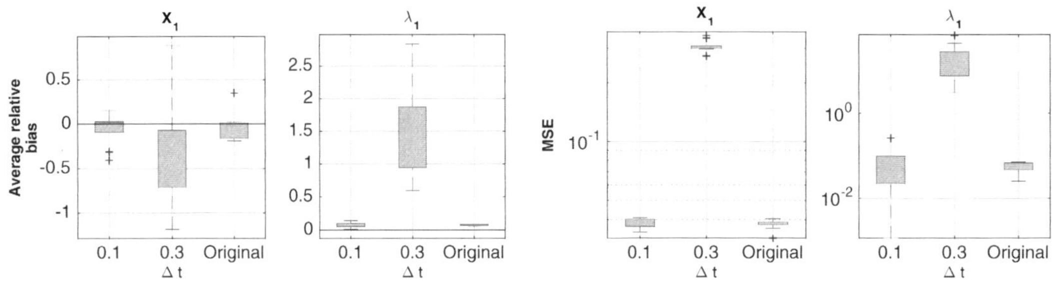

Figure 8. **Van der Pol system**: observation sample size. Boxplots of relative bias (*Left*) and MSE on a log scale (*Right*) obtained over 10 data sets for the three scenarios considered in Table 2.

of the van der Pol system, with different initialisations. UKF estimation is quite robust to initialization for the van der Pol system, even for 150% offset, with the results indicating the bias for all states is, on average, close to 0 for all initialisations. For the parameter λ_1 the estimation differs only by 1 to 2% on average. The MSE is similar in terms of median and range across offsets, indicating that reliable estimates can be obtained from a big range of starting points.

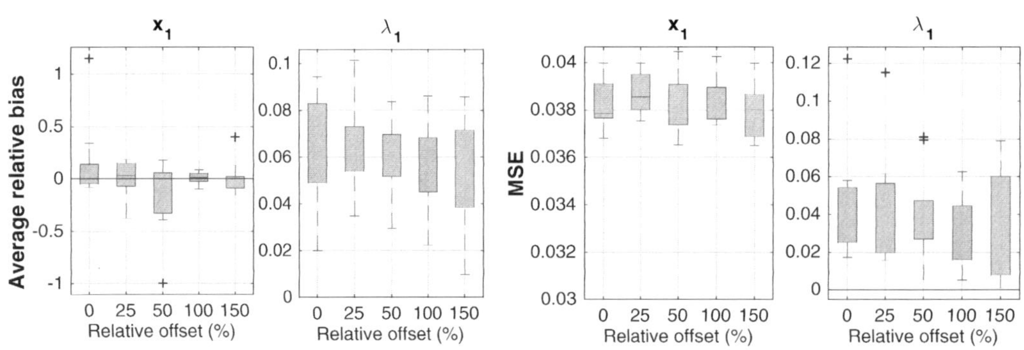

Figure 9. **Van der Pol system**: initialisation. Boxplots of relative bias, zoomed-in to a more informative scale (*Left*), and MSE(*Right*). Estimates obtained over 10 data sets with relative offsets $\{0\%, 25\%, 50\%, 100\%, 150\%\}$.

6 Discussion

Motivated by recent developments in mathematical biology, our study has focused on the accuracy of parameter inference in dynamical systems approximated by a UKF. We have selected three systems with features that are representative of many complex systems, including chaotic attractors and intrinsic stochasticity. The relative simplicity of these systems has allowed us to run hundreds of independent simulations. This has enabled us to obtain the distributions of two figures of merit, quantifying the distance from the known ground truth both in parameter and function space.

First, we looked at the dependence of the parameter estimation on the observation sample size, and we have done this by comparing the models with the original settings as implement in [7] with two different scenarios that contained just 30% of the original sample size. We have shown that for some models, the UKF requires a larger sample size (Lorenz attractor, Lotka-Volterra), or larger frequency (van der Pol) to produce reliable estimates, and this is specific to the particular dynamics involved (stochasticity, periodicity etc.).

Additionally, our particular focus has been on the dependence of the parameter estimation on the initialisation. We have shown that for an initialisation close to the ground truth the results of the seminal study in [7] can be reproduced. Nonetheless, even for perturbations in the order of 100% (a factor of 2), the results noticeably deteriorate. Our plots allow an objective quantification of this deterioration as a function of the initial deviation. The poorer performance for less informative initialisations suggests that the UKF is susceptible to local optima in the likelihood landscape. This appears to be an intrinsic feature of the sequential parameter updating scheme, whereby the relative weight of the corrector step, which updates the parameter distribution by assimilation of new data, exhibits a monotonic decrease proportional to the inverse of the number of data instances. This can render it difficult for the UKF to escape from a local attractor state in the vicinity of a poor initialisation.

In conclusion, our study provides a more cautionary picture than the original publication in [7] and suggests that a complementary fast preliminary parameter scanning scheme, as e.g. proposed in [8], appears to be indispensable to make the method applicable to complex systems for which little knowledge is available a priori.

Bibliography

[1] Baker S.M., Poskar C.H., and Junker, B.H. (2011) *Unscented Kalman filter with parameter identifiability analysis for the estimation of multiple parameters in kinetic models* EURASIP Journal on Bioinformatics and Systems Biology, **2011**:7

[2] Cohen J.E. (2004) *Mathematics Is Biology's Next Microscope, Only Better; Biology Is Mathematics' Next Physics, Only Better* PLoS Biology, **2**(12)

[3] Gao, H., Li, W. G., Cai,L., Berry, C. and Luo, X. Y. (2015) *Parameter estimation in a Holzapfel-Ogden law for healthy myocardium* Journal of Engineering Mathematics **95**(1), 231–248

[4] Julier, S. J. and Uhlmann, J. K., (1997), *New extension of the Kalman filter to nonlinear systems.* AeroSense'97, 182–193.

[5] Macdonald, B., Higham, C., and Husmeier, D. (2015) *Controversy in mechanistic modelling with Gaussian processes* Journal of Machine Learning Research: Workshop and Conference Proceedings **37**, 1539–1547.

[6] Murphy, K.P. (2012) *Machine learning: a probabilistic perspective*, MIT Press, Cambridge.

[7] Sitz, A., Schwarz, U., Kurths, J. and Voss, H. U. (2002) *Estimation of parameters and unobserved components for nonlinear systems from noisy time series* , Phys. Rev. E, **66**(1), 016 – 210

[8] Strebel O. (2013) *A preprocessing method for parameter estimation in ordinary differential equations* Chaos, Solitons & Fractals 57 **93** –104

[9] Xia,J., Lamataa,J., Leeb,J., Moireauc,P., Chapellec,D. and Smith, N., (2011)*Myocardial transversely isotropic material parameter estimation from in-silico measurements based on a reduced-order unscented Kalman filter* Journal of Mechanical Behaviour of Biomechanical Materials **4**, 1090–1102

Using sparse kernels to design computer experiments with tunable precision

Guillaume Sagnol, *Zuse Institute Berlin*, sagnol@zib.de
Hans-Christian Hege, *Zuse Institute Berlin*, hege@zib.de
Martin Weiser, *Zuse Institute Berlin*, weiser@zib.de

Abstract. Statistical methods to design computer experiments usually rely on a Gaussian process (GP) surrogate model, and typically aim at selecting design points (combinations of algorithmic and model parameters) that minimize the average prediction variance, or maximize the prediction accuracy for the hyperparameters of the GP surrogate. In many applications, experiments have a *tunable precision*, in the sense that one software parameter controls the tradeoff between accuracy and computing time (e.g., mesh size in FEM simulations or number of Monte-Carlo samples). We formulate the problem of allocating a budget of computing time over a finite set of candidate points for the goals mentioned above. This is a continuous optimization problem, which is moreover convex whenever the tradeoff function accuracy vs. computing time is concave. On the other hand, using non-concave weight functions can help to identify sparse designs. In addition, using sparse kernel approximations drastically reduce the cost per iteration of the multiplicative weights updates that can be used to solve this problem.

Keywords. Optimal design of computer experiments, Gaussian process, Sparse kernels

1 Introduction

We consider a computer code taking an input $\vec{x} \in \mathcal{X}$ (called a *design point*) in a compact set $\mathcal{X} \subset \mathbb{R}^d$ and a parameter τ specifying the time allowed for the computation, and returning an output of the form

$$Y(\vec{x}, \tau) = \eta(\vec{x}) + \epsilon(\vec{x}, \tau), \tag{1}$$

where $\eta(\cdot)$ is an unknown function and $\epsilon(\vec{x}, \tau)$ represents uncorrelated errors: $\mathbb{E}[\epsilon(\vec{x}, \tau)] = 0$, $\vec{u} \neq \vec{v} \Rightarrow \mathbb{E}[\epsilon(\vec{u}, \tau_u)\epsilon(\vec{v}, \tau_v)] = 0$ for all $\tau_u, \tau_v > 0$, where $\mathbb{E}[X]$ stands for the expectation of X. We assume that the experiments have a *tunable precision*, in the sense that the variance of the error $\mathbb{V}[\epsilon(\vec{x}, \tau)]$ is a decreasing function of the time τ spent to compute an approximation of $\eta(\vec{x})$. Specifically, we assume that there is a *known* parameter σ_N^2 (where the subscript N stands for *noise*) and a *known* differentiable, nondecreasing function $w : \mathbb{R}_+ \mapsto \mathbb{R}_+$ satisfying $w(0) = 0$, such that $\mathbb{V}[\epsilon(\vec{x}, \tau)] = \sigma_N^2 w(\tau)^{-1}$. Let the experimental design (or simply *the design*) be represented by $\xi = \{\vec{x}_i, \tau_i\}_{i \in \{1,\dots,n\}}$, where the *given candidate points* $\vec{x}_i \in \mathcal{X}$ are distinct, and $\tau_i \geq 0$ is the computing time spent on design point $\vec{x}_i$. Note that $\tau_i = 0$ means that no computation is carried out at $\vec{x}_i$, and formally we have $\mathbb{V}[\epsilon(\vec{x}_i, \tau_i)] = +\infty$. Candidate points $\vec{x}_i$ with $\tau_i > 0$ are called *support points* and are those which are actually selected for the design.

We assume that a two-stage approach is used, and observations have already been collected during the initial stage from a design $\xi_{\text{init}} = \{\vec{x}_i^0, \tau_i^0\}_{i=1,\dots,n_0}$, with $\vec{x}_i^0 \neq \vec{x}_j$ for all i, j. The purpose of this article is to develop efficient algorithms for the computation of near-optimal computing times τ_i, subject to a constraint on the total computing time allowed for the second stage: $\sum_{i=1}^n \tau_i = T$. In other words, we search for a (near-)optimal design within the class of all designs that assign a total computing time of T, and whose support is a subset of the $\vec{x}_i$'s. In practice, this two-stage approach can be turned into a sequential one, as follows. Given an optimized design $\vec{\tau}^*$, select one support point $\vec{x}_i$ and compute $Y(\vec{x}_i, \tau_i^*)$. Then, append $\{\vec{x}_i, \tau_i^*\}$ to ξ_{init}, decrement T by τ_i^*, remove $\vec{x}_i$ from the list of candidate points, update the surrogate model for $\eta(\cdot)$, compute the next design $\vec{\tau}^*$, and iterate. This procedure can also be generalized to work on a parallel architecture, where several design points can be processed simultaneously.

Our assumption about the existence of an information function $w()$ is not common. Most authors focus on the search for *exact designs*, i.e. $w_i = w(\tau_i) \in \{0, 1\}$, and $w_i = 1$ indicates that the design point $\vec{x}_i$ belongs to ξ. We refer the reader to [15] for a comprehensive review on exact designs for computer experiments. For the standard linear model, a popular technique is to relax the integer constraint on w_i, which led to the success story of the *theory of approximate designs* [12, 16]. Approximate designs are used most often as a heuristic to find good exact designs, typically by rounding. For computer experiments, however, the total computing time is of much more importance than the number of design points, which motivates to study the tradeoff between tunable accuracy and computing time in more detail. We give two examples:

- In the case of Monte-Carlo simulations, the variance is inversely proportional to the number of samples, and hence $w(\tau) = \tau$. In fact, τ can take integer values only, but we expect the approximation to be good enough for a large number of simulation runs.
- The standard a priori error estimate for finite element solutions of ansatz order p and mesh width h for sufficiently regular stationary elliptic problems in $\Omega \subset \mathbb{R}^d$ is $\|u - u_h\|_{H^1(\Omega)} = \mathcal{O}(h^p)$. The optimal computational complexity is $\tau = \mathcal{O}(h^{-d})$, which means $\tau = \mathcal{O}(\|u - u_h\|_{H^1(\Omega)}^{-d/p})$, see, e.g., [3]. We could thus model the error by a noise of variance $\text{Var}[\epsilon(\vec{x}, \tau)] = \mathcal{O}(\tau^{-2p/d})$, i.e. $w(\tau) = \mathcal{O}(\tau^{2p/d})$.

Note, however, that errors at two design points might be correlated, as is usually the case in finite element models. An alternative could be to use a Gaussian process to model the error process. We leave this for future research, and simply assume as an approximation that the $\epsilon(\vec{x}_i, \tau_i)$ are uncorrelated, which might be acceptable if the $\vec{x}_i$'s are far enough from each other.

Following the kriging methodology, we assume that $\eta(\vec{x}) = f(\vec{x})^T \vec{\beta} + Z(\vec{x})$, where $f : \mathbb{R}^d \mapsto \mathbb{R}^m$ is a spatial regression function (generally a polynomial), $\vec{\beta} \in \mathbb{R}^m$ is an unknown vector of parameters, and Z is a Gaussian random field with zero mean and *known correlation structure*,

$$\mathbb{E}[Z(\vec{x})] = 0, \qquad \mathbb{E}[Z(\vec{u})Z(\vec{v})] = \sigma_Z^2\, C(\vec{u}, \vec{v}),$$

where $C : \mathbb{R}^d \times \mathbb{R}^d \mapsto \mathbb{R}$ is a positive semidefinite kernel satisfying $C(\vec{u}, \vec{u}) = 1$ for all $\vec{u} \in \mathcal{X}$.

Denote by $Y(\xi) = [Y(\vec{x}_1), \ldots, Y(\vec{x}_n)]^T$ the vector of observations associated with the design ξ. We have

$$Y(\xi) \sim \mathcal{N}(\vec{F}^T \vec{\beta},\, \sigma_Z^2 \vec{C} + \vec{\Delta}), \tag{2}$$

where $\vec{F} = [f(\vec{x}_1), \ldots, f(\vec{x}_n)] \in \mathbb{R}^{m \times n}$, $\{\vec{C}\}_{ij} := C(\vec{x}_i, \vec{x}_j)$. and $\vec{\Delta} = \sigma_N^2 \,\text{Diag}(w(\vec{\tau}))^{-1}$. We also define the *signal-to-noise ratio* $\gamma = (\sigma_Z/\sigma_N)^2$ and $\vec{\Sigma} = \vec{C} + \text{Diag}(\gamma w(\vec{\tau}))^{-1}$, so that $\text{Var}[Y(\xi)] = \sigma_Z^2 \vec{\Sigma}$. The best linear unbiased estimator (BLUE) of $\vec{\beta}$ and its variance are given by

$$\hat{\vec{\beta}} = \left(\vec{F}\vec{\Sigma}^{-1}\vec{F}^T\right)^{-1} \vec{F}\vec{\Sigma}^{-1} Y(\xi), \qquad \text{Var}[\hat{\vec{\beta}}] = \sigma_Z^2 \left(\vec{F}\vec{\Sigma}^{-1}\vec{F}^T\right)^{-1}.$$

Then, the best linear unbiased predictor (BLUP) of the unknown function η at $\vec{x} \in \mathcal{X}$ based on the observations $Y(\xi)$ is given by

$$\hat{\eta}(\vec{x}|\xi) = f(\vec{x})^T \hat{\vec{\beta}} + c(\vec{x})^T \Sigma^{-1} \left(Y(\xi) - \vec{F}^T \hat{\vec{\beta}}\right),$$

where $\{c(\vec{x})\}_i := C(\vec{x}, \vec{x}_i)$ is the vector of cross-covariances between $\vec{x}$ and the design points, and the mean-squared prediction error (MSPE) is

$$\begin{aligned}\rho(\vec{x}) &:= \mathbb{E}[(\hat{\eta}(\vec{x}|\xi) - \eta(\vec{x}))^2] \\ &= \sigma_Z^2 \left\{1 - c(\vec{x})^T \vec{\Sigma}^{-1} c(\vec{x}) + (f(\vec{x}) - \vec{F}\vec{\Sigma}^{-1}c(\vec{x}))^T \left(\vec{F}\vec{\Sigma}^{-1}\vec{F}^T\right)^{-1} (f(\vec{x}) - \vec{F}\vec{\Sigma}^{-1}c(\vec{x}))\right\}.\end{aligned}$$

The above expression reduces to $\rho(\vec{x}) = \sigma_Z^2\big(1 - c(\vec{x})^T\vec{\Sigma}^{-1}c(\vec{x})\big)$ when the trend parameter $\vec{\beta}$ is known. Note that $\rho(\vec{x})$ depend on the design ξ through $\vec{\Sigma}$. A standard approach is to choose ξ so as to minimize the integrated mean squared error (IMSE):

$$\mathrm{IMSE}(\xi) := \int_{\mathcal{X}} \rho(\vec{x})\, d\mu(\vec{x}).$$

The IMSE criterion depends on a measure μ on $\mathcal{X}$, which can be used to weigh the interest of the experimenter for knowing the value of η at $\vec{x}$. E.g., if the goal is to minimize $\eta(\vec{x})$ over $\mathcal{X}$, or to estimate the probability that $\eta(\vec{x})$ lies below some threshold, μ should weigh regions of $\mathcal{X}$ such as to balance the exploration/exploitation tradeoff; see, e.g., [9, 1].

It was shown (for the standard case where the variance of ϵ is not a function of τ) in [4, 5] that model (1) can be approximated arbitrarily well by a Bayesian linear model of the form

$$Y(\vec{x}) \simeq [f(\vec{x})^T, g(\vec{x})^T] \begin{bmatrix} \vec{\beta} \\ \vec{\alpha} \end{bmatrix} + \epsilon(\vec{x}), \tag{3}$$

where $\vec{\alpha}$ is a random regression parameter with prior $\vec{\alpha} \sim \mathcal{N}(0, \sigma_Z^2 \vec{I}_s)$, $\vec{I}_s$ is the $s \times s$–identity matrix, and the function $g : \mathbb{R}^d \mapsto \mathbb{R}^s$ can be obtained by truncating the Mercer's expansion of the kernel $C(\cdot,\cdot)$. In our case, recall that observations have already been collected during an initial stage at $\xi_{\text{init}} = \{\vec{x}_i^0, \tau_i^0\}_{i=1,\ldots,n_0}$, $\mathrm{Var}[\epsilon(\vec{x},\tau)] = \sigma_N^2 w(\tau)^{-1}$ and the noise is uncorrelated. Then, by using standard results from the literature on Bayesian designs, see, e.g., [13], one obtains the following approximation of the Kriging variance for the design $\xi = \{\vec{x}_i, \tau_i\}_{i=1,\ldots,n}$:

$$\tilde{\rho}(\vec{x}) \simeq \sigma_Z^2\, h(\vec{x})^T \vec{M}(\xi)^{-1} h(\vec{x}),$$

where $h(\vec{x}) = [f(\vec{x})^T, g(\vec{x})^T]^T$ and $\vec{M}(\xi)$ is the (scaled) Fisher information matrix for $(\vec{\beta}, \vec{\alpha})$:

$$\vec{M}(\xi) := \sum_{i=1}^{n} \gamma\, w(\tau_i) \begin{bmatrix} f(\vec{x}_i) \\ g(\vec{x}_i) \end{bmatrix} \begin{bmatrix} f(\vec{x}_i) \\ g(\vec{x}_i) \end{bmatrix}^T + \underbrace{\sum_{i=1}^{n_0} \gamma\, w(\tau_i) \begin{bmatrix} f(\vec{x}_i^0) \\ g(\vec{x}_i^0) \end{bmatrix} \begin{bmatrix} f(\vec{x}_i^0) \\ g(\vec{x}_i^0) \end{bmatrix}^T + \begin{bmatrix} \vec{0} & \vec{0} \\ \vec{0} & \vec{I}_s \end{bmatrix}}_{\vec{\Gamma}}. \tag{4}$$

Further, the IMSE criterion can be approximated by a criterion of Bayesian A-optimality:

$$\widetilde{\mathrm{IMSE}}(\xi) = \int_{\mathcal{X}} \tilde{\rho}(\vec{x})\, d\mu(\vec{x}) = \operatorname{trace} \vec{M}(\xi)^{-1} \vec{L}$$

with a coefficient matrix $\vec{L} = \sigma_Z^2 \int_{\mathcal{X}} h(\vec{x}) h(\vec{x})^T d\mu(\vec{x})$. We restrict our attention to the situation in which $\vec{\beta}$ is estimable from the observations collected with the initial design ξ_{init}, which ensures that $\vec{\Gamma}$ is positive definite, and $\vec{M}(\xi)$ is invertible for all designs.

This technique was recently used by [20, 8], who compute approximate designs by using standard algorithms of Bayesian A-optimality, and use rounding heuristics to find exact designs. One disadvantage is that it requires the knowledge of a Mercer's expansion of the kernel. To tackle this problem, a polar spectral approximation of the kernel has been used [20, 21], but it is not clear whether this can be generalized for parameter spaces of dimension $d > 2$. In [8] it is assumed that μ has a finite support containing the candidate points $\vec{x}_i$, so the computation of the $g(\vec{x}_i)$ reduce to a standard matrix eigenproblem. In Section 3 we establish a link between this approach and the class of SOR kernels commonly used in machine learning.

In this article, we focus on two classes of sparse kernels, which are commonly used in Machine Learning, and for which there is a simple, finite Mercer's expansion. We will show in Section 3 that using sparse kernels with s inducing points reduce the cost per iteration of the multiplicative weights update algorithm from $\mathcal{O}(n^3)$ to $\mathcal{O}(ns^2)$, when the goal is to compute the weights of a design minimizing the IMSE over n *predefined candidate points*. This reduction is crucial for computer experiments with parameter dimension $d \geq 4$, where a very large number n of candidate points is required to fill the space of parameters. We show further in Section 4 that the same complexity reduction can be achieved for the search of optimal designs for the prediction of hyperparameters of the kernel. However, we point out that the optimization problems we consider are, in general, not convex. Hence, global optimality cannot be guaranteed, but one can implement standard strategies to try and escape local optima. One exception are the sparse IMSE-optimal design problems studied in Section 3, which are convex when the information function $w()$ is concave. Finally, some numerical experiments illustrate our method in Section 5.

The second goal of this article, covered in the next section, is to extend the theory of approximate optimal designs to the situation in which the weights w_i depend on the true design parameters τ_i via an information function $w()$.

2 Approximate designs in presence of an information function

For a design $\xi = \{\vec{x}_i, \tau_i\}_i$, define $\vec{M}(\xi) := \sum_{i=1}^n w(\vec{x}_i, \tau_i) h(\vec{x}_i) h(\vec{x}_i)^T + \vec{\Gamma}$. Throughout this section, we assume that we are given a set $X = \{\vec{x}_1, \ldots, \vec{x}_n\} \subseteq \mathcal{X}$ of candidate points, and we consider a design problem of the form

$$\min_{\vec{\tau} \in \Delta_T} \Phi(\xi_{\vec{\tau}}) := \operatorname{trace} \vec{M}(\xi_{\vec{\tau}})^{-1} \vec{L}, \tag{5}$$

where $\xi_{\vec{\tau}}$ represents the design $\{\vec{x}_i, \tau_i\}_{i=1,\ldots,n}$. The matrix $\vec{L}$ is positive semidefinite, and the computing times are constrained in the set $\Delta_T := \{\vec{\tau} \in \mathbb{R}_+^n : \sum_{i=1}^n \tau_i = T\}$. For the sake of generality, the information function w is allowed to depend on the design point $\vec{x}_i$. For all $\vec{x} \in \mathcal{X}$, the function $w(\vec{x}, \cdot)$ is assumed to be continuously differentiable and nondecreasing on $\mathbb{R}_+$. We restrict our attention to the case of a positive definite $\vec{\Gamma}$ for the sake of simplicity (so that $\vec{M}(\xi_{\vec{\tau}})$ is invertible for all $\vec{\tau} \in \Delta_T$), but we stress that the results presented here can be extended to the case of a positive semidefinite $\vec{\Gamma}$.

It is well known that $\Phi(\xi_{\vec{\tau}})$ is a convex function of the vector of design weights $\vec{w} = [w(\vec{x}_1, \tau_1), \ldots, w(\vec{x}_n, \tau_n)]^T$. Using the fact that $\Phi(\xi_{\vec{\tau}})$ is a nonincreasing function of $w(\vec{x}_i, \tau_i)$, standard composition theorems yield the following:

Proposition 2.1.
If the information function $\tau \mapsto w(\vec{x}, \tau)$ is nondecreasing and concave for all $\vec{x} \in \mathcal{X}$, then the function $\vec{\tau} \mapsto \Phi(\xi_{\vec{\tau}})$ is convex over Δ_T.

The success of the theory of approximate designs is largely due to equivalence theorems such as the Kiefer-Wolfowitz theorem [10], that give simple means to check the optimality of a design.

First we state the Karush-Kuhn-Tucker (KKT) necessary optimality conditions for a vector of design weights $\vec{\tau}$ over Δ_T, which are valid even if the $w(\vec{x}_i, \cdot)$ are not concave:

Proposition 2.2.
Let $\vec{L} = \vec{K}\vec{K}^T$ and $\vec{x}_1, \ldots, \vec{x}_n$ be given candidate points in $\mathcal{X}$. For all $i \in \{1, \ldots, n\}$, define

$$d_i(\vec{\tau}) = \frac{\partial w(\vec{x}_i, \tau_i)}{\partial \tau_i} \| h(\vec{x}_i)^T \vec{M}(\xi_{\vec{\tau}})^{-1} \vec{K} \|^2.$$

If $\vec{\tau}$ is a local minimizer of $\Phi(\xi_{\vec{\tau}})$ over Δ_T, then we have: $\forall i \in \{1, \ldots, n\}$, $d_i(\vec{\tau}) \leq \frac{1}{T} \sum_{k=1}^n \tau_k d_k(\vec{\tau})$. Moreover, the inequality becomes an equality for support points of $\xi_{\vec{\tau}}$, i.e. for all i such that $\tau_i > 0$.

Proof. Observe that $\frac{\partial \Phi(\xi_{\vec{\tau}})}{\partial \tau_i} = -\frac{\partial w(\vec{x}_i, \tau_i)}{\partial \tau_i} \operatorname{trace} \vec{M}(\xi_{\vec{\tau}})^{-1} h(\vec{x}) h(\vec{x})^T \vec{M}(\xi_{\vec{\tau}})^{-1} \vec{L} = -d_i(\vec{\tau})$. Then, the dual feasibility and complementary slackness KKT-conditions of the optimization problem $\min\{\Phi(\xi_{\vec{\tau}}) : \forall i \in \{1, \ldots, n\}, \tau_i \geq 0, \sum_i \tau_i = T\}$ can be expressed as follows:

$$\exists \lambda \geq 0 : \forall i \in \{1, \ldots, n\}, \quad \Big((\tau_i = 0 \text{ and } d_i(\vec{\tau}) \leq \lambda) \quad \text{or} \quad (\tau_i \geq 0 \text{ and } d_i(\vec{\tau}) = \lambda) \Big).$$

Moreover, the Lagrange multiplier λ must satisfy $T\lambda = \sum_i \tau_i \lambda = \sum_i \tau_i d_i(\vec{\tau})$. Substituting the value of λ in the KKT conditions yields the proposition. □

If the information functions are concave, we obtain a much stronger result. For the next theorem we temporarily drop the assumption that the $\vec{x}_i$'s are given. We characterize optimal designs over the set

$\Xi = \Big\{\xi = \{\vec{x}_i, \tau_i\}_{i=1,\ldots,n} : n \in \mathbb{N},\ \forall i \in \{1,\ldots,n\},\ \vec{x}_i \in \mathcal{X}, \tau_i \geq 0,\ \sum_i \tau_i = T\Big\}$ of all designs with support points in $\mathcal{X}$:

Theorem 2.1. *Let $\vec{L} = \vec{K}\vec{K}^T$, and assume that the condition of Proposition 2.1 is satisfied. For all $\vec{x} \in \mathcal{X}$ define*

$$d(\xi;\vec{x}) = \frac{\partial w(\vec{x},\tau(\vec{x}))}{\partial \tau}\|h(\vec{x})^T \vec{M}(\xi)^{-1}\vec{K}\|^2,$$

where $\tau(\vec{x})$ is the computing time spent on the design point $\vec{x}$ (i.e., $\tau(\vec{x}) = 0$ if $\vec{x} \notin \operatorname{supp}(\xi)$ and $\tau(\vec{x}) = \tau_i$ if $\vec{x} = \vec{x}_i \in \operatorname{supp}(\xi)$). Then, $\xi^* = \{\vec{x}_i, \tau_i\}_{i=1,\ldots,n}$ minimizes Φ over Ξ if and only if

$$\forall \vec{x} \in \mathcal{X},\ d(\xi^*;\vec{x}) \leq \frac{1}{T}\sum_{i=1}^{n} \tau_i\, d(\xi^*, \vec{x}_i).$$

Moreover the above inequality becomes an equality for all support points of ξ^*.

Proof. First note that the *only if* part of the theorem is a simple consequence of Proposition 2.2. For the *if* part, assume that the condition of the theorem holds. It implies

$$\forall \vec{\tau'} \in \mathbb{R}^n_+ \quad \text{such that} \ \sum_i \tau'_i = T, \qquad \sum_{i=1}^{n} \tau'_i\, d(\xi^*, \vec{x}_i) \leq \sum_{i=1}^{n} \tau_i\, d(\xi^*, \vec{x}_i).$$

Using the fact that $\frac{\partial \Phi(\xi^*)}{\partial \tau_i} = -d(\xi^*, \vec{x}_i)$, this can be rewritten as:

$$\forall \xi' = \{\vec{x}_i, \tau'_i\} \in \Xi, \qquad \left.\frac{\partial \Phi((1-\alpha)\xi^* + \alpha\xi')}{\partial \alpha}\right|_{\alpha=0} \geq 0.$$

Finally, consider an arbitrary design $\xi' \in \Xi$, and define the function $\psi : \alpha \mapsto \Phi((1-\alpha)\xi^* + \alpha\xi')$. By proposition 2.1, ψ is convex on $[0,1]$, and we know that $\psi'(0) \geq 0$. So we have $\Phi(\xi') = \psi(1) \geq \psi(0) + \psi'(0)(1-0) \geq \psi(0) = \Phi(\xi^*)$, which proves the optimality of ξ^*. □

We next adapt the multiplicative weights update algorithm of Titterington [19] for Problem (5). The multiplicative algorithm was originally presented in the general setting in which a function f must be minimized over a unit simplex $\{\vec{w} \in \mathbb{R}_n^+ : \sum_i w_i = 1\}$, so it can be adapted in a straightforward manner to the case of a design problem with information functions, i.e., $w_i = w(\vec{x}_i, \tau_i)$. Given an exponent $q > 0$ and an initial vector $\vec{\tau}^{(0)} > \vec{0}$, the iterations are:

$$\forall i \in \{1,\ldots,n\}, \quad \tau_i^{(k+1)} \leftarrow T\frac{\tau_i^{(k)}\ d_i(\vec{\tau}^{(k)})^q}{\sum_{j=1}^{n} \tau_j^{(k)}\ d_j(\vec{\tau}^{(k)})^q}. \tag{6}$$

In its standard version, that is, when $w(\vec{x}_i, \tau_i) = \tau_i$, this algorithm converges monotonically towards an $A-$optimal design when $q = \frac{1}{2}$. The process can be accelerated by pruning candidate points with a sufficiently low weight, which ensures that they do not belong to the support of any optimal design [14]. Convergence for a variety of optimality criteria was shown in [25].

Consider now the general setting of a function f to be minimized over Δ_T, with $d_i(\vec{\tau}) = \frac{\partial f(\vec{\tau})}{\partial \tau_i}$. If the iterations (6) converge, then the limit point $\vec{\tau}^*$ must satisfy the necessary condition of Proposition 2.2, under some mild conditions [6]. In practice, we experienced numerical convergence of the above algorithm towards *local minima* of $f : \vec{\tau} \mapsto \Phi(\xi_{\vec{\tau}})$ when q is well chosen, even in the cases where the information functions $w(\vec{x}_i, \cdot)$ are not concave.

3 Sparse covariance kernels

Here we consider two classes of sparse kernel functions commonly used in machine learning for Gaussian process regression with a large number n of samples. These approximations rely on a small set of *inducing points*, $\{\vec{u}_1,\ldots,\vec{u}_s\} \subset \mathcal{X}$, and assume that the covariance $\operatorname{cov}(Z(\vec{x}), Z(\vec{y}))$ of the process between the points $\vec{x}$ and $\vec{y} \in \mathcal{X}$ only depends on the covariances between the $\vec{u}_i$, $\vec{x}$ and the $\vec{u}_i$, and $\vec{y}$ and the $\vec{u}_i$.

This reduces the complexity of training a Gaussian process on a dataset with n samples from $\mathcal{O}(n^3)$ to $\mathcal{O}(ns^2)$. We refer to [17] for a comprehensive review.

SOR-kernels. The *Subset of Regressors* (SOR) approximation consists in replacing the correlation function $C(\vec{x},\vec{y})$ by a low-rank kernel,

$$C_{\text{SOR}}(\vec{x},\vec{y}) = c_u(\vec{x})^T \vec{K}_{u,u}^{-1} c_u(\vec{y}),$$

where $\{\vec{K}_{u,u}\}_{i,j} = C(\vec{u}_i,\vec{u}_j)$ is the $s\times s$ matrix of correlations between the $Z(\vec{u}_i)$, and $\{c_u(\vec{x})\}_i = C(\vec{u}_i,\vec{x})$ is the s-dimensional vector of correlations between $Z(\vec{x})$ and the $Z(\vec{u}_i)$. Hence, if we let $\vec{J}_u$ be any matrix satisfying $\vec{J}_u\vec{J}_u^T = \vec{K}_{u,u}^{-1}$, then the function $g:\vec{x}\mapsto \vec{J}_u^T c_u(\vec{x})$ satisfies

$$\forall(\vec{x},\vec{y})\in\mathcal{X}\times\mathcal{X},\quad C_{\text{SOR}}(\vec{x},\vec{y}) = g(\vec{x})^T g(\vec{y}).$$

Hence, for a SOR-kernel the observation model (1) is equivalent to model (3). Indeed, $Z(\xi) = [Z(\vec{x}_1),\ldots,Z(\vec{x}_n)]^T \sim \mathcal{N}(0,\sigma_Z^2\vec{C})$ has the same distribution as $[g(\vec{x}_1),\ldots,g(\vec{x}_n)]^T\vec{\alpha}$, where $\vec{\alpha}\sim\mathcal{N}(0,\sigma_Z^2\vec{I}_s)$. To put it in other words, if we assume that the true kernel is C_{SOR}, then model (3) is exact, so that $\text{IMSE}(\xi_{\vec{\tau}}) = \widetilde{\text{IMSE}}(\xi_{\vec{\tau}}) = \operatorname{trace}\vec{M}(\xi_{\vec{\tau}})^{-1}\vec{L}$, and we can use the multiplicative weights update (6) to compute an optimal design. The complexity of computing $\text{IMSE}(\xi_{\vec{\tau}})$ and its gradient $[d_1(\vec{\tau}),\ldots,d_n(\vec{\tau})]^T$ is $\mathcal{O}(n(s+m)^2)$, which is $\mathcal{O}(ns^2)$ because the dimension m of the regression parameter $\vec{\beta}$ is a small constant. So the cost of one iteration (6) is $\mathcal{O}(ns^2)$. In contrast, for a full kernel the computation involves $\vec{\Sigma}^{-1}$ and takes $\mathcal{O}(n^3)$ operations.

There is a vast literature on the selection of inducing points $\vec{u}_i$ to approximate a kernel $C(\cdot,\cdot)$ by a SOR kernel [22, 2, 24]. For example, [24] uses a regular grid to exploit the Kronecker structure of $\vec{K}_{u,u}$ when C is a product of one-dimensional kernels, and to speed-up the computations by using fast Fourier transforms. Note that if the points $\vec{u}_1,\ldots,\vec{u}_s$ are sampled randomly and independently from the probability measure μ, the approximation $C(\vec{x},\vec{y})\simeq C_{\text{SOR}}(\vec{x},\vec{y})$ can be regarded as an expansion of the form $C(\vec{x},\vec{y}) = \sum_{i=1}^s \lambda_i\phi_i(\vec{x})\phi_i(\vec{y})$, where the λ_i and $\phi_i(\cdot)$ are the solutions of the Nystrm approximation of the eigenproblem

$$\int_{\mathcal{X}} C(\vec{x},\vec{y})\phi(\vec{y})d\mu(\vec{y}) = \lambda\phi(\vec{x}).$$

This approximation consists in replacing the integral by $\frac{1}{s}\sum_{i=1}^s C(\vec{x},\vec{u}_i)\phi(\vec{u}_i)$, and reduces the infinite-dimensional eigenproblem to a standard $s\times s$–matrix eigenproblem [23]. If we choose $\vec{J}_u = \vec{U}\vec{\Lambda}^{-\frac{1}{2}}$, where $\vec{U}\vec{\Lambda}\vec{U}^T$ is a spectral decomposition of $\vec{K}_{u,u}$, this is equivalent to the approach of [7, 8], where μ is approximated by a discrete measure $\hat{\mu}$ supported by the $\vec{u}_i$'s (i.e., the IMSE is approximated by a quadrature). In [7], the authors further suggest to choose the candidates $\vec{x}_i$ in the support of $\hat{\mu}$, i.e. $n\leq s$. Then, g must only be evaluated at the $\vec{u}_i$'s, and the vectors $g(\vec{u}_i)$ $(i=1,\ldots,s)$ are the columns of $\vec{\Lambda}^{\frac{1}{2}}\vec{U}^T$, hence they are orthogonal. This contrasts with our approach, where we generate a large number of candidate points in order to fill the design space, but use a small number of inducing points for the sake of computation ($n\gg s$).

FITC-kernels. The FITC approximation (*Fully Independent Training Conditional*) is very similar to SOR, but a diagonal noise is added to the kernel, to ensure that $C_{\text{FITC}}(\vec{x},\vec{x}) = C(\vec{x},\vec{x}) = 1$:

$$C_{\text{FITC}}(\vec{x}_i,\vec{x}_j) = C_{\text{SOR}}(\vec{x}_i,\vec{x}_j) + \big(1 - C_{\text{SOR}}(\vec{x}_i,\vec{x}_i)\big)\delta_{ij},$$

If we define as before the function $g:\vec{x}\mapsto\vec{J}_u^T c_u(\vec{x})$, where $\vec{J}_u\vec{J}_u^T = \vec{K}_{u,u}^{-1}$, we obtain:

$$C_{\text{FITC}}(\vec{x}_i,\vec{x}_j) = g(\vec{x}_i)^T g(\vec{x}_j) + (1 - \|g(\vec{x}_i)\|^2)\delta_{ij}.$$

It follows that for a FITC kernel, the observation model is equivalent to

$$Y(\vec{x},\tau) = [f(\vec{x})^T, g(\vec{x})^T]\begin{bmatrix}\vec{\beta}\\ \vec{\alpha}\end{bmatrix} + \nu(\vec{x}) + \epsilon(\vec{x},\tau), \tag{7}$$

where $\vec{\alpha}$ is a random regression parameter with prior $\vec{\alpha} \sim \mathcal{N}(0, \sigma_Z^2 \vec{I}_k)$, and $\nu(\vec{x})$ is an unbiased and uncorrelated noise, which is heteroschedastic with $\mathrm{Var}[\nu(\vec{x})] = \sigma_Z^2(1 - \|g(\vec{x})\|^2)$. Under this model, the Fisher information matrix for $(\vec{\beta}, \vec{\alpha})$ becomes (up to a scaling factor σ_Z^2):

$$\vec{M}(\xi) := \sum_{i=1}^{n} w(\vec{x}_i, \tau_i)\, h(\vec{x}_i)h(\vec{x}_i)^T + \vec{\Gamma}, \qquad \text{where } w(\vec{x}, \tau) := \frac{\gamma w(\tau)}{1 + (1 - \|g(\vec{x})\|^2)\gamma w(\tau)}. \tag{8}$$

Note that $\vec{M}(\xi)$ has the form of the Fisher information matrix of the problem studied in Section 2. Moreover, elementary calculus shows that $\tau \mapsto w(\vec{x}, \tau)$ is concave if $\tau \mapsto w(\tau)$ is concave. In this situation, Proposition 2.1 shows that (5) is a convex optimization problem for FITC kernels.

4 Optimal designs for the estimation of kernel hyperparameters

Until now, we have assumed that the kernel function $C(\cdot, \cdot)$ was known. In practice however, the kernel depends on a set of hyperparameters $\vec{\theta} \in \mathbb{R}^p$, which must be estimated by maximum likelihood from the set of observations $Y(\xi)$. Recall that $Y(\xi) \sim \mathcal{N}\big(\vec{F}^T\vec{\beta},\, \sigma_Z^2 \vec{\Sigma}_\theta\big)$, where $\vec{\Sigma}_\theta = \vec{C}_\theta + \vec{D}^{-1}$, $\vec{D} = \mathrm{Diag}(\gamma w(\vec{\tau}))$, and we have inserted the symbol θ as subscript to stress the dependency on the hyperparameters. Then, the $p \times p$ Fisher information matrix for the vector of parameters $\vec{\theta}$ can be derived from standard formulas:

$$\{\vec{M}_\theta(\xi)\}_{ij} = \frac{1}{2} \operatorname{trace} \vec{\Sigma}_\theta^{-1} \frac{\partial \vec{C}_\theta}{\partial \theta_i} \vec{\Sigma}_\theta^{-1} \frac{\partial \vec{C}_\theta}{\partial \theta_j}. \tag{9}$$

Given a current estimate of $\vec{\theta}$, we propose to search a design ξ maximizing the criterion of $D-$optimality, $\log\det \vec{M}_\theta(\xi)$. Here, note that we assume that $\vec{\beta}$ and σ_Z^2 are known. We refer to [15] for a review on approaches to deal with a total Fisher information matrix for the set of parameters $(\vec{\beta}, \sigma_Z^2, \vec{\theta})$. In particular, Mller and Stehlík proposed a compound criterion with a weighing factor that balances the goals of estimating $\vec{\beta}$ and estimating $\vec{\theta}$ [11].

We want to optimize the computing times τ_i associated with a large number of candidate points $\vec{x}_i$. This is a hard optimization problem, since here the $D-$criterion is not convex with respect to $\vec{\tau}$. Nevertheless we propose to use the multiplicative update iterations (6), where $d_i(\vec{\tau}) := \frac{\partial \log\det \vec{M}_\theta(\xi_{\vec{\tau}})}{\partial \tau_i}$, in order to identify good local optima. However, if n is very large, the computation of $\vec{M}_\theta(\xi_{\vec{\tau}})$ and $d_i(\vec{\tau})$ is extremely time-consuming. It involves the inversion of the $n \times n$ matrix $\vec{\Sigma}_\theta$, and many $n \times n$ matrix-matrix multiplications.

In this section we show that this computational burden can be reduced if sparse kernels are used. From now on, we assume that $\vec{C}_\theta = \vec{G}_\theta \vec{G}_\theta^T$, where $\vec{G}_\theta = [g_\theta(\vec{x}_1), \ldots, g_\theta(\vec{x}_n)]^T \in \mathbb{R}^{n \times s}$. As is previous section, the function g_θ is defined by $g_\theta(\vec{x}) = \vec{J}_u^T c_u(\vec{x})$, where $\vec{J}_u \vec{J}_u^T = \vec{K}_{u,u}^{-1}$. From now on we set $\vec{J}_u$ to the Cholesky factor of $\vec{K}_{u,u}^{-1}$, because this choice yields compact formulas.

First note that the low-rank decomposition makes it possible to use the Woodbury matrix identity:

$$\vec{\Sigma}_\theta^{-1} = (\vec{C}_\theta + \vec{D}^{-1})^{-1} = \vec{D} - \vec{D}\vec{G}_\theta(\vec{I}_s + \vec{G}_\theta^T \vec{D}\vec{G}_\theta)^{-1}\vec{G}_\theta^T \vec{D}. \tag{10}$$

Then, we also need to compute derivatives of g_θ with respect to θ. This is possible thanks to the following lemma:

Lemma 4.1. *Define the function Φ which returns the lower triangle and half the diagonal of a square matrix:*

$$\forall \vec{M} \in \mathbb{R}^{n \times n}, \quad \{\Phi(\vec{M})\}_{ij} = \begin{cases} M_{ij} & \text{if } i > j \\ \frac{1}{2} M_{ij} & \text{if } i = j \\ 0 & \text{if } i < j. \end{cases}$$

Then, we have: $\forall \vec{x} \in \mathcal{X}$,

$$\frac{\partial g_\theta(\vec{x})}{\partial \theta_i} = \vec{J}_u^T \frac{\partial c_u(\vec{x})}{\partial \theta_i} - \Phi\left(\vec{J}_u^T \frac{\partial \vec{K}_{u,u}}{\partial \theta_i} \vec{J}_u\right)^T \vec{J}_u^T c_u(\vec{x}).$$

Proof. A formula for the derivative of the Cholesky decomposition $\vec{X} = \vec{J}\vec{J}^T$ can be found in [18, Theorem A.1], and can be proved by implicit differentiation:

$$\frac{\partial \vec{J}}{\partial \theta} = \vec{J}\Phi(\vec{J}^{-1}\frac{\partial \vec{X}}{\partial \theta}\vec{J}^{-T}).$$

The formula of the lemma can now be obtained, by applying standard formulas for the differentiation of products and matrix inverse. □

We can use this lemma to compute the matrices $\vec{G}_i := \frac{\partial \vec{G}_\theta}{\partial \theta_i}$. Now, we also define $\vec{G}_0 := \vec{G}_\theta$ to simplify the notation. Substituting $\frac{\partial \vec{C}_\theta}{\partial \theta_i} = \vec{G}_i \vec{G}_0^T + \vec{G}_0 \vec{G}_i^T$ and (10) into (9) yields an expression for $\{\vec{M}_\theta(\xi)\}_{ij}$ that depends on $\vec{G}_0, \vec{G}_1, \ldots, \vec{G}_p$. After some simplifications, we obtain

$$\{\vec{M}_\theta(\xi)\}_{ij} = \operatorname{trace} \vec{A}_{0i}\vec{A}_{0j} + \vec{A}_{00}\vec{A}_{ij},$$

where for all $k, l \in \{0, \ldots, p\}$, $\vec{A}_{kl} := \vec{B}_{kl} - \vec{B}_{k0}(\vec{I}_s + \vec{B}_{00})^{-1}\vec{B}_{0j}$ and $\vec{B}_{kl} := \vec{G}_k^T \vec{D} \vec{G}_l$. From these expressions, it is easy to see that $\vec{M}_\theta(\xi)$ can be computed in $\mathcal{O}(ns^2)$, which is a great improvement compared to $\mathcal{O}(n^3)$ for a full kernel.

Similarly, we can compute $\vec{\nabla}_\tau \{\vec{M}_\theta(\xi)\}_{ij} = \left[\frac{\partial \{\vec{M}_\theta(\xi)\}_{ij}}{\tau_1}, \ldots, \frac{\partial \{\vec{M}_\theta(\xi)\}_{ij}}{\tau_n}\right]^T$ in $\mathcal{O}(ns^2)$. For all $k \in \{0, \ldots, p\}$, define $\vec{P}_k := \vec{G}_k - \vec{G}_0(\vec{I}_s + \vec{B}_{00})^{-1}\vec{B}_{0k}$. Then, we can show that (details omitted):

$$\vec{\nabla}_\tau \{\vec{M}_\theta(\xi)\}_{ij} = \gamma \operatorname{Diag}(w'(\vec{\tau})) \operatorname{Diag}\left(\vec{P}_i \vec{A}_{0j} \vec{P}_0^T + \vec{P}_j \vec{A}_{0i} \vec{P}_0^T + \vec{P}_0 \vec{A}_{ij} \vec{P}_0^T + \vec{P}_j \vec{A}_{00} \vec{P}_i^T\right).$$

Finally, the gradient of the criterion is obtained by $d_i(\xi) = \operatorname{trace} \vec{M}_\theta(\xi)^{-1} \frac{\partial \vec{M}_\theta(\xi)}{\partial \tau_i}$. Hence, we have shown that the gradient of the criterion can be computed in $\mathcal{O}(ns^2)$ for a sparse kernel with s inducing points. In contrast, for a full kernel one requires $\mathcal{O}(n^3)$ operations.

5 Numerical Experiments

We consider the Ishigami-like function η to illustrate the effect of the information functions:

$$\forall \vec{x} \in \mathcal{X} = [0,1] \times [0,1], \qquad \eta(\vec{x}) = 1.1 \sin(\pi(2x_1 - 1)) + 7 \sin^2(\pi(2x_2 - 1)).$$

First, 8 noisy observations of the function $\eta(\vec{x})$ are taken, with $\sigma_N^2 = 0.05$, and $\tau_{init} = \frac{1}{8}$ at each of the 8 locations indicated by yellow dots in Figure 1. These initial values are used to estimate, by maximum likelihood, σ_Z^2 and the hyperparameters (ℓ_1, ℓ_2) of the Gaussian kernel

$$C(\vec{x}, \vec{y}) = e^{-\frac{1}{2}\left[\left(\frac{x_1 - y_1}{\ell_1}\right)^2 + \left(\frac{x_2 - y_2}{\ell_2}\right)^2\right]}$$

The plots of Figure 1 show some designs for the distribution of $T = 1$ additional hour of computing time over a regular grid of $n = 31^2 = 961$ candidate points. The size of the red dots indicate the time to spend on a design point, and the color in the background indicates the prior Kriging variance (after the initial 8 observations; blue: small variance, red: high variance), according to the considered covariance model: In Plot (a) and (b), we respectively used the SOR and FITC Kernel associated with $C(\cdot, \cdot)$, for $s = 12$ inducing points (marked with black squares), that were generated in a space-filling fashion with a Sobol sequence; the number $s = 12$ is rather small, on purpose, to illustrate the effect of sparsity. Plots (c)-(g) rely on a SOR kernel with $s = 60$ inducing points (not marked for the sake of visibility); with that many inducing points, the relative errors between IMSE(ξ) and $\widetilde{\text{IMSE}}(\xi)$ were in the order of 0, 1% for the designs we computed. Also, different information functions were used, cf. Plot (h).

The plots (a),(b),(c),(e),(g) show (near-)optimal weights τ_i for the IMSE criterion at the specified n locations of the $\vec{x}_i$'s, while (d),(f) are nearly $D-$optimal weights for the estimation of $\vec{\theta} = (\ell_1, \ell_2)$. For

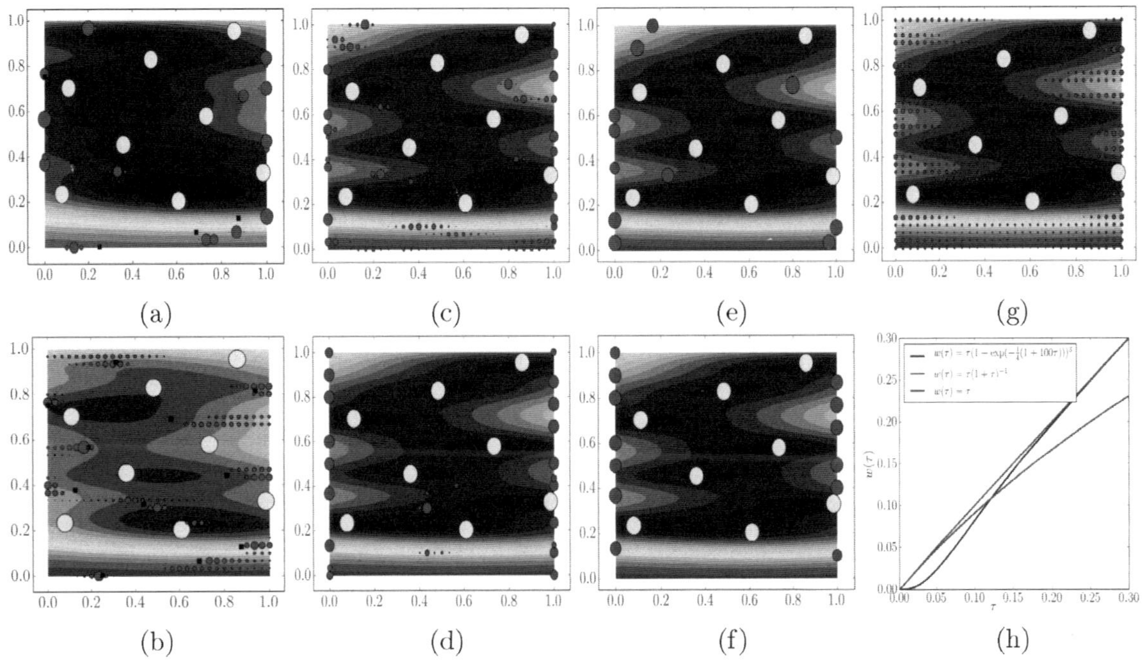

Figure 1. Near-optimal design weights for the test function. Kernel: C_{SOR} with $s = 12$ inducing points in (a); C_{FITC} with $s = 12$ inducing points in (b), and C_{SOR} with $s = 60$ inducing points in (c)-(g). Optimality criterion: IMSE in (a),(b),(c),(e),(g) and D-criterion for $\vec{\theta} = (\ell_1, \ell_2)$ in (d),(f). Information function: $w(\tau) = \tau$ in (a)-(d); $w(\tau) = \tau(1 - \exp(-\frac{1}{4}(1 + 100\tau)))^3$ in (e),(f); $w(\tau) = \tau(1 + \tau)^{-1}$ in (g). These information functions are plotted in (h).

all computations, the matrix $\vec{L} = \sigma_Z^2 \int_{\mathcal{X}} h(\vec{x})h(\vec{x})^T d\mu(\vec{x})$ was computed with a Monte-Carlo method with $N = 10^5$ samples, with μ the Lebesgue measure over $[0, 1]^2$. The stopping criterion for the multiplicative update iterations was

$$\max_{i=1,\dots,n} d_i(\vec{\tau}) \leq \frac{1}{T} \sum_{k=1}^{n} \tau_k d_k(\vec{\tau}) + \varepsilon, \tag{11}$$

where $\varepsilon = 10^{-9}$. Note that the design weights plotted in (a)-(c) and (g) are provably optimal (up to the tolerance ε), because the considered optimization problems are convex. This is not the case for the designs shown in Plots (d), (e) and (f). Here, the multiplicative update algorithm is likely to fall in local optima, so we performed several restarts and kept the best design.

Plots (a) and (b) show the effect of using a sparse kernel, and are to be compared with Plot (c), which can be considered as the optimal design for the full kernel C when $w(\tau) = \tau$. Observe that the kriging variance tends to be underestimated with the SOR kernel (a), while it is overestimated with the FITC kernel (b). As a consequence of the (strict) concavity of $w(\vec{x}, \tau)$, the FITC design is more spread out than the SOR design. Also, the FITC design has a slightly better efficiency than the SOR design (73% vs. 69%, cf. Formula (12)).

The effect of the information function can be seen by comparing the second column (standard case $w(\tau) = \tau$) to the third and fourth columns. In the second column, the information function is convex near 0, so that we need some minimal amount of computing time to get some information; see blue curve in plot (f). As a consequence, the IMSE-optimal design for this information function are very sparse, a feature that can be very valuable for the experimenter. In contrast, a concave information function is used in the third column (green curve in plot (f)). This incentivizes designs with many design points spread out over regions with a high variance.

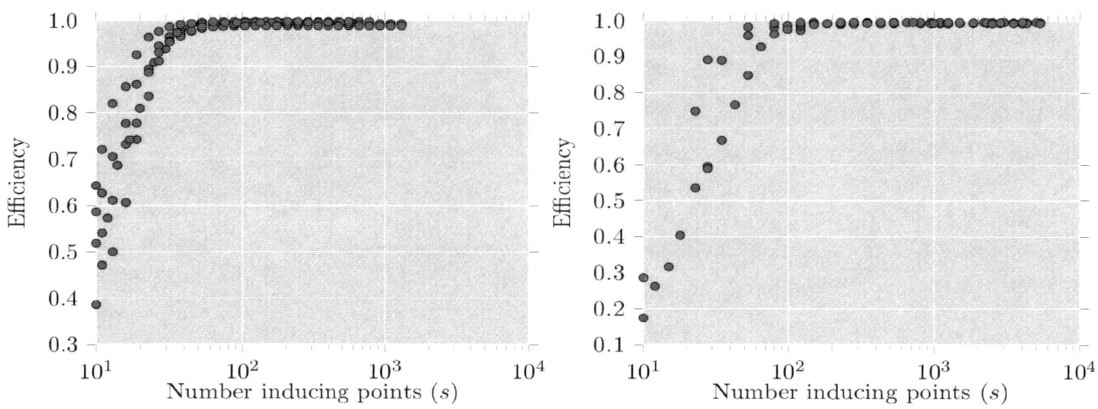

$d = 4$ $d = 7$

Figure 2. Efficiency of IMSE-optimal design, for an approximation of the kernel based on a SOR kernel with s inducing points.

Next, we show some results in higher dimensional spaces to illustrate the importance of using sparse approximations of the kernel. We report results for tests in dimension $d = 4$ and $d = 7$. In each case, we considered 10 instances, corresponding to different functions $\eta(\cdot)$; These functions were dummy rational functions, in which we have selected the coefficients at random.

Our experiments used $T_{\text{init}} = 1$ hour of computing time distributed uniformly over $n_{\text{init}} = 50$ initial observations, and aimed at distributing $T = 1$ additional hour of computing time over $n = 1500$ randomly generated candidate points for the problems in dimension 4, and $n = 5300$ points for the problem in dimension 7. The function $w()$ was set to the identity: $w(\tau) = \tau$.

For each problem, we have computed the *true* optimal design ξ^* (within the subset of designs supported over the given candidate points), by using the multiplicative update iterations (6) with a formula for the derivative of the true criterion: $d_i(\xi) = \frac{\partial \text{IMSE}(\xi)}{\partial \tau_i}$. The efficiency of a design was evaluated by the following formula:

$$\text{efficiency}(\xi) = \frac{\text{IMSE}(\xi)^{-1} - \text{IMSE}(\xi_{\text{init}})^{-1}}{\text{IMSE}(\xi^*)^{-1} - \text{IMSE}(\xi_{\text{init}})^{-1}}. \tag{12}$$

Here, ξ_{init} denotes the initial design supported by the n_{init} initial observation points, so the numerator expresses the gain of information provided by ξ, compared to the situation where no additional measurement is performed. Figure 2 shows the efficiency of designs computed by using a SOR approximation of the kernel, for 10 instances with $d = 4$ (left) and $d = 7$ (right). In both cases, we observe an excellent efficiency when $s \geq 100$, and even for $s \geq 70$ for the instances in a 4-dimensional space. In terms of computing time, the speed-up was on the order of x200 on average for $s = 70$ and $d = 4$, and even of x350 for $s = 100$ and $d = 7$. For the latter instances, the computations took more than 36 hours on a Linux PC with 8 cores at 3.60 GHz, while with the SOR kernel a solution was found within 6 minutes (with a tolerance parameter ε set to 10^{-4} in the stopping criterion (11)).

Acknowledgments. The authors want to thank the two anonymous reviewers for their valuable comments and suggestions that significantly improved the presentation.

Bibliography

[1] Bect, J., Ginsbourger, D., Li, L., Picheny, V., and Vazquez, E. (2012). *Sequential design of computer experiments for the estimation of a probability of failure.* Statistics and Computing, **22**(3):773–793.

[2] Cao, Y., Brubaker, M., Fleet, D., and Hertzmann, A. (2015). *Efficient optimization for sparse gaussian process regression.* IEEE transations on Pattern Analysis and Machine Intelligence, **37**(12):2415–2427.

[3] Deuflhard, P. and Weiser, M. (2012). *Adaptive numerical solution of PDEs.* Walter de Gruyter.

[4] Fedorov, V. (1996). *Design of spatial experiments: Model fitting and prediction.* Handbook of Statistics, **13**:515–553.

[5] Fedorov, V. and Flanagan, D. (1997). *Optimal monitoring network design based on Mercer's expansion of covariance kernel.* Journal of Combinatorics, Information and System Sciences, **23**:237–250.

[6] Gaffke, N. and Mathar, R. (1992). *On a class of algorithms from experimental design theory.* Optimization, **24**(1-2):91–126.

[7] Gauthier, B. and Pronzato, L. (2014). *Spectral approximation of the IMSE criterion for optimal designs in kernel-based interpolation models.* SIAM/ASA Journal on Uncertainty Quantification, **2**(1):805–825.

[8] Gauthier, B. and Pronzato, L. (2016). *Optimal design for prediction in random field models via covariance kernel expansions.* In Proceedings of the 11th Workshop on Model-Oriented Data Analysis and Optimum Designs (mODa'11).

[9] Jones, D., Schonlau, M., and Welch, W. (1998). *Efficient global optimization of expensive black-box functions.* Journal of Global optimization, **13**(4):455–492.

[10] Kiefer, J. and Wolfowitz, J. (1960). *The equivalence of two extremum problems.* Canadian Journal of Mathematics, **12**:363–366.

[11] Müller, W. and Stehlík, M. (2010). *Compound optimal spatial designs.* Environmetrics, **21**(3-4):354–364.

[12] Pázman, A. (1986). *Foundations of optimum experimental design.* D. Reidel Dordrecht.

[13] Pilz, J. (1991). *Bayesian estimation and experimental design in linear regression models*, volume 212. John Wiley & Sons Inc.

[14] Pronzato, L. (2013). *A delimitation of the support of optimal designs for Kiefer's ϕ_p-class of criteria.* Statistics & Probability Letters, **83**(12):2721–2728.

[15] Pronzato, L. and Müller, W. (2012). *Design of computer experiments: space filling and beyond.* Statistics and Computing, **22**(3):681–701.

[16] Pukelsheim, F. (1993). *Optimal Design of Experiments.* Wiley.

[17] Quiñonero-Candela, J. and Rasmussen, C. (2005). *A unifying view of sparse approximate gaussian process regression.* The Journal of Machine Learning Research, **6**:1939–1959.

[18] Särkkä, S. (2013). *Bayesian filtering and smoothing*, volume 3. Cambridge University Press.

[19] Silvey, S., Titterington, D., and Torsney, B. (1978). *An algorithm for optimal designs on a finite design space.* Communications in Statistics - Theory and Methods, **7**(14):1379–1389.

[20] Spöck, G. and Pilz, J. (2010). *Spatial sampling design and covariance-robust minimax prediction based on convex design ideas.* Stochastic Environmental Research and Risk Assessment, **24**(3):463–482.

[21] Spöck, G. and Pilz, J. (2015). *Incorporating covariance estimation uncertainty in spatial sampling design for prediction with trans-gaussian random fields.* Frontiers in Environmental Science, **3**(39).

[22] Titsias, M. (2009). *Variational learning of inducing variables in sparse gaussian processes.* In International Conference on Artificial Intelligence and Statistics, pages 567–574.

[23] Williams, C. and Seeger, M. (2001). *Using the Nyström method to speed up kernel machines.* In Proceedings of the 14th Annual Conference on Neural Information Processing Systems, number EPFL-CONF-161322, pages 682–688.

[24] Wilson, A. and Nickisch, H. (2015). *Kernel interpolation for scalable structured gaussian processes (KISS-GP).* In Proceedings of The 32nd International Conference on Machine Learning, volume 37 of *JMLR: W&CP*, pages 1775–1784, Lille, France.

[25] Yu, Y. (2010). *Monotonic convergence of a general algorithm for computing optimal designs.* The Annals of Statistics, **38**(3):1593–1606.

Author Index

Adachi, Kohei, 25
Afreixo, Vera, 255
Alibrandi, Angela, 231

Bartlett, Thomas Michael, 349
Bastos, Carlos, 255
Boccato, Levy, 349
Breiteneder, Christian, 183
Brito, Paula, 255
Brodinova, Sarka, 183
Bry, Xavier, 169, 195

Caeiro, Frederico, 279
Cardot, Hervé, 49
Chauvet, Jocelyn, 169
Collado, Ricardo A., 111
Corbellini, Aldo, 13
Creamer, Germán G., 111

De Moliner, Anne, 49

Einarsson, Baldvin, 243

Ferreira, Paulo, 255
Filzmoser, Peter, 183
Fujimiya, Hitoshi, 339

García-Díaz, J. Carlos, 149
Geniaux, Ghislain, 1
Ghattas, Badih, 327
Giacalone, Massimiliano, 231
Giorgio, Massimiliano, 99
Giurghita, Diana, 383
Goga, Camelia, 49
Gomes, M. Ivette, 279
Goto, Masashi, 123
Grossi, Luigi, 13
Guida, Maurizio, 99
Guin, Jayanta, 243

Hege, Hans-Christian, 397
Hitaj, Asmerilda, 159
Hubalek, Friedrich, 159
Husmeier, Dirk, 383

Ishioka, Fumio, 85

Kidé, Saikou Oumar, 37
Koskinen, Lasse, 361
Kubota, Takafumi, 291, 339
Kurihara, Koji, 85

Laurini, Fabrizio, 13
Lee, Sharon X., 137
Luoma, Arto, 361

Manté, Claude, 37
Martinetti, Davide, 1
Maté, Carlos G., 303
Mavrogonatou, Lida, 73
McLachlan, Geoffrey J., 137
Mercuri, Lorenzo, 159
Michel, Pierre, 327
Moleti, Mariacarla, 231
Mortier, Frédéric, 169
Murdoch, W. James, 217

Nakamura, Masatoshi, 123
Neves, M. Manuela, 279

Ochi, Yoshimichi, 123
Ortner, Thomas, 183

Penalva, Helena, 279
Philipp, Michel, 315
Pinho, Armando, 255
Postiglione, Fabio, 99
Pulcini, Gianpaolo, 99
Pulido-Rojano, Alexander, 149

Redondo, Javier, 303
Rodrigues, João, 255
Rroji, Edit, 159

Sagnol, Guillaume, 397

Sakamoto, Wataru, 267
Sakurai, Hirohito, 205
Salminen, Tommi, 361
Simac, Théo, 195
Strobl, Carolin, 315

Taguri, Masaaki, 205
Takebayashi, Yoshitake, 291
Tavares, Ana, 255
Torii, Hiroyuki A., 339
Trendafilov, Nickolay, 25
Trottier, Catherine, 169
Tsubaki, Hiroe, 291

Víšek, Jan Ámos, 59
Verron, Thomas, 195
Vyshemirsky, Vladislav, 73

Wójcik, Rafał, 243
Weiser, Martin, 397

Yamada, Sanetoshi, 373
Yamamoto, Yoshiro, 373

Zaharieva, Maia, 183
Zeileis, Achim, 315
Zhu, Mu, 217
Zirilli, Agata, 231

International Statistical Institute (ISI)
P.O. Box 24070
2490 AB The Hague
The Netherlands

ISBN 978-1-5108-3759-1

CURRAN ASSOCIATES INC.
proceedings.com